Lipid Oxidation

Challenges in Food Systems

Lipid Oxidation
Challenges in Food Systems

EDITORS

Amy Logan
Uwe Nienaber
Xiangqing (Shawn) Pan

AOCS PRESS
Urbana, Illinois

AOCS Mission Statement

AOCS advances the science and technology of oils, fats, surfactants and related materials, enriching the lives of people everywhere.

ISBN 978-0-12-810229-9 (print)

Library of Congress Cataloging-in-Publication Data
Lipid oxidation : challenges in food systems / editors, Amy Logan, Uwe Nienaber, Xiangqing Pan.
pages cm
ISBN 978-0-12-810229-9
1. Oils and fats, Edible—Deterioration. 2. Lipids—Oxidation. 3. Food—Quality. I. Logan, Amy, 1977– II. Nienaber, Uwe, 1962– III. Pan, Xiangqing, 1970–
TX560.O3L57 2013
664'.3—dc23

2012044763

17 16 15 14 13 5 4 3 2 1

Contents

Preface vii

CHAPTER 1 1

Challenges in Elucidating Lipid Oxidation Mechanisms: When, Where, and How do Products Arise?
K. M. Schaich

CHAPTER 2 53

Challenges in Analyzing Lipid Oxidation: Are One Product and One Sample Concentration Enough?
K. M. Schaich

CHAPTER 3 129

Oxidation in Different Food Matrices: How Physical Structure Impacts Lipid Oxidation in Oil-in-Water Emulsions and Bulk Oils
Bingcan Chen, D. Julian McClements, and Eric A. Decker

CHAPTER 4 155

Substrate and Droplet Size: Important Factors for Understanding Aqueous Lipid Oxidation
Kazuo Miyashita and Masashi Hosokawa

CHAPTER 5 177

The Role of the Interfacial Layer and Emulsifying Proteins in the Oxidation in Oil-in-Water Emulsions
Claude Genot, Claire Berton, and Marie-Hélène Ropers

CHAPTER 6 211

Oxidative Stability of Enzymatically Processed Oils and Fats
Derya Kahveci, Zheng Guo, Ling-Zhi Cheong, Mia Falkeborg, Warowan Panpipat, and Xuebing Xu

CHAPTER 7 243

The Polar Paradox: How an Imperfect Conceptual Framework Accelerated Our Knowledge of Antioxidant Behavior
Roger Nahas and Donald Berdahl

CHAPTER 8 261

Role of Hydrophobicity on Antioxidant Activity in Lipid Dispersions: From the Polar Paradox to the Cut-Off Theory
Mickaël Laguerre, Ann-Dorit Moltke Sørensen, Christelle Bayrasy, Jérôme Lecomte, Charlotte Jacobsen, Eric Andrew Decker, and Pierre Villeneuve

CHAPTER 9 297

Understanding Antioxidant and Prooxidant Mechanisms of Phenolics in Food Lipids
Lisa Zhou and Ryan J. Elias

CHAPTER 10 323

Antioxidant Evaluation and Antioxidant Activity Mechanisms
Dejian Huang and Ik Chian Wong

CHAPTER 11 345

Strategies to Minimize Oxidative Deterioration in Aquatic Food Products: Application of Natural Antioxidants from Edible Mushrooms
Huynh Nguyen Duy Bao and Toshiaki Ohshima

CHAPTER 12 381

The Natural Antioxidant Ergothioneine: Resources, Chemical Characterization, and Applications
The Han Nguyen, Reiko Nagasaka, and Toshiaki Ohshima

CHAPTER 13 417

Rosemary and Green Tea Extracts as Natural Antioxidants: Chemistry, Technology, and Applications
S. P. J. Namal Senanayake

CHAPTER 14 439

Using Natural Plant Extracts to Delay Lipid Oxidation in Foods
Kristen Robbins and William Schroeder

CHAPTER 15 457

Strategies to Prevent Oxidative Deterioration in Oil-in-Water Emulsion Systems: Canola-Based Phenolic Applications
Amy Logan and Peter Fagan

Editors and Contributors 485

Index 501

Preface

Lipid oxidation in food systems is one of the most important factors that affect food quality, nutrition, safety, color and consumers' acceptance. The control of lipid oxidation remains an ongoing challenge as most foods constitute very complex matrices. Lipids are mostly incorporated as emulsions, and chemical reactions occur at various interfaces throughout the food matrix. Incorporation of healthy lipids into food systems to deliver the desired nutrients has become a popular topic in the food industry. Many food ingredients contain a vast array of components, several of them unknown or constituting diverse or undefined molecular structures. It is both a challenge and a huge need in the food industry to develop effective approaches to mitigate lipid oxidation in food systems. Many approved and safe synthetic food additives that function as very effective antioxidants are perceived by consumers as artificial and potentially detrimental to their health, whereas naturally derived compounds are readily accepted. Although a number of naturally found antioxidants are widely used, they often lack the specificity and effectiveness of their synthetic counterparts, especially in the application of metal chelators.

The motivation behind this book came about as a result of several sessions organized by the Lipid Oxidation and Quality Division for the AOCS Annual Meeting in 2011. These sessions were entitled "A Re-examination of the Antioxidant 'Polar Paradox' Paradigm," "Lipid Oxidation Challenges and Potential Solutions in Food Systems," and "Antioxidants and Oxidation Control: Analytical Methodologies and Efficacies."

It is the objective of this book to provide recent perspectives aimed at a better understanding of lipid oxidation mechanisms and strategies to improve the oxidative stability of food systems. We were fortunate to have an international group of leading researchers from academic, industrial, and governmental entities share their knowledge and latest research findings. The first two chapters provide a thorough overview and discussion of the complexities of lipid oxidation mechanisms and the appropriate techniques to analyze intermediate and end-products of lipid oxidation. Chapters 3 through 5 focus on emulsion systems which are very susceptible to rapid lipid oxidation. Here, the role of interfaces and the effects of emulsifiers are discussed in much detail. Chapter 6 addresses the oxidative stability of enzymatically produced oils and fats, which is of growing importance as an alternative venue to produce customized products for the food industry. The following Chapters 7 and 8 review the antioxidant "polar paradox," discuss studies that contradict the current understandings and introduce the concept of the cut-off theory. Chapter 9 covers antioxidant and prooxidant effects of phenolics, while Chapter 10 reviews antioxidant assays with special emphasis on redox sensitive fluorescent probes to evaluate antioxidant capacity in lipid systems. The latter section of the book (Chapters 11 through 15) discuss various naturally-derived antioxidants with a focus on applications within food systems.

We are grateful to all authors who contributed to this book and invested valuable time to prepare extensive chapters far beyond their original oral presentations. We also would like to thank AOCS Press for the opportunity to execute this book project and kept it on track from beginning to end.

Amy (Richards) Logan
Uwe Nienaber
Xiangqing (Shawn) Pan

CHAPTER

1

Challenges in Elucidating Lipid Oxidation Mechanisms

When, Where, and How Do Products Arise?

K. M. Schaich[1]

Introduction

More than a century ago, researchers first noted that lipids oxidize by a free-radical mechanism, and more than 60 years ago, they identified the three stages (that is, initiation, propagation, and termination) of the chain reactions involved (Farmer et al., 1943; Boll and, 1945; Gunstone & Hilditch, 1945; Farmer, 1946; Holman & Elmer, 1947; Bolland, 1949). Several decades of intense research on both practical and theoretical aspects of lipid oxidation then developed current knowledge about lipid oxidation kinetics, processes, mechanisms, and products. This knowledge is used throughout the food industry to stabilize oils and foods, and it has also been applied to research on aging, cancer, and many other diseases. A number of excellent references have provided general overviews of lipid oxidation as well as detailed coverage of specific stages of the reactions (Swern, 1961; Frankel, 1962; Lea, 1962; Lundberg, 1962; Scott, 1965; Brodnitz, 1968; Frankel, 1980; Porter et al., 1981; Frankel, 1982, 1984; Porter et al., 1984; Porter & Wujek, 1984; Frankel, 1985; Porter, 1986; Chan, 1987; Porter & Wujek, 1987a; Porter, 1990; Frankel, 1991; Porter et al., 1995; Frankel, 2005; Schaich, 2005, 2006).

Despite this experience and research, preventing oxidation and maintaining the desired shelf life of foods formulated with high polyunsaturated fatty acids are still significant challenges. In fact, stability is often impossible to control or predict, even under established production protocols, oxidation kinetics are

[1]Dept. of Food Science, Rutgers University, 65 Dudley Rd., New Brunswick, NJ 08901-8520.

frequently inconsistent with expected reactions or product patterns, and preservation approaches based on the traditional free-radical chain reaction do not always work as expected. In addition, many routinely observed oxidation products cannot be explained by standard radical recombinations or alkoxyl radical scissions. The very fact that this book revisits the challenges of lipid oxidation demonstrates that we still have critical gaps in our understanding of lipid oxidation reactions and mechanisms.

The intent of this chapter, therefore, is not to rehash established material but rather to challenge anyone concerned with lipid oxidation—whether in fundamental chemical research, food quality deterioration, pathology or toxicology, cell signaling, or any other application—to think beyond hydrogen abstraction in established free-radical chain reactions and to consider other mechanisms that may alter the course, kinetics, and product distribution of lipid oxidation.

According to traditional explanations, hydrogen abstraction controls propagation, hydroperoxides are always the key intermediate, and products do not form until after hydroperoxides decompose. However, fundamental free radical chemistry shows that organic peroxyl radicals (ROO•) and alkoxyl radicals (RO•) also undergo independent rearrangement, addition, and scission or elimination reactions in competition with hydrogen abstractions in radical chain reactions. This means that end products other than hydroperoxides can accumulate in early stages of lipid oxidation, and secondary products can be generated without going through hydroperoxides. In lipid oxidation, these alternate reactions can have very important consequences for the course, kinetics, and analysis of lipid oxidation, as well as for the deterioration of food quality associated with lipid oxidation. Products of the various pathways—including alcohols, epoxides, dimers, ketones, and aldehydes—have different effects on the flavors and browning potential of food and, perhaps more importantly, different abilities to react with other molecules and broadcast oxidative damage. While all these product classes are well recognized in lipid oxidation, that they arise separately and from different mechanisms is not. Hence, multiple classes of oxidation products are seldom measured and important components of overall system oxidation (lipids plus other molecules) may be missed.

This chapter, thus, introduces the chemistry of alternate reactions to show that lipid oxidation is much more complex than the simplistic radical chain reactions normally presented, and to encourage more detailed thinking about how and when products are generated in oxidizing lipids and where these products have important effects in foods. The emphasis here will be on propagation processes. Initiation reactions and terminations have been discussed in detail elsewhere (Schaich, 2005, 2006; Schaich et al., 2013), so will not be covered here.

Radical Reactions Involved in Lipid Oxidation

Any discussion of lipid oxidation must begin by considering the classical free-radical chain reaction. All versions of traditional reaction emphasize that repeated hydrogen abstractions exclusively drive the reaction chain to pass radicals along to new molecules. Fig. 1.1 shows one version that is more complete than the abbreviated sequences included in most papers. In addition to the most familiar reactions, this version

- includes branching reactions that amplify propagation
- differentiates products in multiple hydroperoxide decomposition reactions (such as heat and light vs. oxidizing metals vs. reducing metals)
- includes hydrogen abstraction from hydroperoxides, which is ten times faster than abstraction from allylic sites (Howard & Ingold, 1967)
- adds termination from scission reactions, which actually produce more rancid flavors and off-odor compounds than radical recombinations

While not departing drastically from traditional representations, this version does present lipid oxidation as more complex than generally portrayed. Nevertheless, all versions of traditional free-radical chains have several problems:

1. All products are presented as deriving from hydroperoxides, directly or downstream.
2. Hydroxy lipids or lipid alcohols should be major intermediate products of each alkoxyl radical on par with hydroperoxides from peroxyl radicals.
3. Secondary products form entirely by radical recombination and only at the very end of the reaction sequence.

In practice, none of these conditions hold in either model systems or foods. The formation of many well-recognized lipid oxidation products, such as epoxides, cannot be explained by the classical chain reactions. Some products form independently of hydroperoxides, lipid alcohols (when reported) are seldom present at significant levels, product distributions are dominated by scission products (Frankel, 1982, 1984, 1987, 2005), and all the various products accumulate at different rates, some even before or simultaneously with hydroperoxides, depending on reaction conditions. It seems clear, then, that some reactions are missing, incorrect, or incomplete in current understanding of lipid oxidation.

It is always intriguing when data has been available and recognized for decades in the literature, yet has not been connected to related research. One such case appears in the extensive body of data[2] compiled by reknown free-radical

---

[2]Compiled and integrated in volumes 1 and 2 of *Free Radicals* (Wiley: New York, 1973).

CLASSICAL FREE RADICAL CHAIN REACTION MECHANISM OF LIPID OXIDATION

Initiation *(formation of ab initio lipid free radical)*

$$L_1H \xrightarrow{k_i} L_1\bullet \qquad (1)$$

Propagation

Free-radical chain reaction established

$$L_1\bullet + O_2 \underset{k_\beta}{\overset{k_0}{\rightleftarrows}} L_1OO\bullet \qquad (2)$$

$$L_1OO\bullet + L_2H \xrightarrow{k_{p1}} L_1OOH + L_2\bullet \qquad (3)$$

$$L_2OO\bullet + L_3H \xrightarrow{k_{p1}} L_2OOH + L_3\bullet \text{ etc.} \longrightarrow L_nOOH \qquad (4)$$

Free-radical chain branching (initiation of new chains)

$$L_nOOH \xrightarrow{k_{d1}} L_nO\bullet + OH^- \text{ (reducing metals)} \qquad (5)$$

$$L_nOOH \xrightarrow{k_{d2}} L_nOO\bullet + H^+ \text{ (oxidizing metals)} \qquad (6)$$

$$L_nOOH \xrightarrow{k_{d3}} L_nO\bullet + \bullet OH \text{ (heat and UV)} \qquad (7)$$

$$\left.\begin{array}{c} L_nO\bullet \\ L_nOO\bullet \\ HO\bullet \end{array}\right\} + L_4H \xrightarrow[k_{p1}\; k_{p3}]{k_{p2}} \left.\begin{array}{c} L_nOH \\ L_nOOH \\ HOH \end{array}\right\} + L_4\bullet \qquad \begin{array}{c} (8a) \\ (8b) \\ (8c) \end{array}$$

$$L_1OO\bullet + L_nOOH \xrightarrow{k_{p4}} L_1OOH + L_nOO\bullet \qquad (9)$$

$$L_1O\bullet + L_nOOH \xrightarrow{k_{p5}} L_1OH + L_nOO\bullet \qquad (10)$$

Termination *(formation of non-radical products)*

$$\left.\begin{array}{c} L_n\bullet \\ L_nO\bullet \\ L_nOO\bullet \end{array}\right\} + \left.\begin{array}{c} L_n\bullet \\ L_nO\bullet \\ L_nOO\bullet \end{array}\right\} \xrightarrow[k_{t2}\; k_{t3}]{k_{t1}} \begin{array}{l} \text{Radical recombinations} \\ \text{polymers, non-radical monomer} \\ \text{products (ketones, ethers, alkanes,} \\ \text{aldehydes, etc.)} \end{array} \qquad \begin{array}{c} (11a) \\ (11b) \\ (11c) \end{array}$$

$$\left.\begin{array}{c} LOO\bullet \\ LO\bullet \end{array}\right\} \xrightarrow[k_{ts2}]{k_{ts1}} \begin{array}{l} \text{Radical scissions} \\ \text{non-radical products (aldehydes,} \\ \text{ketones, alcohols, alkanes, etc.)} \end{array} \qquad \begin{array}{c} (12a) \\ (12b) \end{array}$$

i—initiation; o—oxygenation; β—O_2 scission; p—propagation; d—dissociation; t—termination; ts—termination/scission

Fig. 1.1. Free-radical chain reaction of lipid oxidation as normally presented with propagation by a series of hydrogen abstractions, expanded version.

From Schaich (2005), used with permission.

chemist Jay Kochi, which document a number of alternate reactions of organic peroxyl and alkoxyl radicals that occur in competition with hydrogen abstraction:

Peroxyl radicals	**Alkoxyl radicals**
atom or group transfer (H-abstraction)	atom or group transfer (H-abstraction)
rearrangement/cyclization to epidioxides	rearrangement/cyclization to epoxides
addition to double bonds	addition to double bonds
disproportionation (LOO• + LOO•)	α- and β-scission (fragmentation)
recombination	recombination
β-scission	
e^- transfer (LOO• + e^- → LOO^-)	

The dominance of or balance between these reactions controls the overall progress and final effects of lipid oxidation. For both peroxyl and alkoxyl radicals, hydrogen abstraction, internal rearrangement, and addition reactions contribute to chain propagation, though under different conditions. For peroxyl radicals, disproportionation leads to branching and a shift in kinetics without the formation of hydroperoxides, radical recombination and electron transfer terminate radical chains, and β-scission mediates *cis* to *trans* isomerization. Electron transfer is an active antioxidant mechanism that converts radicals to ions, particularly in the presence of active redox agents such as metals. For alkoxyl radicals, fragments from α- and β-scission generate many of the most recognized oxidation products, but the intermediate radicals can also contribute to propagation. The mechanisms occurring in any given system are determined by the availability of H donors, lipid concentration, solvent, double bond structure and steric factors in the target molecule, and reaction conditions, particularly temperature (Kochi, 1973b).

Instances of these alternate reactions have been reported for lipid oxidation (Bors et al., 1987; Gardner, 1989; Porter, 1990), but not systematically. Only recently has evidence for these alternate reactions in lipid oxidation been compiled and integrated to allow the visualization of relationships between the pathways (Schaich, 2005). Fig. 1.2 shows one version of how alternate pathways may fit together in an integrated reaction scheme for lipid oxidation. The traditional reaction sequence involving hydrogen abstractions is presented vertically in boxes down the center of the scheme as the core of the oxidation process. Alternate pathways are added as competing reaction cycles that divert peroxyl and alkoxyl radicals, and yield different intermediate products and secondary radicals at new

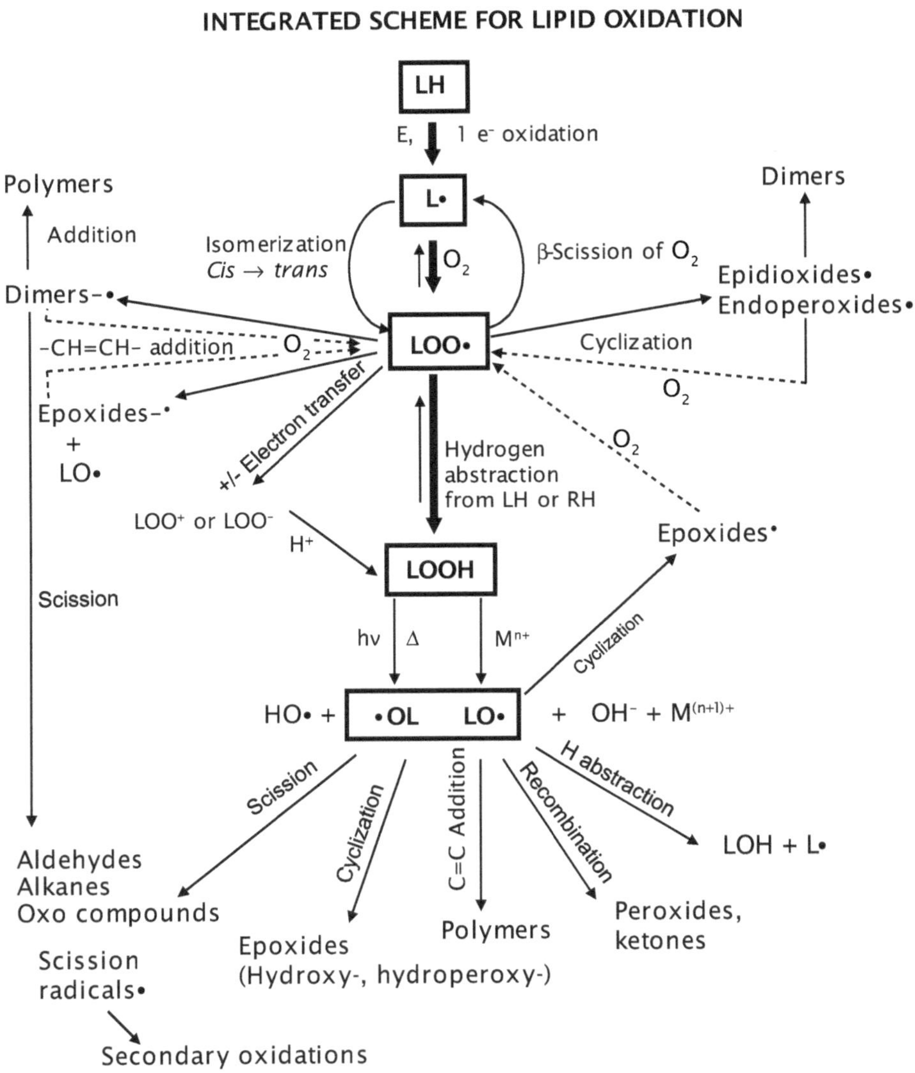

Fig. 1.2. Integration of multiple alternate reactions of lipid oxidation. Vertical boxed reactions in center denote the classical chain reaction driven by hydrogen abstraction. Alternate reactions are added to the lipid species involved (LOO•, LOOH, or LO•). Dotted lines indicate paths for oxygen addition to secondary radicals formed in cyclic and addition products, with formation of new peroxyl radicals.

From Schaich (2005), used with permission.

sites. These secondary radicals add oxygen and form peroxides that can either reenter the traditional H abstraction flow, designated by the dotted lines, or undergo further reaction outside the core process. When the reaction is drawn this way, it is clear just how extensively alternative reaction paths increase the complexity of both kinetics and product mixes of lipid oxidation.

This chapter proposes such an integrated scheme as a first step toward broader recognition of the complexities of lipid oxidation. The scheme is intended to augment and complement, rather than replace, the traditional radical chain reaction. Including documented alternate pathways does not provide just another version of standard lipid oxidation. Rather, accounting for alternate reactions requires that we dramatically rethink how the oxidation process is organized, how products are generated, and how lipid oxidation should be measured beyond hydrogen abstractions.

The discussion that follows provides documentation to support individual alternate reactions in lipid oxidation. The challenge now will be to test the integration of alternate reactions and document how they interact in lipid molecules alone, in foods, and in other systems. The author hopes that this chapter will stimulate broader recognition of lipid oxidation's complexity, rethinking of lipid oxidation reaction sequences and mechanisms, and a parallel shift in analysis of lipid oxidation from only hydroperoxides and hexanal to coordinated measurements of conjugated dienes, hydroperoxides, epoxides, alcohols, carbonyls, and dimers simultaneously in test materials (see also Chapter 2 of this book). Even more information about multiple mechanisms can be obtained by creatively applying chemometrics and advances in GC and HPLC with mass spectrometric detection to identify individual volatile and non-volatile products and determine how they change under different conditions. Only with such integration and analytical detail can we eventually construct a complete and accurate picture of lipid oxidation.

Reactions of L•

Ab initio lipid alkyl radicals (L•) are formed when initiators such as metals, preformed free radicals, light, and heat react with lipids by electron transfer or H abstraction. Although usually ignored in considerations of lipid oxidation mechanisms, the presence of L• must be considered ubiquitous unless all initiation sources are eliminated. L• are reasonably reactive themselves (Pryor, 1986) but they react faster with oxygen than with lipids and their redox potentials are low enough that they are inefficient oxidizers (Buettner, 1993).

L• half-life with typical substrate, 10^{-3} M, 37 °C	10^{-8} s	(Pryor, 1986)
Average reaction rate, k (L mol^{-1} s^{-1})	10^4–10^8	(Pryor, 1986)
Reaction rate with O_2, k (L mol^{-1} s^{-1})	10^8–10^{10}	(Ingold, 1969)
E° (mV)	600	(Koppenol, 1990)

Consequently, under normal conditions (i.e., in the presence of oxygen), L• is preferentially converted to LOO•, and peroxyl radicals become the active chain carriers for oxidations. Nevertheless, even L• have some alternate reactions that must be considered when evaluating the overall course of lipid oxidation.

L• reactions become very important at low oxygen pressures (<100 mm Hg) or under vacuum (Labuza, 1971). Under these conditions, the dominant propagation reactions of L• are additions to double bonds because their abstraction reactions are slower than addition. This leaves a new free radical on the adjacent carbon:

$$L\bullet + \text{-}CH_2\text{-}CH{=}CH\text{-}CH_2\text{-} \rightarrow \text{-}CH_2\text{-}\underset{\displaystyle L}{\underset{|}{C}}H\text{-}\dot{C}H\text{-}CH_2\text{-} \qquad (1.1)$$

Thus, although removing oxygen blocks the production of oxygenated products responsible for off-flavors and odors, this "protective" practice shifts pathways to addition, with generation of a variety of alkanes and potential impacts on textures.

L• are also important in termination reactions under low p_{O2}, again by addition, but this time to other radicals:

$$L\bullet + \left\{ \begin{array}{l} L\bullet \\ LO\bullet \\ LOO\bullet \end{array} \right. \rightarrow \begin{array}{l} L\text{-}L \\ LOL \\ LOOL \end{array} \qquad (1.2)$$

Importantly, in addition to being responsible for many characteristic lipid oxidation products, including alkanes, alcohols, and ketones, L• recombinations remove other oxygenated intermediates from the reaction system. This shifts the spectrum of products away from those normally measured to compounds seldom analyzed. As a consequence, a system evaluated by hydroperoxides or volatile aldehydes alone may appear to have low oxidation when instead reactions have been transferred to less oxygenated pathways for which products remain invisible (not measured).

Reactions of Peroxyl Radicals, LOO•

Peroxyl radicals that are traditionally written as the exclusive main chain propagators are relatively slow in their hydrogen abstraction reactions (k = 36–62 L

mol^{-1} s^{-1}) (Howard & Ingold, 1967; Gebicki & Bielski, 1981); they are also selective, abstracting only hydrogens with low bond energy (e.g., doubly allylic -CH_2-, thiols, phenols) (Sheldon & Kochi, 1981). As a result, there is plenty of time for alternate reaction pathways to compete and change the direction of oxidation (Simic, 1975), yielding distinctly different products at different rates with significant consequences for the ultimate mixture of products. β-Elimination of oxygen intervenes to slow the initial accumulation of LOO• and to isomerize double bonds.

β-Elimination of Oxygen

β-Elimination cleaves the C-O bond in L-OO• and releases O_2, leaving behind an alkyl radical:

$$LOO\bullet \leftrightarrow L\bullet + O_2 \qquad (1.3)$$

Because oxygen adds to L• almost instantaneously, a kind of cyclic oxygenation of L• can be established under some conditions. In linoleic acid, the rate of β-elimination at room temperature (27–430 M^{-1} s^{-1}) (Porter et al., 1995; Tallman et al., 2001) is competitive with H abstraction from allylic positions (1–400 M^{-1} s^{-1}) (Ingold, 1969; Howard, 1973; Porter et al., 1981). Thus, when abstractable hydrogens are limited (e.g., in dilute solution, in aprotic solvents, and as temperature increases), β-elimination becomes more important.

Reversible addition of oxygen to the pentadienyl system was first documented by kinetic (Howard & Ingold, 1968), EPR (Janzen et al., 1967), and ^{17}O evidence (Campbell et al., 1974; Chan et al., 1979). However, most of the evidence for the phenomenon and its consequences to lipid oxidation comes from Porter's group at Vanderbilt University (Porter & Nixon, 1978; Porter et al., 1980a, 1980b; Porter et al., 1981; Weenan & Porter, 1982; Roe et al., 1983; Porter et al., 1984; Porter & Wujek, 1984; Porter, 1986; Porter & Wujek, 1987a, 1987b; Porter, 1990; Mills et al., 1992; Porter et al., 1994a, 1994b; Porter et al., 1995; Tallman et al., 2001; Pratt et al., 2003).

β-Elimination of oxygen from LOO• has three important consequences to lipid oxidation: prevention of LOOH formation, with prolongation of the induction period; positional isomerisation along the acyl chain, which affects final product distributions; and geometric isomerisation of hydroperoxides, which affects product viscosity and its susceptibility to secondary oxidation. Basically, the process involves removal of the peroxyl oxygen, bond rotation, and/or migration of the free radical within the resonance system of the fatty acid, then readdition of the oxygen in a different orientation or at a new carbon position, respectively. This is quite similar to shifts and rotation of bonds during hydrogenation of double bonds. As shown in Fig. 1.3, interconversion of peroxyl position and orientation can continue

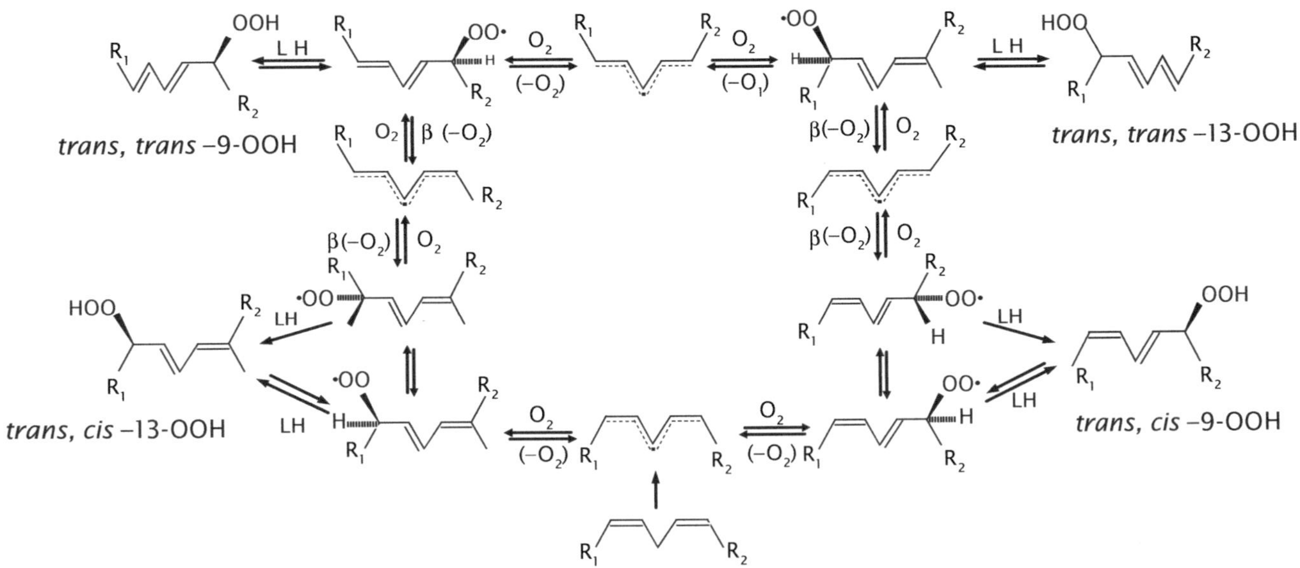

Fig. 1.3. Reaction scheme for positional isomerization of double bonds and formation of *trans,trans*-hydroperoxides during oxidation of linoleic acid via reversible β-scission of oxygen.

Data from Porter (1980, 1984); Weenan and Porter (1982).

From Schaich (2005), used with permission.

indefinitely as long as the radical is in the resonance manifold. The peroxyl radical becomes fixed as a hydroperoxide once it is protonated in reaction, but it can return to the manifold if the LOOH hydrogen is abstracted. For example,

$$\text{LOO}\bullet \text{ or LO}\bullet + \text{L}_1\text{OOH} \rightarrow \text{LOOH or LOH} + \text{L}_1\text{OO}\bullet \qquad (1.4)$$

Because β-elimination of oxygen mediates *cis-trans* and positional interconversions, it can critically influence the final product mixtures in ways that previous research has largely ignored or misattributed. The course of modification is driven by a kinetic–thermodynamic balance in which hydrogen abstraction kinetically competes with the thermodynamics that drive molecular rotation and transformation to isomeric products (Porter et al., 1980b; Porter & Wujek, 1987a; Tallman et al., 2001). *Trans,cis* isomers form rapidly and are the first products when good H donors from protic solvents, antioxidants, co-substrates, or the allylic hydrogens of the fatty acid chains themselves are present. In general, *trans,cis* isomer formation is favored in oriented systems or at high concentrations that increase interchain contact, while *trans,trans* isomers are favored in dilute solutions, aprotic solvents, and at elevated temperatures where there is less interchain contact and decreased H availability (Weenan & Porter, 1982). Contrary to earlier reports, the *cis-trans* ratio does not vary with the extent of oxidation, unless reaction conditions change or H abstraction from LOOH occurs, allowing LOO• to undergo β-scission. β-Elimination of oxygen is likely to be involved in interconversions between C-9 and C-13 hydroperoxides of linoleic acid at elevated temperatures, but the controlling factors and the direction of isomerizations remain controversial.

Hydrogen Abstraction

As shown in Fig. 1.1, hydrogen abstraction drives classical free-radical chain reactions. Peroxyl radicals initially formed at any site on a fatty acid abstract hydrogens from an allylic position (next to a double bond) or from a hydroperoxide on adjacent lipid molecules to transfer the unpaired electron and extend the chain. The process repeats itself indefinitely until the chain is intercepted.

$$\text{LOO}\bullet + \text{LH} \rightarrow \text{LOOH} + \text{L}\bullet \qquad (1.5)$$

$$\text{LOO}\bullet + \text{L'OOH} \rightarrow \text{LOOH} + \text{L'OO}\bullet \qquad (1.6)$$

Hydroperoxides were the first lipid oxidation products discovered (Lundberg & Chipault, 1947), and they remain the cornerstone of lipid oxidation analysis.

Position: Hydrogen abstraction by most oxyl radicals is quite specific, occurring preferentially at allylic hydrogens with the weakest C-H bond energies. In fatty acids, allylic hydrogens have the lowest C-H bond energies (Kerr, 1966), so they are the preferred sites for H removal and formation of a free radical. The -CH_2- groups between two double bonds in lipids (called doubly allylic) are doubly activated so C-H bond energies drop dramatically (Scott, 1965; Kerr, 1966), as shown in the structure below. Numbers in parentheses are bond energies in kcal/mol.

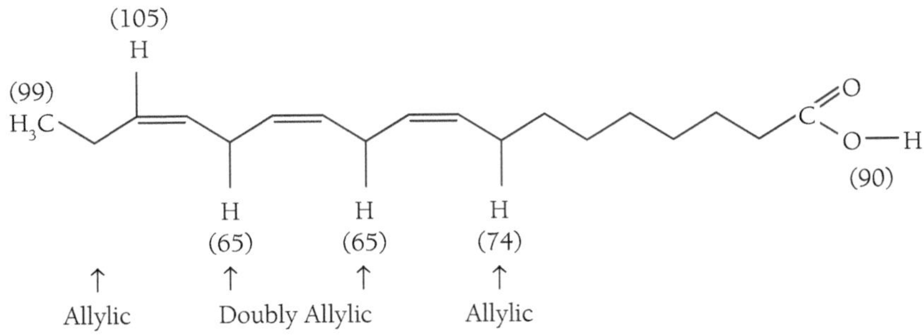

The order of reactivity, then, is doubly allylic H between two double bonds > singly allylic H next to double bonds >> H α to the -COOH group > H on methylene groups farther down the acyl chains. There is one exception to this rule. The hydroxyl radical, HO•, is so strongly electrophilic and reactive that it can abstract H indiscriminately from all positions along the acyl chain, and the new radicals formed then migrate to the acyl carbon with the weakest bonding (allylic H) or abstract allylic hydrogens from a neighboring lipid molecule (Patterson & Hasegawa, 1978).

What happens after the initial H abstraction is an interesting quandary. Older literature presents radicals as distributing over double bond systems to form resonance structures with multiple sites, each with equal probability of forming hydroperoxides (Frankel, 1962; Scott, 1965). The radicals predicted were at the two external carbons and all the methylene groups between double bonds. According to current understanding, however, the free electron becomes distributed across a resonance-stabilized double bond system following hydrogen abstraction from the allylic hydrogens. Because H abstraction occurs at either side of a double bond, each isolated double bond has two resonance systems, with the electron concentrated on the center carbon. For polyunsaturated fatty acids, resonance systems of individual double bonds overlap, so electron density doubles on allylic carbons between double bonds. In both of these systems, concentration of electron density on the center carbon makes the outside carbons relatively electron-deficient (denoted with arrows) and, hence, susceptible to oxygen addition (see Reactions 1.7 and 1.8).

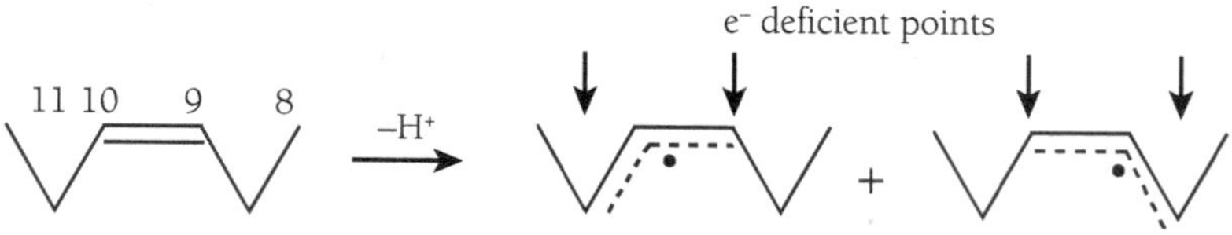

(1.7)

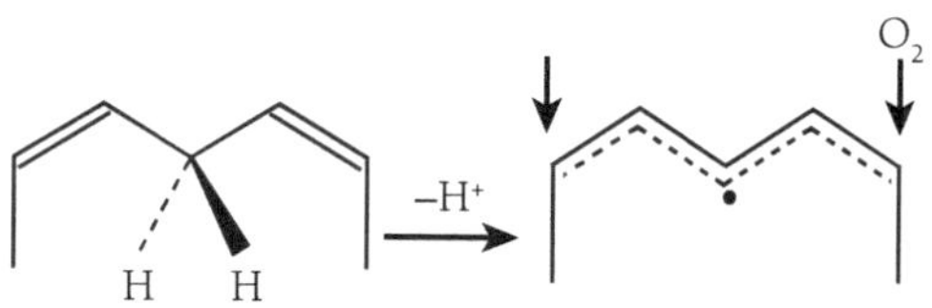

(1.8)

Table 1.A. gives the H abstraction positions predicted by extending these overlapping and 1,4-diene patterns to higher polyunsaturated fatty acids. These positions are important to recognize because they provide markers for formation of hydroperoxides and derivative secondary products. Despite the large number of theoretically possible positions for radical formation, it is now well established that only minor amounts of the corresponding product hydroperoxides are detected at internal positions in practice and the dominant hydroperoxides of autoxidizing fatty acids are always found at the external positions, regardless of the number of double bonds. This product pattern develops for two reasons. First, in 1,4-diene systems, radical formation on external hydrogens leads to isomerization of non-conjugated to conjugated double bonds:

$$\text{-CH=CH-CH}_2\text{-CH=CH-} \rightarrow \text{-}\dot{\text{C}}\text{H-CH=CH-CH=CH-} + \text{-CH=CH-CH=CH-}\dot{\text{C}}\text{H-} \quad (1.9)$$

Conjugation energy facilitates oxygen addition by stabilizing the C-OO• bond. Radicals formed at internal positions lack this stabilization and, hence,

Table 1.A. Carbons at Which Hydrogen Abstraction Occurs and Hydroperoxides Subsequently Form in Unsaturated Fatty Acids

Oleic acid	C-8,9,10,11
Linoleic acid	C-9,11,13
Linolenic acid	C-9,12,13,16
Arachidonic acid	C-5,8,9,11,12,15
EPA	C-5,8,9,11,12,14,15,18
DHA	C-4,7,8,10,11,13,14,16,17,20

Data from Frankel (2005).

undergo rapid β-elimination of the oxygen to regenerate the original dienyl radical (Tallman et al., 2001):

$$\text{-CH=CH-}\overset{\text{OO}\bullet}{\overset{|}{\text{C}}}\text{H-CH=CH-} \rightarrow O_2 + \text{-CH=CH-}\overset{\bullet}{\text{C}}\text{H-CH=CH-} \tag{1.10}$$

That a radical attack does indeed occur at the doubly allylic position has been verified by recent EPR and HPLC detection of doubly allylic radicals and non-conjugated hydroperoxides in methyl linoleate and higher PUFAs (Brash, 2000; Pratt et al., 2011), and resulting products can be isolated when high concentrations of abstractable hydrogens are present (e.g., with high added tocopherols). However, the dienyl radical is unstable and more rapidly isomerizes or undergoes β-scission (Pratt et al., 2011). The actual fate is influenced by the extent of double bond conjugation, solvent polarity (Jha & Pratt, 2008; Xu et al., 2009), solvent viscosity, temperature, and other factors.

Internal peroxyl radicals also have a very strong tendency towards cyclization (Pryor & Stanley, 1975; Porter et al., 1976; Coxon et al., 1981; Neff et al., 1981), so initial products rapidly transform into epidioxides and secondary products.

Why is the position of radical attack of more than academic importance? Current theories of lipid oxidation were established primarily from studies in purified fatty acids and their esters, where products are relatively simple and kinetics are easy to follow. However, in complex materials such as foods, cosmetics, and biological tissues, lipid orientations and molecular environments change and products arise that cannot be explained by radical attack only at the prescribed external positions. As Chapter 2 will discuss, strategies for detecting and quantitating lipid oxidation have traditionally been based on detecting expected products from predicted H abstraction sites only, and this artificially restricts what will be detected. The challenge then becomes which alternate pathways and products must be followed to determine the full extent of oxidation. Non-conjugated LOO• and LOOH may be too unstable for normal detection, but must be considered when evaluating products. Products observed in our recent work cannot be explained by C9- and C13-OOH alone.

Rates: H abstraction by peroxyl radicals (Reactions 1.5 and 1.8) is slow at room temperature,

$k \leq 1\ M^{-1}\ s^{-1}$ for most sites

$k = 62\ M^{-1}\ s^{-1}$ for dienes (Howard & Ingold, 1967)

with a strong preference for doubly allylic hydrogens (Bors et al., 1987; Bockman et al., 1998). This partially accounts for the slow oxidation during induction periods, where LOO• are the main chain carriers. H abstraction from hydroperoxides (Reaction 1.6) is ten times faster (Ingold, 1969) but, paradoxically, can slow oxidation by reducing the number of new LOOH and new chains formed. In some cases a steady state can develop in which hydroperoxides recycle each other, and propagation plateaus.

Three factors govern H abstraction rates by LOO•:

a) Availability of low bond-energy H sources in solvent and lipids, which provide targets. Chain propagation is facilitated in neat lipids where the lipid allyls are the only source of hydrogens, at high lipid concentrations where fatty acid chains come in closer contact, in aprotic solvents that do not compete with lipids as targets, and in higher polyunsaturated fatty acids with multiple bisallylic hydrogens. In contrast, when the solvent or other components in the system have H sources, competitive abstraction from non-lipid sites occurs, consequently quenching the radical and interrupting the chain, rather than propagating it.

b) Viscosity of the medium, which modulates accessibility (Walling & Padwa, 1963). H abstraction increases in low viscosity media but is impeded in viscous solvents (Factor et al., 1965; Porter et al., 1995).

c) Temperature, which provides thermal energy to reduce bond dissociation energy. The H abstraction rate increases to 10^3–10^4 M^{-1} s^{-1} at 65 °C (Ingold, 1969). However, any increases in H abstraction at elevated temperatures compete against reaction rates that are also increasing for other pathways, and the final balance is not always predictable.

Internal Rearrangement (Cyclization to Epidioxides)

When insufficient H_2 gas is present in the hydrogenation of unsaturated oils, radicals formed at the double bond find alternate pathways for reaction to reduce the system energy, and geometric and positional isomers form in competition with double bond saturation. Similarly, when abstractable hydrogens are not immediately available during oxidation of unsaturated lipids, peroxyl radicals find alternate pathways, the most facile of which is addition to a neighboring double bond to generate a cyclic product.

To accommodate the two oxygens, cyclization requires the presence of a *cis* double bond two carbons away from the LOO• (Baldwin, 1975; Porter et al., 1976). LOO• adds to the first carbon of this double bond to form a 5-exo ring, leaving a radical on the distal carbon of the double bond (Reaction 1.11). Addition of oxygen to this radical generates a second peroxyl radical (new position) which, in turn, abstracts a hydrogen from a neighboring lipid molecule to propagate the chain and form a hydroperoxy epidioxide product (Reaction 1.11a) (Chan, 1980).

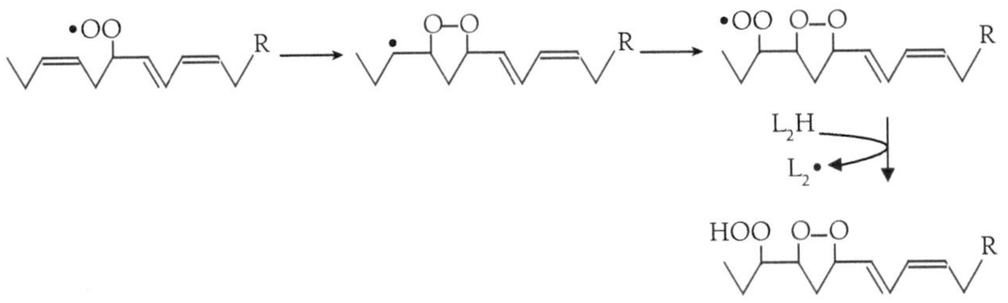

(1.11)

(1.11a)

The β-double bond structural requirement means that cyclization of peroxyl radicals occurs preferentially at internal positions. Peroxyl radicals of autoxidizing linoleic acid do not undergo cyclization because the requisite structure is not present. However, internal 10- and 12-hydroperoxides from $^1O_2$-photosensitized oxidation do have the required β-*cis* double bond. LOO• generated by H abstraction from these internal hydroperoxides rearrange to form propagating hydroperoxy epidioxide radicals (Reaction 1.12) and their corresponding products (Reaction 1.12a) in high proportions (Frankel et al., 1979; Mihelich, 1980; Frankel et al., 1982).

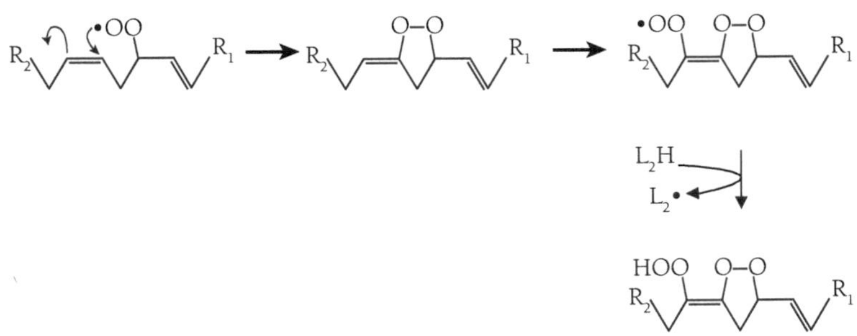

(1.12)

(1.12a)

As the number of overlapping 1,4-diene systems increases in higher polyunsaturated fatty acids (HPUFAs), the sensitivity to H abstraction and formation of LOO• at internal positions increases:

Linolenic acid	25% internal LOO•
Arachidonic acid	33
DHA[3]	66

As a result, HPUFAs are particularly prone to formation of cyclic peroxides. Intramolecular cyclization rates ($k{\sim}10^3\ s^{-1}$) (Porter, 1990) are faster than both β-scission of oxygen ($27\text{–}430\ s^{-1}$) (Porter et al., 1995) and H abstraction (<1–400

[3]Docosahexaenoic acid

$M^{-1}\,s^{-1}$) (Ingold, 1969; Howard, 1973; Porter et al., 1981), so cyclization should be able to compete with these two reactions as an initial process, especially in fatty acids with three or more double bonds.

Addition to Double Bonds

A second reaction that becomes competitive when abstractable hydrogens are limited (by aprotic solvents or low temperatures, for example) is addition of LOO• to double bonds to form an initial dimer complex. A radical simultaneously formed on the second carbon of the double bond then adds oxygen to form a peroxyl radical that becomes the propagating species.

$$\text{LOO}\bullet + \text{-CH=CH-} \rightarrow \text{-}\overset{\text{LOO}}{\overset{|}{\text{C}}}\text{H-}\overset{\bullet}{\text{C}}\text{H-} \xrightarrow{O_2} \text{-}\overset{\text{LOO}}{\overset{|}{\text{C}}}\text{H-}\overset{\text{OO}\bullet}{\overset{|}{\text{C}}}\text{H-} \qquad (1.13)$$

The dimer peroxy complex is unstable and reacts further to generate both monomer products (epoxides and epidioxides) and dimers or polymers. The ultimate products depend on the nature of the target double bond.

In early stages of oxidation, LOO• add to isolated or non-conjugated double bonds, though not with facility due to steric hindrance from the long acyl chain (Elson, 1975), and the adduct undergoes 1,3-cyclization to form an epoxide and an allylic radical, eliminating LO• in the process. The allylic radical then adds oxygen to form a new peroxyl radical (Reaction 1.14) (Gardner, 1989). This is a true branching reaction since two new propagating radicals (LO• and epoxy OO•) with increased reactivities are generated from the initial LOO•.

$$R_1\text{-CH}_2\text{-}\underset{\text{OOL}}{\underset{|}{\underset{\bullet}{\text{C}}}}\text{H-CH-}R_2 \rightarrow \text{LO}\bullet + R_1\text{-HC}\bullet\text{-}\underbrace{\text{CH-CH}}_{\text{O}}\text{-}R_2 \rightarrow R_1\text{-H}\overset{\text{OO}\bullet}{\overset{|}{\text{C}}}\bullet\text{-}\underbrace{\text{CH-CH}}_{\text{O}}\text{-}R_2 \qquad (1.14)$$

However, peroxyl radicals prefer to add to conjugated, terminal, and 1,1-disubstituted double bonds (Hiatt & McCarrick, 1975), as well as *trans* double bonds (Lewis & Mayo, 1958), at least in part because these configurations counterbalance steric impediments. The probability of LOO• addition increases in later stages of oxidation because doubly allylic hydrogens have been removed during initial reactions, the double bond system has been shifted to conjugated, and terminal double bonds then arise as intermediates in alkoxyl radical scission reactions. These conditions limit competition by hydrogen abstraction and greatly facilitate LOO• addition, which can then account for 80–100% of the products (Mayo, 1958a, 1968).

The complexity of addition products also increases in later stages of lipid oxidation, including not only the characteristic oxypolymers that increase the viscosity of very oxidized oils (Witting et al., 1957; Privett, 1959; Sims & Hoffman, 1962) but also hydroperoxyepoxides (A), dihydroperoxides (B) (Miyashita et al., 1982), and alkoxyl radicals (C), as shown in Reaction 1.15. Note that even in advanced oxidation, the addition product still retains a propagating free radical, distinguishing this addition from peroxyl radical recombinations that yield peroxo dimers without radicals in termination reactions.

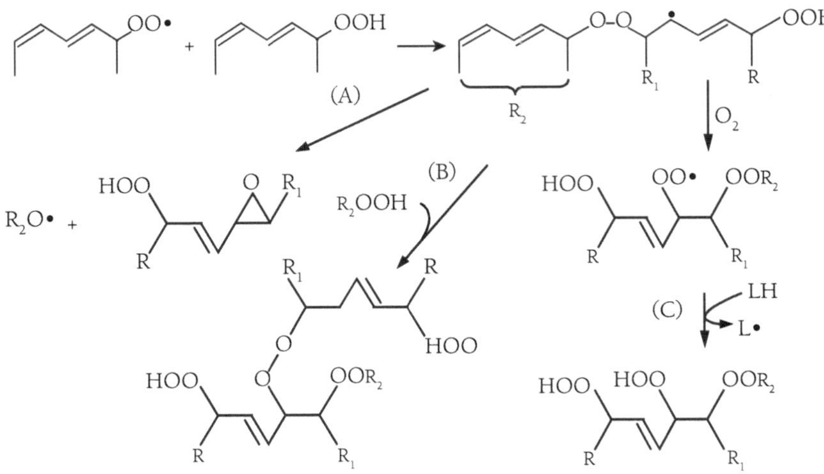

(1.15)

LOO• additions increase with heat (Mounts et al., 1970), solvent polarity (Haynes & Vonwiller, 1990), and extent of oxidation, which alters double bond structures and generates more reactive species. Epidioxide peroxyl radicals—secondary products of LOO• internal rearrangement—exhibit a particularly strong tendency to add to double bonds, with greater than 90% dimerization at 40 °C (Frankel et al., 1988). Heat not only increases addition rates but also changes the form of linkages. Dimer linkages were mostly C-O-O-C at lower temperatures, but shifted to C-C and C-O-C as the temperature and LOO• additions increased (Neff et al., 1988). Copper salts catalyze addition reactions of peroxyl radicals (Kochi, 1962a), which has interesting implications for food systems.

Disproportionation

Although peroxyl radical recombinations are most often written as termination reactions that generate lipid peroxides (ROOR) (Reaction 1.16a), an alternate propagation reaction occurs in lipids oxidized neat or in aprotic solvents where radical recombination forms new alkoxyl radicals rather than stable products (Reaction 1.16) (Thomas, 1965; Lindsay et al., 1973).

$$R_1OO\bullet + R_2OO\bullet \rightarrow [R_1OOOOR_2] \rightarrow R_1O\bullet + \bullet OOOR_2 \rightarrow R_1O\bullet + O_2 + \bullet OR_2 \quad (1.16)$$

$$[R_1OOOOR_2] \rightarrow R_1OOR_2 + O_2 \quad (1.16a)$$

(1.16) $2k = 9.5 \times 10^3\ M^{-1}\ s^{-1}$ (1.16a) $2k = 3.9 \times 10^2\ M^{-1}\ s^{-1}$

A unique and notable aspect of both dismutation and recombination of LOO• is the regeneration of molecular oxygen that can then continue to support oxidation with two important consequences: O_2 release makes oxygen consumption appear to slow in analyses, but prolongs oxidation in oxygen-limited systems. We have detected both actions in methyl linoleate and oils, particularly at elevated temperatures (Schaich, unpublished data).

Although disproportionation rates vary with the nature of the R groups (Adamic et al., 1969), dismutation to alkoxyl radicals (Reaction 1.16) is about 25 times faster than peroxide formation (Reaction 1.16a) (Thomas, 1965) in neat lipids (e.g., oils). This, plus the rapid reactions of RO•, leads to greatly accelerated oxidation (Traylor & Russell, 1965; Lindsay et al., 1973). Therefore, LOO• accumulation and disproportionation by Reaction 1.16 in later stages of lipid oxidation is probably a significant driver of the marked acceleration of oxidation during the bimolecular rate period.

In contrast, in polar solvents or aqueous solutions, the preferred reactions of LOO• shift to β-scission, the rate of LOO• decomposition increases dramatically, and termination by peroxide formation (Reaction 1.16a) becomes dominant (Walling et al., 1970; Heijman et al., 1985). The rate constant for LOO• recombination by Reaction 1.16a is ($2k = 2 \times 10^7\ L\ mol^{-1}\ s^{-1}$) in aqueous solution at pH 10.5 (Hasegawa & Patterson, 1978).

Reactions of Alkoxyl Radicals, LO•

As discussed above, LOO• are the radical chain carriers in early stages of oxidation. However, as LOO• react and form LOOH that subsequently decompose (Reactions 5–7 in Fig. 1.1), LO• radicals ascend to take over chain propagation as well as to direct product formation by multiple pathways. Because LO• react faster than LOO• by several orders of magnitude, one hallmark of secondary oxidation stages is very rapid oxidation that takes off after very slow oxidation in the induction period.

Described below are four major mechanisms for radical chain propagation by alkoxyl radicals. The mechanism dominating in a given system is determined largely by double bond structure, solvent conditions, and steric factors (Kochi, 1973b).

Hydrogen Abstraction

As discussed above, H abstractions by LOO• are relatively slow and selective for bis-allylic hydrogens alone (Bors et al., 1987). In contrast, because the redox potential of LO• is considerably higher than for LOO•—1.6 V vs. 1.0 V (NHE), respectively (Merenyi et al., 1994)—LO• can abstract hydrogens from positions with higher bond energies (both allylic and bis-allylic), and the reactions are very fast ($k{\sim}10^7$–10^8 $M^{-1}\,s^{-1}$) (Erben-Russ et al., 1987). Secondary alkoxyl radicals, including lipid alkoxyl radicals, are particularly effective in abstracting allylic hydrogens (Kochi, 1973b), so the following should be a major reaction in lipid oxidation (only one of the two possible abstraction positions is shown):

$$\underset{\displaystyle O\bullet}{R_1\text{-}\overset{}{\underset{|}{CH}}\text{-}R_2} + \text{-}CH_2\text{-}CH{=}CH\text{-}CH_2\text{-} \rightarrow \underset{\displaystyle OH}{R_1\text{–}\underset{|}{CH}\text{–}R_2} + \text{-}\dot{C}H\text{-}CH{=}CH\text{-}CH_2\text{-} \quad (1.17)$$

However, lipid alcohols are seldom more than trace components of product mixtures, so cyclization and scission reactions of LO• must compete with H abstraction and limit the role of this reaction in chain propagation.

H abstraction from other lipid chains by LO• is most effective in neat lipids where allylic groups are plentiful and the only source of hydrogens. The rates of H abstraction from different fatty acids by LO• are approximately proportional to the number of allylic or doubly allylic hydrogens (Table 1.B.) (Bors et al., 1984; Erben-Russ et al., 1987) but independent of position (Small et al., 1979; Bascetta et al., 1982). When bent chain configurations in HPUFAs align opposite ends of the acyl chains, the proximity of allylic hydrogens on the two parallel chains enhances *internal* H abstraction by lipid alkoxyl radicals. This generates radicals in positions that may not be normal for autoxidation and that have previously (Frankel, 1982) been attributed to HO• attack. At the same time, alignment of double

Table 1.B. Effects of Fatty Acid Structure and Solvent on Rates of H Abstraction by Lipid Alkoxyl Radicals

	$k \times 10^{-6}$ L mol^{-1} s^{-1}	
Fatty Acid	**Nonpolar Solvent**	**Aqueous Solution**
Oleic	3.8	68
Linoleic	8.8	130
Linolenic	13.0	160
Arachidonic	20.5	180

From Schaich (2005) used with permission. Data from Bors et al. (1984).

bonds also facilitates competing cyclization and epoxide formation (Hiatt & Szilagyi, 1970; Hiatt & Zigmund, 1970; Gilbert et al., 1977).

H abstraction by LO• to propagate free-radical chains is facile in nonpolar aprotic solvents when lipids are at high concentrations. However, at moderate lipid concentrations, H abstraction must compete with internal rearrangements and scission (Walling & Padwa, 1963), and at low concentrations it may become insignificant (Tsentalovich et al., 1998).

Surprisingly, the most important H abstraction by LO• may well be from lipid hydroperoxides as they form (Reaction 1.18), thus regenerating LOO•.

$$\text{LO}\bullet + \text{LOOH} \rightarrow \text{LOH} + \text{LOO}\bullet \tag{1.18}$$

The bond dissociation energy of the hydroperoxide hydrogen is higher than that of the allylic hydrogens (90 vs. 65–85 kcal mol^{-1}, respectively), but hydrogen bonding between the LO• and LOOH greatly decreases the E_a for the abstraction (Avila et al., 1995). In mixtures of fatty acids and their hydroperoxides, *tert*-butoxyl radicals abstract hydrogens almost exclusively from the hydroperoxides (Porter et al., 1976) at diffusion controlled rates ($k = 2.5 \times 10^8 M^{-1}\ s^{-1}$) (Paul et al., 1978). Similarly, cumylalkoxyl radicals abstract H from hydroperoxides faster than they do from alkyl substrates (Avila et al., 1995). This reaction eliminates the need for other catalysts to decompose the hydroperoxides and begin chain branching, but it also paradoxically slows down overall oxidation rates.

In aqueous and protic solvents where H sources are plentiful, hydrogen abstractions by LO• are faster as a reaction (Table 1.B.) but less effective in chain propagation for two reasons: (1) selectivity of H abstraction decreases, and H abstraction must compete with increasing rate of LO• scissions (Kochi, 1962b); and (2) increased H abstractions from water, solvents, or dissolved solutes terminate rather than propagate lipid radical chains (Walling & Wagner, 1964; Simic, 1981; Simic et al., 1992).

Finally, the effects of temperature on rates of H abstraction by LO• must be considered. In polar and aqueous solvents, the activation energy for H abstraction is low, so there is little thermal enhancement of abstraction rates (Walling & Wagner, 1964). In contrast, the rate of H abstraction by LO• increases notably with temperature in lipids oxidized neat or in nonpolar solvents where the only H sources are fatty acids. Up to about 80 °C, this leads to marked acceleration of oxidation and enhancement of LOOH formation over cyclization (Neff et al., 1978; Selke et al., 1978; Neff et al., 1981; Neff et al., 1988). However, at T > 100 °C the relative importance of H abstraction in propagation diminishes as LO• scission is enhanced more than abstractions, and secondary processes begin to dominate (Mayo, 1958a, 1958b).

Internal Rearrangement (Cyclization)

As with LOO•, LO• seek alternate pathways for reaction under conditions where abstractable hydrogens are not immediately available. One major option that is important in terms of products but often overlooked is internal cyclization, in which the alkoxyl –O• adds to the α carbon of the immediately adjacent double bond to form an epoxide, with transfer of the free electron to the β carbon to generate a new epoxyallylic radical (Reaction 1.19).

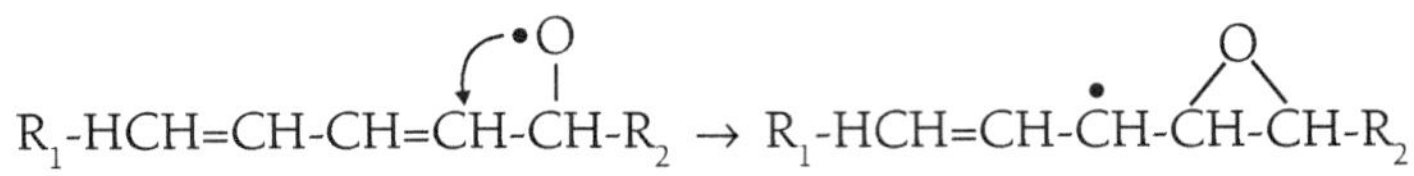

(1.19)

Under some conditions, this fast reaction can even exceed the rate of H abstraction (Kochi, 1962b). As with LOO•, LO• has a strong tendency to cyclize from internal positions, so $^1O_2$-photosensitized oxidations yield high concentrations of cyclic products (Neff, 1980). Indeed, the levels and positional distribution of these products are characteristic markers distinguishing autoxidation from photosensitized oxidation.

Cyclization of LO• to epoxides is the dominant reaction in aprotic solvents (including neat lipids), when lipids are at low concentration (Van Sickle et al., 1967) or highly dispersed on a surface (Wu et al., 1977, 1978), at room temperature (Acott & Beckwith, 1964; Dix & Marnett, 1981; Bors et al., 1984), and at low oxygen pressures (Mayo, 1958b; Van Sickle et al., 1967). We have found epoxides to be the dominant products in methyl linoleate and food oils incubated neat (Xie & Schaich, 2012). The reaction accelerates with increasing polarity of aprotic solvents (Kochi, 1962b; Walling & Wagner, 1964; Bors et al., 1984), but counterbalancing this is considerably reduced stability of LO• in polar solvents and greatly enhanced H abstraction and scission reactions in protic solvents (Kochi, 1962b; Walling & Wagner, 1964). Thus, although short-lived epoxyallylic radicals have been observed in aqueous solutions by pulse radiolysis (Bors et al., 1984), they become progressively less detectable as solvent polarity increases and other products become favored (Table 1.C.).

Temperature has relatively little effect on cyclization because the activation energy for the rearrangement is low. Cyclization thus dominates in neat lipids at room temperature, but as the temperature increases, H abstraction and scission become more important directors of propagation. At high temperatures (>100 °C) rearrangement is a relatively minor process (Mayo, 1958b; Acott & Beckwith, 1964).

Cyclization is activated by metals, particularly Fe and Cu. The ubiquitous presence of metals in foods thus would imply that epoxides should always be a major product in oxidizing lipids. However, these metals also direct isomerization

Table 1.C. Effect of Solvent Polarity on Balance of LO• Internal Cyclization with Other Products in Reactions of Methyl Linoleate Hydroperoxides

		Product Distribution (%)					
Solvent and System	**Catalyst**	**LOH/ LOOH**[a]	**Cyclic**[b]	**Scission**	**Other**	**Unknown**	**Reference**
CH_2Cl_2	$FeCl_3$		100				Haynes & Vonwiller, 1990
MeOH	$FeCl_3$/ cysteine	3–8	75–80	13–15			Gardner & Kleiman, 1981
Cyclohexane	UV Light	15	68		18[c]		Schieberle et al., 1988
80% EtOH	$FeCl_3$/ cysteine	30	11	—	7[d]	7[e]	Gardner et al., 1974

[a]Total of all H abstraction products, all isomers
[b]Total of all products that had any cyclic component
[c]Oxo dienes
[d]Hydroxyl ethoxylated products from reaction with solvent radicals
[e]Unidentified soluble products and volatile scission products

to multiple positional dihydroxy and hydroxyene isomers (Acott and Beckwith, 1964) and conversion of HO-epoxides to ketols (Gardner and Kleiman, 1979), and competition from these reactions reduces epoxide yields. Epoxides are rarely mentioned as products in research papers, in large part because epoxides are seldom measured due to lack of sensitive methodology and difficulty in detecting epoxides. Transformations of epoxides in the presence of metals, instability of epoxides in water and other polar solvents noted above, volatility of epoxides (Schaich, unpublished data), and very rapid reactions of epoxides with proteins (Schaich, 2008) all remove epoxides from the product mix in foods and complex model systems. Thus, lack of epoxide detection does not mean that internal rearrangement to epoxides is not an active pathway; it only indicates that downstream products must be measured to track and verify intermediate epoxide formation.

Addition

The high redox potential of LO• makes them strongly electrophilic, so alkoxyl radicals have unusually strong preference for allylic attack. Thus, intermolecular H abstraction or internal cyclization dominate as long as allylic hydrogens are present. Addition of LO• to double bonds occurs only under selective conditions, particularly, in the absence of allylic hydrogens and conjugation (Table 1.D). These conditions develop as oxidation progresses. As a consequence, propagation by LO• addition is most active in catalyzing chain branching during secondary stages of oxidation.

Table 1.D. Effect of Lipid Structure on Competition between H Abstraction and Addition by RO•. Both product distributions (top) and thermodynamics (bottom) show facilitation of addition by trans and conjugate double bonds.

Alkene	Abstraction (%)	Addition (%)	Reference
R-CH=CH_2	97	3	Kochi, 1973
R_2-C=CH_2	83	17	
R-CH_2-CH=CH_2-R (*trans*)	95	3–4	
R-CH_2-CH=CH_2-R (*cis*)	83	17	
	ΔH Abstraction kcal/mol	**ΔH Addition kcal/mol**	
RCH_2CH=$CHCH_2R'$	−15	−8	Mayo, 1968
RCH=CH-CH=CHR'	−19	−20	
RCH=CH-CH_2-CH=CHR'	−26	−8	

Note that a free radical forms on the opposite carbon in the double bond, and this becomes the new propagating species (Reaction 1.20). In contrast to LOO• additions, LO• addition increases with *cis* configuration and asymmetrical substitution on double bonds (Walling & Thaler, 1961; Kochi, 1973b). Because H abstraction transforms adjacent double bonds from *cis* to *trans* conformation, LO• preferentially adds to the unoxidized double bonds in polyunsaturated fatty acids:

(1.20)

Given the reaction requirements, propagation by LO• addition is most important in neat lipids (e.g., oils) and organic solvents (Bors et al., 1984). It occurs rapidly but is not competitive with scission and rearrangement reactions in aqueous systems. Heat catalyzes additions by decomposing LOOH to LO• (Mounts et al., 1970).

α- and β-Scission

Scission of the C-C bond on either side of the LO• group yields a mixture of carbonyl products and free radicals—typically aldehydes, alkanes, and oxo-esters from the initial alkoxyl radicals (Chan et al., 1976).

The α and β notations in the fragmentation refer to the position of chain scission relative to the -COOH on the fatty acid: α indicates scission of the alkoxyl radical occurs on the side closest to -COOH, while β designates scission on the chain terminus side (Reaction 1.21).

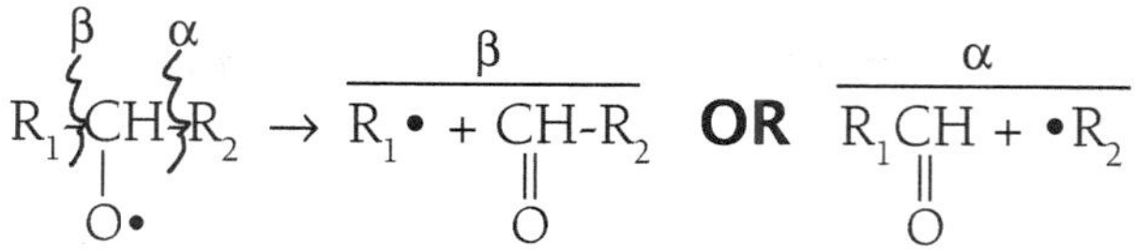
(1.21)

The radical fragments add oxygen and abstract hydrogens to propagate the radical chain, a process that is often overlooked. Fragments containing the oxygen atom become the volatiles and flavor compounds most strongly associated with rancidity. With two possible scissions at each LO• site, the distribution of products can become quite complex in secondary stages of oxidation, especially in higher polyunsaturated fatty acids (Figs. 1.4–1.6 on pages 26–31). Indeed, α and β scissions have often been presented as the most important reactions of alkoxyl radicals in complex systems, such as foods (Frankel, 1982, 1984, 1985, 1987), perhaps due to their sensory impact and perhaps because their products have been much more extensively identified than in other products.

What determines whether a scission will be α or β relative to the -COOH? A number of guiding rules have been suggested for this question, although none have been fully substantiated:

a) Scissions should occur between the alkoxyl radical and the double bond (Frankel, 1982, 1987). For example,

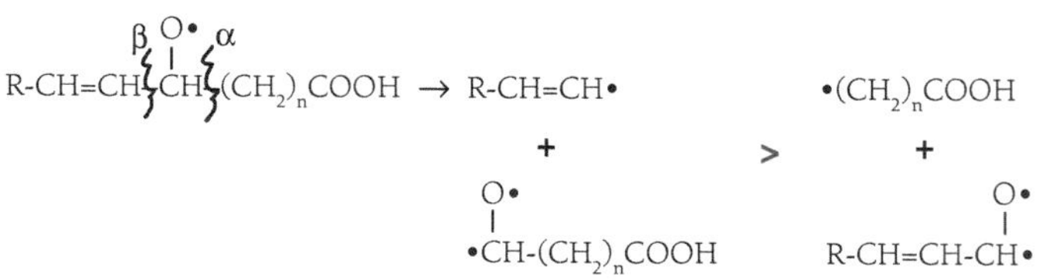
(1.22)

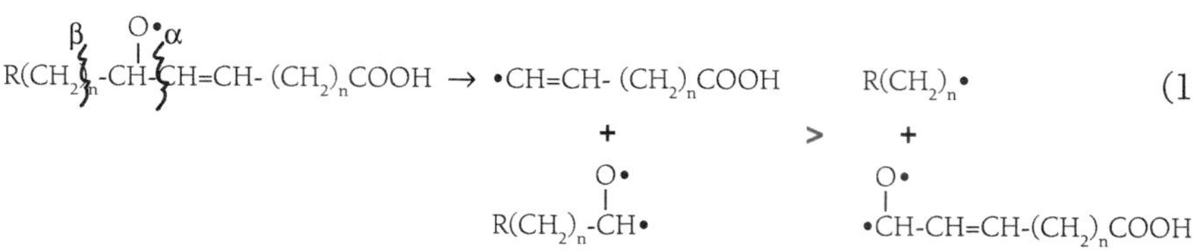
(1.23)

This rule has been questioned on energetic grounds because the dissociation energy for vinyl bonds is 109 kcal, but is 60 kcal for allylic bonds (Nawar, 1969).

b) Scission should occur preferentially at the site that generates the product most thermodynamically stable; for example, saturated aldehydes are more stable than unsaturated aldehydes, as is the case in Reactions 1.22 and 1.23. However, it is not clear whether the determining force is the stability of the radical, as in

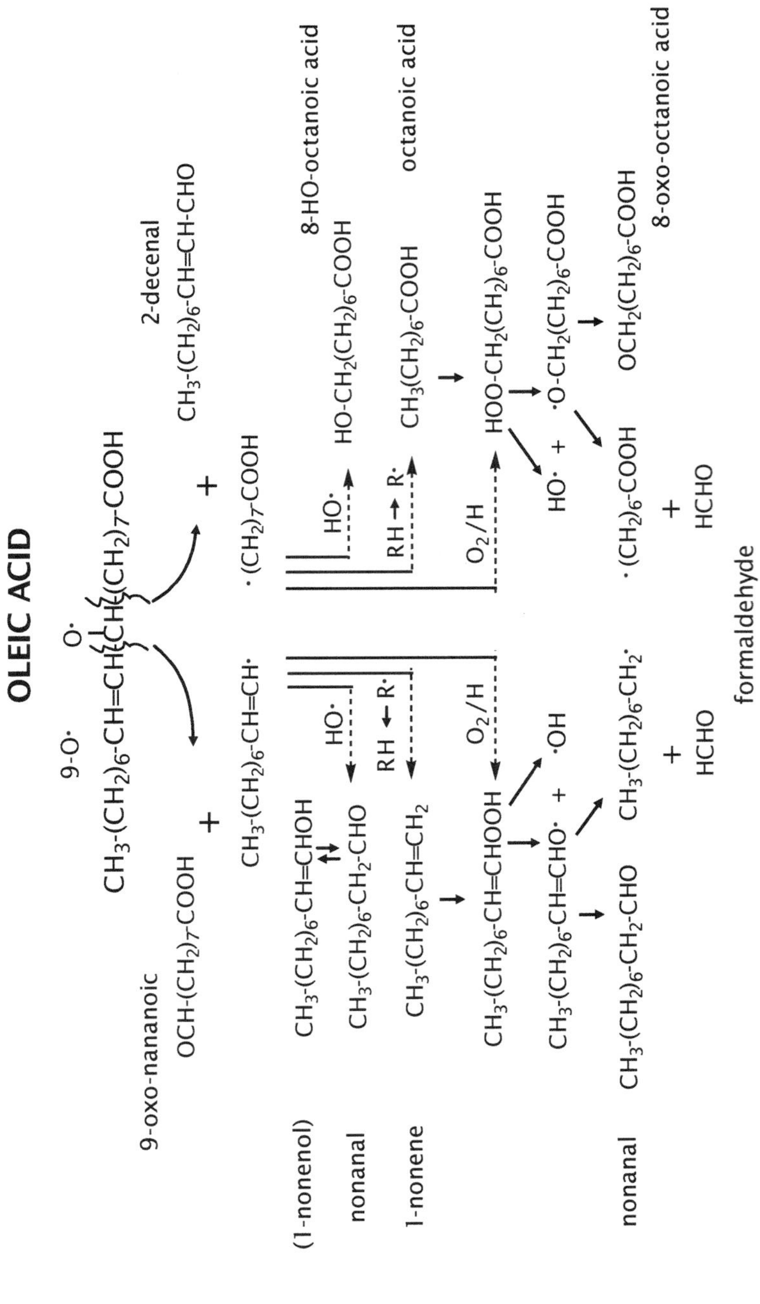
OLEIC ACID
9-O·
O·
CH3-(CH2)6-CH=CH-CH-CH-(CH2)7-COOH
9-oxo-nananoic
OCH-(CH2)7-COOH
+
CH3-(CH2)6-CH=CH·
2-decenal
CH3-(CH2)6-CH=CH-CHO
+
·(CH2)7-COOH
(1-nonenol)
CH3-(CH2)6-CH=CHOH
nonanal
CH3-(CH2)6-CH2-CHO
HO·
RH R·
1-nonene
CH3-(CH2)6-CH=CH2
O2/H
CH3-(CH2)6-CH=CHOOH
CH3-(CH2)6-CH=CHO· + ·OH
nonanal
CH3-(CH2)6-CH2-CHO
CH3-(CH2)6-CH2·
+
HCHO
formaldehyde
8-HO-octanoic acid
HO-CH2(CH2)6-COOH
octanoic acid
CH3(CH2)6-COOH
HOO-CH2(CH2)6-COOH
HO· + ·O-CH2(CH2)6-COOH
·(CH2)6-COOH
OCH2(CH2)6-COOH
8-oxo-octanoic acid

Following the same fragmentation pattern—

	β-scission	*α-scission*
8–O• →	8-oxo-octanoic acid + decanal 1-decene [nonanol nonane nonanal formaldehyde]	2-undecenal + 7-HO-heptanoic acid heptanoic acid 7-oxo-heptanoic acid [6-HO-hexanoic acid hexanoic acid 6-oxo-hexanoic acid formaldehyde]
10–O• →	10-oxo-8-decenoic acid + octanol octane octanal [heptanol heptane heptanal formaldehyde]	nonanal + 9-oxo-nonanoic acid 8-nonenoic acid [octanol octane octanal formaldehyde]
11–O• →	11-oxo-9-undecenoic acid + heptanol heptane heptanal [hexanol hexane hexanal formaldehyde]	octanal + 10-oxo-decanoic acid 9-decenoic acid [nonanal nonane nonal formadehyde]

Fig. 1.4. Typical scission products of oxidizing oleic acid. Parentheses indicate unstable intermediates; brackets denote products from secondary scissions.

Data from Frankel (1987) and Grosch (1987). Modified from Schaich (2005), used with permission.

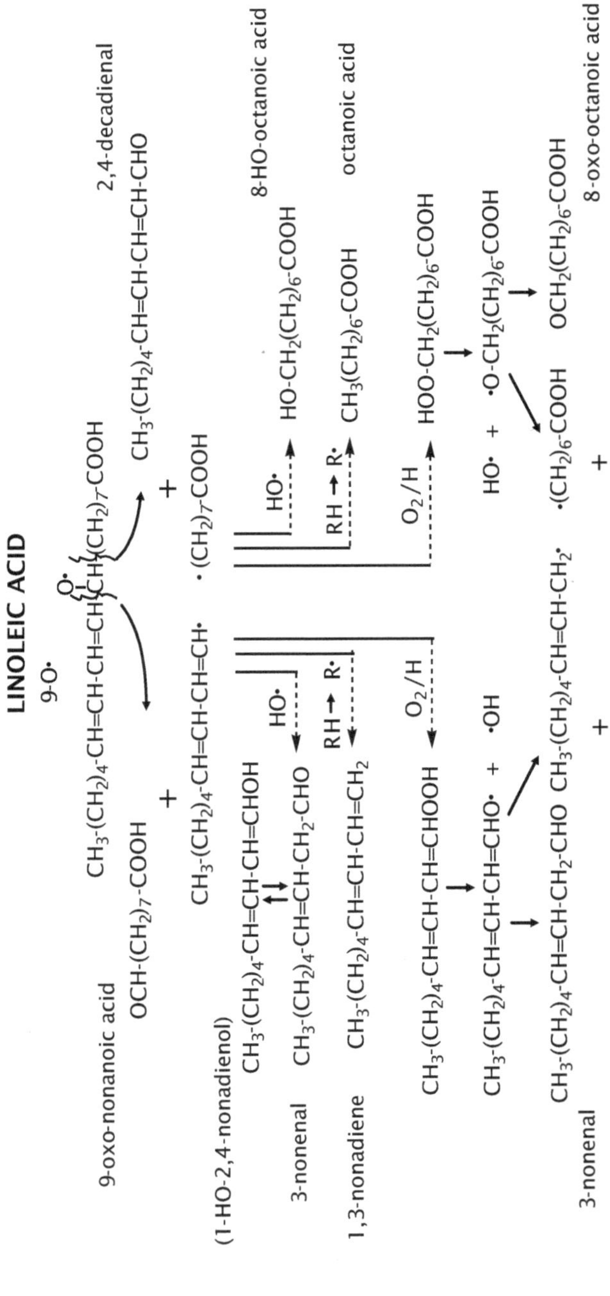
LINOLEIC ACID
9-O•
9-oxo-nonanoic acid
2,4-decadienal
(1-HO-2,4-nonadienol)
3-nonenal
1,3-nonadiene
8-HO-octanoic acid
octanoic acid
8-oxo-octanoic acid
formaldehyde
3-nonenal

Following the same fragmentation pattern—

13–O→ 13-oxo-9,11-tridecadienoic acid + pentanol, pentane, pentanal, [butanol, butane, butanal, formaldehyde]

hexanal + 12-oxo-9-dodecenoic acid, 9,11-dodecadinoic acid, [11-HO-9-undecenoic acid, 9-undecenoic acid, 11-oxo-9-undecenoic acid, formaldehyde]

Fig. 1.5. Typical scission products of oxidizing linoleic acid. Parentheses indicate unstable intermediates; brackets denote products from secondary scissions.

Data from Frankel (1987) and Grosch (1987). Modified from Schaich (2005), used with permission.

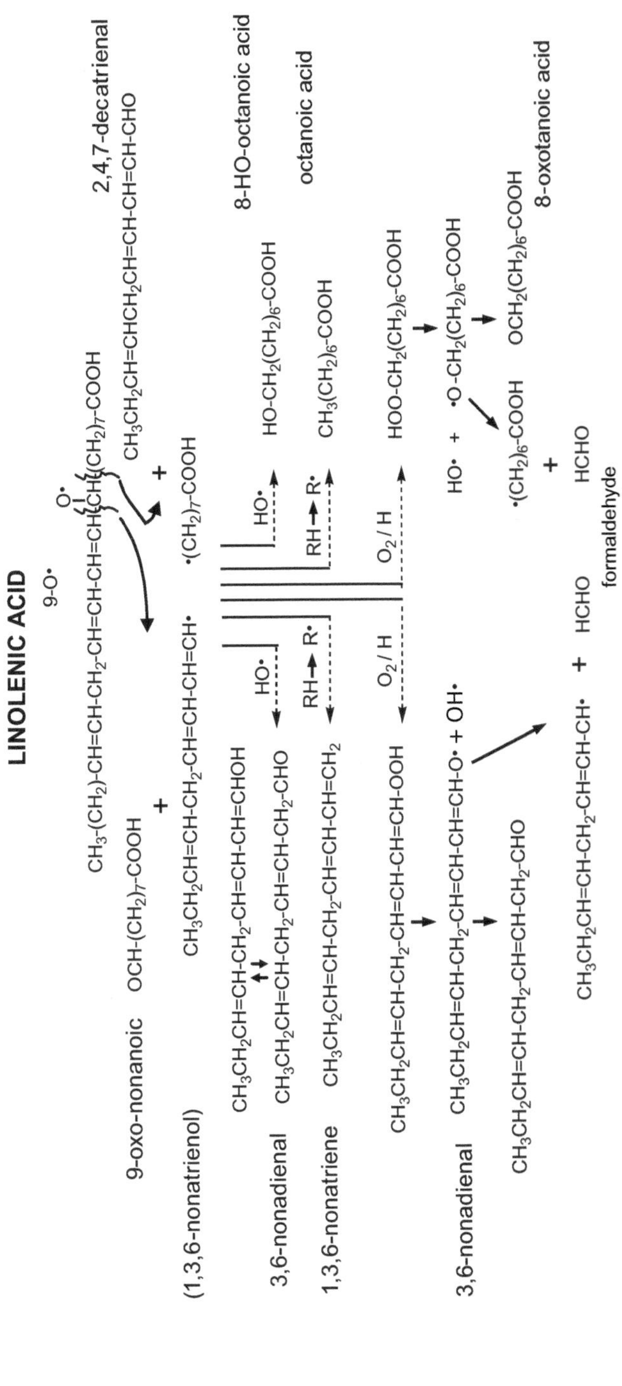
LINOLENIC ACID
9-O•
9-oxo-nonanoic
(1,3,6-nonatrienol)
3,6-nonadienal
1,3,6-nonatriene
3,6-nonadienal
formaldehyde
2,4,7-decatrienal
8-HO-octanoic acid
octanoic acid
8-oxotanoic acid

Following the same fragmentation pattern—

	β-scission	*α-scission*
16–O• →	16-oxo-9,12,14-hexadecadienoic acid + ethane	propanal + 15-oxo-9,12-pentadecadienoic acid 9,12,14-pentadecatrienoic acid [14-HO-9,12-tetradecadeinoic acid 9,12-tetradecadeinoic acid 14-oxo-9,12-tetradecadienoic acid formaldehyde]
13–O• →	13-oxo-9,11-tetradecadienoic acid + 2-pentenol 2-pentene 2-pentenal [butanal butene formaldehyde]	3-hexenal + 12-oxo-9-dodecenoic acid [9,11-dodecadienoic acid 11-HO-9-undecenoic acid 9-undecenoic acid 11-oxo-9-undecenoic acid formaldehyde]
12–O• →	12-oxo-9-dodecenoic acid + 3-hexenal 3-hexene [2-pentenol 2-pentene 2-pentenol formaldehyde]	2,4-heptadienal + 11-HO-9-undecenoic acid 9-undecenoic acid [11-oxo-9-undecenoic acid 10-oxo-decanoic acid 9-decenoic acid formaldehyde]

Fig. 1.6. Typical scission products of oxidizing linolenic acid. Parentheses indicate unstable intermediates; brackets denote products from secondary scissions.

Data from Frankel (1987) and Grosch (1987). Modified from Schaich (2005), used with permission.

Reaction 1.22 (Russell, 1956; Fukuzumi & Kochi, 1981; Wilsey et al., 1999; Baciocchi et al., 2002), or the carbonyl products produced in Reaction 1.23 (Frankel, 1987; Avila et al., 1993).

c) The presence of a free -COOH group increases the tendency toward α scission by inductive effects, but also favors selective cleavage between LO• and the double bond (α or β, depending on LO• position) (Kimoto & Gaddis, 1974).

d) The tendency for α cleavage increases as the alkoxyl radical position moves away from the carboxylic acid, consistent with preferential scission between the alkoxyl radical and the double bond as well as the formation of saturated aldehydes (Selke et al., 1978).

It is clear from these "rules" that many of the factors influencing scission points have opposing effects. In practice, scission patterns reported for oxidizing fatty acids have generally been mixed and inconsistent, varying with the product structure and reaction conditions. Thus, the issue remains unresolved for the time being.

It should be noted that most of these guidelines were developed from GC analysis of volatiles, where methodology and instrumental sensitivity play critical roles in what is detected. Different analytical approaches may provide new insights into scission preferences. For example, in recent research on alternate pathways using cryotrapping to ensure detection of all low molecular-weight products, we found that C-13 attack on linoleic acid was dominant in all systems studied, and that pentane from β-scission was the dominant product, present at levels an order of magnitude or more higher than the expected hexanal from α-scission (Bogusz & Schaich, 2012a, 2012b). Therefore, to explain lipid oxidation more fully and accurately, we also must actively apply contemporary analytical methods to gain sensitivity and clarity in analysis of lipid oxidation products, in addition to thinking beyond H abstraction in traditional radical chain reactions. This issue will be addressed in more detail in Chapter 2.

For scission to go forward kinetically, its products must be stabilized thermodynamically to make the process energetically favorable. The system increases in polarity as nonpolar alkoxyl radicals transform to polar cleavage products, as shown in Reaction 1.24 (Kochi, 1973b).

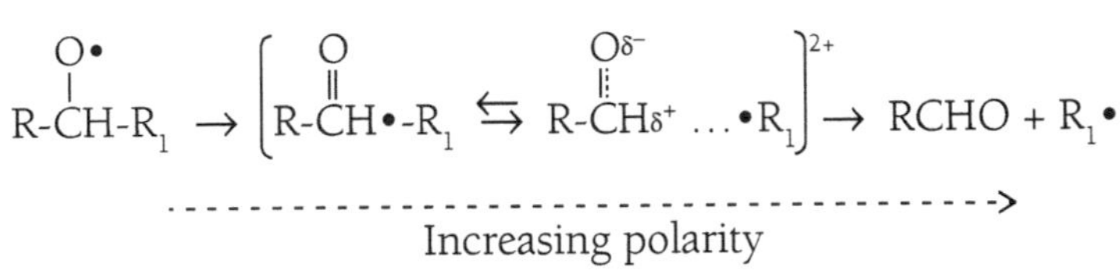

(1.24)

Thus, stabilization for scission is provided by two polar processes: (1) solvation and hydrogen bonding by water and polar protic solvents to support transition states and reduce the activation energy for bond rupture (Walling & Padwa, 1963; Walling & Wagner, 1963; Kochi, 1973; Avila et al., 1993; Tsentalovich et al., 1998), and (2) H^+ from solvents that add immediately to the scission radicals to provide the driving force for the reaction (Schauenstein, 1967).

Polar protic solvents provide critical stabilization at each stage. Scission of alkoxyl radicals in aqueous solution ($k_s = 10^6 – 10^7\ s^{-1}$) (Farmer & Sutton, 1943; Gilbert et al., 1981; Neta et al., 1984) is 10–100 times faster than in nonpolar organic solvents with dielectric constants comparable to fatty acid methyl esters (Baignee et al., 1983; Neville et al., 1989; Baciocchi et al., 2002). Even though this rate is still slower than H-abstraction (Table 1.E), scission usually competes effectively and can account for high proportions of products in polar media. For example:

- ~48% fragmentation, ~48% H abstraction in aqueous solutions of *tert*-butyl-O• on a pulse radiolysis time scale (ms to s) (Bors et al., 1984).
- scission fragment 2-hydroxyheptanal as the sole product of n-6 fatty acids oxidized in Tris-KCl + $FeSO_4$ /ascorbic acid for up to 24 h (Loidl-Stahlhofen & Spiteller, 1994).
- 7–10% scission oxidation products in neat triolein shifted to 66–75% in the presence of acid (Kimoto & Gaddis, 1974).
- proportion of scission products of cumene-O• increased from 24 to 81% as solvent polarity increased (solvent series in Table 1.E); H abstraction was able to compete only when scission was less than 35% (Boto et al., 2003).

Table 1.E. Effects of Solvent Polarity on Rates of H Abstraction versus β-Scission of Cumene Alkoxyl Radicals

Solvent	Polarity Index[a]	k_a, H Abstraction[b] $10^{-6}\ M^{-1}s^{-1}$	k_β β-Scission[b] $10^{-5}s^{-1}$	Ratio[b] $k_a/k_\beta\ M^{-1}$
CCl_4	0.259	1.1	2.6	4.5
C_6H_6	0.111	1.2	3.7	3.2
C_6H_5Cl	0.188	1.1	5.5	2.0
$(CH_3)_3COH$	0.389	1.3	5.8	2.3
CH_3CN	0.460	1.2	6.3	1.9
CH_3COOH	0.648	1.3	19.0	0.7

[a]Reichardt (2003)
[b]Avila et al. (1993)

LO• scission is particularly important as a propagation process in all solvents at elevated temperatures (Kochi, 1973b) where the thermal energy helps to overcome large E_a and log A (Arrhenius factor) for alkoxyl radical scissions (Horner et al., 2000) and reduces viscosities of the solvent and lipid (Hiatt & Traylor, 1965). However, reactions become increasingly difficult to track as product patterns shift with increasing temperature. At room temperature, H abstraction and cyclization dominate, but primary scission of the alkoxyl radicals (Reaction 1.21) increases disproportionately with temperature (Richardson et al., 1973), and secondary scissions also occur. At processing and frying temperatures greater than 100 °C, scission can become the major contributor to propagation (Mayo, 1958b). Shifts in scission products at different temperatures have been reviewed in detail by Grosch (1987).

Scission is not a favored process in neat lipids or nonpolar aprotic solvents, yet everyone working with lipid oxidation routinely detects scission products. Scission is facilitated in aprotic environments under two conditions: (1) when lipids are in dilute solution, there is less competition from hydrogen abstraction; and (2) double bonds in unsaturated fatty acids and oils provide sufficient polarity to allow scission to proceed (Farmer, 1946; Kimoto & Gaddis, 1974). Dipole-dipole interactions between alkoxyl radicals and double bonds on an adjacent unsaturated fatty acid form a charge transfer transition state in which electrons and charges are redistributed, thus promoting scission (Reaction 1.25) (Walling & Padwa, 1963).

$$R_1(R_2)HC\text{-}O\bullet + \begin{matrix} CR_2 \\ \| \\ CR_2 \end{matrix} \rightarrow R_1(R_2)HC\text{-}O^- \begin{matrix} {}^{+}CR_2 \\ | \\ \bullet CR_2 \end{matrix} \rightarrow R_1HC{=}O + R_2\bullet + \begin{matrix} CR_2 \\ \| \\ CR_2 \end{matrix} \qquad (1.25)$$

Reactions of Lipid Hydroperoxides, LOOH

Hydroperoxides are usually considered to be rather innocuous, stable intermediates that only decompose to provide propagating radicals—mainly, LO• or scission products. This perspective greatly underestimates the role of hydroperoxides in orchestrating the progress of lipid autoxidation by alternate reactions of their own.

Perceptible oxidation does not progress as long as peroxyl radicals remain in Porter's β-scission manifold (Fig. 1.3), with O_2 continuously added to or eliminated from lipid alkyl radicals. In the presence of abstractable hydrogens and under appropriate conditions, however, production of lipid hydroperoxides by the pathways described above drives the oxidation reaction forward. Along with conjugated dienes, LOOH are the first stable products of lipid oxidation, and in the absence of prooxidant heat, metals, hemes, UV light, peroxyl radicals, antioxidant acids, or nucleophiles, they might be expected to accumulate indefinitely and stall the progress of oxidation. In reality, this does not occur because

a) Decomposing factors are nearly always present to catalyze the scission of low-energy O-O and O-H bonds, (re)generating reactive radicals
b) At high enough concentrations, LOOH react with each other, undergo disproportionation, and again regenerate reactive radicals (Sliwiok et al., 1974; Hiatt & McCarrick, 1975).
c) LOOH can also undergo induced decomposition in cage reactions with other molecules (Pryor, 1966; Schaich & Karel, 1975).

Rather, hydroperoxides provide a reservoir of potential radicals waiting to be generated. Indeed, all but traces of the LOO• and all of the LO• involved in propagation arise from some form of LOOH decomposition rather than direct initiations by external agents (Bateman et al., 1953).

Decomposition reactions of LOOH, as depicted in Reactions 5–7 in the classical radical chain reaction (Fig. 1.1), have been extensively documented; yet, they are persistently considered equivalent, to the detriment of understanding oxidation kinetics. Metal-mediated decompositions of LOOH are heterolytic, producing one radical and one ion per event. Reactions of LO• generated from Fe^{2+} reactions (Reaction 5) are orders of magnitude faster than precursor LOO•, so the oxidation rate increases; slower reacting Fe^{3+} recycles stable LOOH to more reactive LOO• to restart abstraction chains (Reaction 6) (Schaich, 2005). Even with only single radicals generated, the metals are very effective catalysts because they are always present in foods and biological systems.

Heterolytic reactions of metals, however, must be distinguished from homolytic decomposition by heat and UV light (Reaction 7, Fig.1.1), which breaks the O-O bond by energy deposition rather than chemical reaction. This process yields two radical species that are considerably more reactive and more strongly oxidizing than the original LOO• (E° = 1 V): alkoxyl (LO•) and hydroxyl radicals (•OH) (E° = 1.600 and 2.32 V vs. NHE, respectively) (Wood, 1988; Koppenol, 1990). An obvious result of homolytic decomposition, then, is a dramatic acceleration of oxidation and destabilization of the system. The O-O in organic hydroperoxides (bond dissociation energy = 25–38 kcal/mol) begins decomposing at about 50 °C and is completely decomposed at 160 °C (Chan et al., 1976). This suggests that all food processing should markedly accelerate lipid oxidation while, paradoxically, low LOOH levels are commonly observed due to enhanced decomposition. Therefore, oxidation must be followed during and immediately after processing by a range of secondary products rather than only peroxide values, or reactions can be missed. Rates for decomposition of LOOH by UV light appear to be unavailable, but they are fast enough that hydroperoxides generated by lipoxygenase cannot be collected at high levels when enzyme generation and isolation of LOOH fractions are performed under normal lab lights (Schaich, unpublished data). This marked instability to light and heat underlies standard protocols that keep lipids cold and in the dark to limit oxidation.

Perhaps of greater consequence to overall lipid oxidation, however, is the co-generation of HO•, which are much more reactive and less selective than even LO• in their attack on lipids and other molecular components. Thus, the product mix becomes increasingly complicated and difficult to track. In addition to boosting kinetics, HO• and LO• together direct a transition in mechanism that greatly amplifies initiation, expands chain branching, and broadcasts oxidation far beyond lipids in any system. Multiple alternate pathways are clearly involved under such conditions and must be considered when planning analyses and interpreting kinetics and product distributions in oxidation.

The monomolecular decompositions just described dominate at low LOOH concentrations during early stages of lipid oxidation. However, when catalysts exceed decomposers, LOOH accumulate to levels at which the molecules can interact with each other at rates competitive with Reactions 5–7 in Fig.1.1. Labuza sets this concentration at greater than 1% oxidation (Labuza, 1971). Traditional explanations for bimolecular LOOH decomposition involve hydrogen bonding between the two LOOH to form transition dimers, followed by bimolecular decomposition (Sliwiok et al., 1974; Hiatt & McCarrick, 1975):

$$2\ LOOH \rightarrow LO\boxed{OH \ldots H}OOL \rightarrow LO\bullet + H_2O + \bullet OOL \qquad (1.26)$$

Bimolecular decomposition has the same effect as heat and UV light: generation of two reactive radicals from a relatively stable LOOH. With faster reactions, LO• dominates kinetically, but LOO• still produces secondary chains. The net result should be a dramatic increase in new chain initiation, detectable in accelerated oxygen consumption rates, but whether a corresponding increase in LOOH occurs depends on relative rates of all the reactions involved. We have observed cycling of hydroperoxides in a number of systems, which we believe is due to sequential repetitive local accumulation of hydroperoxides, bimolecular decomposition, regeneration of new chain reactions, then secondary accumulation of LOOH (Rossi Olson, 2011; Bogusz & Schaich, 2012; Ravi, 2012; Xie & Schaich, 2012; Bogusz & Schaich, 2012a). An example of such cycling is shown in Fig. 1.7.

This explanation for bimolecular decomposition in Reaction 1.26 is well established in the literature and supported by experimental data (Walling & Heaton, 1965; Hicks & Gebicki, 1993), but it remains somewhat controversial due to the

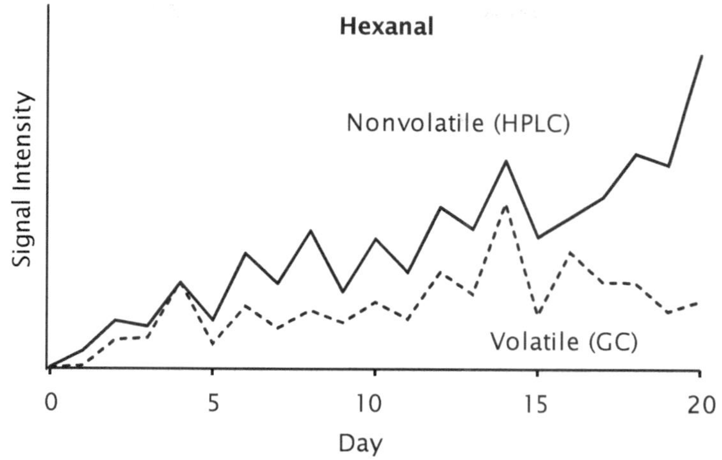

Fig. 1.7. Example of product cycling (hexanal) during lipid oxidation in a closed system. Oxygen was not limiting and intensity excursions were orders of magnitude larger than noise and replicate variability. Conjugated dienes, hydroperoxides, and other aldehydes presented comparable cycling patterns.

From Bogusz and Schaich (2012a).

poor fit of oxidation kinetics for some compounds (Brodnitz, 1968; Howard & Bennett, 1972). Two alternative reaction pathways appear to be active, in addition to Reaction 1.26, and account for different kinetic patterns.

a) Hydroperoxides can be "recycled" via H abstraction by peroxyl radicals (see also Reaction 1.6). The net effect is a plateau in hydroperoxide concentrations and a stall in oxidation progress.

$$\text{L'OOH} + \text{LOO}\bullet \rightarrow \text{LOOH} + \text{L'OO}\bullet \quad (1.27)$$

b) Thermodynamic considerations and EPR evidence for free-radical intermediates suggest that bimolecular decomposition leading to acceleration of oxidation may be specific to allylic hydroperoxides and proceed not by dissociation of hydrogen-bonded dimers (Reaction 1.26) but via LOO• radical–induced decomposition in which one slowly reacting radical adds to an allylic, non-propagating hydroperoxide to generate *three* very reactive radicals—one LO•, one epoxy-LO•, and one •OH (Brodnitz, 1968; Elson, 1975). This is shown in Reaction 1.28, with the heavy arrows indicating the favored pathway.

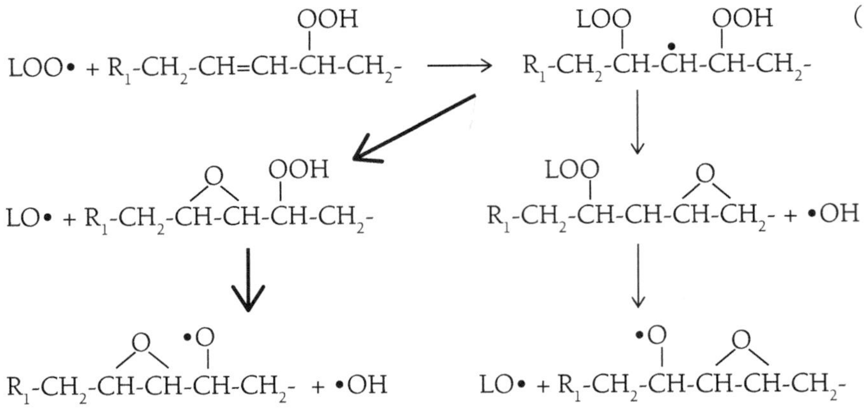
(1.28)

Radical-induced LOOH decomposition is thermodynamically favorable (E_a = 37.5 kcal). It is also more consistent with the characteristics of bimolecular initiation by hydroperoxides originally proposed by Russell (1959), the kinetics measured in lipid oxidation systems, and significant epoxide products reported in many studies. Most notably, the radical-induced decomposition described in Reaction 1.28 provides a powerful cascade of reactive radicals to fuel the very rapid increase in oxidation during the bimolecular rate period.

Conditions Favoring Specific Pathways

Rate constants for alternate competing reactions of LOO• and LO• (Table 1.F) are surprisingly comparable, showing that the multiple pathways discussed in this chapter should indeed be competitive and yield mixed products in lipid oxidation. The similarities in rates, however, also mean that reaction conditions usually determine which pathway(s) dominate. Substantial shifts in product distributions can be effected simply by modifying the reaction's environment.

Solvent proticity, temperature, and lipid concentration and conformation are key factors that control the rate, direction, and balance of alternate reaction pathways in lipid oxidation. Some factors affecting rates of individual reactions were discussed above. Major characteristics that influence reaction rates and dominant pathways may be summarized as follows (Schaich, 2012):

H abstraction from other lipids is facilitated

a) in neat lipids and aprotic solvents where lipids are the sole source of H
b) at high lipid concentrations where fatty acid chains come in closer contact

Table 1.F. Comparison of Rate Constants for Competing Reactions of Lipid or Related Peroxyl and Alkoxyl Radicals

	Solvent	LOO•	LO•	
H abstraction, LH	nonpolar organic	<1–400 M^{-1} s^{-1}	10^4–10^7 M^{-1} s^{-1}	
	polar, aqueous		10^6–10^8 M^{-1} s^{-1}	
H abstraction, LOOH	nonpolar organic	600 M^{-1} s^{-1}	2.5 x 10^8 M^{-1} s^{-1}	
	polar, aqueous	NA	NA	
Cyclization	nonpolar organic	10^1–10^3 s^{-1}	10^4–10^5 s^{-1}	
	polar, aqueous	NA	NA	
Addition	nonpolar organic	NA	10^4–10^8 M^{-1} s^{-1}	
	polar, aqueous	NA	NA	
β-scission	oleate	1–8 s^{-1}	10^3–10^5 s^{-1}	organic
	linoleate	27–430 s^{-1}	10^4–10^5 s^{-1}	polar organic
			10^6 – 10^7 s^{-1}	aqueous
Dismutation	nonpolar organic	10^6–10^9 M^{-1} s^{-1}	10^9–10^{10} M^{-1} s^{-1}	
	polar, aqueous	10^7–10^8 M^{-1} s^{-1}	NA	
	oleate-OO•	10^6 M^{-1} s^{-1}		

Data included authentic fatty acids whenever possible, plus primarily cumyl, tetralinyl, and *tert*-butyl peroxyl and alkoxyl radicals.

NA: data not available

Full references available in Schaich (2005).

c) in higher polyunsaturated fatty acids with multiple bisallylic hydrogens and bent chain conformations that bring opposite ends of fatty acid chains into close proximity
d) in low-viscosity media, which allow close lipid chain alignment and association
e) at high temperatures, which provide energy to facilitate bond dissociation

In competition with hydrogen abstraction, rearrangement to epidioxides and epoxides is facilitated

a) as the number of double bonds increases
b) at moderately elevated temperatures (rearrangement > H abstraction at 40 °C)
c) at low oxygen pressure
d) in aprotic solvents and dilute solutions at room temperature where external abstractable hydrogens are absent or limited

e) in neat lipids (abstraction and cyclization compete to generate mixed products)
f) by lipid chain orientation, as in membranes or monolayers on surfaces

Addition is facilitated by

a) aprotic solvents (in which abstractable hydrogens are limited), particularly when polar
b) the presence of conjugated double bonds (in secondary stages of oxidation, for example)
c) high lipid concentration (addition competes with increased rates of H abstraction)
d) heat (addition competes with increased rates of H abstraction)

Scission, disproportion, and elimination reactions are facilitated by

a) the presence of water, polar protic solvents, or other strong proton donors (which stabilize scission products)
b) high oxygen (which adds to scission radicals and prevents reassociation)
c) high levels of LOO• (disproportionation)
d) low solvent viscosity

Consequences of Multiple Pathways of Lipid Oxidation

As the food industry strives to meet the challenges of stabilizing foods with healthful high PUFA lipids, old habits impose obstacles to development of innovative and effective new strategies. Viewing lipid oxidation as simple chain reactions will not solve today's problems because the mechanism is incomplete; therefore, conclusions drawn from classical theory may be incorrect.

The alternate reactions and products of lipid oxidation described in the discussion above are not new; they were documented as long ago as the 1940s. What is new is:

- the recognition that pathways other than hydrogen abstraction are involved in lipid oxidation and that these alternate pathways can change the course of oxidation by competing with hydrogen abstraction for the propagation of the free-radical chain and the generation of oxidation products
- the association of specific products with individual pathways and oxidation conditions
- the identification of where specific products are generated in the sequence of lipid oxidation rather than the expectation that they amass at the end of the process

- the integration of multiple competing reactions into a coherent reaction scheme of interrelated alternate pathways that together may more fully explain complicated oxidation kinetics, account for complex product mixes, enable more accurate evaluation of the extent of oxidation, and facilitate the design of more effective antioxidant strategies

A number of consequences result from considering lipid oxidation as a process with multiple interrelated pathways rather than a single radical chain based on hydrogen abstractions alone. Perhaps most obvious is the problem of analyzing lipid oxidation. Standard approaches monitor only one or a few products in oils or extracts—typically, conjugated dienes, hydroperoxides, and perhaps also hexanal (which would form if the only reaction sequence was hydrogen abstraction at C13 of linoleic acid, LOOH decomposition, and α-scission of the resulting alkoxyl radical). Analyzing limited products, however, greatly underestimates lipid oxidation or misses it altogether when alternate pathways, particularly internal rearrangements, are also active. Similarly, the balance between pathways changes with reaction conditions, so dominant degradation pathways and products should shift dramatically when lipid oxidation occurs in bulk oils vs. emulsions vs. multiphase foods with different moisture contents. With the current analytical practices of measuring limited standard products, a shift to rearrangements would not be detected because epoxides are seldom measured and would, in fact, be interpreted as decreased lipid oxidation. Avoiding this error requires use of a combination of product analyses to fingerprint the most likely alternate pathways and to more accurately measure the full extent of lipid oxidation.

A second problem is kinetics, both in terms of time course of the reaction and the creation of rate equations for quantitating or modeling oxidation. Obviously, multiple competing reactions alter propagation sequences and change kinetics, dominant products, and detectability, making the overall process much more complex to follow and describe than in previous practice. In particular, when multiple LOO• and LO• reaction pathways occur simultaneously and in competition with each other, product mixtures can form at all stages, not just at the ends of chains. We view the course of oxidation in time curves of products. One diagram that has been reproduced extensively in different forms depicts the induction period with little or no oxidation, increasing hydroperoxides, and then all secondary products developing after hydroperoxide formation peaks (Fig. 1.8). This pattern is set in our thinking about lipid oxidation; yet, it may not accurately describe all products or situations. For example, lipid oxidation based on LOO• abstractions is considered to be a relatively slow process, so analyses in shelf life studies are often conducted weekly or even less often. This practice misses a tremendous amount of chemistry that could provide intriguing and challenging insights into degradation

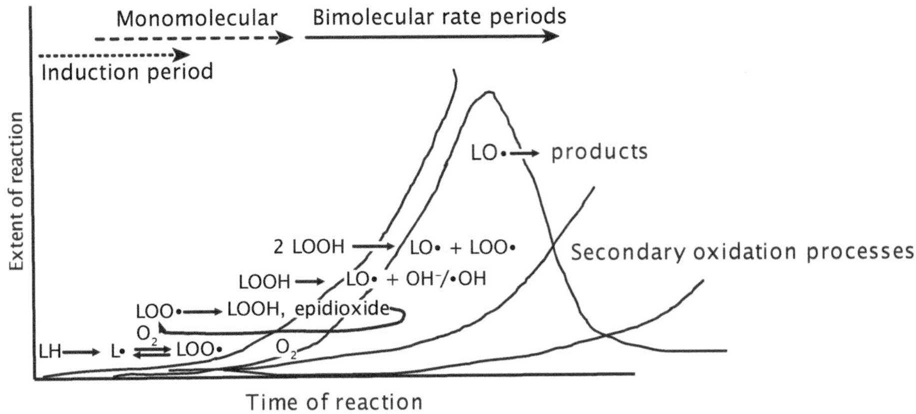

Fig. 1.8. Conceptual diagram of time course of product development in lipid oxidation as has been traditionally depicted.

From Schaich (2005), used with permission.

processes. Following multiple products, we have found that epoxides accumulate in methyl linoleate faster than LOOH under some conditions. With daily analyses, we have also found that some products repeatedly cycle (that is, increase and then decrease), while others do not, in closed but not oxygen-limited systems. Furthermore, more aldehydes are generated than can be explained by α- and β-scissions of 9 and 13 alkoxyl radicals of linoleic acid, and cycling aldehydes appear to be related to LOOH cycling, while non-cycling aldehydes are independent of LOOH (Bogusz & Schaich, 2012a, 2012b; Xie & Schaich, 2012). Although only initial attempts to map the existence and effects of multiple alternate pathways, these studies clearly show that our current expectations for the behavior of lipid oxidation only account for very limited aspects of the overall degradation. Most of the active chemistry remains unidentified, undocumented, or even unknown. This untracked chemistry very likely contributes substantially to difficulties in stabilizing high PUFA foods as well as to pathological processes involving lipid oxidation in biological systems.

Another common assumption about lipid oxidation is that reactions in accelerated shelf life studies will be the same at 60 °C as at room temperature, merely accelerated by higher thermal energy according to Arrhenius kinetics. In contrast, the theoretical considerations of alternate pathways presented in this chapter suggest that as temperature increases, major changes should occur in the mechanism, caused by shifts in the balance between pathways. Tracking multiple products in both model systems and foods validates this prediction (Schaich, unpublished data).

Calculating the kinetics of oxidation in complex systems is challenging enough when assumptions are based on straightforward radical chain reactions with LOOH formation and decomposition. Modeling lipid oxidation as a complex process with multiple reactions whose rates and balance change differentially with conditions (Tappel et al., 1989) will be an even more daunting task. At a minimum, more complete rate data for individual pathways is necessary, but more basic research on the relation between and interdependence of alternate pathways will also be critical for developing accurate kinetic models of lipid oxidation.

Lipid oxidation products strongly impact food's properties and quality, generating flavors, off-odors, browning, texture alterations, and color loss, as well as mediating cooxidation of non-lipid molecules. Each of these effects derives from specific products. Shifts away from H abstraction replace hydroperoxides and alcohols with epoxides, aldehydes, dimers, and other products that markedly alter physical properties, chemical reactivity, and stability patterns in foods. Aldehydes are largely responsible for the off-flavors associated with rancidity (Grosch, 1987); they also contribute to browning (Pokorny, 1976; Hidalgo et al., 1999) as well as protein crosslinking and modification (Schaich, 2008). Carbonyl condensations of aldehydes are likely important contributors to viscosity increases in aging oils. Epoxides are 1,000 times more reactive than aldehydes with proteins (Schaich, 2008) and rapidly transform into other largely unidentified products that have no reported sensory impact. Tracking and elucidating causal relationships between lipid oxidation and food properties will thus require consideration of the full range of lipid oxidation pathways and products.

Finally, alternate pathways of lipid oxidation can seriously impact toxicity and safety. More than a sensory issue, foods with oxidized lipids contain lipid radicals, aldehydes, epoxides, dimers, and potentially also polymers, all of which are known to have some toxicity or even weak carcinogenicity (Fukuzumi, 1970; Sanders, 1989). Shifts from volatile scission products to non-volatile internal rearrangement and addition products reduce off-odors, which is desirable. However, off-odors are critical cues to consumers that foods are rancid and should not be eaten. Epoxides are more toxic than aldehydes and react more rapidly with DNA and proteins, while dimers and polymers have reduced digestibility and can damage the intestinal mucosa (Bird et al., 1981). These products have no sensory warning signals for consumers. Obviously, stopping oxidation altogether would eliminate these toxicity issues. However, knowing more about the balance between alternate lipid oxidation pathways in foods may also make it possible to design formulations or processing that preferentially favor specific oxidation pathways to enhance some effects and minimize others.

Conclusion

Hydrogen abstraction by lipid radicals to propagate radical chains and generate stable lipid hydroperoxides has long been considered the anchoring process in lipid oxidation. However, β-scission of oxygen, internal rearrangement to epidioxides, addition, and disproportionation of lipid peroxyl radicals, as well as internal rearrangement to epoxides, addition, and α- or β-scissions of alkoxyl radicals, are also active and compete with hydrogen abstraction in propagation to alter the kinetics, course, and products of lipid oxidation.

This chapter described these alternate reactions, provided basic information about how their kinetics and balance change with reaction environment, noted data supporting the existence of these pathways,[4] and discussed the consequences of alternate pathways on the progress and effects of lipid oxidation as starting points for rethinking the chemistry of lipid oxidation. A scheme (Fig. 1.2) was presented showing how alternate pathways of LOO•, LOOH, and LO• may be integrated and interrelated to more fully account for the total products of lipid oxidation. Developed from data available in the literature (Schaich, 2005), the scheme may or may not be completely accurate as proposed, and it is open to revision as new information dictates. For now, the integrated scheme challenges all lipid oxidation investigators to think more broadly than hydrogen abstractions in a simple radical chain reaction. It offers a starting point for systematic studies to more fully elucidate the conditions that drive oxidation reactions and associated products in different directions, and it illustrates the need to monitor multiple products when tracking lipid oxidation in future studies, a topic that Chapter 2 will further address.

[4]Additional documentation is available from Schaich (2005).

References

Acott, B.; Beckwith, A. L. J. Reactions of alkoxy radicals. IV. Intramolecular hydrogen-atom transfers in the presence of cupric ion: A novel directive effect. Austral. J. Chem. 1964, 17, 1342–1353.

Adamic, K.; Howard, J. A.; Ingold, K. U. Absolute rate constants for hydrocarbon autoxidation. XVI. Reactions of peroxy radicals at low temperatures. Can. J. Chem. 1969, 47, 3803–3808.

Avila, D. V.; Brown, C. E.; Ingold, K. U.; Lusztyk, J. Solvent effects on the competitive β-scission and hydrogen atom abstraction reactions of the cumyloxyl radical. Resolution of a long-standing problem. J. Am. Chem. Soc. 1993, 115, 466–470.

Avila, D. V.; Ingold, K. U.; Kusztyk, J. Dramatic solvent effect on the absolute rate constant for abstraction of the hydroxylic hydrogen atom from *tert*-butyl hydroperoxide and phenol

by the cumyloxyl radical. The role of hydrogen bonding. J. Am. Chem. Soc. 1995, 117, 2929-2930.
Baciocchi, E.; Bietti, M.; Salamone, M.; Steenken, S. Spectral properties and absolute rates constants for β-scission of ring-substituted cumyloxyl radicals. A laser flash photolysis study. J. Org. Chem. 2002, 67, 2266–2270.
Baignee, A.; Howard, J. A.; Scaiano, J. C.; Stewart, L. C. Absolute rate constants for reations of cumyloxy in solution. J. Am. Chem. Soc. 1983, 105, 6120–6123.
Baldwin, J. E. Rules for ring closure. J. Chem. Soc. Chem. Commun. 1975, 734–736.
Bascetta, E.; Gunstone, F. D.; Scrimgeour, C. M.; Walton, J. C. ESR observation of pentadienyl and allyl radicals on hydrogen abstraction from unsaturated lipids. J. Chem. Soc. Chem. Commun. 1982, 110–112.
Bateman, L.; Hughes, H.; Morris, A. L. Hydroperoxide decomposition in relation to the initiation of radical chain reactions. Discuss. Faraday Soc. 1953, 14, 190–199.
Bird, R. P.; Basrur, P. K.; Alexander, J. C. Cytotoxicity of thermally oxidized fats. In Vitro 1981, 17, 397–404.
Bockman, T. M.; Hubig, S. M.; Kochi, J. K. Kinetic isotope effects for electron-transfer pathways in the oxidative C-H activation of hydrocarbons. J. Am. Chem. Soc. 1998, 120, 2826–2830.
Bolland, J. L. The course of autoxidation reactions in polyisoprenes and allied compounds. Part IX. The primary thermal oxidation products of ethyl linoleate. J. Chem. Soc. 1945, 445–447.
Bolland, J. L. Kinetics of olefin oxidation. Quart. Rev. 1949, 3, 1–21.
Bors, W.; Erben-Russ, M.; Saran, M. Fatty acid peroxyl radicals: Their generation and reactivities. J. Electroanal. Chem. 1987, 232, 37–49.
Bors, W.; Michael, C.; Saran, M. Inhibition of the bleaching of the carotenoid crocin. A rapid test for quantifying antioxidant activity. Biochim. Biophys. Acta. 1984, 796, 312–319.
Bors, W.; Tait, D.; Michel, C.; Saran, M.; Erben-Russ, M. Reactions of alkoxyl radicals in aqueous solutions. Isr. J. Chem. 1984, 24, 17–24.
Boto, A.; Hernandez, D.; Hernandez, R.; Suarez, E. β-Fragmentation of primary alkoxyl radicals versus hydrogen abstraction: Synthesis of polyols and α,γ-differently substituted cyclic ethers from carbohydrates. J. Org. Chem. 2003, 68, 5310–5319.
Brodnitz, M. H. Autoxidation of saturated fatty acids. A review. Agric. Food Chem. 1968, 16, 994–999.
Buettner, G. R. The pecking order of free radicals and antioxidants: Lipid peroxidation, α-tocopherol, and ascorbate. Arch. Biochem. Biophys. 1993, 300, 535–543.
Campbell, I. M.; Caton, R. B.; Crozier, D. N. Complex formation and reversible oxygenation of free fatty acids. Lipids 1974, 9, 916–920.
Chan, H. W.-S., Ed. Autoxidation of Unsaturated Lipids; Academic Press: London, 1987.
Chan, H. W.-S.; Levett, G.; Matthew, J. A. The mechanism of the rearrangement of linolenate hydroperoxides. Chem. Phys. Lipids 1979, 24, 245–256.
Chan, H. W.-S.; Matthew, J. A.; Coxon, D. T. A hydroperoxy-epidioxide from the autoxidation of a hydroperoxide of methyl linolenate. J. Chem. Soc. Chem. Commun. 1980, 235–236.
Chan, H. W.-S.; Prescott, F. A. A.; Swoboda, P. A. T. Thermal decomposition of individual positional isomers of methyl linoleate hydroperoxide: Evidence of carbon-oxygen bond scission. J. Am. Oil Chem. Soc. 1976, 53, 572–576.

Coxon, D. T.; Price, K. R.; Chan, H. W.-S. Formation, isolation and structure determination of methyl linolenate diperoxides. Chem. Phys. Lipids 1981, 28, 365–378.

Dix, T. A.; Marnett, L. J. Free radical epoxidation of 7,8-dihydroxy-7,8-dihydrobenzo[*a*]pyrene by hematin and polyunsaturated fatty acid hydroperoxides. J. Am. Chem. Soc. 1981, 103, 6744–6746.

Elson, I. H.; Mao, S. W.; Kochi, J. K. Electron spin resonance study of addition of alkoxy radicals to olefins. J. Am. Chem. Soc. 1975, 97, 335–341.

Erben-Russ, M.; Michael, C.; Bors, W.; Saran, M. Absolute rate constants of alkoxyl radical reactions in aqueous solution. J. Phys. Chem. 1987, 91, 2362–2365.

Factor, A.; Russell, C. A.; Traylor, T. G. Bimolecular combination reactions of oxy radicals. J. Am. Chem. Soc. 1965, 87, 3692–3696.

Farmer, E. H. Peroxidation in relation to olefinic structure. Trans. Faraday Soc. 1946, 42, 228–236.

Farmer, E. H.; Koch, H. P.; Sutton, D. A. The course of autoxidation reactions in polyisoprenes and allied compounds. VII. Rearrangement of double bonds during autoxidation. J. Chem. Soc. 1943, 541–547.

Farmer, E. H.; Sutton, D. A. Course of autoxidation reactions in polyisoprenes and allied compounds. IV. The isolation and constitution of photochemically-formed methyl oleate peroxide. J. Chem. Soc. 1943, 119–122.

Frankel, E. N. Hydroperoxides. Symposium on Foods: Lipids and Their Oxidation; Schultz, H.; Day, E. D.; Sinnhuber, W. O., Eds.; AVI Publishing: Westport, CT, 1962; 51–76.

Frankel, E. N. Lipid oxidation. Prog. Lipid Res. 1980, 19, 1–22.

Frankel, E. N. Volatile lipid oxidation products. Prog. Lipid Res. 1982, 22, 1–33.

Frankel, E. N. Lipid oxidation: Mechanisms, products, and biological significance. J Amer. Oil Chem. Soc. 1984, 61, 1908–1917.

Frankel, E. N. Chemistry of free radical and singlet oxidation of lipids. Prog. Lipid Res. 1985, 23, 197–221.

Frankel, E. N. Secondary products of lipid oxidation. Chem. Phys. Lipids 1987, 44, 73–85.

Frankel, E. N. Recent advances in lipid oxidation. J. Sci. Food Agric. 1991, 54, 495–511.

Frankel, E. N. Lipid Oxidation, 2nd edition; The Oily Press: Bridgewater, UK, 2005.

Frankel, E. N.; Neff, W. E.; Bessler, T. R. Analysis of autoxidized fats by gas chromatography-mass spectrometry: V. Photosensitized oxidation. Lipids 1979, 14, 961–967.

Frankel, E. N.; Neff, W. E.; Selke, E.; Brooks, D. D. Analysis of autoxidized fats by gas chromatography-mass spectrometry: X. Volatile thermal decomposition products of methyl linolenate dimers. Lipids 1988, 23, 295–298.

Frankel, E. N.; Neff, W. E.; Selke, E.; Weisleder, D. Photosensitized oxidation of methyl linoleate: Secondary and volatile thermal decomposition products. Lipids 1982, 17, 11–18.

Fukuzumi, K. Significance of lipoperoxides in gastric cancer. Sonderdruck aus Fette 1970, 72, 853–855.

Fukuzumi, S.; Kochi, J. K. Electrophilic additions to olefins. A new approach to unifying the mechanisms of bromination and oxymercuration. J. Am. Chem. Soc. 1981, 103, 2783–2791.

Gardner, H. W. Oxygen radical chemistry of polyunsaturated fatty acids. Free Radic. Biol. Med. 1989, 7, 65–86.

Gardner, H. W.; Kleiman, R. Lack of regioselectivity in formation of oxohydroxyoctadecenoic acids from the 9- or 13-hydroperoxide of linoleic acid. Lipids 1979, 14, 848–851.

Gardner, H. W.; Kleiman, R. Degradation of linoleic acid hydroperoxides by a cysteine-$FeCl_3$ catalyst as a model for similar biochemical reactions. II. Specificity in formation of fatty acid epoxides. Biochim. Biophys. Acta 1981, 665, 113–125.
Gardner, H. W.; Kleiman, R.; Weisleder, D. Homolytic decomposition of linoleic acid hydroperoxide: Identification of fatty acid products. Lipids 1974, 9, 696–706.
Gebicki, J. M.; Bielski, B. H. J. Comparison of the capacities of the perhydroxyl and the superoxide radicals to initiate chain oxidation of linoleic acid. J. Am. Chem. Soc. 1981, 103, 7020–7022.
Gilbert, B. C.; Holmes, R. G. G.; Norman, R. O. C. Electron spin resonance studies. Part LII. Reactions of secondary alkoxyl radicals. J. Chem. Res. 1977, 1, 101–113.
Gilbert, B. C.; Marshall, P. D. R.; Norman, R. O. C.; Pineda, N.; Williams, P. S. Electron spin resonance studies. Part 61. The generation and reactions of the t-butoxyl radical in aqueous solution. J. Chem. Soc. Perkin II 1981, 1392–1400.
Grosch, W. Reactions of Hydroperoxides—Products of Low Molecular Weight. Autoxidation of Unsaturated Lipids; Chan, H. W.-S., Ed.; Academic Press: London, 1987; 95–139.
Gunstone, F. D.; Hilditch, T. P. The union of gaseous oxygen with methyl oleate, linoleate, and linolenate. J. Chem. Soc. 1945, 836–841.
Hasegawa, K.; Patterson, L. K. Pulse radiolysis studies in model lipid systems: Formation and behavior of peroxy radicals in fatty acids. Photochem. Photobiol. 1978, 28, 817–823.
Haynes, R. K.; Vonwiller, S. C. Iron(III) and copper(II) catalyzed transformation of fatty acid hydroperoxides: Efficient generation of peroxy radicals with Cu(II) trifluoromethane sulfonate. J. Chem. Soc. Chem. Commun. 1990, 1102–1104.
Heijman, M. G. J.; Nauta, H.; Levine, Y. K. A pulse radiolysis study of the dienyl radical in oxygen-free linoleate solutions: Time and linoleate concentration dependence. Radiat. Phys. Chem. 1985, 26, 73–82.
Hiatt, R.; McCarrick, T. On "bimolecular initiation" by hydroperoxides. J. Am. Chem. Soc. 1975, 97, 5234–5237.
Hiatt, R.; Szilagyi, S. Homolytic decomposition of di-sec-butyl peroxide. Can. J. Chem. 1970, 48, 615–627.
Hiatt, R.; Traylor, T. G. Cage recombination of t-butoxy radicals. J. Am. Chem. Soc. 1965, 87, 3766–3768.
Hiatt, R.; Zigmund, L. Interaction of s-butyl peroxy and alkoxy radicals. Can. J. Chem. 1970, 48, 3967–3969.
Hicks, M.; Gebicki, J. M. Continuous measurement of oxygen consumption by linoleic acid membranes exposed to free radicals generated by g-radiation. Int. J. Radiat. Biol. 1993, 64, 143–148.
Hidalgo, F. J.; Alaiz, M.; Zamora, R. Effect of pH and temperature on comparative nonenzymatic browning of proteins produced by oxidized lipids and carbohydrates. J. Agric. Food Chem. 1999, 47, 742–747.
Holman, R. T.; Elmer, O. C. The rates of oxidation of unsaturated fatty acids and esters. J. Amer. Oil Chem. Soc. 1947, 24, 127–129.
Horner, J. H.; Choi, S.-Y.; Newcomb, M. Laser flash photolysis studies of alkoxyl radical kinetics using 4-nitrobenzenesulfenate esters as radical precursors. Org. Lett. 2000, 2, 3369–3372.
Howard, J. A. Homogeneous Liquid-Phase Autoxidations. Free Radicals; Kochi, J. K., Ed.; Wiley-Interscience: New York, 1973; Vol. II, 3–62.

Howard, J. A.; Bennett, J. E. The self-reaction of *sec*-alkylperoxy radicals: A kinetic electron spin resonance study. Can. J. Chem. 1972, 50, 2374–2377.

Howard, J. A.; Ingold, K. U. Absolute rate constants for hydrocarbon autoxidation. VI. Alkyl aromatic and olefinic hydrocarbons. Can. J. Chem. 1967, 45, 793–802.

Howard, J. A.; Ingold, K. U. Absolute rate constants for hydrocarbon oxidation. XI. The reactions of tertiary peroxy radicals. Can. J. Chem. 1968, 46, 2655–2660.

Ingold, K. U. Peroxy radicals. Acc. Chem. Res. 1969, 2, 1–9.

Janzen, E. G.; Johnston, F. J.; Ayers, C. A. The reversible thermal decomposition of triphenylmethylperoxyl radical to triphenylmethyl radical and oxygen. J. Am. Chem. Soc. 1967, 89, 1176–1183.

Kerr, J. A. Bond dissociation energies by kinetic methods. Chem. Rev. 1966, 66, 465–500.

Kimoto, W. I.; Gaddis, A. M. Monocarbonyl compounds from catalytic decomposition of autoxidized unsaturated fatty acid esters. J. Amer. Oil Chem. Soc. 1974, 51, 307–311.

Kochi, J. K. Addition of peroxides to conjugated olefins catalyzed by copper salts. J. Am. Chem. Soc. 1962a, 84, 2785–2793.

Kochi, J. K. Chemistry of alkoxyl radicals. J. Am. Chem. Soc. 1962b, 84, 1193–1197.

Kochi, J. K. Free Radicals; John Wiley and Sons: New York, 1973a; Vol. 1.

Kochi, J. K. Oxygen Radicals. Free radicals; Kochi, J. K., Ed.; John Wiley: New York, 1973b; Vol. 2, 665–710.

Koppenol, W. Oxyradical reactions: From bond-dissociation energies to reduction potentials. FEBS Letters 1990, 264, 165–167.

Labuza, T. P. Kinetics of lipid oxidation in foods. CRC Crit. Rev. Food Sci. Nutr. 1971, 2, 355–405.

Lea, C. H. The Oxidative Deterioration of Food Lipids. Symposium on Foods: Lipids and Their Oxidation; Schultz, H. W.; Day, E. A.; Sinnhuber, R. O., Eds.; AVI Publishing: Westport, CT, 1962; 3–30.

Lewis, S. E.; Mayo, F. R. Copolymerization. IX. A comparison of some *cis* and *trans* isomers. J. Am. Chem. Soc. 1958, 70, 1533–1536.

Lindsay, D. A.; Howard, J. A.; Horswill, E. C.; Iton, L.; Ingold, K. U.; Cobbley, T.; Li, A. The bimolecular self-reaction of secondary peroxy radicals. Product studies. Can. J. Chem. 1973, 51, 870–880.

Loidl-Stahlhofen, A.; Spiteller, G. a-Hydroxyaldehydes, products of lipid peroxidation. Biochim. Biophys. Acta 1994, 1211, 156–160.

Lundberg, W. O., Ed. Autoxidation and Antioxidants; Interscience: New York, 1962; Vol. 1–2.

Lundberg, W. O.; Chipault, J. R. The oxidation of methyl linoleate at various temperatures. J. Am. Chem. Soc. 1947, 69, 833–836.

Mayo, F. R. The oxidation of unsaturated compounds. VIII. The oxidation of aliphatic unsaturated compounds. J. Am. Chem. Soc. 1958a, 80, 2497–2500.

Mayo, F. R. The oxidation of unsaturated compounds. IX. The effects of structure on the rates and products of oxidation of unsaturated compounds. J. Am. Chem. Soc. 1958b, 80, 2500–2507.

Mayo, F. R. Free radical autoxidation of hydrocarbons. Acc. Chem. Res. 1968, 1, 193–201.

Merenyi, G.; Lind, J.; Engman, L. One- and two-electron reduction potentials of peroxyl radicals and related species. J. Chem. Soc. Perkin Trans. 2 1994, 2551–2553.

Mihelich, E. D. Structure and stereochemistry of novel endoperoxides isolated from the sensitized photooxidation of methyl linoleate. Implications for prostaglandin biosynthesis. J. Am. Chem. Soc. 1980, 102, 7143–7145.

Mills, K. A.; Caldwell, S. E.; Dubay, G. R.; Porter, N. A. An allyl radical-dioxygen caged pair mechanism for *cis*-allylperoxyl rearrangements. J. Am. Chem. Soc. 1992, 114, 9689–9691.

Miyashita, K.; Fujimoto, K.; Kaneda, T. Formation of dimers during the initial stage of autoxidation in methyl linoleate. Agric. Biol. Chem. 1982, 46, 751–755.

Mounts, T. L.; McWeeny, D. J.; Evans, C. D.; Dutton, H. J. Decomposition of linoleate hydroperoxides: Precursors of oxidative dimers. Chem. Phys. Lipids 1970, 4, 197–202.

Nawar, W. W. Thermal degradation of lipids. J. Agric. Food Chem. 1969, 17, 18–21.

Neff, W. E. Quantitative analyses of hydroxystearate isomers from hydroperoxides by high pressure liquid chromatography of autoxidized and photosensitized-oxidized fatty esters. Lipids 1980, 15, 587–590.

Neff, W. E.; Frankel, E. N.; Fujimoto, K. Autoxidative dimerization of methyl linolenate and its monohydroperoxides, hydroperoxy epidioxides and dihydroperoxides. J. Amer. Oil Chem. Soc. 1988, 65, 616–623.

Neff, W. E.; Frankel, E. N.; Schofield, C. R.; Weisleder, D. High-pressure liquid chromatography of autoxidized lipids: I. Methyl oleate and linoleate. Lipids 1978, 13, 415–421.

Neff, W. E.; Frankel, E. N.; Weisleder, D. High-pressure liquid chromatography of autoxidized lipids: II. Hydroperoxy-cyclic peroxides and other secondary products from methyl linolenate. Lipids 1981, 16, 439–448.

Neta, P.; Dizdaroglu, M.; Simic, M. Radiolytic studies of the cumyloxyl radical in aqueous solutions. Isr. J. Chem. 1984, 24, 25–28.

Neville, A. G.; Brown, C. E.; Rayner, D. M.; Lusztyk, J.; Ingold, K. U. Direct measurement of the rate constant for β-scission of the cumyloxyl radical by laser flash photolysis with time-resolved IR detection. J. Am. Chem. Soc. 1989, 111, 9269–9270.

Patterson, L. K.; Hasegawa, K. Pulse radiolysis studies in model lipid systems. The influence of aggregation on kinetic behavior of OH induced radicals in aqueous sodium linoleate. Ber. Bunsenges. Phys. Chem. 1978, 82, 951–956.

Paul, H.; Small, R. D., Jr.; Scaiana, J. C. Hydrogen abstraction by *tert*-butoxy radicals. A laser photolysis and electron spin resonance study. J. Am. Chem. Soc. 1978, 100, 4520–4527.

Pokorny, J.; El-Zeany, B. A.; Luan, N.; Janicek, G. Nonenzymic browning. XV. Effect of unsaturation on browning reactions of oxidized lipids with proteins. Z. Lebensm. Unters-Forsch. 1976d, 161, 271–272.

Porter, N. A. Mechanisms for the autoxidation of polyunsaturated lipids. Acct. Chem. Res. 1986, 19, 262–268.

Porter, N. A. Autoxidation of Polyunsaturated Fatty Acids: Initiation, Propagation, and Product Distribution (Basic Chemistry). Membrane Lipid Oxidation; Vigo-Pelfrey, C., Ed.; CRC Press: Boca Raton, FL, 1990; Vol. I, 33–62.

Porter, N. A.; Nixon, J. R. Stereochemistry of free radical substitution on the peroxide bond. J. Am. Chem. Soc. 1978, 100, 7116–7117.

Porter, N. A.; Wujek, D. G. Autoxidation of polyunsaturated fatty acids, an expanded mechanistic study. J. Am. Chem. Soc. 1984, 106, 2626–2629.

Porter, N. A.; Wujek, D. G. The Autoxidation of Polyunsaturated Fatty Acids. Reactive Oxygen Species in Chemistry, Biology, and Medicine; Quintanilla, A., Ed.; Plenum Publishing Corp.: New York, 1987a; 55–79.

Porter, N. A.; Wujek, J. S. Allylic hydroperoxide rearrangement: β-Scission or concerted pathway? J. Org. Chem. 1987b, 52, 5085–5089.

Porter, N. A.; Caldwell, S. E.; Mills, K. A. Mechanisms of free radical oxidation of unsaturated lipids. Lipids 1995, 30, 277–290.

Porter, N. A.; Funk, M. O.; Gilmore, D.; Isaac, R.; Nixon, J. The formation of cyclic peroxides from unsaturated hydroperoxides: Models for prostaglandin biosynthesis. J. Am. Chem. Soc. 1976, 98, 6000–6005.

Porter, N. A.; Lehman, L. S.; Weber, B. A.; Smith, K. J. Unified mechanism for polyunsaturated fatty acid autoxidation. Competition of peroxy radical hydrogen atom abstraction, β-scission, and cyclization. J. Am. Chem. Soc. 1981, 103, 6447–6455.

Porter, N. A.; Lehman, L. S.; Wujek, D. G. Oxidation Mechanisms of Poly-Unsaturated Fatty Acids. Oxygen Radicals in Chemistry and Biology; Bors, W.; Saran, M.; Tait, D., Eds.; Walter de Gruyter: Berlin, 1984; 235–237.

Porter, N. A.; Mills, K. A.; Caldwell, S. E.; Dubay, G. R. The mechanisms of the [3,2]-allylperoxyl rearrangement. A radical-dioxygen pair reaction that proceeds with stereochemical memory. J. Am. Chem. Soc. 1994a, 116, 6697–6705.

Porter, N. A.; Mills, K. A.; Carter, R. L. A mechanistic study of oleate oxidation: Competing peroxyl H-atom abstraction and rearrangement. J. Am. Chem. Soc. 1994b, 116, 6690–6696.

Porter, N. A.; Roberts, D. H.; Ziegler, C. B., Jr. A new route to lipid hydroperoxides: Orbital symmetry controlled ring opening of vinylcyclopropyl bromides. J. Am. Chem. Soc. 1980a, 102, 5912–5913.

Porter, N. A.; Weber, B. A.; Weenan, H.; Kahn, J. A. Autoxidation of polyunsaturated lipids. Factors controlling the stereochemistry of product hydroperoxides. J. Am. Chem. Soc. 1980b, 102, 5597–5601.

Pratt, D. A.; Mills, J. H.; Porter, N. A. Theoretical calculations of carbon-oxygen bond dissociation enthalpies of peroxyl radicals formed in the autoxidation of lipids. J. Amer. Chem. Soc. 2003, 125, 5801–5810.

Pratt, D. A.; Tallman, K. A.; Porter, N. A. Free radical oxidation of polyunsaturated lipids: New mechanistic insights and the development of peroxyl radical clocks. Acc. Chem. Res. 2011, 44, 458–467.

Privett, O. S. Autoxidation and autoxidative polymerization. J. Amer. Oil Chem. Soc. 1959, 36, 507–512.

Pryor, W. A. Free Radicals; McGraw-Hill: New York, 1966.

Pryor, W. A. Oxy-radicals and related species: Their formation, lifetimes and reactions. Ann. Rev. Physiol. 1986, 48, 657–667.

Pryor, W. A.; Stanley, J. P. A suggested mechanism for the production of malonaldehyde during the autoxidation of polyunsaturated fatty acids: Nonenzymatic production of prostaglandin endoperoxides during autoxidation. J. Org. Chem. 1975, 40, 3615–3617.

Reichardt, C. Solvents and Solvent Effects in Organic Chemistry, 3rd edition; Wiley-VCH Publishers: Weinheim, DE, 2003.

Richardson, W. H.; Yelvington, M. B.; Andrist, A. H.; Ertley, E. W.; Smith, R. S.; Johnson, T. D. Thermal decomposition of tertiary alkyl peroxides. Substituent effects in peroxide bond homolysis and β-scission of alkoxyl radicals. J. Org. Chem. 1973, 38, 4219–4225.

Roe, A. N.; McPhail, A. T.; Porter, N. A. Serial cyclization: Studies in the mechanism and stereochemistry of peroxy radical cyclization. J. Am. Chem. Soc. 1983, 105, 1199–1203.

Russell, G. A. The rates of oxidation of aralkyl hydrocarbons. Polar effects in free radical reactions. J. Am. Chem. Soc. 1956, 78, 1047–1054.

Russell, G. A. Fundamental processes of autoxidation. J. Chem. Ed. 1959, 36, 111–118.

Sanders, T. A. B. Nutritional Aspects of Rancidity. Rancidity in Foods; Allen, J. C.; Hamilton, R. J., Eds.; Elsevier Applied Science: London, 1989; 125–137.
Schaich, K. M. Lipid Oxidation in Fats and Oils: Theoretical Aspects. Bailey's Industrial Fats and Oils, 6th ed.; Shahidi, F., Ed.; John Wiley: New York, 2005; 2681–2767.
Schaich, K. M. Lipid Oxidation in Specialty Oils. Nutraceutical and Specialty Lipids and Their Coproducts; Shahidi, F., Ed.; CRC Press: Boca Raton, FL, 2006; 401–448.
Schaich, K. M. Cooxidations of Oxidizing Lipids: Reactions with Proteins. Lipid Oxidation Pathways; Kamal-Eldin, A.; Min, D. B., Eds.; CRC Press: Boca Raton, FL, 2008; Vol. 2, 183–274.
Schaich, K. M. Thinking outside the classical chain reaction box of lipid oxidation. Lipid Technology 2012, 24, 55–58.
Schaich, K. M.; Karel, M. Free radicals in lysozyme reacted with peroxidizing methyl linoleate. J. Food Sci. 1975, 40, 456–459.
Schaich, K. M.; Shahidi, F.; Zhong, Y.; Eskin, N. A. M. Lipid Oxidation. Biochemistry of Foods, 3rd ed.; Eskin, N. A. M., Ed.; Elsevier: London, 2013; 419–478.
Schauenstein, E. Autoxidation of polyunsaturated esters in water: Chemical structure and biological activity of the products. J. Lipid Research 1967, 8, 417–428.
Schieberle, P.; Trebert, Y.; Firl, J.; Grosch, W. Photolysis of unsaturated fatty acid hydroperoxides. 4. Fatty acid products from the aerobic decomposition of methyl 13(S)-hydroperoxy-9(Z), 11(E)-octadecadienoate dissolved in cyclohexane. Chem. Phys. Lipids 1988, 48, 281-288.
Scott, G. Atmospheric Oxidation and Antioxidants; Elsevier: London, 1965.
Selke, E.; Frankel, E. N.; Neff, W. E. Thermal decomposition of methyl oleate hydroperoxides and identification of volatile components by gas chromatography-mass spectrometry. Lipids 1978, 13, 511–513.
Sheldon, R. A.; Kochi, J. K. Metal-Catalyzed Oxidations of Organic Compounds; Academic Press: New York, 1981.
Simic, M. Fast Processes in Radiation Chemistry and Biology; Adams, G. E.; Fielden, E. M.; Michael, B. D., Ed.; Wiley: New York, 1975; 162–179.
Simic, M. G. Free radical mechanisms in autoxidation processes. J. Chem. Ed. 1981, 58, 125–131.
Simic, M. G.; Jovanovic, S. V.; Niki, E. Mechanisms of Lipid Oxidative Processes and Their Inhibition. Lipid Oxidation in Food; St. Angelo, A. J., Ed.; Amer. Chem. Soc.: Washington, D.C., 1992; 14–32.
Sims, R. P.; Hoffman, W. H. Oxidative Polymerization. Autoxidation and Antioxidants; Lundberg, W. O., Ed.; Interscience: London, 1962; Vol. II, 629–694.
Sliwiok, J.; Kowalska, T.; Kowalski, W.; Biernat, A. The influence of hydrogen-bond association on the destruction of hydroperoxides in the autoxidation process of oleyl alcohol, oleic acid, and methyl oleate. Microchem. J. 1974, 19, 362–372.
Small, R. D., Jr.; Scaiano, J. C.; Patterson, L. K. Radical processes in lipids. A laser photolysis study of t-butoxy radical reactivity toward fatty acids. Photochem. Photobiol. 1979, 29, 49–51.
Swern, D. Primary Products of Olefinic Autoxidation. Autoxidation and Antioxidants; Lundberg, W. O., Ed.; Interscience: New York, 1961; Vol. 1, 1–54.
Tallman, K. A.; Pratt, D. A.; Porter, N. A. Kinetic products of linoleate peroxidation: Rapid β-fragmentation of non-conjugated peroxyls. J. Am. Chem. Soc. 2001, 123, 11827–11828.
Tappel, A. L.; Tappel, A. A.; Fraga, C. A. Application of simulation modeling to lipid peroxidation processes. Free Radic. Biol. Med. 1989, 7, 361–368.

Thomas, J. R. The self-reaction of *t*-butylperoxy radicals. J. Am. Chem. Soc. 1965, 87, 3935–3940.

Traylor, T. G.; Russell, C. A. Mechanism of autoxidation. Terminating radicals in cumene autoxidation. J. Am. Chem. Soc. 1965, 87, 3698–3706.

Tsentalovich, Y. P.; Kulik, L. V.; Gritsan, N. P.; Yurkovskaya, A. V. Solvent effect on the rate of β-scission of the tert-butoxyl radical. J. Phys. Chem. 1998, 102, 7975–7980.

Van Sickle, D. E.; Mayo, F. R.; Gould, E. S.; Arluck, R. M. Effects of experimental variables in oxidations of alkenes. J. Am. Chem. Soc. 1967, 89, 977–984.

Walling, C.; Heaton, L. Hydrogen bonding and complex formation in solutions of *t*-butyl hydroperoxide. J. Amer. Oil Chem. Soc. 1965, 87, 48–51.

Walling, C.; Padwa, A. Positive halogen compounds. VI. Effects of structure and medium on the β-scission of alkoxyl radicals. J. Am. Chem. Soc. 1963, 85, 1593–1597.

Walling, C.; Thaler, W. Positive halogen compounds, III. Allyl chlorination with *t*-butyl hypochlorite. The stereochemistry of allylic radicals. J. Am. Chem. Soc. 1961, 83, 3877–3884.

Walling, C.; Wagner, P. J. Effects of solvents on transition states in the reactions of *t*-butoxy radicals. J. Am. Chem. Soc. 1963, 85, 2333–2334.

Walling, C.; Wagner, P. J. Positive halogen compounds. X. Solvent effects in the reactions of *t*-butoxy radicals. J. Am. Chem. Soc. 1964, 86, 3368–3375.

Walling, C.; Waits, H. P.; Milanovic, J.; Pappiaonnou, C. G. Polar and radical paths in the decomposition of diacyl peroxides. J. Am. Chem. Soc. 1970, 92, 4927–4932.

Weenan, H.; Porter, N. A. Autoxidation of model membrane systems: Cooxidation of polyunsaturated lecithins with steroids, fatty acids, and a-tocopherol. J. Am. Chem. Soc. 1982, 104, 5216–5221.

Wilsey, S.; Dowd, P.; Houk, K. N. Effect of alkyl substituents and ring size on alkoxy radical cleavage reactions. J. Org. Chem. 1999, 64, 8801–8811.

Witting, I. A.; Chang, S. S.; Kummerow, F. A. The isolation and characterization of the polymers formed during the autoxidation of ethyl linoleate. J. Am. Oil Chem. Soc. 1957, 34, 470–473.

Wood, P. M. The potential diagram for oxygen at pH 7. Biochem. J. 1988, 253, 287–289.

Wu, G.-S.; Stein, R. A.; Mead, J. F. Autoxidation of fatty acid monolayers adsorbed on silica gel: II. Rates and products. Lipids 1977, 12, 971–978.

Wu, G.-S.; Stein, R. A.; Mead, J. F. Autoxidation of fatty acid monolayers adsorbed on silica gel: III. Effects of saturated fatty acids and cholesterol. Lipids 1978, 13, 517–524.

Xu, L.; Davis, T. A.; Porter, N. A. Rate constants for peroxidation of polyunsaturated fatty acids and sterols in solution and in liposomes. J. Am. Chem. Soc. 2009, 131, 13037–13044.

CHAPTER 2

Challenges in Analyzing Lipid Oxidation

Are One Product and One Sample Concentration Enough?

K. M. Schaich[1]

Fundamental Requirements of Lipid Oxidation Analyses

Students learn about monitoring reactions in the context of straightforward reactions with fixed stable products that form in high concentrations; for example,

$$A + B \rightarrow C \tag{2.1}$$

$$\text{or} \quad A + B \rightarrow C + D \tag{2.1a}$$

This situation directly contrasts with lipid oxidation, which has multiple possible pathways that change with conditions and over time. In these reactions, products are actually intermediates that degrade or transform into other compounds, as shown in Fig. 1.2 of Chapter 1, and product yields are quite low; a total reaction of less than 1% renders foods inedible. Thus, lipid oxidation is generally considered to be the greatest analytical challenge in food science. Similar arguments could be made for cosmetics and many personal care products, as well as for biological tissues in which lipid oxidation and its effects are complex, even when total oxidation levels are low.

One question that persists in lipid oxidation analysis is which product is best to monitor? The answer is strongly influenced by the either quantitative or qualitative end goals of analyses. Industry rejects batches and makes decisions about

[1]Dept. of Food Science, Rutgers University, 65 Dudley Rd., New Brunswick, NJ 08901-8520.

formulations based on quantitation, which requires accuracy and reproducibility, and product class analysis (that is, for all hydroperoxides or all epoxides, regardless of structure) is usually adequate. In contrast, basic research seeks to elucidate reaction mechanisms, and oxidation sequences need as much quantitative and qualitative detail about individual products as possible. Oxygen consumption is often preferred for determining oxidation kinetics when starting from fresh products because it is independent of product transformation, but it cannot be used for spot analyses of products off the production line or storage shelf. For the latter, the question becomes, does one measure hydroperoxides because they form first (even though they decompose) or secondary products to detect hydroperoxide breakdown and because consumers can smell and taste them? Does a researcher analyze only volatiles because gas chromatographs are readily available, only nonvolatile products because they remain in the product, or both? The picture of lipid oxidation that we construct depends on the analytical strategy used.

A wide variety of assays for various lipid oxidation products has been developed; many methods have been standardized, and more are used simply by following (and modifying) procedures published in research papers. Some of the most common assays for various lipid oxidation products are listed in Table 2.A. A number of excellent reviews and books have presented details of individual methods (McDonald & Mossoba, 1997; Shahidi, 1998; Dobarganes & Velasco, 2002; Yildiz et al., 2003; Kamal-Eldin & Pokorny, 2005). Therefore, it is not the goal of this chapter to provide protocols for analyses or to recommend specific assays. Rather, this chapter argues for new approaches and seeks to change how lipid oxidation analyses are viewed and used to attain more complete and accurate information. First and foremost, the author challenges readers to "think chemistry" in all lipid oxidation analyses, rather than seeing black boxes in which procedures are accepted and applied blindly; instead, they should look beyond peroxide values to track multiple products and attempt mass balance between products.

Lipid oxidation presents several distinctive challenges for analysis:

1. *Providing quantitative, reproducible analyses that accurately reflect oxidation chemistry.* As noted above, this is mandatory when decisions about formulating or discarding products are based on the analyses, as on industrial production lines or in storage facilities. It should also be the first priority in most analyses, even in research where relative levels are often used to follow reaction progress or compare oxidation levels. Accuracy here refers both to the individual assays and product pathways monitored. Hence, if a single sample size is used and the product levels fall outside the product detection range, the assay itself may work, but it does not correctly document oxidation. Similarly, an assay may accurately measure a

Table 2.A. Examples of Analytical Methods Used to Determine Oxidation in Lipids

Lipid Oxidation Product	Standard Method	Reference
Conjugated dienes	AOCS Th 1a-64, Ti 1a-64 AOCS Ch 5-91 cyclohexane	White, 1995; AOCS, 2011b, 2011c
	various other methods	de Andrade et al., 2010 (review)
Hydroperoxides		Dobarganes & Velasco, 2002 (review)
Iodometric titration	AOCS 8b-90, JOCS 2-4 12-71 isooctane	JOCS, 2009; AOCS, 2011i
	AOCS Cd 8-53; IUPAC 2.501 $CHCl_3$	IUPAC, 1992a; AOCS, 2011h
	AOAC 41.1.16	AOAC, 2000
Xylenol orange—direct		Jiang et al., 1990, 1991; Jiang et al.,1992; Wolff, 1994; Nourooz-Zadeh et al., 1995; Bou et al., 2008
Xylenol orange—PeroxySafe™	AOAC alternate method 030405	Osawa et al., 2007; Biomedicals, 2012
Xylenol orange—Pierce kit		Pierce, 2011
Ferrithiocyanate—direct	IDF Standard Method 74A:1991	Shantha & Decker, 1994; Mihaljevic et al., 1996
Ferrithiocyanate—Cayman kit		Cayman, 2011
Triphenyl phosphine		Nakamura & Maeda, 1991; Akasaka & Ohrui, 2000; Talpur et al., 2010; Gotoh et al., 2011
Fourier transform infrared spectroscopy		Van de Vort et al., 1994; Sedman et al., 1997; Guillen and Cabo, 2002; Yu et al., 2007
NMR		Hamalainen & Kamal-Eldin, 2005
RP-HPLC—234 nm		Bauer-Plank & Steenhorst-Slikkerveer, 2000
post-column reaction		Yang, 1992

(continued)

Table 2.A. *Continued*

Lipid Oxidation Product	Standard Method	Reference
Alcohols (hydroxylated products)	AOCS Cd 4-40, Cd 13-60, Tx 1a-66	AOCS, 2011d; 2011e; 2011f
GC-MS of TMS ether derivatives		Guido et al., 1993
Epoxides		
HBr titration	AOCS Cd 9-57	Durbetaki, 1956; Maerker, 1965; AOCS, 2011g
4-(*p*-Nitrobenzyl)pyridine		Hammock et al., 1974; Agarwal et al., 1979
Diethyldithiocarbamate/ HPLC		Dupard-Julien et al., 2007
Tetrafluorobenzenethiol/ GC-MS		Newman & Hammock, 2001
NMR		Hamalainen & Kamal-Eldin, 2005
HPLC-MS		Sjovall et al., 2001
Fourier transform infrared spectroscopy		Patterson, 1954; Bomstein, 1958; George, 1975
Carbonyls (soluble)		
Dinitrophenylhydrazones —optical	JOCS 2.5.4	White, 1995; JOCS, 1996
—HPLC		Seppanen & Csallany, 2001
—chemiluminescence		Townshend & Wheatley, 1998
Indole		Nagawade & Shinde, 2006
p-Anisidine—chemical	AOCS Cd 18-90; IUPAC 2.504	IUPAC, 1987; White, 1995; AOCS, 2011a
FTIR		Dubois, 1996
Infrared spectroscopy		Moya Moreno et al., 1999; Mobaraki & Hemmateenejad, 2011
NMR		Haywood et al., 1995; Moya Moreno et al., 1999; Hamalainen & Kamal-Eldin, 2005
Volatile products—gas chromatography		Christie, 1989
Static headspace		Przybylski & Eskin, 1995

Table 2.A. *Continued*

Lipid Oxidation Product	Standard Method	Reference
Volatile products—gas chromatography *(continued)*		
Solid phase microextraction (SPME)		Pawliszyn, 1999; Lee et al., 2007
Purge and trap/thermal desorption		Hartman et al., 1991; Hartman et al., 1993
Dimers/Polymers—HPLC-SEC	AOCS Cd 22-91, IUPAC 2.508	IUPAC, 1992a; AOCS, 2011j; Márquez-Ruiz & Dobarganes, 1997; 2006;
Maldi-MS		Fuchs et al., 2010

specific product (or class of products), but if that product is in a minor or absent pathway, the assay does not accurately reflect oxidation of the system.

2. *Tracking multiple pathways to avoid missing oxidation and to gain a more complete picture of the full extent and effects of oxidation.* As discussed in Chapter 1, conjugated dienes, hydroperoxides, and hexanal are not the only products of lipid oxidation. Dienes and hydroperoxides decompose under many conditions, and hexanal is not always present or may remain a minor product. Multiple reaction pathways need to be monitored to map oxidation, particularly under conditions favoring hydroperoxide decomposition or internal radical rearrangement to epidioxides or epoxides, or when proteins or other compounds are present to react with oxidation products and remove them from the analysis stream. At a minimum, these include hydroperoxides (H abstraction), epoxides (internal rearrangement), and multiple secondary products (scission), both volatile and non-volatile. Full tracking also requires adding lipid alcohols and dimers to this list.

3. *Obtaining maximum detailed information about specific products, their formation and degradation rates, and their potential transformations.* Formerly the purview of fundamental research, the challenge here is to resolve and isolate individual reactions and identify individual products. While this approach is necessary to elucidate oxidation mechanisms, it is also critical for determining potential problem points in processing or handling and distinguishing combinations of products that will accurately follow oxidation in quality control. Contemporary instrumentation now makes it possible to achieve detailed qualitative mapping of oxidation without sacrificing absolute quantitation.

4. *Tracking the footprints of oxidation in cooxidation products to avoid underestimating lipid oxidation and to assess the full range of oxidative deterioration in the system.* This is particularly important in systems with proteins, pigments, or other molecules that readily provide abstractable hydrogens, double bonds susceptible to lipid oxyl radical addition, or groups that react with lipid carbonyls (for example, amines and other carbonyls) (Schaich, 2008). Here, interactions broadcast damage beyond lipids and substantially degrade product quality while reducing detectable lipid oxidation.

5. *Data mining: gleaning maximum information from large volumes of data.* Given the substantial challenges of elucidating lipid oxidation mechanisms in greater detail and tracking complex reaction pathways, we cannot afford to throw away data. For example, GC chromatograms from oxidizing samples may have tens to hundreds of peaks, only a small fraction of which are identified and even fewer of which are used for routine monitoring of lipid oxidation; yet, these ignored peaks contain a wealth of information about oxidation processes and scission points. Similarly, analyzing products from multiple pathways provides more complete insight into the directions of oxidation, but integrating and correlating the masses of data can be overwhelming. To manage and exploit complex data obtained in lipid oxidation analyses, we need to embrace the application of chemometric approaches and other advanced statistical analyses to detect kinetic and product relationships that are not immediately obvious.

6. *Integrating multi-pathway data to develop a new understanding* of lipid oxidation, which can be applied creatively to better stabilize foods and food ingredients, cosmetics and personal care products, and other lipid-based materials. The information gained should also be useful for elucidating lipid oxidation in living tissues where reactions may be even more complex.

Let's consider each of these challenges in more detail. Some of the points made in the following discussion may seem to be obvious examples of good laboratory practices, but they are presented as reminders of the attention to detail that is required to make lipid oxidation analyses reproducible and accurate. Other points provide new perspectives that will hopefully stimulate industrial and academic researchers to look beyond traditional lipid oxidation reactions and analyses to develop a more complete picture of oxidative degradation in foods, biological cells and tissues, personal care products, and other materials. The first part of the chapter presents more experimental detail because more research has taken place in this area. Later sections of the chapter introduce more concepts and possibilities because these represent the directions we need to move in the development of lipid oxidation assays.

Providing Quantitative, Reproducible Analyses That Accurately Reflect Oxidation Chemistry

This seemingly straightforward challenge is not so easy to meet in practice because experimenters must address several main conditions. First, lipids must be extracted and oils must be handled in a manner that does not increase oxidation or change the distribution of oxidation products. Highly variable, off-target results more often reflect extraction or handling inconsistencies than actual oxidation.

A second requirement is that assays must accurately measure the desired chemistry. Once again, this concept may seem obvious, but it is often not achieved because many factors contribute to getting the chemistry right, including:

1. assay specific for the measured product, providing accurate and full detection of measured product independent of oxygen
2. response linear with concentration in the range that the product occurs
3. constant stoichiometry that involves only the test compound and is independent of molecular structure beyond the measured functional group
4. results reproducible when handling and conditions are appropriately controlled

Assuring accurate detection is particularly problematic with lipid oxidation because product levels and distributions change with time, introducing both quantitative and qualitative complications. A very sensitive assay that is appropriate for the early stages of oxidation may saturate as oxidation progresses, and assays capable of detecting high oxidation levels are useful in later stages of oxidation but may be inaccurate until products sufficiently develop.

A third requirement is that the analyses applied actually cover the products present. The presence of alternate reaction pathways for peroxyl radicals, alkoxyl radicals, and hydroperoxides, as well as shifts from hydroperoxides in early oxidation to a variety of secondary products in the later stages, certainly present challenges in accurately tracking product pathways. Which product(s) should be analyzed? Limiting assays to hydroperoxides and/or volatile products such as hexanal grossly underestimates or may even completely miss oxidation when conditions favor internal rearrangement to epoxides. Similarly, monitoring only volatile products or only hexanal in solid matrices that bind aldehydes and ketones yields erroneously low values of oxidation.

In practice, the same one or two assays are nearly always used for all stages of oxidation, and only single concentrations of samples are analyzed. Both of these practices can introduce considerable inaccuracies in quantitating and interpreting the extent of lipid oxidation.

The use of standardized assays does not ensure accuracy. Standardized assays that have been tested extensively are available for most lipid oxidation products (Table 2.A.). However, assays do not apply to all circumstances just because they are standardized, particularly when sample and assay concentration ranges do not match and the experimenters have not considered stoichiometry. Standard assays developed for use in the oil industry, where large volumes of material are always available, must be micronized for use in research labs, which have limited sample availability. Standard assays based on traditional technologies often lack the sensitivity needed to detect trace levels of products that develop early in oxidation or are unstable. In addition, standard methods have not been established with the newest and most sensitive instrumentation. Typically, each lab uses its own version of an analysis, modified from a standard assay or adopted from a literature report. These conditions present no guarantee that any of the three requirements outlined previously are met.

Satisfying these requirements is actually a challenge in "thinking chemistry" and applying it in detail, rather than blindly following procedures. The subsections that follow address some of the chemical details that are critical for accurate, reproducible assays of lipid oxidation.

Limiting Oxidation Artifacts during Extraction

More often than not, analyses of lipid oxidation focus on foods or other products, rather than oils. In these cases, the first challenge in ensuring the accurate analysis of oxidation must be developing extraction procedures that limit artifacts. Though simple in concept, limiting oxidation during extraction is no mean feat in practice because nearly every action destabilizes lipids. The common practice of handling and storing samples in plastic bags or containers before and during analyses allows oxygen to slowly diffuse into test materials. Extraction pretreatments disrupt the phase and cellular isolation of lipids, release catalysts, and expose the system to oxygen. Grinding to reduce particle size and enhance exposure to solvents also heats materials, at least locally. Unless sparged with inert gas (preferably argon), extraction solvents carry reasonable concentrations of oxygen, and sample transfers without inert gas protection further increase oxygen exposure. All these actions enhance oxidation and increase the formation of hydroperoxides. In counterbalance, any heating above 40 °C or exposure to laboratory lights decomposes hydroperoxides and stimulates the formation of secondary products. As a consequence, measured lipid oxidation too often reflects the artifacts of handling procedures more than actual oxidation levels. This situation can be disastrous and very costly for decisions about retaining production lots, which depend on absolute values of some oxidation products, such as peroxide values.

To limit oxidation during extractions, standard practices require constant protection of all samples from air and light. Glass containers provide the first line of defense against air in the laboratory because they can be sealed, as opposed to plastic materials, which are oxygen-permeable (Table 2.B). The second line of defense is keeping test materials under inert gas at all times. Of the inert gases, nitrogen is most commonly used due to habit and its lower cost, but argon is preferred. Since argon is heavier than air, it does not readily escape from solutions or vessels. Also, it does not contain the traces of oxygen present in nitrogen, and argon solubility in both water (Figure 2.1) and oil is double (or more) that of nitrogen (Table 2.C).

Solvents themselves carry catalytic levels of oxygen (Table 2.D), so they must be freshly distilled or sparged with inert gas. In water, argon solubility is slightly higher than oxygen, but about 2.5 times the solubility of nitrogen over the temperature range 0–50 °C (Battino & Clever, 1966); the solubility of argon in lipids is twice that of nitrogen (Behnke & Yarbrough, 1939). The order of gas solubility in chloroform is CO_2 > Ar > O_2 > H_2 > N_2 > He, while in dichloromethane (a solvent increasingly used to replace $CHCl_2$ in lipid extractions) it is CO_2 > O_2 > Ar >H_2 > N_2 > He (Shirono et al., 2008). Hence, the common assumption that nitrogen sparging or blanketing prevents lipid oxidation does not always hold. To be most effective, gas sparging (especially with nitrogen) must be alternated with vacuum and carried out for at least three 15–30 minute cycles (Rollie et al., 1987).

Solvent stability is another critical but often-overlooked issue in lipid extractions and oxidation analyses. Solvents that degrade over time (such as chloro-

Table 2.B. Oxygen and Water Vapor Permeability of Common Packaging Plastics

Polymer	Oxygen ml/100 sq. inch	Water g/100 sq. inch
Low density polyethylene	300–400	1.0–1.5
High density polyethylene	100–200	0.3–0.5
Polypropylene	150–200	0.2–0.5
Polystyrene	300–400	5–10
Polycarbonate	200–300	3–8
Poly(vinyl) chloride	5–10	0.9–2.0
Polyethylene terephthalate	10	0.9
Nylon 6	1–3	6–22
Ethylene vinyl alcohol, dry	0.1	6

Excerpted from Massey (2003), used with permission.

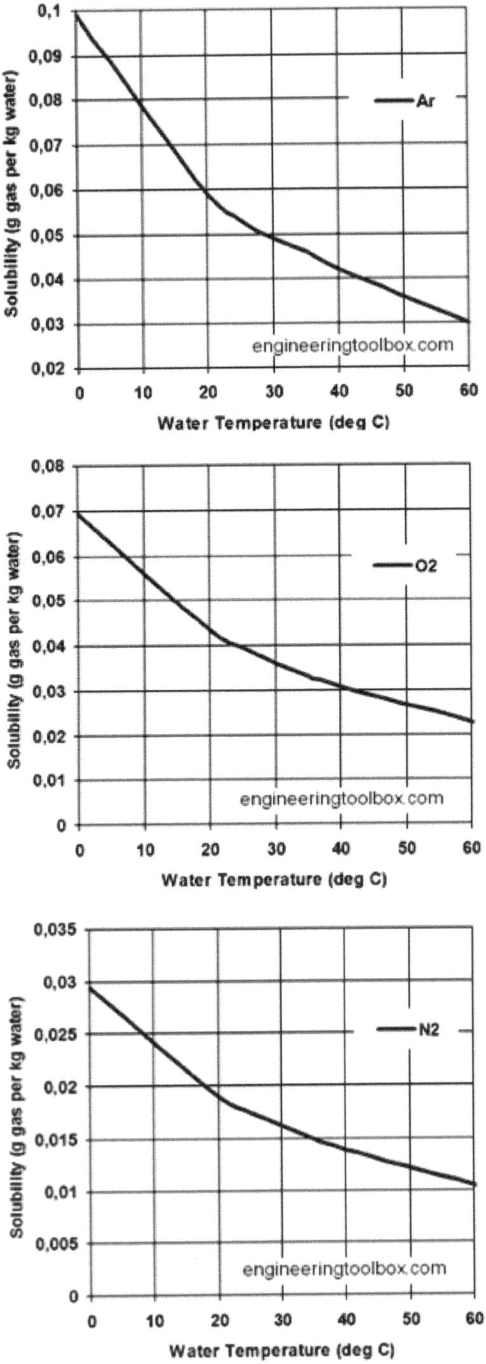

Fig. 2.1. Solubility of inert gases argon and nitrogen relative to oxygen in water.

Graphs excerpted from http://www.engineeringtoolbox.com/gases-solubility-water-d_1148.html. Used with permission.

Table 2.C. Solubility of Oxygen and Nitrogen in Oils (Data Based on Hardened Lard and Olive Oil)

	Solubility at 1 bar (mg/kg oil)		Dissolved in oil saturated with air (mg/kg oil)	
Temp. °C	O_2	N_2	O_2	N_2
0	170	80		
25	180	85	38	66
50	185	90		
75	190	95		
100	200	105		
125	—	110		
150	—	115		

Adapted from Gunstone and Padley (1997).

Table 2.D. Solubility of Oxygen, Nitrogen, and Argon in Common Lipid Solvents at Room Temperature (293 K)

	Mol fraction × 10^4		
Solvent	**Oxygen**	**Nitrogen**	**Argon**
Octane	21.22		
Iso-octane		14.13[2]	29.2[3]
Heptane	19.98		25.0[3]
Hexane	19.60	14.0[3]	25.3[3]
Cyclohexane	12.31	7.55[3]	14.9[3]
Toluene	9.17		10.95[3]
Acetone	8.34	5.19[2]	
Tetrahydrofuran	8.03		
Butanol	8.03	4.52[2]	
Chloroform	7.23[1]	7.89[1]	3.82[1]
Dichloromethane	5.61[1]	3.27[1]	5.02[1]
Ethanol	5.87	3.55[2]	
Methanol	4.15	2.75[2]	
Olive oil	0.51		

[1]Shirono et al. (2008)
[2]Kretschmer et al. (1946)
[3]Jolley & Hildebrand (1958)

Unless otherwise noted, data compiled from Battino et al. (1983).

form and methylene chloride) usually have some stabilizer added (Honeywell, Burdick, Jackson, 2008). Chloroform, in particular, forms toxic phosgene ($COCl_2$) over time, so it must be handled and stored with care (preferably under argon after opening) (Sigma-Aldrich, 2012). From a toxicity standpoint, ethanol may be the preferred chloroform stabilizer, but it must be added at levels higher than 1% (Honeywell, Burdick, Jackson, 2008; Sigma-Aldrich, 2012). This can introduce problems because ethanol is susceptible to radical attack by oxidizing lipids, forming reactive radicals that accelerate oxidation (Schaich et al., 1993), and it also complexes with lipids and other radicals (Wu et al., 1977; Rockenbauer & Tudos, 1979; Gardner et al., 1985). Cyclohexene, another stabilizer, can similarly become a source of radicals and also interfere with the GC analysis of products (Sigma-Aldrich, 2012). Amylene (2-methyl-2-butene) is a free-radical scavenger that limits adventitious lipid oxidation during extraction (Yao & Schaich, 2012a), but Richards and Feng (2000) reported high backgrounds in the ferric thiocyanate hydroperoxide assay with this stabilizer, and amylene appears to be less effective in preventing phosgene formation over time (Turk, 1998). Awareness of stabilizer effects in assay chemistry has recently risen, and considering that all stabilizers are chemically reactive, it seems prudent at this point to recommend that all lipid oxidation assays be thoroughly reevaluated to determine solvent effects, degradation products, and stabilizers.

A second problem with solvent stability is the formation of solvent peroxides. Furans and ethers are notable peroxide-formers, as are isopropanol (Redemann, 1942), ethyl acetate, and butanol (Navas et al., 2004), solvents commonly used in lipid extractions and oxidation assays. My research group has found significant interference in both peroxide and carbonyl analyses when isopropanol oxidizes. To limit solvent peroxide formation, researchers should keep all solvent bottles sparged (liquid and headspace) with argon and tightly sealed after opening (Clark, 2001), protect them from light and maintain minimum headspace (Clark, 2001), and add peroxide decomposers, such as XPell commercial pellets, which are inert and act like molecular sieves in solvents (XploSafe, 2012).

Some rethinking of extraction methods is required also to minimize oxidation artifacts during extraction. A battle constantly exists between maximizing yield and minimizing oxidation in extractions, and the two objectives are not always compatible because the extreme conditions required to disrupt difficult matrices also degrade lipids. Even with rigorous protection in the simplest extractions (using traditional manual methods), it is difficult to avoid exposing macerated materials to oxygen and light during multiple handling steps and long incubations (24 to 72 hours). This is especially true in matrices where lipids are bound or entrapped and, thus, require extreme solvents (such as hot water–saturated butanol) and extraction conditions for removal.

Soxhlet and solvent reflux methods were developed to maximize lipid yields, and they have been used extensively for total lipid extractions from complex cellular matrices, such as microbes (Liu et al., 2010; McNichol et al., 2012), muscle tissue (Sahasrabudhe & Smallbone, 1983; Kazala et al., 2006), and plant materials (Melton et al., 1979; Shen & Shao, 2005). Unfortunately, these methods have been applied with little concern for associated degradation, and since lipid oxidation products are present at trace levels (< 1% total lipid), they are obscured in reduced yields. In actuality, Soxhlet extractions have serious limitations for applications in which oxidation is an endpoint. Solvents strongly affect what is extracted and, hence, the picture of lipid oxidation portrayed. Ethers, the most common solvents for Soxhlet, preferentially solubilize polar lipids (including oxidation products) and underextract triacylglycerols, so they can lead to overestimation of lipid oxidation levels. Hexane is a poor solvent for polar lipids (McNichol et al., 2012), and because phospholipids are particularly sensitive to oxidation, this solvent underestimates oxidation in Soxhlet extracts. In contrast, chloroform-methanol extracts high levels of extraneous, non-saponifiable material (McNichol et al., 2012); if this is not removed or accounted for, erroneously low lipid oxidation levels are calculated. Regardless of solvent, the extreme conditions (for example, the high temperature and long time) of Soxhlet extractions induce *trans* isomerization (Pérez-Serradilla et al., 2007). Most important, the elevated temperatures employed are even worse than lab lights for accelerating oxidation and changing product levels and distributions. Studies with extruded total-nutrition products that require very low oxidation found that Soxhlet extracts exhibited unacceptably high levels of both peroxides and carbonyl secondary products (Strange & Schaich, 2000). Similarly, extractions of soybean oil by Soxhlet and two related solvent reflux methods gave peroxide values of about 7 meq/kg oil (Nikolić et al., 2009), which is above the normal industry reject value of 5, far too oxidized to be useful for most applications. Thus, Soxhlet is a poor extraction choice when analyzing oxidized lipids.

A number of alternate extraction procedures have been proposed as replacements for Soxhlet extraction. Indeed, these new methods generally enhance lipid extractions from complex matrices where highly cross-linked molecules entrap lipids and make them difficult to disrupt or digest. They also facilitate extraction in highly compartmentalized tissue structures, which release catalysts when homogenized for extraction. Unfortunately, the elevated temperatures that increase extraction efficiency in most of these methods (Raynie, 2006) also cause significant complications when the end analyses will be lipid oxidation products rather than fatty acid or lipid class composition.

Microwave-assisted extractions utilize focused, localized heating under pressure to break down tissues and release lipids (García-Ayuso et al., 2000; Lee et al.,

2009). However, microwave extractions typically raise temperatures to between 60 and 120 °C for many minutes (Turner, 2006; Boldor et al., 2010). This heat accelerates hydrolysis and lipid oxidation, breaking down pre-formed hydroperoxides and increasing both radical chain propagation and transformation to secondary products. For example, even exposure times as short as 10–15 seconds (65 °C to 85 °C) generated up to 11 nmol/mg malonaldehyde and formaldehyde in polyunsaturated methyl esters (Wong et al., 1991). Ten minutes at high microwave power (210 °C final temperature) induced significant increases in all oxidation products, plus increased hydrolysis of triacylglycerols and loss of polyunsaturated fatty acids (Cossignani et al., 1998). In addition, microwave heating does not inactivate all lipases so it allows increased hydrolysis and fatty acid production in fresh materials (Ponne et al., 1996).

Ultrasound-assisted extractions use high frequency acoustic waves to disrupt cells and molecular matrices. Unfortunately, the method is highly problematic when oxidation is the endpoint because the cavitation caused by collapsing bubbles forms free radicals in any solvents present. HO• and H• radicals form during the ultrasonication of water in solution or tissues (Riesz et al., 1990; Riesz & Kondo, 1992; Price et al., 1997; Somaglino et al., 2011). Subsequent reactions of these radicals are responsible for the cell killing associated with ultrasound and also many forms of molecular degradation, including lipid oxidation, in cells and foods. Ultrasound also induces decomposition of organic solvents, particularly chlorinated hydrocarbons (Kruus et al., 1997). Unlike other extraction methods, ultrasound is one case in which sparging with inert gas does not protect against oxidation. Paradoxically, argon increases radical production from cavitation because it is more soluble than air so more active bubbles form, friction between bubbles generates higher local temperatures and enhances bubble collapse, and the fragments become nuclei for further cavitation (Okitsu et al., 2006). Therefore, as with microwave heating and Soxhlet, this method of facilitating lipid extraction is not appropriate when analysis of lipid oxidation is the end point.

Supercritical fluid extraction (SFE) of lipids utilizes CO_2 under high pressure above its critical temperature of 31 °C, under which conditions it flows and dissolves compounds like a liquid solvent (Clifford, 1993; Peter, 1996). Because of its low viscosity and small size, CO_2 can penetrate matrices more efficiently than other solvents and methods. For oxidation analyses, however, the elevated temperatures of SFE are a distinct disadvantage. When searching for efficient extraction methods for extruded materials, my research group found that SFE could indeed remove lipids inaccessible to standard extraction methods, including water-saturated butanol. However, temperatures exceeding 60 °C (depending on the material) caused substantial degradation of the lipid extracts (Chaudry, 1998; House, 1999). In SFE of olive husks, peroxide values decreased with extraction temperature from 40 to 60 °C, and all were higher than acceptable for human consumption (43 to 8 meq/kg oil, respectively) (de Lucas et al., 2002); free fatty acids were also unacceptably high (14 to 43%). In oxidized walnuts extracted at 80 °C, the quality of the CO_2 had marked effects on lipid oxidation, with welding-grade CO_2 accelerating oxidation more than Coleman grade; all volatiles were lost during the extraction (Crowe et al., 2002). The drying required before SFE of fresh materials such as meats (King et al., 1989) also decreases lipid stability. One additional drawback is that CO_2 is very hydrophobic, so it does not extract phospholipids without the addition of water or ethanol as modifiers (Temelli, 1992; Cocero & Calvo, 1996; Montarini et al., 1996). Because phospholipids are highly susceptible to oxidation—oxidizing even before triacylglycerols—SFE without ethanol can significantly underestimate oxidation in extracts.

Accelerated solvent extraction (ASE, Dionex trademark), also known as pressurized fluid extraction (PFE) or pressurized solvent extraction (PSE), was developed to circumvent both the high temperatures and the restrictor clogging that has plagued SFE (Singleton & Stikeleather, 1999). Several manufacturers now market instruments for this extraction. During static steps, pressure at elevated temperatures forces solvent penetration into sample matrices to dissolve lipids. Lipids and solvent are then washed out during dynamic (rinse) steps using single solvents or solvent mixtures, as necessary, to solubilize the lipid classes present or desired. Surface lipids are usually removed in a single cycle; lipids in complex matrices more typically require two or more static-dynamic cycles for complete extraction. Important for lipid oxidation, the entire process is performed under inert gas—usually, nitrogen—and in the dark, thus reducing degradation. Multisample systems also provide automation, which reduces handling exposure and oxidation further.

ASE has gained increasing attention over the past decade because it efficiently removes lipids from difficult matrices, including muscle tissue, fish tissue (Dodds et al., 2004), poultry meat (Toschi et al., 2003), microbial cells (Liu et al., 2010; Wang et al., 2012), corn and oats (Moreau et al., 2003), and oilseeds (Singleton & Stikeleather, 1999). As with other extraction methods, however, little attention has been given to issues of oxidative stability during the extraction when developing applications.

My research program has been particularly interested in determining both the completeness of ASE (so that we can determine where oxidation occurs) and its effects on oxidation (so that we can determine the oxidation actually present in the sample instead of that induced by extraction handling). Overall, we have found that ASE can effectively replace manual and other methods, providing comparable or higher yields with lower oxidation and contaminant levels in short extraction times, 10–40 minutes depending on the matrix (Table 2.E) (Yao & Schaich, 2012a, 2012b). Blank studies with corn oil show that, unlike other methods, ASE

Table 2.E. Comparison of Yields and Lipid Oxidation in Manual Extractions vs. ASE from Extruded Food Product. Values reported are averages of triplicate analyses.

	Yield, % lipid	PV[a]	CD[a]
Manual—24 hrs	17.33	4.47	1.04
Manual—48 hrs	19.34	4.08	0.94
ASE—10 min, 4 cycles	16.98	2.56	0.10
Acid hydrolysis	16.82	—	—

[a]mmol/mol TAG

procedures alone produce little change in conjugated dienes and hydroperoxides (Fig. 2.2). Most important for this discussion, the ASE of lipids from food matrices can be accomplished at 40 °C, where hydroperoxides are still stable, and under nitrogen or argon, which limits oxidation. At higher extraction temperatures, oxidation increases (Fig. 2.2).

Obviously, extraction details must be optimized for each type of material and lipid composition, and ASE still requires protection of samples during all handling steps, particularly during pre-extraction grinding, cell loading, and post-extraction solvent evaporation and storage. As long as these stages are controlled, ASE can produce extracts with significantly less extraneous oxidation than manual or Soxhlet extractions. With continued evaluation and validation, ASE may well become the method of choice for lipid extractions when oxidation is the main analytical goal.

Accurately Measuring Target Chemistry

It is easy to assume that no assay would be run if the chemistry was not correct. However, in recent reevaluations of lipid oxidation assays, my research group has found many factors that reduce the accuracy and sometimes even the validity of results. True, we are pushing the limits of assays because we seek absolute quantitation rather than relative data to determine mass balance between pathways. However "black-box" chemistry, where procedures are followed by rote with little or no accounting for underlying chemistry, should no longer be accepted. The following questions must be addressed for all lipid oxidation assays:

1. **Is the assay specific for the measured product?**

Specificity includes considerably more elements than just the reagent reacting with the target compound. Loss of specificity results from at least three sources that are not easy to correct in blanks. First, extraneous compounds whose opti-

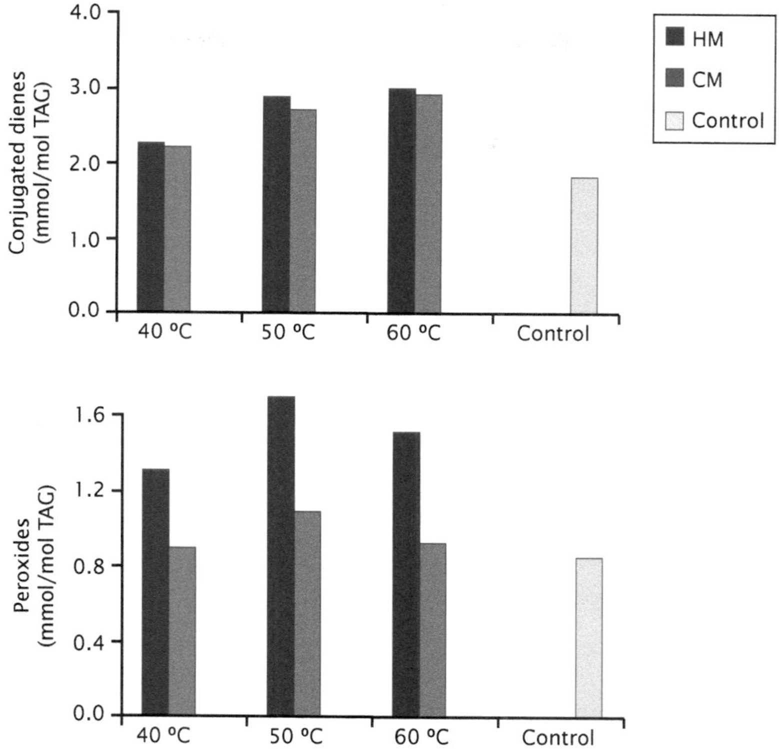

Fig. 2.2. Corn oil exposed to ASE demonstrates very low levels of oxidation ASE induced by heat and handling during ASE. ASE conditions: 1500 psi, 4 ten-minute cycles. Control was corn oil analyzed before ASE. HM: hexane methanol 2:1; CM: chloroform methanol 2:1.

cal spectra overlap absorbance wavelengths used in lipid oxidation assays add to calculated product concentrations. This can happen even in an assay as simple as conjugated dienes (CDs), in which the absorption of antioxidant additives or their oxidation products, such as tocopheryl hydroperoxides in oils or extracts (Gruszka et al., 2008), overlap the 231–235 nm region of CDs. We have observed similar interference with a number of antioxidants. Similarly, carotenoids add to visible range absorbance in xylenol orange and ferric thiocyanate optical assays of lipid hydroperoxides, forcing erroneously high peroxide values (Hornero-Méndez et al., 2001). Pigments (Dobarganes & Velasco, 2002) and lipid-soluble food colors can have similar effects if not removed from extracts.

A second loss of accuracy arises when the assay is broader in specificity than expected, skewing the interpretation of oxidation progression or active pathways involved. An example here is the Alde-Safe™ assay for aldehydes. Based on the indole assay, this test was designed by Virginia Gordon as a replacement for the toxic *p*-anisidine assay, which is specific for 2-alkenals. However, reported correlations of the Alde-Safe™ assay with anisidine analyses of frying oils are unimpressive ($R^2 = 0.74$) (Osawa et al., 2007), probably because the indole assay detects saturated aldehydes even more strongly than alkenals (Peng, 2010). This may be a moot point if the goal is only to record some value for comparison of oxidation levels. However, when the measured value is used to make decisions about product retention or formulation, or to understand why and how fast a product deteriorates, such inaccuracy can lead to incorrect assessment of both the assay's validity and the extent of oxidation.

Finally, side reactions, perhaps unrecognized, can introduce serious errors in assay specificity. A prime example is when alkoxyl radicals generated from hydroperoxide decomposition react with solvents as well as reporter compounds (Steltzer, 2012). The net effect on the assay depends on whether the solvent involved generates an oxidizing or reducing radical, as will be discussed in greater detail in the stoichiometry section.

2. Is the assay independent of oxygen?

Measurement of nearly all lipid oxidation products involves redox reactions, so it should not be surprising that oxygen can present major problems for quantitation. Oxygen has long been recognized as causing critical interference in lipid hydroperoxide (LOOH) assays, particularly the classical iodometric titration in which dissolved or atmospheric oxygen oxidizes released iodide ions and increases the apparent LOOH concentration (Lea, 1946, 1952). Similarly, oxygen oxidizes ferrous iron in xylenol orange and ferrithiocyanate assays for LOOH (Kolthoff & Medalia, 1949), and it forms new peroxides from any radicals generated during the reactions (Kolthoff & Medalia, 1951; Mihaljevic et al., 1996); both effects lead to high peroxide values (see a more detailed discussion in the following section).

Epoxide assays are also sensitive to oxygen. HBr oxidizes in the standard HBr titration of epoxides, yielding low epoxides. We have found consistently lower values with standard epoxides when titrating with HBr in air than under argon; the differences increase as concentrations and epoxide chain length decrease (Liao, 2013). Nitrobenzyl pyridine (NBP) assays for epoxides are more variable when oxygen is present.

To avoid interference from oxygen and to protect both the lipids and the reaction, a good general rule is to run all lipid oxidation assays under argon (sparging both solvent and headspace).

3. **Does the assay have full product reactivity and constant stoichiometry that involves only the test compound?**

Although stoichiometry is introduced in general chemistry courses, it is seldom considered in lipid oxidation analyses. Stoichiometry is an indication of the directness and cleanness of a reaction. In the reaction,

$$T + L_{ox} \rightarrow \text{T-Product} + L_{ox}\text{-Product} \tag{2.2}$$

the stoichiometry is 1:1, and concentrations of target products (T) should exactly match lipid oxidation levels (L_{ox}). If the stoichiometry is less than one, full detection of lipid oxidation is not achieved. If the stoichiometry is greater than one, side reactions are involved and oxidation is overestimated. Stoichiometry should be calculated for all assays and conditions to ensure that the reaction is complete and constant. Indeed, known stoichiometry is an absolute requirement for accurate quantification, and poorly reproducible stoichiometry is a serious limitation to quantitation, whether absolute or relative.

Assays that involve ferrous iron reduction of hydroperoxides have serious stoichiometric issues. Both the xylenol orange and ferrithiocyanate assays nominally have a stoichiometry of 2:1 (two mol Fe^{3+} generated per mol LOOH) because both the hydroperoxide and the alkoxyl radical resulting from its reduction oxidize Fe^{2+} to Fe^{3+} (Wolff, 1994; Dobarganes & Velasco, 2002):

$$LOOH + Fe^{2+} \rightarrow LO\bullet + OH- + Fe^{3+} \tag{2.3}$$

$$LO\bullet + Fe^{2+} + H^{+} \rightarrow LOH + Fe^{3+} \tag{2.3a}$$

In actuality, however, the stoichiometry is quite variable (Mihaljevic et al., 1996) and can range as high as 4:1, depending on the solvent. In chloroform:methanol 2:1, for example, alkoxyl radicals attack solvent chloroform, producing trichloromethyl radicals ($Cl_3C\bullet$). These then add oxygen, become strongly oxidizing trichloroperoxyl radicals, and continue to oxidize ferrous ions directly:

$$LO\bullet + CHCl_3 \rightarrow LOH + \bullet CCl_3 \tag{2.4}$$

$$Fe^{2+} + \bullet CCl_3 + H^{+} \rightarrow CHCl_3 + Fe^{3+} \tag{2.4a}$$

$$\bullet CCl_3 + O_2 \rightarrow \bullet OOCCl_3 \tag{2.4b}$$

$$\bullet OOCCl_3 + Fe^{2+} \rightarrow Fe^{3+} + {}^{-}OOCCl_3 \tag{2.4c}$$

Counterbalancing this excess generation of Fe^{3+}, alkoxyl radical attack on methanol produces hydroxymethyl radicals, which are strong reducing agents. They recycle Fe^{3+} to Fe^{2+} and decrease the apparent LOOH detected.

$$LO\bullet + CH_3OH \rightarrow LOH + \bullet CH_2OH \tag{2.5}$$

$$\bullet CH_2OH + Fe^{3+} \rightarrow HCHO + Fe^{2+} + H^+ \tag{2.5a}$$

These are all competing reactions, so the net amount of Fe^{3+} generated (hence, the LOOH levels reported) in a given assay will be influenced strongly by the nature of the solvent, levels and type of hydroperoxide, iron-binding constants of the trapping agent (xylenol orange, ferrithiocyanate), oxygen availability, and pH (Mihaljevic et al., 1996; Bou et al., 2008). However, the major controlling factors are the redox potentials of individual hydroperoxides (Reaction 2.3) and the resulting alkoxyl radicals relative to the secondary targets—iron in Reaction 2.3a or solvents in Reactions 2.4 and 2.5.

For example, differences in hydroperoxide redox potentials alter the rate of Fe^{2+} reduction of hydroperoxides, as well as final product yields and stoichiometry in any assay (Orr & Williams, 1953). They also account for variations in extinction coefficients measured for hydroperoxides of different structures (and redox potentials) in the xylenol orange assay (Jiang et al., 1990, 1991a, 1992; Gay et al., 1999a, 1999b). Differences in hydroperoxide response can be normalized by running standard curves (the basis for the PeroxySafe™ assay) or calculating extinction coefficients for known hydroperoxides. However, standards of lipid hydroperoxides are rarely available, and standard curves for available pure hydroperoxides are not applicable to ROOH with other structures, particularly lipids. For example, since reactivity generally increases with R chain length in ROOH, using cumene hydroperoxide as a standard greatly overestimates lipid hydroperoxides in most iron-based assays. Likewise, significant differences in the excess equivalents of Fe^{3+} observed for *t*-butyl (2.5), cumene (3), and oleic or linoleic (~4) hydroperoxides in the ferrithiocyanate assay (Barthel & Grosch, 1974) parallel differences in corresponding RO• redox potentials and hence capability to oxidize solvents and other adventitious molecules. Hence, stoichiometry greater than 2 mol Fe^{3+} generated per mol of hydroperoxide in iron-based assays strongly indicates that side reactions are actively adding to apparent LOOH concentrations.

In addition to these compositional complications, we have found that response stoichiometry also varies with hydroperoxide concentration, particularly in the ferrithiocyanate reaction (Table 2.F) (Steltzer, 2012). Such variation can occur when a reaction is diffusion-controlled and/or when competition between potential targets (including oxygen) is active.

Table 2.F. Effect of Hydroperoxide Concentration on Reaction Stoichiometry in the Ferrithiocyanate Assay for Hydroperoxides

[ROOH] Reacted (μM)	[ROOH] Detected (μM)	Stoichiometry
Cumene Hydroperoxide		
32.25	71.20	2.21
24.81	55.19	2.22
19.85	43.49	2.19
12.4	25.40	2.05
7.12	13.53	1.90
6.10	11.78	1.93
4.07	6.11	1.50
2.03	3.08	1.51
1.22	1.08	0.89
0.81	0	0
tert-*Butyl Hydroperoxide*		
62.81	99.48	1.58
41.87	79.13	1.89
33.5	63.44	1.89
20.94	38.28	1.83
12.56	23.06	1.84
7.15	13.24	1.85
3.58	6.32	1.77
2.15	3.15	1.47
1.43	1.37	0.96
0.36	0	0

Data from Steltzer (2012), used with permission.

Overall, the primary message for this section is: stoichiometry that changes with solvent or analyte composition or concentration compromises any assay. We have found variable stoichiometry to be a serious shortcoming of all optical assays for lipid oxidation products (that is, lipid oxidation products react with a target to generate a colored agent with an extinction coefficient). Iodometric titration of LOOH is the only assay we have found with constant stoichiometry that is independent of R group and peroxide concentration.

Inconsistent stoichiometry affects the accuracy of results and can cause erratic and irreproducible results, especially when comparing different materials or when studying complex materials with many unsaturated fatty acids. This stoichiometry issue needs to be addressed with all assays for lipid oxidation products to ensure accuracy in the detection, interpretation, and application of results.

4. **Is the assay response independent of structure beyond the measured functional group?**

This is a particularly important issue with lipids because hydroperoxides, epoxides, alcohols, and aldehydes each form at multiple sites along the acyl chain, and the position of the functional group can alter its reactivity. For accurate quantitation, assays for individual classes of products must be specific for the functional group but independent of the chain structure (R-) attached to the functional group.

When reevaluating lipid oxidation assays to track all of the alternate pathways described in Chapter 1, we found structure effects on reactivity to be a serious issue limiting the utility of several assays. Surprisingly, iodometric titration of hydroperoxides, much maligned due to difficulties in reproducible handling, was completely blind to molecular structure, giving the same response regardless of hydroperoxide. In contrast, responses of both xylenol orange and ferric thiocyanate assays varied with structure (Steltzer, 2012) (Fig. 2.3a). HBr and NBP assays for epoxides are even more sensitive to molecular structure, with their reaction rates and extents increasing dramatically as R chain length increases (Fig. 2.3b) (Liao, 2013). Agarwal and colleagues (1979) and Hammock and colleagues (1974) also reported large differences in extinction coefficients for different epoxides reacted with NBP, even with similarities in structure.

Why is structure-dependent response important? Lipids in foods and biological tissues have mixed compositions with a range of fatty acid chain lengths and number of double bonds. Mixed fatty acids generate hydroperoxides, aldehydes, and epoxides with many different structures. When the assay response varies with product structure, it is extremely difficult to determine whether the responses measured reflect product structure or product levels. When response varies with structure, standards must be matched to each individual material and it will be difficult, if not impossible, to find appropriate standards for all products. If the sample has an unknown composition or contains mixed hydroperoxides in unknown proportions, approximate ROOH concentrations can be calculated by using an appropriate general molar absorption coefficient or by expressing the concentrations as hydrogen peroxide equivalents (or equivalents of hydroperoxide used as standard). Such relative values are useful for following oxidation of a single type

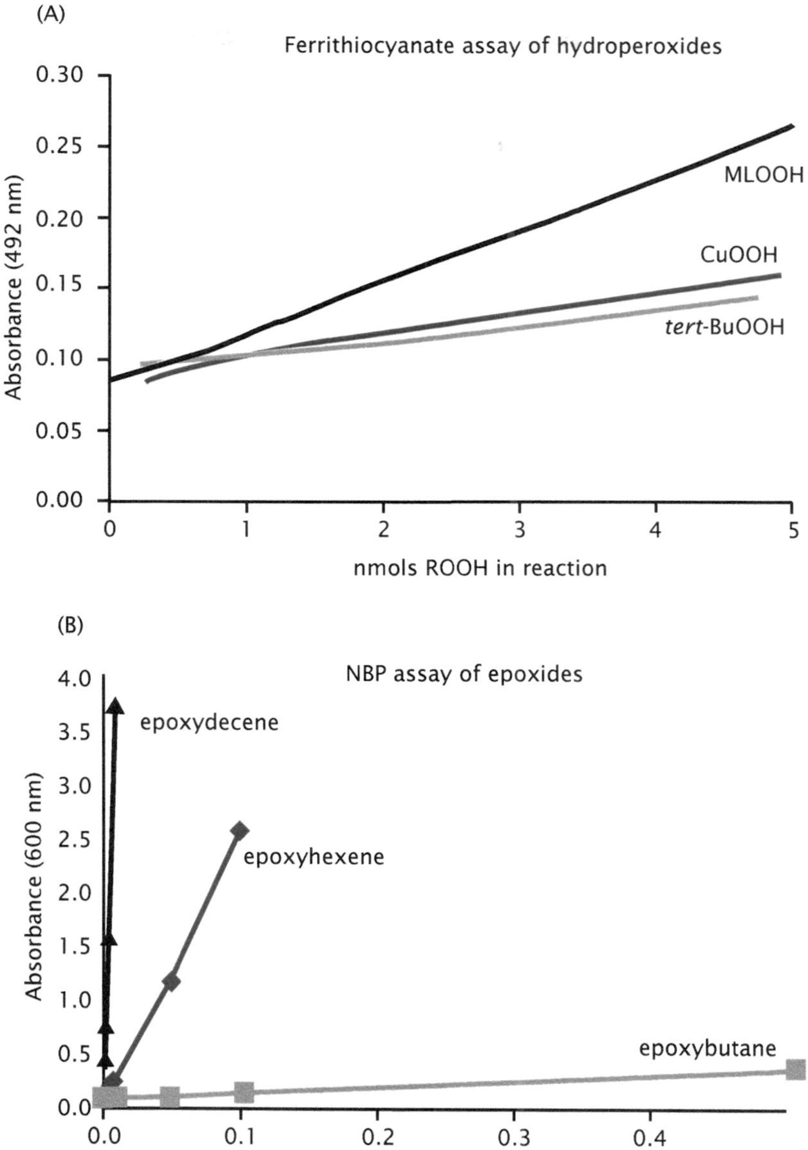

Fig. 2.3. Effect of molecular structure on reaction responses of lipid oxidation products. (A) Ferrithiocyanate assay for lipid hydroperoxides. MLOOH, methyl linoleate hydroperoxide; CuOOH, cumene hydroperoxide; *tert*-BuOOH, *tert*-butyl hydroperoxide. (B) Nitrobenzylpyridine (NBP) assay of lipid epoxides.

of material over time, comparing like materials, or when precision is not necessary. However, these values are not absolute, which means they are not appropriate for mechanism studies, and they definitely should not be the basis for acceptance or rejection of a commercial product.

Matching Product Concentrations in Samples to Assay Detection Ranges

Because each assay has its own accurate detection range, the stage of oxidation and product concentrations of samples must be matched to the sensitivity of each assay. Unfortunately, oxidation levels are not often known before analyses; they may be suspected from experience, but that is no guarantee. Therefore, accurately matching assay sensitivity requires the routine analysis of multiple sample concentrations, especially with "high-sensitivity" assays. Accurate matching may also require shifting from one assay with micromolar detection ranges in the early stages of oxidation to a second assay that works in the millimolar and higher concentration range as oxidation progresses. Analyzing outside the active detection ranges means that reported concentrations are inaccurate (usually too low), and in worst cases, products may go undetected altogether.

Matching product concentrations to assays (and vice versa) addresses the following question: Is the assay response linear with analyte concentration in the range that the product is expected? This actually combines two questions—is the assay linear and are the product concentrations "in range"? Assays that were originally intended to detect trace levels of peroxides and other oxidation products in solvents (Young et al., 1936; Kolthoff & Medalia, 1949, 1951) or living tissues (Wolff, 1994) have increasingly been adapted for oil and food applications. Examples include xylenol orange, ferrithiocyanate, and triphenyl phosphine assays, which detect hydroperoxides at nanomolar levels (Dobarganes & Velasco, 2002; Steltzer, 2012) (Table 2.G). These reactions have responses that increase linearly with analyte concentration, but the slope of the response curve often changes with ROOH concentration, creating several discrete detection ranges (Fig. 2.4). Reactions may saturate at higher analyte concentrations due to competition for reaction sites, inhibition of molecular diffusion, and an excess of analyte over reporter molecules, or they may even reverse due to secondary reactions. The challenge, then, is to determine which detection range is most appropriate for the material being tested and to match the test concentrations to that range.

Unfortunately the micromolar to nanomolar levels of hydroperoxides for which these assays are designed are well below the levels normally found even in fresh foods. Hence, super-sensitive assays nearly always require that oils and lipid extracts be routinely diluted or that very small samples be used. Testing a series of sample

Table 2.G. Comparison of Detection Ranges for Various Hydroperoxide Assays

Assay	ROOH Detection Range
Iodometric Titration	2-200 µmoles
PeroxySafe™	2–12 nmoles
Xylenol Orange—Pierce	0.8–50 nmoles
Ferric Thiocyanate	5–90 nmoles
Cayman Assay (Ferric Thiocyanate)	0.5–5 nmoles

Data from Steltzer, (2012) used with permission.

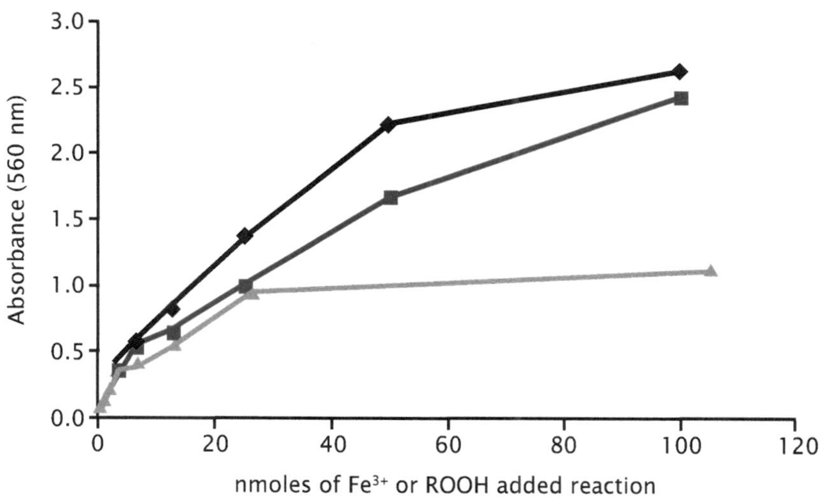

Fig. 2.4. Examples of non-linear response and saturation in lipid oxidation assays. Curves are from a commercial xylenol orange assay kit for lipophilic hydroperoxides. Response curves technically fit polynomial equations with high R^2 values but depart from theoretical shape in the higher concentration ranges. Hence, concentrations calculated from polynomial standard regressions will be overestimated. Alternatively, several discrete linear response regions with different slopes can be identified and used for calculations. Accurate determination of product concentrations from either approach requires separate standard curves developed from authentic compounds for each product to be analyzed, as well as analysis of a range of concentrations for each test sample.

dilutions is particularly important when product concentrations are near either extreme of the calibration range or are calculated from regression equations without referring to physical graphs that reveal plateau points. Matching oil sample dilution or size to assay detection range becomes an empirical process when LOOH concentrations are unknown and only single concentrations of samples are analyzed.

My research group has found that analyzing only undiluted samples has serious drawbacks in both xylenol orange and ferrithiocyanate assays for lipid hydroperoxides. In both of these assays, high concentrations of hydroperoxides oxidize and bleach the detection complex, making the assay appear erroneously as if no peroxides were present:

$$\text{LOOH (or LO}\bullet\text{)} + Fe^{3+}\text{-XO (purple/blue)} \rightarrow Fe^{3+}\text{-XO}_{ox}\text{ (yellow to light orange)} \quad (2.6)$$

$$\text{LOOH (or LO}\bullet\text{)} + Fe^{3+}\text{-SCN (dark red)} \rightarrow Fe^{3+}\text{-SCN}_{ox}\text{ (light red)} \quad (2.7)$$

In fact, these assays are counterintuitive. Expectations of low oxidation boost the tendency to increase sample size or concentration in order to increase sensitivity, but paradoxically, this can put the sample concentration beyond the linear response range and into the bleaching range. Without dilution until consistent response is obtained, samples can erroneously be accepted as having very low or no oxidation while peroxide levels are actually quite high. To avoid this problem and ensure that responses fall into linear ranges of the assays, we recommend diluting samples serially by factors of 10 over at least three orders of magnitude (more is often needed).

Keeping a sample and assay matched in range also means that appropriate sample size or dilution is very likely to change as oxidation progresses, as in shelf life tests or when samples are pulled out of extended storage. This behavior can be seen readily in the dramatic reduction in methyl linoleate required to keep the sample in reading range and avoid bleaching Fe^{3+}-SCN as incubation time progresses from 4 to 13 days (Fig. 2.5). When sample sizes for ultrasensitive assays must be reduced to the point that measurement accuracy suffers, the assay applied should be shifted to one with a high detection range, such as iodometric titration for hydroperoxides.

Combining and Integrating Multiple Assays to Track Alternate Oxidation Pathways

Analyze Multiple Products

When multiple or alternate pathways are active, all of the handling precautions and reaction considerations discussed above become moot if the measured products

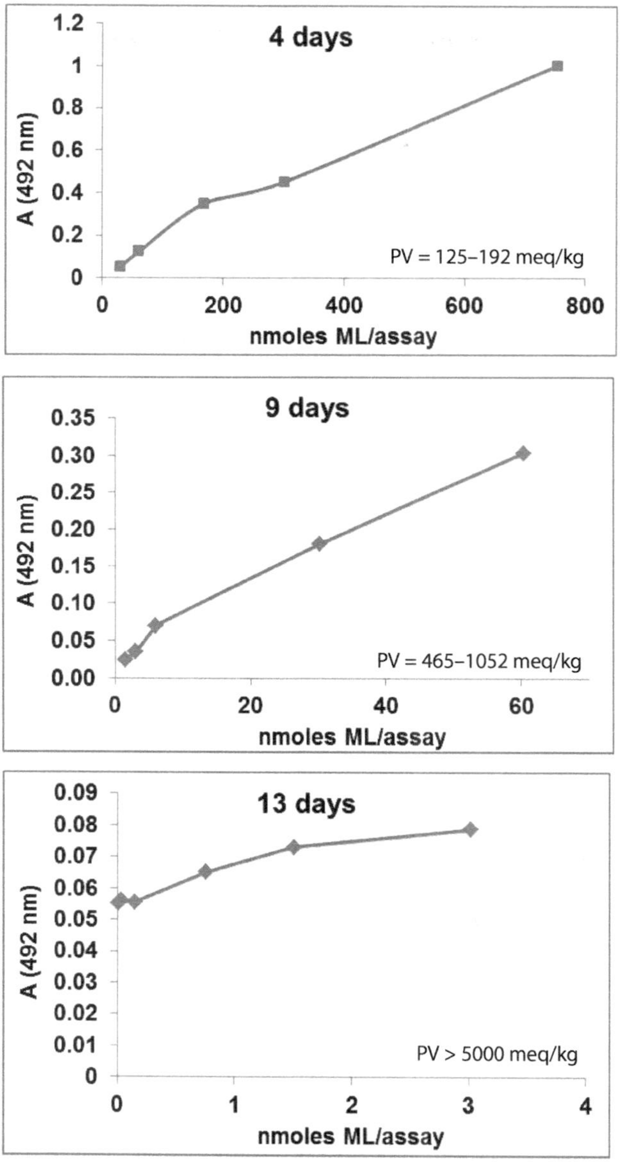

Fig. 2.5. For many assays, sample size required to match response to detection range decreases as oxidation increases. Methyl linoleate was oxidized at 40 °C and hydroperoxides were analyzed 4, 9, and 13 days by the ferrithiocyanate assay. In this assay, excess LOOH bleaches detection complex, so the same sample size cannot be used at all oxidation levels.

are in trace pathways and the dominant product pathway is missed. Thus, the question regarding analysis accuracy includes whether the analytical approach actually measures the major oxidation products being generated.

The need for analyzing multiple products is more than a theoretical issue of determining reaction pathways. The sequence of product generation has always been accepted as radicals → hydroperoxides → decomposition products (mostly carbonyls), so hydroperoxides are almost always measured and volatiles (usually hexanal) are sometimes measured as routine analyses of lipid oxidation (reactions under broad arrow pathway in Fig. 2.6). There should be common recognition that hydroperoxides decompose under a variety of conditions (especially under laboratory lighting), so peroxide values alone do not accurately reflect the state of oxidation. Nevertheless, a large proportion of studies reported in the literature do not include at least one additional analysis to distinguish when low peroxide values result from low versus advanced oxidation and to track breakdown products. Despite documentation of dozens of compounds isolated from lipid oxidation degradation pathways (Schieberle et al., 1979; Grosch & Megele, 1984; Grosch, 1987; Gardner, 1989; Frankel, 2005; Belitz et al., 2009), few secondary products are regularly followed. These are usually only volatiles generated from alkoxyl radical scission reactions, and even then, the customary practice is to quantitate limited numbers of products, most commonly hexanal. The argument often given is that products known to be associated with off-flavors or odors are the most important cues of rancidity for consumers, and other products are only of academic or theoretical interest.

Two fallacies emerge from this line of thought. First, one size does not fit all in lipid oxidation analyses; the same volatiles are not produced by lipid oxidation in every material or under all conditions. Hexanal is not always the main product (indeed, is not always generated), product distributions shift as oxidation progresses, and intermediate products degrade or transform. Hence, in isolated analysis (such as sample pulled off a production line or from storage with no previous analyses or history), there is a large chance that oxidation will be underestimated or missed altogether when limited numbers of fixed products are measured and other pathways are more active. Because lipid oxidation is a dynamic process, recording low values for any one product does not distinguish whether the material is just beginning to oxidize or, rather, has already progressed into secondary stages in which the analyzed compound is degrading, reacting, or transforming. This is particularly true when proteins are present because these molecules react with nearly every intermediate and product of lipid oxidation (Schaich, 2008). Monitoring multiple products of lipid oxidation is, therefore, critical in both routine quality control analyses and in research to reveal when low measures of one product result from active alternate pathways rather than low oxidation.

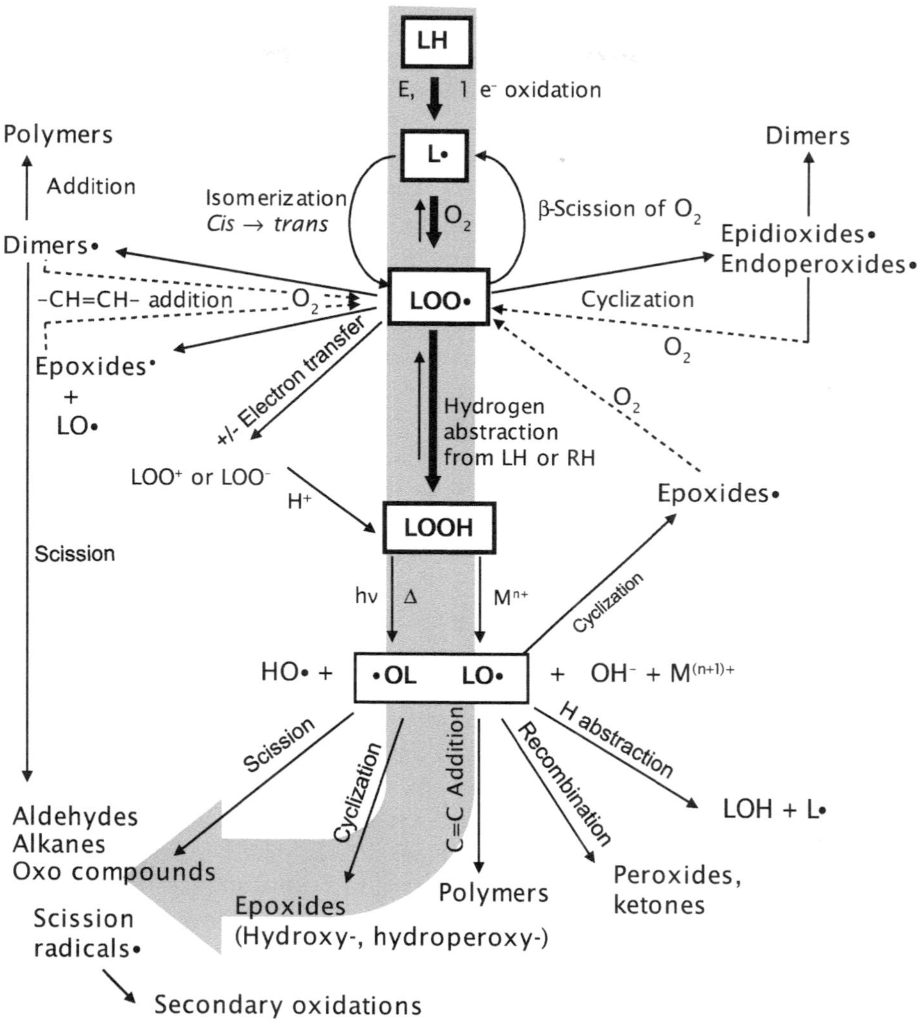

Figure 2.6. Typical lipid oxidation products analyzed (broad underlying gray arrow), in context of multiple alternate pathways.

Adapted from Schaich (2005), used with permission.

The second fallacy is that only sensory cues are important. Lipid oxidation generates toxic products (Cortesi & Privett, 1972; Kubow, 1992; Esterbauer, 1993), and many (if not most) of these have no documented flavor or odor or are toxic at levels much lower than sensory thresholds. Epoxides, for example, are considerably more reactive than aldehydes with proteins and have shown cytotoxicity (Vock et al., 1999) as well as carcinogenicity (Van Duuren, 1966; Chung et al., 1993; Blair, 2001; Lee et al., 2001), genotoxicity (Ehrenberg & Hussain, 1981; von der Hude et al., 1992; Hemminki et al., 1994), and mutagenicity (Lee et al., 2002). Epoxides have been identified as dominant or sole products in neat oils or aprotic solvents (Gardner et al., 1974, 1978, 1985; Gardner, 1989; Haynes & Vonwiller, 1990), and we have observed the same in methyl linoleate model systems (Xie & Schaich, 2012). Even so, lipid epoxides are seldom monitored in routine analyses. 4-Hydroxynonenal, which has no documented flavor, is highly cytotoxic at 100 μM (Esterbauer et al., 1990) and causes necrosis of the liver, thymus, and kidney *in vivo* (Esterbauer, 1993). Lipid dimers and polymers also have documented toxicity (Bird et al., 1981; Izaki et al., 1984). While acute oral toxicity appears to be low, chronic ingestion of lipid oxidation products over time has been associated with increased incidence of gastrointestinal cancers (Fukuzumi, 1970), hepatotoxicity and fatty liver disease (Comporti, 1985), modification of low-density lipoproteins and development of atherosclerosis (Heinecke, 1987; Stringer et al., 1989), enhancement of tumor growth and enhanced mutagenesis (Esterbauer et al., 1990), carcinogenesis (Orengo et al., 1989), and altered signal transduction (Esterbauer et al., 1990; Suzuki et al., 1997; Leonarduzzi et al., 2000). Consequently, the effects of lipid oxidation must be considered beyond their sensory impact to ensure that products are safe. Then, it is necessary to monitor multiple products for a second reason: to reveal the presence of potentially toxic products.

Finally, current understanding of lipid oxidation developed over decades as data documenting products in model systems and intact materials accumulated. Today's instrumentation, which is far superior in sensitivity and analytical capability, offers unique opportunities to obtain more detailed and perhaps more accurate information about the extent of oxidation in individual samples. In addition, monitoring multiple products while taking advantage of improved instrumentation is necessary to begin developing an expanded database from which enhanced understanding of lipid oxidation can be derived. Measuring a fingerprint of products, rather than single compounds will allow researchers to document the presence of multiple pathways and determine how the pathways balance and interact under various conditions. Such information will be required to improve control over lipid oxidation, guiding not only analysis strategies but also antioxidant and processing design. Learning which pathways are dominant may eventually make it possible to direct oxidation to produce particular products for flavors or reactivity or to avoid the generation of toxic products.

It is easy to recommend analyses of multiple products; it is far more difficult to accomplish this goal practically. Doing so presents several technical challenges for lipid oxidation analysis: (1) the development of accurate, rapid chemical assays with high sensitivity for all the key classes of products: conjugated dienes, hydroperoxides, alcohols, epoxides, carbonyls, dimers, and polymers, and (2) the coordination of assays to measure multiple products shortly after extraction in multiple assays or, optimally, in the same analysis. Both issues currently present obstacles to analyzing multiple products.

Which assays should be employed? Protocols exist for measuring all of the products noted above, but many assays lack the sensitivity to detect low levels of products in early oxidation. On the other hand, xylenol orange and ferrithiocyanate reactions are designed specifically to detect nano- to micromolar hydroperoxides, but they lack stoichiometric accuracy and consistency, and high peroxide concentrations normally found in foods bleach the detection product, as discussed above. Methods with comparable sensitivity for lipid hydroxyl groups, a measure of H atom transfer in the chain reaction, are not available, but are necessary for determining mass balance between products. HBr (Maerker, 1965) and nitrobenzylpyridine (NBP) (Hammock et al., 1974; Agarwal et al., 1979) assays detect epoxides in ranges of 0.01–0.1% and 10^{-3}–10^{-5} M, respectively. However, the reaction response varies considerably with R structure for both assays, and the NBP assay has temperature-time issues for full reaction and reproducibility (Liao, 2013), so these assays are useful only with lipids that are extensively and selectively oxidized. Diethyldithio-carbamate forms adducts with epoxides that can be separated by HPLC (Dupard-Julien et al., 2007), but adducts with mid-chain epoxides on unsaturated fatty acids and triacylglycerols undergo hydrolysis, complicating quantitation (Shallcross & Schaich, unpublished data). HPLC (IUPAC, 1992) and size exclusion chromatography (HPLC and column) methods are available for dimers and polymers (Márquez-Ruiz & Dobarganes, 2006; Dobarganes & Márquez-Ruiz, 2007), but they are seldom used, except in analyses of thermal degradation. Sensitive methods for dimers are needed to track addition reactions in early oxidation.

Volatile carbonyl products can be readily measured using gas chromatography, but static headspace analyses have poor sensitivity in early oxidation. Solid phase microextraction (SPME) techniques increase sensitivity but do not extract analytes quantitatively from matrices (Pawliszyn, 1999). In addition, selective adsorption of higher molecular weight compounds skews the product distribution detected (Povolo & Contarini, 2003; Mallia et al., 2005) and can compromise accurate quantitation. Purge and trap/thermal desorption methods can detect the lowest levels of products by continually concentrating collected fractions, but they are more time consuming and can have leak problems.

In terms of tracking oxidation, it is important to remember that volatile products are secondary products generated in scission and other decomposition reactions. The binding of volatiles by complex food matrices limits release, compromises quantitation, and underestimates the extent of oxidation. It can also skew the interpretation of reaction pathways. In addition, the other halves of the molecular scission pairs remain in the solid or liquid phase, perhaps in core fragments still attached to acylglycerols. Identifying them could help to elucidate details of the degradation processes. Thus, a critical need for additional analyses is the development of HPLC-MS methods with the sensitivity and ability to identify multiple classes of individual, non-volatile oxidation products. HPLC-MS can then be combined with GC-MS and chemical analyses of oxidation products to provide a more accurate and complete assessment of oxidation in any sample.

When should assays be performed? A significant impediment to the routine application of multiple assays is the time required to perform the analyses and the short period available after extraction to accurately detect products. Clearly, a single operator working alone cannot run more than a few assays per day. However, the old adage "time is of the essence" could not be more true than with lipid oxidation analyses. It is common practice to extract samples one day, store extracts frozen under inert gas, then analyze oxidation products the next day. In tracking how long samples can be held without change, we have found that even with very careful handling and storage at –80 °C, oxidation levels more than doubled during the first overnight, and after that, competition between degradation of products and continued oxidation led to increasingly erratic results. Thus, samples cannot be accumulated after extraction until researchers have the time to analyze them; lipid oxidation must be determined within hours of extraction for accurate results. This requirement imposes practical limitations on how many products can be measured.

Combine Chemical and Physical Analyses

One way in which multiple product analyses can be accomplished without long reaction times is to combine chemical assays with instrumental analyses that can be performed rapidly on intact foods or extracts, are automatable, and detect multiple products in each run. Gas chromatography (GC) analyses of volatile oxidation products are perhaps the most familiar, having been used for decades to detect flavors and odors associated with lipid oxidation. Methods for improving GC sensitivity will be discussed later. Instrumental analyses are also needed to detect and quantitate non-volatile products that contribute to flavor, nutrition, digestibility, and cooxidations in oils and foods, and are ingested when foods are consumed. High-pressure liquid chromatography (HPLC), IR (infrared), and NIR (near infrared) are obvious candidates to meet this challenge. Recent developments in IR and NIR merit more detailed consideration here. In the past, IR and NIR have been considered to be only semiquantitative methods. However, with contemporary Fourier transform instrumentation, IR and NIR can be performed on solid or fluid samples in seconds. The wavelengths of key oxidation products are well documented (George, 1975; Chapman et al., 1994), and very small differences in oxidation product regions can be discriminated (Lam et al., 2001; Lerma-Garcia et al., 2011; Maggio et al., 2011). With appropriate use of calibrators and application of chemometrics, IR and NIR methods can now compete with chemical analyses for quantitating oxygen uptake (Makino et al., 2010), hydroperoxides (Ma et al., 1997; Yildiz et al., 2002), alcohols (hydroxylated products) (Marengo et al., 2004; Ferrao et al., 2007), epoxides (Pramanik et al., 2012), carbonyls (Mobaraki & Hemmateenejad, 2011), and free fatty acids (from hydrolysis and the oxidation of aldeyhdes) (Cozzolino et al., 2005). The potential also exists for detecting multiple oxidation products in a single sample, if methods are developed appropriately with multiple calibrators (Schaich, unpublished data).

Despite ease in physical handling and operation, IR and NIR analyses do have specific requirements for obtaining valid data. Sample sizes are very small, and food materials are heterogeneous, so to reduce data scatter in quantitation, many replicates must be analyzed, In addition, each sample should be scanned at many positions (e.g., by rotation during spectral recording), and signal averaged. Perhaps the greatest challenge is correlating the spectra of complex multicomponent materials to quantities and identities of individual components. This requires (1) the development of data libraries from appropriate standards to calibrate both quantity and specific spectral regions for identification of individual products and (2) the application of chemometrics to separate and distinguish small differences between samples. The former can add considerably to the development time of methods, but once available, the standards greatly increase the power and usefulness of IR and NIR assays. In addition, for validation during methods development, oxidation products and levels presumably identified in IR and NIR spectra must be correlated with corresponding chemical analyses of the same samples. These requirements, plus the high cost of instrumentation, provide obstacles that have probably slowed the adoption of infrared methods for routine analyses. Nevertheless, FT-IR and FT-NIR methods deserve more attention because the instrumentation and rapid scanning are easily adapted to on-line monitoring (Reis et al., 2007) and because the application of chemometrics makes it possible to distinguish multiple products and small differences between samples.

Returning to gas chromatography as a physical method, headspace analyses of oils and foods have been conducted in parallel with chemical analyses for some time to provide measures of initial (hydroperoxides) and secondary (aldehydes) products in tracking the progress of lipid oxidation. However, few products are tracked and little integration to elucidate mechanisms has been attempted, in part due to poor sensitivity and difficulty in identifying and quantitating products in standard headspace analyses. Static headspace analyses can only report what has volatilized from samples, and this is limited by two practical factors:

1. Headspace equilibrium: The usual practice is to fill the sample vial as full as possible to concentrate the volatiles. However, smaller headspace volume also saturates more readily, which then impedes the evaporation of additional molecules. Use of larger headspace volumes dilutes oxidation products and requires injection of larger volumes to retain sensitivity. However, injector design limits the sample volumes that can be loaded. Hence, low sensitivity is a critical issue in headspace analyses.

2. Matrix binding of volatiles: "Flavor scalping" (binding of volatiles by macromolecular components such as proteins and starches) is well known; comparable binding occurs with aldehydes and other lipid oxidation products. The physical structures of dense materials also trap volatile oxidation products and impede their release. Both processes limit GC usefulness for quantitative analyses and skew the profiles of volatiles detected.

Two instrumental issues further limit usefulness of headspace analyses for tracking and quantitating lipid oxidation:

a. Small sample sizes accepted by mass spectrometer detectors: Mass spec detection may be necessary for product identification, but typically 100:1 dilutions of the analyte stream are required to avoid overloading the nebulizer. This extreme dilution seriously limits detection. Indeed, peaks present with flame ionization detection are frequently missed when using mass spec detection.

b. Requirement for internal standards: Gas chromatography detection has no absolute quantitative reporting. For absolute quantitation, peak areas must be related to areas of standard compounds spiked into the experimental samples, but addition of standards into samples with reactive components always raises questions of recovery efficiency.

Contemporary approaches that overcome some of these problems are to use dynamic headspace analyses (purge and trap collection of volatiles with thermal desorption to gas chromatography) or DART–MS (Direct Analysis in Real Time–Mass Spectrometry) detection of products. Purge and trap (P/T) collection of volatiles is a well-established methodology (Hartman et al., 1991). P/T increases detection of volatiles in two ways: by continually removing released volatiles from the headspace under inert gas flow, thus preventing saturation, and by cumulatively collecting and concentrating volatiles. The length of collection determines the reaction time window monitored; samples with low concentrations of volatiles can be collected for a long time to increase sensitivity, while traps can be changed every few minutes to follow reactions in rapidly oxidizing samples. Despite such advantages, this methodology is not commonly used in practice, possibly because the instrumentation required costs more than simple headspace sampling, thermal desorption units often leak at the injector, and questions are frequently raised about the binding or modification of volatiles by solid supports used either in the purge cycle or the traps.

While acknowledging criticism of this technique, we have found P/T collection of volatiles with thermal desorption to GC to be extremely useful in following both the kinetics and pathways of thermal degradation in oils (Qin, 2011) and in solid matrices of baked and extruded products (unpublished data). Because P/T increases detection of products (number and concentrations), this methodology has the potential to provide more accurate supporting volatiles data to integrate with chemical analyses than static headspace analyses or SPME, if all peaks are identified and quantitated rather than limiting reporting to one or a few select products. Thus, purge and trap/thermal desorption GC merits increased investigation and application to analysis of lipid oxidation in any materials.

The GC discussion above introduced the problem of overloading nebulizers of MS detectors. This problem can be overcome with DART–MS (Direct Analysis in Real Time–Mass Spectrometry), a relatively new technology that replaces gas or liquid chromatography delivery of volatile analytes and enables the direct analysis of sample composition from nearly all matrices by MS without any sample pretreatment or component separation. Any mass spectrometer can be fitted to DART analyses with appropriate interfaces, although time of flight spectrometers, which provide superior resolution and mass accuracy, are required to unambiguously identify analyte components (Cody et al., 2005a, 2005b). In brief, the DART module is positioned in front of the MS nebulizer (Fig. 2.7) to provide an ion source in both positive and negative modes, and inert gas is routed through it. Inside the

Fig. 2.7. DART ion source module positioned in front of mass spectrometer nebulizer with two different multiple sampling units.

From IonSense (2012), used with permission.

DART chamber, an electrical discharge creates plasma containing ions, electrons, and excited neutral atoms and molecules. The gas is heated, and the ions and electrons are removed as the gas passes through the source (Fig. 2.8). At the DART exit, the gas flow moves neutral excited-state species of the inert gas (helium or nitrogen) toward the mass spectrometer inlet at atmospheric pressure, rather than under vacuum. As they pass over materials positioned between the DART cap and MS, these excited-state species induce the formation of ions in molecules on the sample's surface and stimulate the desorption of these ions; the gas flow then carries the analyte ions into the MS. Because of this induced surface ionization, DART detects most, if not all, components present on the sample's surface, not just volatiles, so it offers considerable advantage over normal headspace analyses. The DART output is presented on molecular ion rather than conventional time basis; mass spectra for each peak are also recorded for compound identification (Hajslova et al., 2011). A variety of sample presentation units are available, providing flexibility in analyzing single, manual multiple, and automated multiple samples (IonSense, 2012). Sample components can be analyzed individually by comparing molecular ions and mass spectra with libraries; chemometrics provide the capability to differentiate and categorize samples.

DART–MS offers new opportunities to detect products at ppb levels and identify them at the same time (Cody et al., 2005; Yu et al., 2009). DART's low tendency to suppress ions increases its sensitivity and is particularly important in facilitating analyses of liquid samples and effluents from high-pressure liquid chromatography (Beismann et al., 2011). Although few studies have been directed at lipids, those published demonstrate that this technology has dramatic potential for differentiating samples based on fatty acid or oil composition. For example, DART

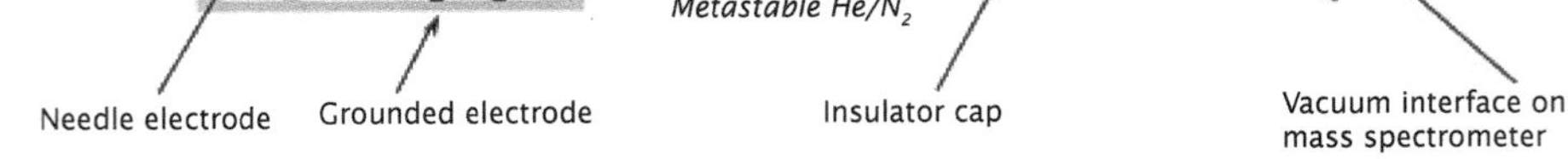

Fig. 2.8. Conceptual diagram of DART chamber design for delivering excited state gas molecules that induce ionization on the surface of analytes. The gas flow stimulates desorption of analyte ions and carries them to the mass spectrometer nebulizer at atmospheric pressure.

Adapted from IonSense (2012), used with permission.

analyses have been used to determine fatty acid, triacylglycerol, and hydroxytyrosol composition of various forms of olive oil, providing a basis for distinguishing olive oil quality levels as well as detecting hazelnut oil adulteration (Vaclavik et al., 2009). DART has been similarly employed to measure the generation of fatty acid methyl esters by bacterial cells (Pierce et al., 2007), to differentiate between beef and pork fats (tallow and lard, respectively), and to detect adulteration by intermixing based on triacylglycerol compositions (Vaclavik et al., 2011). In the latter study, small volatiles were also monitored.

If such sensitivity can be observed in separating large molecules with low volatility like triacylglycerols, it is reasonable to expect that DART may be even more useful for detecting and identifying multiple oxidation products that are small, volatile molecules. We have at least preliminary evidence that this application of DART is very promising. Obviously, quantitative and qualitative analyses will only be as good as the libraries developed for specific oxidation products and for each sample matrix. Assuming that focused research can overcome this hurdle, DART is another technology that merits serious evaluation for analyzing oxidized lipids.

Obtaining Maximum Information about Rates and Products

Traditional assays for quantitation of lipid oxidation have generally measured classes of products containing a single functional group such as hydroperoxides, carbonyls (ketones and aldehydes together), or epoxides. This approach provides comparative numbers for tracking oxidation, but information about individual products is lost and time/labor requirements for the multiple analyses are extensive. While class analyses are arguably sufficient for quality control and detailed product identification is needed only for research, many times results are not consistent or show unexpected degradation patterns. In such cases, identification of individual products within classes can reveal whether the problem is in the analytical technique or chemistry changing in the sample. Furthermore, a clear need to better understand lipid oxidation processes is the impetus for the development of this book. Thus, another challenge in lipid oxidation analysis is abandoning simplistic analytical approaches and learning how to gain as much specific information as possible from each assay.

More fully elucidating lipid oxidation reactions and more accurately tracking oxidation products will require an integrated approach that combines both (1) converting global test-tube assays into more detailed HPLC separations, coupled with product identification by mass spectrometry, and (2) collecting volatile products more effectively and utilizing GC data more fully. The latter involves changing methods for collecting volatiles and developing more extensive identification of products by mass spectrometry, as well as the increased application of chemometrics for identifying product patterns. HPLC offers a better option for detecting a wide range of oxidation products and distinguishing whether these occur as free monomers or are retained on acylglycerols or phospholipids. However, for both routine analyses and fundamental research in oxidation mechanisms, results from the two chromatographies must be integrated to track both sides of scission reactions and secondary product decompositions.

Modifying Traditional Class Assays to Provide Information about Products and Distribution

Traditional class assays tell what products are being generated, but they do not specify where. For example, quantitating the total hydroperoxides of an extract does not distinguish whether the oxidation occurred on triacylglycerols, free fatty acids, phospholipids, or all three; it also does not identify which fatty acids are preferentially oxidized. Similarly, hydroperoxide and carbonyl analyses of oils at elevated temperatures (as found in accelerated shelf-life storage or frying) do not distinguish whether the products occur at the same positions as oxidation at room temperature or are completely different. Meeting the challenge of fully elucidating lipid oxidation processes requires that individual products, rather than classes, be tracked.

A first step in this direction is converting assays such as DNPH (dinitrophenylhydrazine) complexation of carbonyls or DETC (diethylcarbamate) complexation of epoxides from test tube optical analyses to HPLC separation of individual products. Adducts from both reactions are stable, so can survive transit through HPLC columns. Once adducts are separated, HPLC analysis can identify individual products by MS or comparison to standards. HPLC can also quantitate total and invidual products from total peak areas (Fig. 2.9), although conversion of peak area to product concentrations requires development of standard curves using pure products or compounds similar to expected products. The sensitive detection and product differentiation in these HPLC assays often reveals many more products than current theory would predict, and these new product patterns can offer interesting and useful insights into oxidation reactions (Xie & Schaich, May 2012; Liao, 2013). Of course, most chemical assays suffer from the limitations discussed above, particularly reaction response that varies with product structure, and these limitations must be considered when deriving reaction sequences and mechanisms from the HPLC data.

Ultimately, we need to move increasingly toward the development of detailed direct analyses of lipid oxidation products on-line with HPLC. This requires either

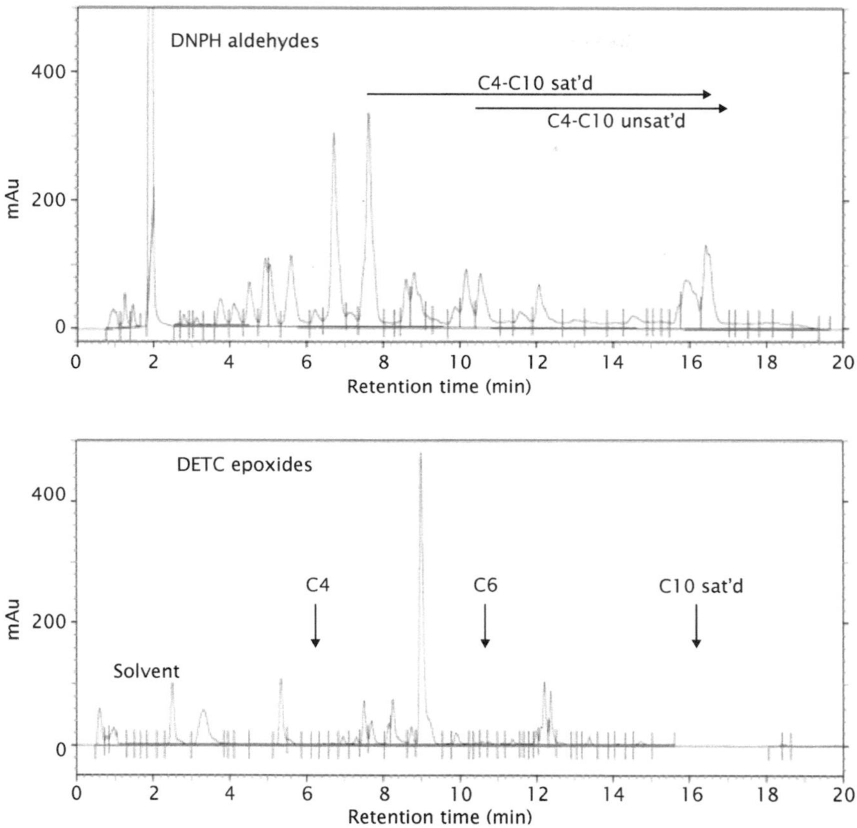

Fig. 2.9. HPLC separation of multiple products from methyl linoleate oxidized at 60 °C for 8 days (DETC-epoxides) or 10 days (DNPH-aldehydes). Retention times do not match standards so will require mass spectrometric determination of structure. Approximate chain lengths of products are shown on scale.

Xie and Schaich, unpublished data.

pre-HPLC on-line separation of total lipid classes as a critical first step, coupled with specific detection methods and/or mass spectrometry, or complex gradient elution, possibly with two columns, to separate both classes and individual products in a single run. In HPLC analyses of total lipids in oils and extracts, any handling to separate fractions before chromatography risks introducing oxidation artifacts, so it should be avoided whenever possible. The variety of HPLC columns currently available, particularly aqueous reversed-phase columns, as well as ultra-high pressure chromatography systems, offer mixed-mode separations with considerably more flexibility in conditions than were previously accessible. Thus, HPLC analysis of lipid classes, as well as individual components and oxidation products within classes in oils and extracts, now appears to be feasible without preliminary separations. However, HPLC applications need some additional methods development as shown in the discussion below.

HPLC separations of total lipid classes are challenging due to the differences in solubility of the various classes. Although a number of experimental methods for HPLC separation of lipids have been published, none have been routinely applied in quality control or oxidation studies. The most common approach has been normal phase HPLC with evaporative light scattering (ELSD) or flame ionization (FID) detection for quantitation. These detectors are necessary to provide a quantitative basis for oxidation products since lipids have no specific chromophores for optical detection. To cite a few examples of approaches, a ternary gradient system (isooctane-isopropanol-water) was applied to normal phase columns to separate fatty acids, acylglycerols, galactolipids, and phospholipids (Moreau et al., 1990). Nine distinct classes of lipids from blue-green algae were separated on a 100×3.0 mm i.d. column packed with a polyvinylalcohol bonded-silica gel stationary phase with elution via a complex gradient moving from hexane through acetonitrile, methanol, and ethyl methyl ketone and ending with a polar mixture or organic solvents containing water, morpholine, and acetic acid (Nordback et al., 1998). Unusual lipids varying from very nonpolar squalene, wax esters, and sterol esters to highly polar phospholipids were separated on a monolithic silica column using a ternary gradient of (isooctane:ethylacetate), (acetone:ethylacetate:acetic acid), and (isopropanol:water with acetic acid and ethanolamine) (Graeve & Janssen, 2009). Seven lipid classes from plasma and animal tissues were separated on normal phase columns in an ultraperformance LC system using a quarternary gradient of (isooctane:tetrahydrofuran), (acetone:dichloromethane), (isopropanol:chloroform), and (isopropanol:water) with ELSD detection (McLaren et al., 2011). Two critical limitations of normal phase HPLC for total lipid analyses are stronger retention of minor components (polar fractions and oxidation products) than triacylglycerols, and marked sensitivity of separations to traces of water in the solvents. Hence, increasing attention has been given to reversed phase applications, which provide more flexibility in separation modes.

Reversed phase separations of total lipid classes have their own challenges arising from the poor solubility of triacylglycerols in the polar solvents normally used and the strong hydrophobicity of the columns, which must be modified to retain polar lipids and oxidation products. However, a variety of new columns that modulate analyte binding and tolerate a broader range of solvents have been developed to overcome these limitations, and these specialty columns offer unique opportunities for tailoring separations of both native and oxidized lipids. For example, we have developed a reversed-phase application that completely separates all lipid classes and some components within classes using an ultra-aqueous C18 column and ternary gradient based on acetonitrile/water/(isopropanol/hexane) (Teng & Schaich, unpublished data).

Once separated on HPLC columns, how can lipid oxidation products be distinguished? Despite being required for quantitation of lipids, neither ELSD or FID alone is adequate when detection of oxidized products or fractions is also desired. To monitor oxidation in class analyses, ELSD/FID quantitation must be supplemented at least with analysis of conjugated diene products at 231–235 nm using UV or diode array detectors placed in-line before the destructive detectors. However, many secondary products lose the conjugated diene structure, so additional analyses are also required to fully detect and quantify fractions with lipid oxidation.

To detect specific lipid oxidation products, at least two options are available that can be integrated into the total HPLC assay. Hydroperoxides can be detected by post-column reaction with luminol, followed by chemiluminescence detection (Miyazawa, 1989). This sensitive assay can detect picomoles of hydroperoxides in individual fractions, but has two shortcomings for quantitative analyses: (1) the intensity of luminol emission varies with the hydroperoxide reacting, and (2) the luminol reaction is destructive, so if ELSD or FID quantitation is also needed, the effluent line must be split or two separate runs must be made with different detection.

Electrochemical detection offers greater flexibility because it is not destructive and can be used in tandem with diode array and ELSD detectors. Single electrodes have been used for hydroperoxide analysis of phospholipids (Yamada et al., 1987), fatty acids (Funk, 1987), and a variety of lipids, including sterols (Korytowski et al., 1995, 1999). The redox potential is 530 mV vs. NHE, and the lower detection limit is low nanograms (Funk, 1987). Since hydroperoxide, alcohol, epoxide, aldehyde, and carboxylic acid products of lipid oxidation all have different redox potentials, electrochemical detectors with multiple channels that can operate at different redox potentials create even more interesting opportunities to simultaneously detect and distinguish multiple oxidation products in the same sample. Thus, this is another technology that merits serious investigation to push the boundaries of lipid oxidation analysis.

Electrochemical detection does have the disadvantage of requiring an electrolyte in the eluting solvent, and very few non-aqueous electrolytes are available. Formerly the electrolyte of choice, tetraethylammonium perchlorate is currently difficult to obtain because of its use in explosives. Tetraethylammonium bromide works reasonably well as an alternative.

The detection methods described above can provide considerably more detailed information about lipid oxidation products than we currently collect.

Ultimately, however, mass spectrometry will be required to verify all oxidation products identified by luminol and electrochemical methods, as well as to identify other products not detected by these methods.

Don't Throw Away Peaks! Utilizing the Wealth of Data from GC-MS Analyses to Best Advantage

When tracking lipid oxidation progress and pathways, the complement of HPLC detection of non-volatile products is, of course, analysis of volatile products by GC. Volatility, however, introduces complex analytical issues that differ from those seen with HPLC and chemical assays. Critical considerations include (1) losing volatiles due to open incubations and lack of cryotrapping in GC injectors, and (2) throwing away data by ignoring peaks other than a few selected products. Both compromise the validity of conclusions that we draw from GC data.

For typical oxidation studies, samples are incubated either open to air or sealed in packages which are opened before transfer to headspace vials for GC analysis. These practices lose the earliest products generated as well as the most volatile products, thus skewing the distribution of products detected and compromising interpretation of results. To illustrate this problem, while working in closed systems that are not oxygen limited, we have observed that pentane and other volatile degradation products from C13 of linoleic acid far exceed those from C9. These results contradict previous reports that hydroperoxide formation at C9 and C13 is equal (Frankel, 2005). One possible explanation for the observed difference is that products from C13 hydroperoxide decomposition are smaller and more volatile than C9 hydroperoxide products; these products are retained in closed systems, but in open systems they are more readily lost. Samples packaged in paper or permeable plastic bags become open systems so that the product distribution detected from either samples or package headspace is not what was originally generated, but instead portrays downstream products that are less volatile. This problem needs to be recognized, especially when sampling products "off the shelf." To detect all products, analysis of volatiles must be conducted in closed systems, as well as open ones.

A second issue for complete detection of products is the requirement for cryotrapping during injection to prevent the escape of highly volatile products with chain length less than six carbons. Whether accomplished by an *in situ* liquid nitrogen cryotrap or dry ice in the GC oven, failure to inject cold results in significant loss of short-chain products. It can be argued that products this volatile are unimportant to long-term product quality, but this fraction includes products that collect in packages and are smelled by consumers when packages are opened.

Thus, they need to be accounted for and identified in all analyses seeking to track the pathways or extent of oxidation.

Problems of volatility limitations and volatiles binding to matrices in GC analyses have already been discussed. Even with these restrictions, however, more oxidation products are usually detected than can be easily identified. Increased efficiency in volatiles collection with the P/T methods discussed above only compound this problem. As a consequence, peaks not arising from products expected to have sensory (such as hexanal) or diagnostic (decadienal) value are routinely ignored. Whether primary or secondary products, peaks that we do not measure or explain provide invaluable information about the specific pathways active in lipid oxidation. They can supplement and explain the results of chemical assays and may even reveal previously undocumented chemistry. Earlier limitations in instrumentation provided excuses for such omissions. However, the more extensive availability of high-sensitivity mass spectrometers has made it easier to identify peaks that do not match standards. The application of chemometrics facilitates the recognition of patterns and relationships among the various peaks, as well as between volatile products and other assays. Combined, these two factors issue new challenges to think beyond a well-recognized few products and identify all other peaks present in GC chromatograms. With this information, we can learn what these peaks reveal about alternate pathways; effects on flavors, odors, and cooxidations; and potentially also toxicity.

Tracking the Footprints of Lipid Oxidation—Identifying and Measuring Changes in Critical Cooxidation Target Molecules

That all lipid oxidation products damage proteins, nucleic acids, vitamins, enzymes, starches, and other molecules has been extensively documented (Schaich, 2008), and some damage products have been identified (Hidalgo & Zamora, 1993; Zamora et al., 1994; Hidalgo & Zamora, 1995; Zamora et al., 1995; Hidalgo & Zamora, 2004; Hidalgo et al., 2005; Zamora & Hidalgo, 2005). Less recognized is the concurrent antioxidant effect these cooxidations have on lipid oxidation: Each radical transfer from oxidizing lipids to non-lipid molecules stops that chain reaction in lipids while propagating it elsewhere, and reaction of aldehydes and epoxides with proteins and other food components removes these products from lipid oxidation reaction and analysis streams. As a result, lipid oxidation may appear to be paradoxically low when cooxidations are active, while in fact, oxidation has been broadcast to other molecules, amplifying degradation in the entire system.

We first noted this paradox in model systems while studying free-radical transfer from oxidizing lipids to proteins. Lipid hydroperoxides were barely detectable, yet the reaction mixtures turned increasingly yellow and intense EPR free-radical signals developed in the proteins (Schaich, unpublished data). A similar loss of measurable lipid oxidation can occur when any lipid oxidation intermediates or products—free radicals, hydroperoxides, epoxides, or aldehydes—react with susceptible target molecules and broadcast oxidation beyond lipids (Borg & Schaich, 1984; Refsgaard et al., 2000). As a consequence, in complex materials such as foods, personal care products, and biological tissues, measuring only lipid oxidation misses a large part of the action and thus can present an erroneous and misleading picture of system oxidation. In addition, cooxidations cause a shift in the system damage profile so that texture changes from molecular crosslinking and scission, new off-flavors and odors, loss of nutritional value, Maillard or other browning, and fading are superimposed on underlying off-flavors, odors, and browning from lipid oxidation (Schaich, 2008). Under these conditions, material quality changes often do not correlate with standard lipid oxidation measures, and the causal connection to lipid oxidation becomes obscured. This situation presents one of the greatest analytical challenges in understanding lipid oxidation: including cooxidations in routine analytical schemes and learning how to track the footprints of lipid oxidation in other molecules.

For decades, researchers have recognized that lipids cause damage to other molecules, yet cooxidation research remains relatively undeveloped. Although considerable data correlating easily detectable behaviors (physical properties such as texture and browning in foods or cell death or function loss in tissues) to lipid oxidation has been amassed (Fig. 2.10.), relatively little is known about the chemistry responsible for the changes. Even when chemical properties such as protein crosslinking are followed, underlying chemical modifications other than disulfide crosslinking are seldom determined. Filling this gap, establishing causal relationships, and developing quantitative assays that link specific non-lipid molecular changes to lipid oxidation will require elucidation of cooxidation chemistry that is as detailed as the lipid oxidation chemistry outlined in Chapter 1.

In the past, tracking individual cooxidation products in complex systems might have been a technically insurmountable task, in terms of both separating damaged molecules and identifying individual reaction products. Tracking specific cooxidation sites in proteins using conventional methods, even in model systems, can be daunting, if not impossible, because acid hydrolysis destroys key amino acids that are modified by lipids. Digestion by proteolytic enzymes avoids this problem, but autolysis of the enzymes complicates analyses. Considerable research is needed to identify proteases that adequately degrade modified proteins, then to immobilize them on supports that are easily separable from the reaction

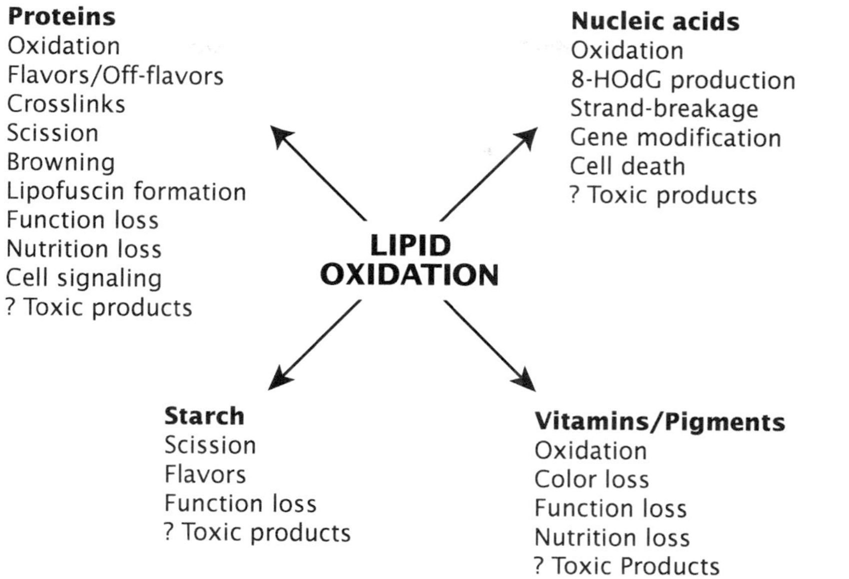

Fig. 2.10. Types of damage that occur when oxidizing lipids cooxidize cellular molecules in foods and living tissues (plant and animal).

From Schaich (2008) used with permission.

mixture and do not bind peptides and amino acids. Agarose, commonly used with commercial immobilized enzymes, does not meet either of these criteria; therefore, more rugged, less reactive supports are needed. Distinguishing lipid cooxidations from damage due to other stressors, such as heat in foods and enzymes or other reactive oxygen species in biological tissues, adds to these analytical challenges. Nevertheless, contemporary instrumentation offers exciting new prospects for establishing quantitative and qualitative roadmaps of oxidative degradation in all kinds of materials.

The most detailed cooxidation studies have been conducted on the pathological effects of lipid oxidation, and the approach used in biological tissues may provide a model for broader applications, particularly foods. First, extensive investigations in model systems with protein targets identified specific cooxidation damage patterns and, most important for analysis, specific products attributable to individual lipid oxidation species. Once identified, these products provided an experimental basis for distinguishing lipid-modified proteins and quantitating lipid cooxidation in proteins isolated from tissues.

An impressive body of research in model systems by Hidalgo and Zamora[2] provides an excellent starting point for applying this approach to cooxidation in foods. Several substantial reviews detail current knowledge about the chemistry involved in protein cooxidation (Davies et al., 1999; Davies, 2004; Schaich, 2008). Key protein oxidation products that may serve as markers for detecting and quantitating protein cooxidation in a broad range of systems have been summarized in Table 2.H. Following these lipid cooxidation footprints will be essential for determining the effects of recently rediscovered healthy lipids in food formulations and designing more effective analytical and stabilization strategies.

Simplifying and Identifying Data Patterns Using Chemometrics or Other Statistical Analyses

When we perform simple experiments in which single molecules generate single analytical data points (for example, peroxide values or a single peak in an IR spectrum or GC chromatograph as a function of incubation time or temperature), it is relatively easy to interpret results and relationships from the shapes of response curves. However, considering only one peak or band in methods that simultaneously detect multiple products in complex molecular systems throws away considerable data. Are product peaks other than target compounds important components of reaction responses or consumer perceptions of rancidity? Similarly, when following reactions, the analysis of single products is relatively straightforward, but determining response patterns arising from multiple pathways generating different products can be daunting.

Chemometrics, sometimes called data mining, is a branch of multivariate statistical analysis that has evolved to address both of these types of challenges in data analysis. The International Union of Pure and Applied Chemistry (IUPAC) defines chemometrics as "the application of statistics to the analysis of chemical data (from organic, analytical or medicinal chemistry) and design of experiments and simulations" (van de Waterbeemd et al., 1997). The goal of chemometrics is to go beyond simple correlations to derive maximum chemical information from complex sets of chemical data (Wold, 1991; 1995; Wold & Sjostrom, 1998) by discerning patterns and relationships among multiple variables and products, components, or

[2]See Hidalgo & Kinsella, 1989; Hidalgo & Zamora, 1993, 1993; Zamora & Hidalgo, 1994; Zamora et al., 1994; Hidalgo & Zamora, 1995a,b; Zamora & Hidalgo, 1995; Hidalgo et al., 1998; Zamora et al., 1999; Hidalgo & Zamora, 2000; Zamora et al., 2000; Hidalgo et al., 2001; Hidalgo & Zamora, 2002, 2004; Hidalgo et al., 2005; Zamora & Hidalgo, 2005; Zamora et al., 2005, 2006, 2007.

Table 2.H. Oxidation Products that May Serve as Markers for Tracking and Quantitating Protein Cooxidation

Lipid Oxidation Product	Protein or Amino Acid	Protein Oxidation Product	Reference
LOOH	LDL, cytochrome c	loss of trp, met, cys, pro, val, leu	Fruebis et al., 1992
		epoxides which cyclize to pyrrolines and tetrahydropyridines	Fruebis et al., 1992
	lupine conglutins	fragmentation and crosslinking	Schaich & Karel, 1975; Lqari et al., 2003
	any protein	(-NO•), (-S•), (-SO•), (-SOO•), S-S crosslinks	Schaich, 2008
Epoxides			
γ-hydroxy-α, β-unsaturated epoxides	lys, cys, his	aminols	Lederer, 1996; Lederer et al., 1998
4,5-epoxy-alkenals	primary amines	N-substituted-2-alkyl pyrroles, alkyl-substituted pyrroles and furans, alkyl pyridines	Zamora et al., 2005
	phenylalanine	N-substituted 2-(1-hydroxyalkyl) pyrroles, phenylacetaldehyde, 2-alkylpyridines	Hidalgo & Zamora, 2004
epoxyoxoenes	phenylalanine	phenylacetaldehyde	Zamora et al., 2005
1,2-epoxides	imines	1,2-dihydropyridines	Brunner et al., 2006
	4-methylimidazole	4,5-epoxy-3-(4-methylimidazol-1-yl) heptanol	Zamora et al., 1999
	N-acetyl-L-histidine methyl ester	N-acetyl-1-[1'-(1",2"-epoxybutyl)-3'-hydroxypropyl]-L-histidine methyl ester N-substituted pyrroles	Zamora et al., 1999

(continued)

Table 2.H. (Continued)

Lipid Oxidation Product	Protein or amino acid	Protein Oxidation product	Reference
Carbonyls			
2,4-decadienal	phenylalanine	phenylacetaldehyde, pyrroles, pyridines	Zamora et al., 2007
methyl-13-oxooctadeca-9,11-dienoate	phenylalanine	phenylacetaldehyde, pyrroles, pyridines	Zamora et al., 2007
4-hydroxy-2-nonenal	phenylalanine	Strecker degradation products	Hidalgo et al., 2005
nonanal, 8-oxononanoic acid	cysteine	thiazolidine dicarboxylic acids	Gardner et al., 1977; Gardner & Jursinic, 1981
alkanals (high concentration)	amino acids	N-substituted quarternary pyridinium salts	Suyama & Adachi, 1979
alkanals	amino acids	alkyl-substituted pyridines	Suyama & Adachi, 1980
	primary amines	imine/enamine Schiff base	McMurray, 2000
	secondary amines	enamine Schiff base	McMurray, 2000
malonaldehyde	one protein-NH_2	imine and enamine Schiff base	Yildiz et al., 1998; Nazir et al., 2000
	two protein-NH_2	enamine or iminopropene crosslinks	Yildiz et al., 1998; Nazir et al., 2000
acrolein	one protein-NH_2	N^{ϵ}-(2-propenal)lysine	Uchida et al., 1997
malonaldehyde (excess) + RCHO	protein-NH_2	1,4-dihydropyridine-3,5-dicarbaldehydes	Nair et al., 1988; Freeman et al., 2005

unsaturated aldehydes	Protein-NH_2	Michael addition products	Esterbauer et al., 1991; Nadkarni & Sayre, 1995
acrolein/crotonaldehyde (2)	lysine (1)[1]	EMP [N^{ϵ}-(5-ethyl-2-methylpyridinium)]	Ichihashi et al., 2001
		FDP [N^{ϵ}-(2,5-dimethyl-3-formyl-3,4-dehydropiperidino)	Ichihashi et al., 2001
4-hydroxynonenal (HNE)	histidine	β-substituted propanals	Uchida & Stadtman, 1992
4-hydroxy-2-alkenals (HNE)/oxononenal (ONE) (4)[1]	excess-NH_2	pyrroles	Sayre et al., 1993
HNE/ONE (4)	protein-NH_2 (1)	2-pentyl-2-hydroxy-1,2-dihydropyrrol-3-one iminium crosslinks	Zhang et al., 2003
4-oxo-2-alkenals	protein-lys	ONE-lys pyrrolidone	Lin et al., 2005
	protein-his	ONE-his-lys pyrrole crosslink	Lin et al., 2005

[1]Numbers in parentheses refer to the moles reacting to generate the corresponding complex.

reactions. Focusing mostly on small differences that are barely discernible to the eye, chemometrics evaluates multiple factors simultaneously to identify qualitative and quantitative similarities and dissimilarities between complex samples, such as in compositions as a function of material origin or product distributions as a function of treatment (Buydens et al., 1999). Chemometrics has also been used to identify flavor patterns most associated with specific sensory or other properties (Chien & Peppard, 1993).

A thorough consideration of chemometrics is beyond the scope of this chapter. For details and background on the statistical bases and applications of chemometrics, the reader should look to several excellent texts (Massart et al., 1998; Otto, 1999; Brereton, 2003; Massart et al., 2003; Brown et al., 2009) and reviews (Lavine, 2000; Doherty & Lange, 2006; Berrueta et al., 2007; Roggo et al., 2007; Pomerantsev & Rodionova, 2010). In addition, a series of studies by Lavine and Workman systematically covers key problem areas in chemometrics (Lavine, 1998; 2000; Lavine & Workman, 2002, 2004, 2006, 2008, 2010). This chapter will introduce the fundamental approaches of chemometrics and propose several ways in which they may apply to analyses of lipid oxidation.

Simplistically, chemometrics seeks to determine the extent and ways in which samples differ by integrating statistical analyses of multiple sets of complex data. Both qualitative and quantitative methodologies must be applied to obtain maximum information (Table 2.I), first to search data for patterns that distinguish similarities or dissimilarities (for example, by principal components analysis), then to determine quantitative relationships between these data sets or clusters using one or more regression analyses. Pattern recognition capabilities combined with quantitation provide powerful tools for discerning small changes between samples, and they have proven to be particularly useful in data-mining complex optical spectra, as well as GC and HPLC chromatographs. Unfortunately, no set or universally applicable guidelines exist that specify which analyses are most appropriate for various applications. Consequently, most cases require trial and error to determine the optimum combination of chemometric assays for each system and data combination studied.

Solving problems in lipid oxidation analysis with chemometrics is a research area in its infancy, with very few applications in the literature. Applications currently reported remain simplistic, but nevertheless demonstrate the usefulness (and challenges) of chemometrics for reducing large amounts of data and deriving relationships. Three potential areas in which chemometrics offers significant promise for gleaning maximum information from lipid oxidation analyses and for elucidating lipid oxidation pathways include:

1. *Discerning small changes in FT-IR and FT-NIR spectra due to fatty acid composition, oxidation, and cooxidation.* Over the past two decades, the application of

Table 2.I. Chemometric Methodologies Commonly Used to Identify Qualitative Patterns and Quantitative Relationships in Complex Spectroscopic, Chromatographic, and Multi-Analysis Data Sets

Qualitative (Classification) Methods

Question asked: Are samples different?

Major analyses: Pattern recognition methods

a. Unsupervised (classifications made directly from data alone; source of cluster differences must then be explained)
   - Principal component analysis
   - Cluster analysis

b. Supervised (classifications developed from applied criteria, e.g., calibration sets with known properties)
   - Correlation coefficient with defined threshold
   - Discriminant analyses—linear (LDA), partial least squares (PLS-DA), quadratic (QDA)—focus on differences between samples; parametric
   - K nearest neighbors (KNN)—differentiates by comparison to nearest neighbors in calibration sets; non-parametric
   - Soft independent modeling of class analogy (SIMCA)—focuses on similarities

Quantitative (Regression) Methods

Questions asked: How much are samples different and in what direction? To what extent are various properties correlated?

Major analyses: Regression analyses

a. Multilinear—multiple measures (e.g., peaks or wavelengths) correlated to specified test property

b. Principal component—principal component scores correlated to specified test property

c. Partial least squares—correlations developed between least squares values for matrices of multiple analytical measures and multiple test properties

chemometrics has become critically important for interpreting Fourier transform infrared (FT-IR) and near infrared (FT-NIR) spectra, discriminating subtle yet key differences in the sharp peaks of FT-IR and broad overtones of NIR and relating them to specific molecular components or functional groups. With chemometrics, both IR methods can distinguish

a. Oils of different fatty acid and triacylglycerols compositions
b. Oils of the same variety but different sources
c. Saturated fats vs. unsaturated oils

d. Oxidized vs. fresh oils, alone and in complex matrices
e. Presence of additional ingredients (additives, contaminants, or food components)

For IR, the challenges are primarily in isolating components from overlapping bands, while for NIR the challenges are in detecting and characterizing small differences in spectra (Fig. 2.11). A common practice has been to record spectra and depend on chemometrics to classify them (unsupervised analyses) with little or no regard for the elucidation of underlying chemistry by auxiliary analyses. Principal components, cluster, or partial least squares analyses can separate samples into groups based on qualitative spectral differences detected, but with unsupervised analyses only the loadings (spectral regions used by the computer to separate groups) provide any clues as to the chemistry underlying the classification (Fig. 2.12). Obtaining maximal information and determining quantitative relationships require development of calibrations to guide regression analyses. The quality and accuracy of calibration sets with known qualitative and quantitative composition are, therefore, key to the success of chemometric analyses. Since such samples require extensive auxiliary analyses and are often difficult to obtain, especially with oxidized materials, the need for valid calibrators also presents a major limitation to the approach.

The most common application of chemometrics to IR and NIR has been the discrimination of oils and adulterants (Armenta et al., 2010; Maggio et al., 2011; Gomez-Caravaca et al., 2012; Longobardi et al., 2012; Rohman & Man, 2012). For example, the C=C bond peak provided the basis for discriminant analysis and canonical variate analysis to discriminate and classify oils and fats based on Fourier transform IR, NIR, and Raman spectra (Yang et al., 2005). With PCA, discriminate analysis, and partial least squares and PC regressions, FT-IR was able to classify Moroccan olive oils according to region or origin and picking times (DeLuca et al., 2011), in addition to determining the adulteration of extra-virgin olive oil with rice bran oil (Rohman & Man, 2012).

As noted above, applications of chemometrics to oxidation analyses have only recently been reported. Analysis of lipid oxidation by traditional IR has been limited by overlapping peaks of many products that absorbed in the carbonyl, -OH, and epoxide regions. Use of extensive training sets and chemometrics enabled Mobaraki and Hemmateenejad (2011) to distinguish between carbonyls in 149 carboxylic acids, 47 aldehydes, 110 esters, and 64 ketones with > 90% predictive accuracy. Although these analyses were performed on pure compounds, the development of appropriate training sets should make this approach applicable to comparing types and levels of carbonyl lipid oxidation products in mixed samples and complex matrices. In unsupervised chemometric analyses, we have

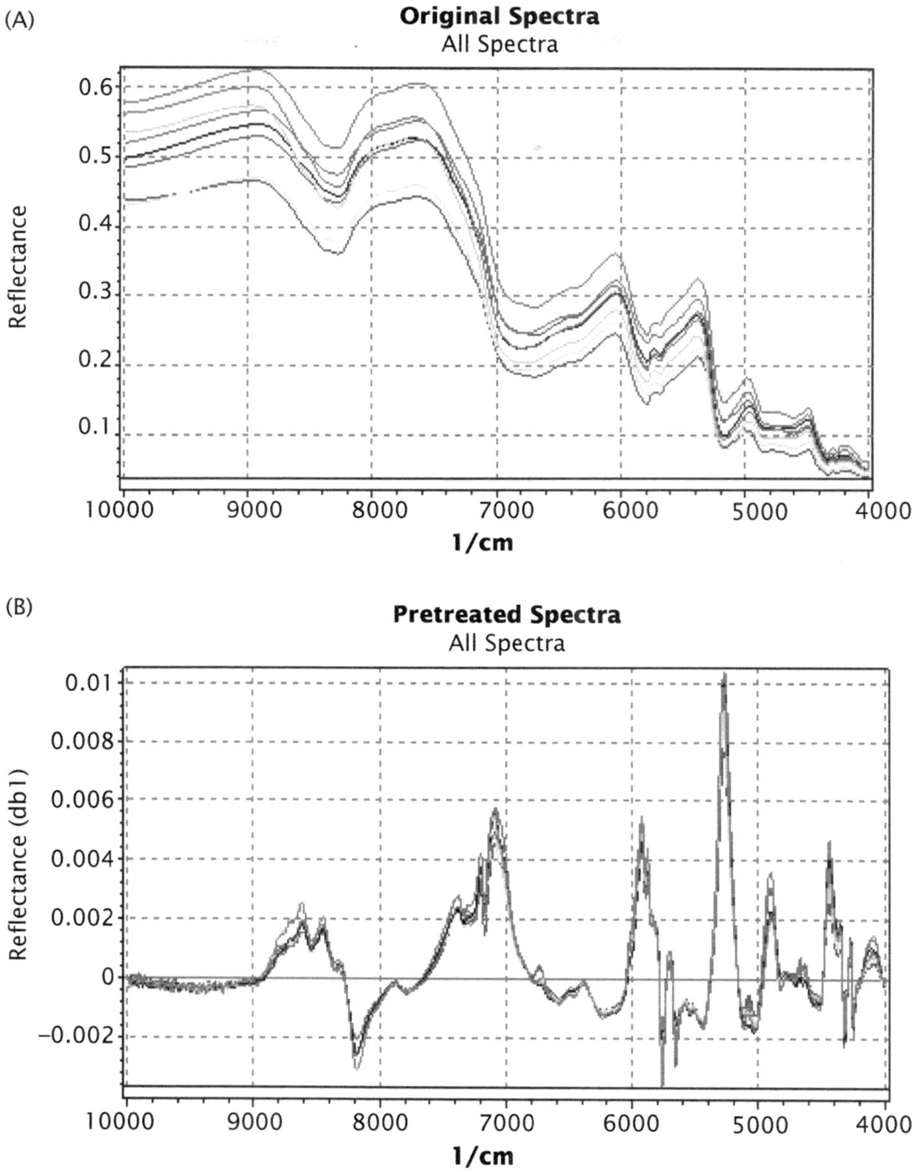

Fig. 2.11. Near infrared gives very little detailed spectral information (A), but pretreatment, for example, by transforming to first derivative curves (B), sharpens spectral features. By application of chemometric analyses, small differences in spectral features can be distinguished.

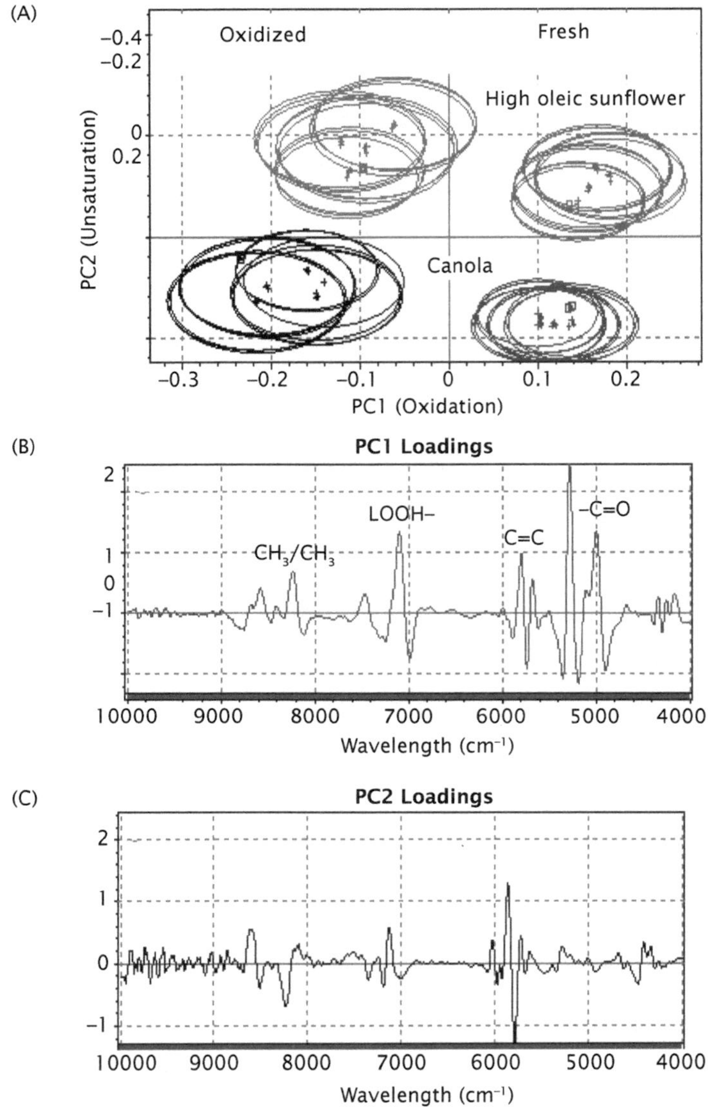

Fig. 2.12. Unsupervised near infrared analysis distinguishes samples, but basis for class separation must be deduced from principal component loadings (spectra of components used by computer to direct the classifications). (A) Model systems of white rice flour mixed 80:20 with soybean oil were oxidized at 60 °C for 48 hrs, NIR spectra were recorded, and principal components (PCA) and cluster analyses were applied to distinguish oxidized components. Each circle in the PCA plot is a separate sample. Loadings showed that presence of hydroperoxides or alcohols, carbonyls, and protein or starch cooxidation products (B) were distinguishing factors used for oxidation, and double bonds plus unidentified factors were used to separate oils.

found that the computer used the loss of double bonds and presence of hydroperoxides or alcohols in distinguishing oxidized oils. It also detected protein or starch cooxidation products as a basis for classifying oxidized extruded products (Figure 2.13).

2. *Distinguishing patterns in chromatographic data and correlating product distributions to traditional chemical analyses.* Success in IR applications suggests that chemometrics should also be able to differentiate key oxidation products, singly or in sets, in complicated gas and liquid chromatograms and relate them to flavor properties or sensory perceptions, processing conditions, or the extent of oxidation, as measured by standard chemical analyses. Chemometrics has indeed been applied effectively to correlate flavor components in headspace analyses with sensory evaluation (Chien & Peppard, 1993), and the approach is now being used extensively in sensory and flavor analysis.

Although more recently developed, methods integrating chemometrics with the GC and HPLC of triacylglycerols are gaining momentum for distinguishing oils and oil sources, as well as detecting adulteration with less costly oils. Approaches for this new combination of analyses have been reviewed and critiqued by Bosque-Sendra and colleagues (2012), with particular reference to olive oils and the shortcomings of most current experimental designs and methods for applying chemometrics. Stepwise linear discriminant analysis successfully used oil chromatograms to cluster the virgin olive oils according to their region of origin (Luna et al., 2006). Principal components analysis of volatiles data from SPME-fast GC-FID, an electronic nose (HS-Enose), and SPME-MS facilitated the detection of olive oil adulteration with hazelnut oil (Mildner-Szkudlarz & Jelen, 2008). Similarly, principal component and cluster analyses of SPME headspace GC and sensory data were able to distinguish rapeseed, soybean, peanut, sunflower, and olive oils, as well as fresh vs. oxidized oils (Mildner-Szkudlarz et al., 2003)

Jahan and colleagues (2005) modeled relationships among quality factors in retailed free-range, corn-fed, organic, and conventional chicken breasts by combining gas chromatography with chemometric analyses. Principal component analysis showed that two multivariate factors of triacylglycerol composition explained 88% of the difference in products detected by flame ionization but only 75% variance in three factors for components revealed by nitrogen-phosphorus detection. Fatty acid compositions and individual antioxidant contents (α-tocopherol, glutathione, catalase, glutathione peroxidase, and reductase) were strongly related ($R^2 > 0.9$) to flavor development in cooked chicken.

Gas chromatograms of lipid oxidation products are often quite complicated, with many products that arise from the native material as well as from degradation. Chemometrics made it possible to differentiate rapeseed oil varieties and the

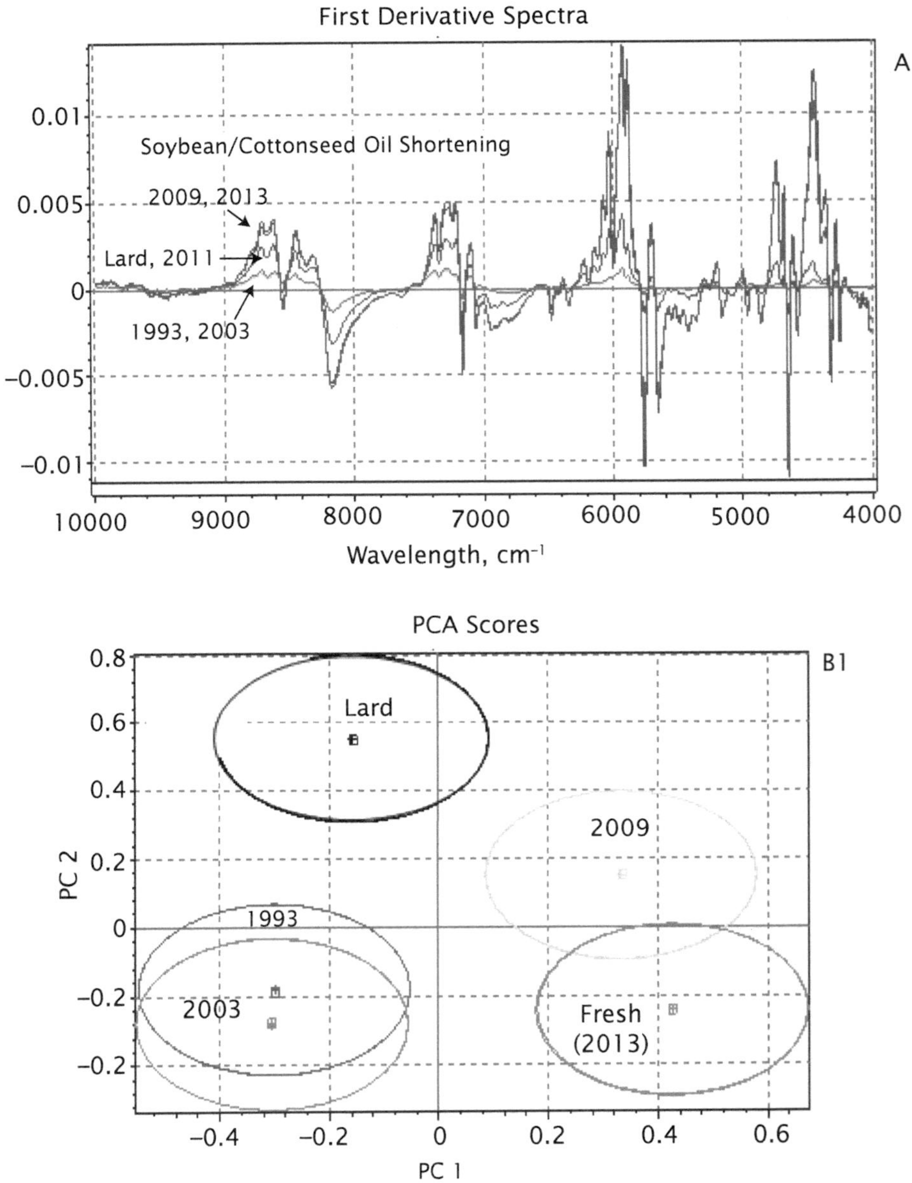

Fig. 2.13. Near infrared analysis of fresh and aged shortenings and lard demonstrating chemometric use of multiple product properties for differentiation. First derivatives of spectra (A) show differences in several spectral regions. PCA analysis (B) identifies unsaturation as the main differentiation (PC 1).

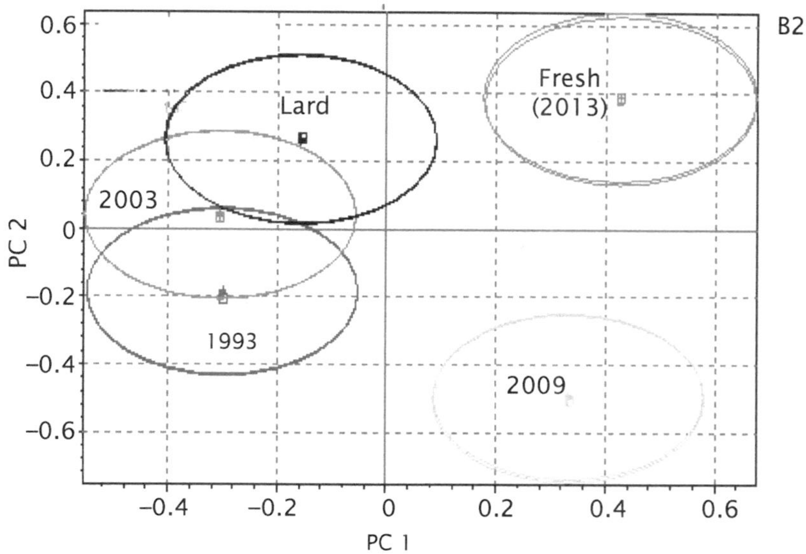

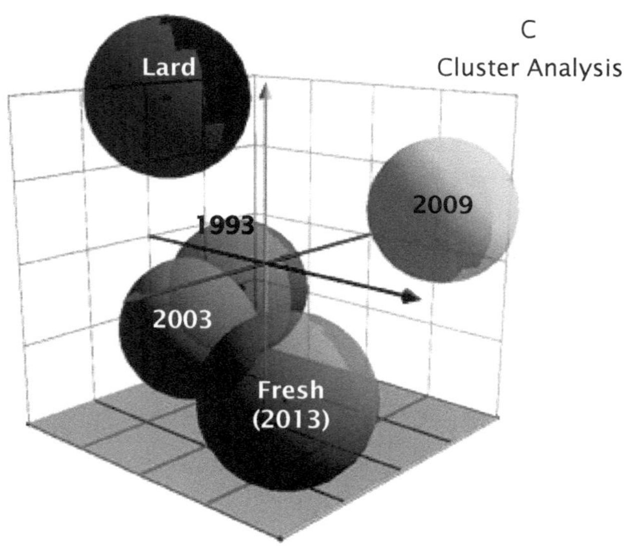

Fig. 2.13. *(Continued)* PC 2 is mostly hydroperoxide and aldehyde oxidation products; the 1993 and 2003 shortening samples were so oxidized that early products had decomposed, and polymerization and secondary products dominated. These two PCs, however, did not completely separate all samples. Loadings of a third component (PC 3) reflected mostly packaging differences in the hydrocarbon regions, that is, solid structure differences. Integrating these three PCs in cluster analysis (C) differentiates the samples in three-dimensional space.

degree of oxidation from SPME-GC chromatograms (Petersen et al., 2012). Octanal and 3-octanone were key components that distinguished different rapeseed varieties, while propanal, t,t-2,4-hexadienal, and t-2-heptenal most accurately sorted oil samples with different extents of oxidation. Clustering of rapeseed oil samples according to their volatile composition correlated strongly with results of sensory duo-trio and paired comparison tests, but the combination of chemometrics with GC detected oxidative modifications earlier.

3. *Determining relationships between multiple products from alternate pathways.* Perhaps one of the greatest challenges in the analysis of lipid oxidation is learning how to integrate complex data from multiple products sometimes arising simultaneously from the multiple pathways shown in Figure 2.6. Chemometric approaches have the potential to elucidate interactions between active pathways under different oxidation conditions as well as to provide respite from tedious individual statistical analyses. Pattern distinction may be relatively straightforward, but quantitation of interactions and effects will be more difficult in the absence of good calibrators with known levels of different oxidation products in food-type matrices. Nevertheless, if multiple products from multiple pathways are analyzed to clearly delineate the progress and extent of lipid oxidation—an absolute requirement for progress—chemometrics must be applied to identify relationships in the resulting complex data matrices.

This approach has been tested in a limited way in the Peterson and colleagues (2012) study cited above. Fifty-five volatile oxidation products of conventional and high oleic rapeseed oil were identified by SPME-GC, and 17 of these were selected as target compounds for comparison with other lipid oxidation measures, including free fatty acids, peroxide values, anisidine values, and Rancimat induction times. Principal component and agglomerative hierarchical cluster analyses were combined to identify octanal and 3-octanone as the most sensitive and accurate volatile markers of lipid oxidation.

A New Understanding of Lipid Oxidation from Multi-Product Data

Obviously, results of the extensive research challenges outlined above will not be accepted or have long-term impact on the field unless the data from multiple analyses is coherently correlated, analyzed, and integrated. Doing so will provide a clearer picture of the lipid oxidation process, which then can be applied to unraveling lipid oxidation and its effects *in vivo* or to improve product stability through modifications in formulation, processing, packaging, and storage. Meeting the challenges of providing this integration and understanding lipid oxidation in a new context will have two requirements: (1) developing methodologies to coordinate large amounts of data from multiple assays that reveal different chemistry in different formats, and (2) looking for solutions beyond traditional chain reactions and expected, accepted mechanisms. Current understanding of lipid oxidation is not incorrect—just incomplete—and accurately filling in the holes may demand that we remain open to different chemistry and approaches.

The integration of multiple forms of lipid oxidation reaction data is a new frontier. Right now, there are no road maps, so we can provide no specific direction for producing new knowledge from data. Certainly, an appropriate starting point would be to mesh chemometrics and computer systems analyses with the manual graphing of reaction pathways suggested by product analyses. Writing out reactions will reveal whether the patterns and correlating products identified by chemometrics make sense in terms of known chemistry or, alternatively, whether accepted reaction sequences need to be reevaluated. When validated, these pathways can be joined to provide a road map for lipid oxidation that parallels the detailed metabolic pathways diagrams so familiar in biochemistry.

It will be just as important to assess how kinetics and balance between various pathways change with conditions. Rate constants for the individual competing pathways shown in Fig. 2.6 need to be determined in defined model systems of increasing molecular complexity. In addition, analyses of lipid oxidation products generated via alternate pathways in model systems and in foods must be followed over time and integrated to develop maps of oxidation under different conditions. Together, these two sets of data can provide a basis for the simulation and modeling of lipid oxidation reactions, such as that attempted by Tappel and colleagues more than 20 years ago (Tappel et al., 1989). The knowledge gained from proving or modifying these models will provide important new molecular-level understandings of how lipid oxidation reactions, mechanisms, kinetics, and product distributions are modified with food system and reaction conditions. This information will also identify the compounds or product classes necessary to accurately track the progress and effects of lipid oxidation in foods, guide the development of improved analytical instrumentation and methodologies that can readily distinguish multiple pathways of lipid oxidation simultaneously, and reveal conditions that enhance the production of compounds with high potential to cooxidize other types of molecules (such as proteins, vitamins, and DNA) or to be toxic.

Indeed, these challenges are serious, and finding solutions will not be simple or fast. Nevertheless, meeting these challenges may radically change how we view, analyze, and control lipid oxidation.

Conclusion

This chapter cannot stress one point enough: Lipid oxidation is a dynamic process—it does not stand still for even a day—and analyses of oxidation only provide a snapshot of the extent of reaction at that specific point in time. Our challenge, therefore, is to change the analytical mindset in lipid oxidation and provide analyses that recognize and embrace these changes. Measuring one product at one sample concentration is not enough. Analyses must accurately portray, both qualitatively and quantitatively, the products actually present. They must also be matched to the level of oxidation in the sample and be flexible enough to shift as oxidation progresses. Quantitatively matching assays means that multiple concentrations of each sample must be analyzed to determine stoichiometry and make sure the sample is within an optimum response range. It means that methods with variable stoichiometry need to be replaced with assays that produce constant responses. It also means that multiple methods with a range of sensitivities are needed so that assays can be changed as samples move from the induction period to extensive oxidation. To match assays qualitatively, fingerprints of analyses that cover multiple products in alternate reaction pathways must be run to ensure detection of products in all active pathways and to determine which pathways are most active. Part of the mindset change is seeking more, rather than less, data and information; we must focus not only on general analyses of product classes but also on the determination of specific non-volatile and volatile products within each class to provide (1) a more detailed and accurate picture of the overall extent of oxidation, and (2) a contemporary database for reevaluating oxidation mechanisms and pathways. Accomplishing either of these goals requires that we identify, rather than ignore, all products, and this demands increased use of mass spectrometry or comparable technology for structural analysis.

"More data" also must include coordinated analyses of non-lipid cooxidation products to identify reactions of lipid oxidation products with other molecules in the system. Cooxidation reactions of lipids broadcast oxidation damage to other sites, greatly expanding the range of deterioration while reducing levels of lipid oxidation products. We cannot determine the full and accurate level and extent of lipid oxidation in any product—whether food, cosmetics, personal care products, or biological systems—without following lipid cooxidations as well as autoxidation.

Finally, tracking lipid oxidation requires that we think in three-dimensional space, where multiple products are changing in different directions at the same time. Integrating multiple products, reconciling disparate reaction kinetics, and organizing the reaction-product space in order to understand oxidation more fully and accurately will require powerful computer and statistical methodologies. The application of chemometrics is a first step in this direction.

Acknowledgments: The author gratefully acknowledges the contributions of graduate students Elah Steltzer, Teng Peng, Chenhsiang Liao, Jia Xie, Brandon Bogusz, Linhong Yao, Xin Tian, and Wan Zunair Wan Ibadullah to the new thinking presented here. Lessons learned from their ongoing research in lipid oxidation reactions and analyses were a major driving force behind this chapter.

References

Agarwal, S. C.; Van Duuren, B. L.; Knei, T. J. Detection of epoxides with 4-(p-nitrobenzyl)pyridine. Bull. Environm. Contam. Toxicol. 1979, 23, 825–829.

Akasaka, K.; Ohrui, H. Development of phosphine reagents for the high-performance liquid chromatographic-fluorometric determination of lipid hydroperoxides. J. Chromatog. A 2000, 881, 159–170.

AOAC. Lipid Hydroperoxides: Method 41.1.16. Official Methods of Analysis of AOAC International, 17th ed.; AOAC International: Gaithersburg, MD, 2000.

AOCS. p-Anisidine Value: AOCS Official Method Cd 18-90 (11). Official Methods and Recommended Practices of the American Oil Chemists' Society, 6th ed.; Firestone, D. Ed.; AOCS Press: Champaign, IL, 2011a.

AOCS. Conjugated Dienoic Acids in Castor Oil and Acids: AOCS Official Method Ti 1a-64 (09). Official methods and recommended practices of the American Oil Chemists' Society, 6th ed.; Firestone, D., Ed.; AOCS Press: Champaign, IL, 2011b.

AOCS. Diene Value of Drying Oils: AOCS Official Method Th 1a-64 (11). Official Methods and Recommended Practices of the American Oil Chemists' Society, 6th ed.; D. Firestone, Ed.; AOCS Press: Champaign, IL, 2011c.

AOCS. Hydroxyl Value of Epoxidized Oils: AOCS Official Method Tx 1a-66 (09). Official Methods and Recommended Practices of the American Oil Chemists' Society, 6th ed.; Firestone, D., Ed.; AOCS Press: Champaign, IL, 2011d.

AOCS. Hydroxyl Value of Fats and Oils (Acetyl Value): AOCS Official Method Cd 4-40 (99). Official Methods and Recommended Practices of the American Oil Chemists' Society, 6th ed.; Firestone, D., Ed.; AOCS Press: Champaign, IL, 2011e.

AOCS. Hydroxyl Values of Fatty Oils and Alcohols: AOCS Official Method Cd 13-60 (09). Official Methods and Recommended Practices of the American Oil Chemists' Society, 6th ed.; Firestone, D., Ed.; AOCS Press: Champaign, IL, 2011f.

AOCS. Oxirane Axygen in Epoxidized Materials: AOCS Official Method Cd 9-57 (09). Official Methods and Recommended Practices of the American Oil Chemists' Society; Firestone, D., Ed.; AOCS Press: Champaign, IL, 2011g.

AOCS. Peroxide Value Using Chloroform: AOCS Official Method Cd 8-53 (03). Official Methods and Recommended Practices of the American Oil Chemists' Society, 6th ed.; Firestone, D., Ed.; AOCS Press: Champaign, IL, 2011h.

AOCS. Peroxide Value Using Isooctane: AOCS Official Method 8b-90. Official Methods and Recommended Practices of the AOCS, 6th ed.; Firestone, D., Ed.; AOCS Press: Champaign, IL, 2011i.

AOCS. Polymerized Triglycerides by Gel-Permeation HPLC: AOCS Official Method Cd 22-91 (09). Official Methods and Recommended Practices of the American Oil Chemists' Society, 6th ed.; Firestone, D., Ed.; AOCS Press: Champaign, IL, 2011j.

Armenta, S.; Moros, J.; Garrigues, S.; de la Guardia Cirugeda, M.; Victor, R. P.; Ronald Ross, W. Determination of olive oil parameters by Near Infrared Spectrometry. Olives and Olive Oil in Health and Disease Prevention; Academic Press: San Diego, CA, 2010; 533–544.

Barthel, G.; Grosch, W. Peroxide value determination: Comparison of some methods. J. Am. Oil Chem. Soc. 1974, 51, 540–544.

Battino, R.; Clever, H. L. The solubility of gases in liquids. Chem. Rev. 1966, 66, 395–463.

Behnke, A. R.; Yarbrough, O. D. Respiratory resistance, oil-water solubility, and mental effects of argon, compared with helium and nitrogen. Am. J. Physiol. 1939, 126, 409–415.

Beißmann, S.; Buchberger, W.; Hertsens, R.; Klampfl, C. W. High-performance liquid chromatography coupled to direct analysis in real time mass spectrometry: Investigations on gradient elution and influence of complex matrices on signal intensities. J. Chromatogr. A 2011, 1218, 5180–5186.

Belitz, H.-D.; Grosch, W.; Scheiberle, P. Food Chemistry; Springer-Verlag: Berlin, 2009; 203–211.

Berrueta, L. A.; Alonso-Salces, R. M.; Heberger, K. Supervised pattern recognition in food analysis. J. Chromatogr. A 2007, 1158, 196–214.

Biomedicals, M. SafTest PEROXYSAFE™ kit product insert. http://images.www.mpbio.com/docs/saftest/Insert-PeroxySafe-07KTPR1010-07KTPR1020-07KTPR1030.pdf (accessed October 2, 2012).

Bird, R. P.; Basrur, P. K.; Alexander, J. C. Cytotoxicity of thermally oxidized fats. In Vitro. 1981, 17, 397–404.

Blair, I. A. Lipid hydroperoxide-mediated DNA damage. Exp. Gerontol. 2001, 36, 1473–1481.

Boldor, D.; Kanitkar, A.; Terigar, B. G.; Leonardi, C.; Lima, M.; Breitenbeck, G. A. Microwave assisted extraction of biodiesel feedstock from the seeds of invasive Chinese tallow tree. Environ. Sci. Technol. 2010, 44, 4019–4025.

Bomstein, J. Infrared spectra of oxirane compounds. Anal. Chem. 1958, 30, 544–546.

Borg, D. C.; Schaich, K. M. Cytotoxicity from coupled redox cycling of autoxidizing xenobiotics and metals. Isr. J. Chem. 1984, 24, 38–53.

Bosque-Sendra, J. M.; Cuadros-Rodriguez, L.; Ruiz-Samblas, C.; de la Mata, A. P. Combining chromatography and chemometrics for the characterization and authentication of fats and oils from triacylglycerol compositional data—A review. Anal. Chim. Acta 2012, 724, 1–11.

Bou, R.; Codony, R.; Tres, A.; Decker, E. A.; Guardiola, F. Determination of hydroperoxides in foods and biological samples by the ferrous oxidation-xylenol orange method: A review of the factors that influence the method's performance. Anal. Biochem. 2008, 377, 1–15.

Brereton, R. G. Chemometrics Data Analysis for the Laboratory and Chemical Plant; John Wiley & Sons: Chichester, UK, 2003.

Brown, S. D.; Tauler, R.; Walczak, B. Comprehensive chemometrics and biochemical data analysis; Elsevier: Amsterdam, Netherlands, 2009; Vol. 4.

Brunner, B.; Stogatis, N.; Lautens, M. Synthesis of 1,2-dihyropyridine using vinyloxiranes as masked dienolates in imino-aldol reactions. Org. Lett. 2006, 8, 3473–3476.

Buydens, L. M. C.; Reijmers, T. H.; Beckers, M. L. M.; Wehrens, R. Molecular data-mining: A challenge for chemometrics. Chemometr. Intell. Lab. 1999, 49, 121–133.

Cayman. Lipid hydroperoxide (LPO) assay kit manual. https://www.caymanchem.com/pdfs/705002.pdf (accessed October 1, 2012).

Chapman, D.; Goni, F. M.; Gunstone, F. D. Physical Properties: Optical and spectral characteristics. Infrared Spectra. The Lipid Handbook, 2nd ed.; Gunstone, F. D.; Harwood, J. L.; Padley, F. B., Eds.; Chapman and Hall: London, 1994; 487–504.

Chaudry, Y. Determination of optimal conditions for extraction of total crude fat from a model extruded pet food system using supercritical carbon dioxide. M.S. Thesis, Rutgers University, New Brunswick, NJ, 1998.

Chien, M.; Peppard, T., Use of Statistical Methods to Better Understand Gas Chromatographic Data Obtained from Complex Flavor Systems. Flavor Measurement; Ho, C.-T.; Manley, C. H., Eds.; Marcel Dekker: New York, 1993; 1–35.

Christie, W. W. Gas Chromatography and Lipids: A Practical Guide; The Oily Press: Dundee, Scotland, 1989.

Chung, F.-L.; Chen, H.-C.; Guttenplan, J. B.; Nishikawa, A.; Hard, G. C. 2,3-Epoxy-4-hydroxynonanal as a potential tumor-initiating agent of lipid peroxidation. Carcinogenesis 1993, 14, 2073–2077.

Clark, D. L. Peroxides and peroxide forming compounds. Chemical Health & Safety 2001, Sept/Oct, 12–22.

Clifford, A. A. Introduction to SFE in Analytical Science. Supercritical Fluid Extraction and Its Use in Chromatographic Sample Preparation; Westwood, S. A., Ed.; CRC Press: Boca Raton, FL, 1993; 1–38.

Cocero, M. J.; Calvo, L. Supercritical fluid extraction of sunflower seed oil with CO2-ethanol mixtures. J. Am. Oil Chem. Soc. 1996, 73, 1573–1578.

Cody, R. B.; Laramée, J. A.; Durst, H. D. Versatile new ion source for the analysis of materials in open air under ambient conditions. Anal. Chem. 2005a, 77, 2297–2302.

Cody, R. B.; Laramée, J. A.; Nilles, J. M.; Durst, H. D. Direct analysis in real time (DARTTM) mass spectrometry. JEOL News 2005b, 40, 8–12.

Comporti, M. Lipid peroxidation and cellular damage in toxic liver injury. Lab Invest. 1985, 53, 599–623.

Cortesi, R.; Privett, O. S. Toxicity of fatty ozonides and peroxides. Lipids 1972, 7, 715–721.

Cossignani, L.; Simonetti, M.; Neri, A.; Damiani, P. Changes in olive oil composition due to microwave heating. J. Am. Oil Chem. Soc. 1998, 75, 931–937.

Cozzolino, D.; Murray, I.; Chree, A.; Scaife, J. R. Multivariate determination of free fatty acids and moisture in fish oils by partial least-squares regression and near-infrared spectroscopy. LWT - Food Sci. Technol. 2005, 38, 821–828.

Crowe, T. D.; Crowe, T. W.; Johnson, L. A. Impact of extraction method on yield of lipid oxidation products from oxidized and unoxidized walnuts. J. Am. Oil Chem. Soc. 2002, 79, 453–456.

Davies, M. J. Markers of protein oxidation: Different oxidants give rise to variable yields of bound and released carbonyl products. Free Radic. Biol. Med. 2004, 36, 1175–1184.

Davies, M. J.; Fu, S.; Wang, H.; Dean, R. T. Stable markers of oxidant damage to proteins and their application in the study of human disease. Free Radic. Biol. Med. 1999, 27, 1151–1163.

de Andrade, D. F.; Fernandes, D. R.; Miranda, J. L. Methods for the determination of conjugated dienes in petroleum products: A review. Fuel 2010, 89, 1796–1803.
De Luca, M.; Terouzi, W.; Ioele, G.; Kzaiber, F.; Oussama, A.; Oliverio, F.; Tauler, R.; Ragno, G. Derivative FTIR spectroscopy for cluster analysis and classification of Morocco olive oils. Food Chem. 2011, 124, 1113–1118.
de Lucas, A.; Rincón, J.; Gracia, I. Influence of operating variables on yield and quality parameters of olive husk oil extracted with supercritical carbon dioxide. J. Am. Oil Chem. Soc. 2002, 79, 237–243.
Dobarganes, C. M.; Márquez-Ruiz, G., Formation and Analysis of Oxidized Nonomeric, Dimeric and Higher Oligomeric Triglycerides. Deep Frying: Chemistry, Butrition and Oractical Applications, 2nd ed.; Erickson, M. D., Ed.; AOCS Press: Champaign, IL, 2007; 87–110.
Dobarganes, M. C.; Velasco, J. Analysis of lipid hydroperoxides. Eur. J. Lipid Sci. Technol. 2002, 104, 420–428.
Dodds, E.; McCoy, M.; Geldenhuys, A.; Rea, L.; Kennish, J. Microscale recovery of total lipids from fish tissue by accelerated solvent extraction. J. Am. Oil Chem. Soc. 2004, 81, 835–840.
Doherty, S. J.; Lange, A. J. Avoiding pitfalls with chemometrics and PAT in the pharmaceutical and biotech industries. Trends Anal. Chem. 2006, 25, 1097–1102.
Dupard-Julien, C.; Kandlakunta, B.; Uppu, R. Determination of epoxides by high-performance liquid chromatography following derivatization with diethyldithiocarbamate. Anal. Bioanal. Chem. 2007, 387, 1027–1032.
Ehrenberg, L.; Hussain, S. Genetic toxicity of some important epoxides. Mutat. Res.-Rev. Genet. 1981, 86, 1–113.
Esterbauer, H. Cytotoxicity and genotoxicity of lipid-oxidation products. Amer. J. Clin. Nutrit. 1993, 57, 779S–785S.
Esterbauer, H.; Eckl, P.; Ortner, A. Possible mutagens derived from lipids and lipid precursors. Mutat. Res.-Rev. Genet. 1990, 238, 223–233.
Esterbauer, H.; Schaur, R. J.; Zollner, H. Chemistry and biochemistry of 4-hydroxynonenal, malonaldehyde, and related aldehydes. Free Radic. Biol. Med. 1991, 11, 81–128.
Ferrao, M. F.; Godoy, S. C.; Gerbase, A. E.; Mello, C.; Furtado, J. C.; Petzhold, C. L.; Poppi, R. J. Non-destructive method for determination of hydroxyl value of soybean polyol by LS-SVM using HATR/FT-IR. Anal. Chim. Acta 2007, 595, 114–119.
Frankel, E. N. Hydroperoxide Formation. Lipid Ixidation, 2nd ed.; Frankel, E. N., Ed.; PJ Barnes & Associates: Bridgewater, UK, 2005; 25–50.
Freeman, T. L.; Haver, A.; Duryee, M. J.; Tuma, D. J.; Klassen, L. W.; Hamel, F. G.; White, R. L.; Rennard, S. I.; Thiele, G. M. Aldehydes in cigarette smoke react with the lipid peroxidation product malonaldehyde to form fluorescent protein adducts on lysines. Chem. Res. Toxicol. 2005, 18, 817–824.
Fruebis, J.; Parsasarathy, S.; Steinberg, D. Evidence for a concerted reaction between lipid hydroperoxides and polypeptides. Proc. Nat. Acad. Sci. USA 1992, 89, 10588–10592.
Fuchs, B.; Suß, R.; Schiller, J. An update of MALDI-TOF mass spectrometry in lipid research. Prog. Lipid Res. 2010, 49, 450–475.
Fukuzumi, K. Significance of lipoperoxides in gastric cancer. Sonderdruck aus Fette 1970, 72, 853–855.
Funk, M. O. The electroanalytical approach to lipid peroxide determinations. Free Radic. Biol. Med. 1987, 3, 319–21.

García-Ayuso, L.; Velasco, J.; Dobarganes, M.; Luque de Castro, M. Determination of the oil content of seeds by focused microwave-assisted soxhlet extraction. Chromatographia 2000, 52, 103–108.
Gardner, H. W. Oxygen radical chemistry of polyunsaturated fatty acids. Free Radic. Biol. Med. 1989, 7, 65–86.
Gardner, H. W.; Jursinic, P. A. Degradation of linoleic acid hydroperoxides by a cysteine-FeCl3 catalyst as a model for similar biochemical reactions. I. Study of oxygen requirement, catalyst and effect of pH. BBA-Lipid Lipid Met. 1981, 665, 100–112.
Gardner, H. W.; Kleiman, R.; Weisleder, D. Homolytic decomposition of linoleic acid hydroperoxide: Identification of fatty acid products. Lipids 1974, 9, 696–706.
Gardner, H. W.; Kleiman, R.; Weisleder, D.; Inglett, G. E. Cysteine adds to lipid hydroperoxide. Lipids 1977, 12, 655–660.
Gardner, H. W.; Plattner, R. D.; Weisleder, D. The epoxyallylic radical from homolysis and rearrangement of methyl linoleate hydroperoxide combines with the thiyl radical of N-acetylcysteine. Biochim. Biophys. Acta 1985, 834, 65–74.
Gardner, H. W.; Weisleder, D.; Kleiman, R. Formation of *trans*-12,13-epoxy-9-hydroperoxy-*trans*-10-octadecadienoic acid catalyzed by either a soybean extract or cysteine-FeCl3. Lipids 1978, 13, 246–252.
Gay, C.; Collins, J.; Gebicki, J. M. Determination of iron in solutions with the ferric-xylenol orange complex. Anal. Biochem. 1999a, 273, 143–148.
Gay, C.; Collins, J.; Gebicki, J. M. Hydroperoxide assay with the ferric-xylenol orange complex. Anal. Biochem. 1999b, 273, 149–155.
George, D. A. Spectral data for aliphatic epoxides. J. Chem. Eng. Data 1975, 20, 445–448.
Gomez-Caravaca, A. M.; Maggio, R. M.; Verardo, V.; Cichelli, A.; Cerretani, L. Fourier transform infrared spectroscopy—Partial least squares (FTIR-PLS) coupled procedure application for the evaluation of fly attack on olive oil quality. LWT - Food Sci. Technol. 2012, 50, 153–159.
Gotoh, N.; Miyake, A.; Takei, H.; Sasaki, K.; Okuda, S.; Ishinaga, M.; Wada, S. Simple method for measuring the peroxide value in a colored lipid. Food Anal. Methods 2011, 4, 525–530.
Graeve, M.; Janssen, D. Improved separation and quantification of neutral and polar lipid classes by HPLC-ELSD using a monolithic silica phase: Application to exceptional marine lipids. J. Chromatogr. B. Analyt. Technol. Biomed. Life Sci. 2009, 877, 1815–1819.
Grosch, W. Reactions of Hydroperoxides—Products of Low Molecular Weight. Autoxidation of Unsaturated Lipids; Chan, H.W.-S., Ed.; Academic Press, London, 1987; 95–139.
Grosch, W.; Megele, J. Preparation, Isomerization, and Breakdown of Hydroperoxides from Methyl Oleate. Oxygen Radicals in Chemistry and Biology; Bors, W.; Saran, M.; Tait, D. Eds.; Walter de Gruyter & Co.: Berlin, 1984; 249–256.
Gruszka, J.; Pawlak, A.; Kruk, J. Tocochromanols, plastoquinol, and other biological prenyllipids as singlet oxygen quenchers—Determination of singlet oxygen quenching rate constants and oxidation products. Free Radic. Biol. Med. 2008, 45, 920–928.
Guido, D. M.; McKenna, R.; Mathews, W. R. Quantitation of hydroperoxy-eicosatetraenoic acids and hydroxy-eicosatetraenoic acids as indicators of lipid peroxidation using gas chromatography-mass spectrometry. Anal. Biochem. 1993, 209, 123–129.
Guillen, M. D.; Cabo, N. Fourier transform infrared spectra data versus peroxide and anisidine values to determine oxidative stability of edible oils. Food Chem. 2002, 77, 503–510.
Gunstone, F. D.; Padley, F. B. Lipid Technologies and Applications; Marcel Dekker: New York, 1997; 181–182.

Hajslova, J.; Cajka, T.; Vaclavik, L. Challenging applications offered by direct analysis in real time (DART) in food-quality and safety analysis. Trends Anal. Chem. 2011, 30, 204–218.
Hamalainen, T. I.; Kamal-Eldin, A. Analysis of Lipid Oxidation Products by NMR Spectroscopy. Analysis of Lipid Oxidation; Kamal-Eldin, A.; Pokorný, J., Eds.; AOCS Press: Champaign, IL, 2005; 70–126.
Hammock, L. G.; Hammock, B. D.; Casida, J. E. Detection and analysis of epoxides with 4-(p-nitrobenzyl)-pyridine. Bull. Environ. Contam. Toxicol. 1974, 12, 759–764.
Hartman, T. G.; Lech, J.; Fukuda, E.; Rosen, R. T. Methodologies for the quantification of purge and trap thermal desorption and direct thermal desorption analyses. The Mass Spec. Source 1991, 14, 18–24.
Hartman, T. G.; Lech, J.; Karmas,, K.; Salinas, J.; Rosen, R. T.; Ho, C.-T. Flavor Characterization Using Adsorbent Trapping-Thermal Desorption or Direct Thermal Desorption-Gas Chromatography and Gas Chromatography-Bass Spectrometry. Flavor Beasurement; Ho, C.-T.; Manley, C. H. Eds.; Marcel Dekker: New York, 1993; 37–60.
Haynes, R. K.;Vonwiller, S. C. Iron(III) and copper(II) catalyzed transformation of fatty acid hydroperoxides: Efficient generation of peroxy radicals with Cu(II) trifluoromethane sulfonate. J. Chem. Soc. Chem. Commun. 1990, 1102–1104.
Haywood, R. M.; Claxson, A. W.; Hawkes, G. E.; Richardson, D. P.; Naughton, D. P.; Coumbarides, G.; Hawkes, J.; Lynch, E. J.; Grootveld, M. C. Detection of aldehydes and their conjugated hydroperoxydiene precursors in thermally stressed culinary oils and fats: Investigations using high-resolution proton NMR spectroscopy. Free Radic. Biol. Med. 1995, 22, 441–482.
Heinecke, J. W. Free radical modification of low-density lipoprotein: Mechanisms and biological consequences. Free Radic. Biol. Med. 1987, 3, 65–73.
Hemminki, A.; Väyrynen T.; Hemminki, K. Reaction kinetics of alkyl epoxides with DNA and other nucleophiles. Chem-Biol. Interact. 1994, 93, 51–58.
Hidalgo, F. J.; Zamora, R. Fluorescent pyrrole products form carbonyl-amine reactions. J. Biol. Chem. 1993a, 268, 16190–16197.
Hidalgo, F. J.; Zamora, R. Non-enzymatic browning and fluorescence development in a (E)-4,5-epoxy-(E)-2-heptenal/lysine model system. J. Food Sci. 1993b, 58, 667–670.
Hidalgo, F. J.; Zamora, R. Characterization of the products formed during microwave irradiation of the nonenzymatic browning lysine/(E)-4,5-epoxy-(E)-2-heptenal model system. J. Agric. Food Chem. 1995a, 43, 1023–1028.
Hidalgo, F. J.; Zamora, R. *In vitro* production of long chain pyrrole fatty esters from carbonyl-amine reactions. J. Lipid Res. 1995b, 36, 725–735.
Hidalgo, F. J; Zamora, R. Modification of bovine serum albumin structure following reaction with 4,5(E)-epoxy-2-(E)-heptenal. Chem. Res. Toxicol. 2000, 13, 501–508.
Hidalgo, F. J.; Zamora, R. Methyl linoleate oxidation in the presence of bovine serum albumin. J. Agric. Food Chem. 2002, 50, 5463–5467.
Hidalgo, F. J.; Zamora, R. Strecker-type degradation produced by the lipid oxidation products 4,5-epoxy-2-alkenals. J. Agric. Food Chem. 2004, 52, 7126–7131.
Hidalgo, F. J.; Alaiz, M.; Zamora, R. A spectrophotometric method for the determination of proteins damaged by oxidized lipids. Anal. Biochem. 1998, 262, 129–136.
Hidalgo, F. J.; Alaiz, M.; Zamora, R. Pyrrolization and antioxidant function of proteins following oxidative stress. Chem. Res. Toxicol. 2001, 14, 582–588.

Hidalgo, F. J.; Gallardo, E.; Zamora, R. Strecker type degradation of phenyl alanine by 4-hydroxy-2-nonenal in model systems. J. Agric. Food Chem. 2005, 53, 10254–10259.
Hidalgo, F. J.; Kinsella, J. E. Changes in β-lactoglobulin B following interactions with linoleic acid 13-hydroperoxide. J. Agric. Food Chem. 1989, 37, 860–866.
Honeywell Burdick & Jackson. Solvent stabilization. http://www51.honeywell.com/sm/rlss/bandj-eu/productsn2/solvents3/solvent-stabilization.html?c=21 (accessed Aug. 1, 2012).
Hornero-Méndez, D.; Pérez-Gálvez, A.; Mínguez-Mosquera, M. A rapid spectrophotometric method for the determination of peroxide value in food lipids with high carotenoid content. J. Am. Oil Chem. Soc. 2001, 78, 1151–1155.
House, K. Comparison of supercritical fluid extraction with solvent extraction and acid hydrolysis in determination of total lipid content of dry extruded pet food. B.S., Rutgers University, New Brunswick, NJ, 1999.
Ichihashi, K.; Osawa, T.; Toyokuni, S.; Uchida, K. Endogenous formation of protein adducts with carcinogenic aldehydes. J. Biol. Chem. 2001, 276, 23903–23913.
IonSense. DARTR technology: A new technology for rapid, non-contact surface sampling of compounds/experiment modules. http://www.ionsense.com/dart_technology (accessed October 1, 2012).
IUPAC. Determination of the p-Anisidine Value, Method 2.504. IUPAC Standard Methods for the Analysis of Oils, Fats and Derivatives, 7th ed.; Alden Press: Oxford, 1987; 210–211.
IUPAC. 2.501: IUPAC Standard Methods for the Analysis of Oils, Fats and Derivatives, 7th ed.; Dieffenbacher, A.; Pocklington, W. D., Eds.; Blackwell: Oxford, 1992.
Izaki, Y.; Yoshikawa, S.; Uchiyama, M. Effect of ingestion of thermally oxidized frying oil on peroxidative criteria in rats. Lipids 1984, 19, 324–330.
Jahan, K.; Paterson, A.; Piggott, J.; Spickett, C. Chemometric modeling to relate antioxidants, neutral lipid fatty acids, and flavor components in chicken breasts. Poult. Sci. 2005, 84, 158–166.
Jiang, Z.; Hunt, J. V.; Wolff, S. P. Ferrous ion oxidation in the presence of xylenol orange for detection of lipid hydroperoxide in low density lipoprotein. Anal. Biochem. 1992, 202, 384–389.
Jiang, Z.; Woollard, A.; Wolff, S. Lipid hydroperoxide measurement by oxidation of Fe^{2+} in the presence of xylenol orange. Comparison with the TBA assay and an iodometric method. Lipids 1991, 26, 853–856.
Jiang, Z.; Woollard, A. C. S.; Wolff, S. P. Hydrogen peroxide production during experimental protein glycation. FEBS Letters 1990, 268, 69–71.
JOCS. 2.5.4-1996 Carbonyl Value. Standard Methods for the Analysis of Fats, Oils, and Related Materials; Japan Oil Chemists' Society: Tokyo, Japan, 1996.
JOCS. 2.5.2.1. (2003): Peroxide Value (Acetic Acid-Isooctane Method). The JOCS Standard Methods for the Analysis of Fats, Oils, and Related Materials; Japan Oil Chemists' Society: Tokyo, Japan, 2009.
Kamal-Eldin, A.; Pokorny, J., Eds. Analysis of Lipid Oxidation; AOCS Press: Champaign, IL, 2005.
Kazala, E. C.; Lozeman, F.; Mir, P.; Aalhus, J.; Schmutz, S.; Weselake, R. Fatty acid composition of muscle fat and enzymes of storage lipid synthesis in whole muscle from beef cattle. Lipids 2006, 41, 1049–1057.
King, J. W.; Johnson, J. H.; Friedrich, J. P. Extraction of fat tissue from meat products with supercritical carbon dioxide. J. Agric. Food Chem. 1989, 37, 951–954.

Kolthoff, I. M.; Medalia, A.I. The reaction between ferrous iron and peroxides. III. Reaction with cumene hydroperoxide, in aqueous solution. J. Am. Chem. Soc. 1949, 71, 3789–3792.
Kolthoff, I. M.; Medalia, A. I. Determination of organic peroxides by reaction with ferrous iron. Anal. Chem. 1951, 23, 595–603.
Korytowski, W.; Geiger, P. G.; Girotti, A. W. High-performance liquid chromatography with mercury cathode electrochemical detection: Application to lipid hydroperoxide analysis. J. Chromatogr. B 1995, 670, 189–197.
Korytowski, W.; Geiger, P. G.; Girotti, A. W. Lipid Hydroperoxide Analysis by High-Performance Liquid Chromatography with Mercury Cathode Electrochemical Detection. Methods in Enzymology; Packer, L., Ed.; Academic Press: San Diego, CA, 1999; 23–33.
Kruus, P.; Burk, R. C.; Entezari, M. H.; Otson, R. Sonication of aqueous solutions of chlorobenzene. Ultrason. Sonochem. 1997, 4, 229–233.
Kubow, S. Routes of formation and toxic consequences of lipid oxidation products in foods. Free Radic. Biol Med. 1992, 12, 63–81.
Lam, H. S.; Proctor, A.; Meullenet, J.-F. Free fatty acid formation and lipid oxidation on milled rice. J. Am. Oil Chem. Soc. 2001, 78, 1271–1275.
Lavine. B. Chemometrics. Anal. Chem. 2000, 72, 91–98.
Lavine, B.; Workman, J. Chemometrics. Anal. Chem. 2004, 76, 3365–3372.
Lavine, B.; Workman, J. Chemometrics. Anal. Chem. 2006, 78, 4137–4145.
Lavine, B.; Workman, J. Chemometrics. Anal. Chem. 2008, 80, 4519–4531.
Lavine, B.; Workman, J. Chemometrics. Anal. Chem. 2010, 82, 4699–4711.
Lavine, B. K. Chemometrics. Anal. Chem. 1998, 70, 209–228.
Lavine, B. K. Chemometrics. Anal. Chem. 2000, 72, 91–98.
Lavine, B. K.; Workman, J. Chemometrics. Anal. Chem. 2002, 74, 2763–2770.
Lea, C. H. The determination of the peroxide values of edible fats and oils: The iodometric method. J. Soc. Chem. Ind. 1946, 65, 286–291.
Lea, C. H. Methods for determining peroxide in lipids. J. Sci. Food Agric. 1952, 3, 586–594.
Lederer, M. O. Reactivity of lysine moieties toward g-hydroxy-a,b-unsaturated epoxides: A model study on protein-lipid oxidation product interaction. J. Agric. Food Chem. 1996, 44, 2531–2537.
Lederer, M. O.; Schuler, A.; Ohmenhäuser, M. Reactivity of lysine moieties toward an epoxyhydroxylinoleic acid derivative: Aminolysis vs hydrolysis. J. Agric. Food Chem. 1998, 47, 4611–4620.
Lee, J.; Kim, D.-H.; Chang, P.-S.; Lee, J. Headspace-solid phase microextraction (HS-SPME) analysis of oxidized volatiles from free fatty acids (FFA) and application for measuring hydrogen donating antioxidant activity. Food Chem. 2007, 105, 414–420.
Lee, J.-Y.; Yoo, C.; Jun, S.-Y.; Ahn, C.-Y.; Oh, H.-M. Comparison of several methods for effective lipid extraction from microalgae. Bioresource Technol. 2009, 101, S75–S77.
Lee, S. H.; Oe, T.; Blair, I. A. Vitamin C-induced decomposition of lipid hydroperoxides to endogeneous genotoxins. Science 2001, 292, 2083–2086.
Lee, S. H.; Oe, T.; Blair, I. A. 4,5-Epoxy-2(E)-decenal-induced formation of 1,N(6)-etheno-2'-deoxyadenosine and 1,N(2)-etheno-2'-deoxyguanosine adducts. Chem. Res. Toxicol. 2002, 15, 300–304.
Leonarduzzi, G.; Arkan, M. C.; Basaga, H.; Chiarpotto, E.; Sevanian, A.; Poli, G. Lipid oxidation products in cell signaling. Free Radic. Biol. Med. 2000, 28, 1370–1378.

Lerma-Garcia, M. J.; Simo-Alfonso, E. F.; Bendini, A.; Cerratani, L. Rapid evaluation of oxidised fatty acid concentration in virgin olive oil using Fourier-transform infrared spectroscopy and multiple linear regression. Food Chem. 2011, 124, 679–684.

Liao, C.-H. Comparison of chemical assays used to determine epoxides in oxidized lipids. M.S. Thesis, Rutgers University, New Brunswick, NJ, 2013.

Lin, D.; Lee, H.; Liu, Q.; Perry, G.; Smith, M. A.; Sayre, L. M. 4-Oxo-2-nonenal is both more neurotoxic and more protein reactive than 4-hydroxy-2-nonenal. Chem. Res. Toxicol. 2005, 18, 1219–1231.

Liu, M.; Shuai, Y.; Liang, H.; Huang, L.; Song, N. Comparative study on extraction methods of microbial lipids. J. Earth Sci. 2010, 21, 300–303.

Longobardi, F.; Ventrella, A.; Casiello, G.; Sacco, D.; Tasioula-Margari, M.; Kiritsakis, A. K.; Kontominas, M. G. Characterisation of the geographical origin of Western Greek virgin olive oils based onrumental and multivariate statistical analysis. Food Chem. 2012, 133, 169–175.

Lqari, H.; Pedroche, J.; Giron-Calle, J.;Vloque, J.; Millan, F. Interaction of Lupinus angustifolius L. a and g conglutins with 13-hydroperoxide-11,9-octadienoic acid. Food Chem. 2003, 80, 517–523.

Luna, G.; Morales, M. T.; Aparicio, R. Characterisation of 39 varietal virgin olive oils by their volatile compositions. Food Chem. 2006, 98, 243–252.

Ma, K.; van de Voort, F.; Sedman, J.; Ismail, A. Stoichiometric determination of hydroperoxides in fats and oils by Fourier transform infrared spectroscopy. J. Am. Oil Chem. Soc 1997, 74, 897–906.

Maerker, G. Determination of oxirane content of derivatives of fats. J. Am. Oil Chem. Soc. 1965, 42, 329–332.

Maggio, R. M.; Valli, E.; Bendini, A.; Gomez-Caravaca, A. M.; Toschi, T. G.; Cerratani, L. A spectroscopic and chemometric study of virgin olive oils subjected to thermal stress. Food Chem. 2011, 127, 216–221.

Makino, Y.; Ichimura, M.; Oshita, S.; Kawagoe, Y.; Yamanaka, H. Estimation of oxygen uptake rate of tomato (lycopersicon esculentum Mill.) fruits by artificial neural networks modelled using near infra-red spectroal absorbance and fruite mass. Food Chem. 2010, 121, 533–539.

Mallia, S.; Fernandez-Garcia, E.; Bosset, J. O. Comparison of purge and trap and solid phase microextraction techniques for studying the volatile aroma compounds of 3 European PDO hard cheeses. Int. Dairy J. 2005, 15, 741–758.

Marengo, E.; Bobba, M.; Robotti, E.; Lenti, M. Hydroxyl and acid number prediction in polyester resins by near infrared spectroscopy and artificial neural networks. Anal. Chim. Acta 2004, 511, 313–322.

Márquez-Ruiz, G.; Dobarganes, M. C. Analysis of Lipid Oxidation Products by Combination of Chromatographic Techniques. New Techniques and Applications in Lipid Analysis; MacDonald, R. E.; Mossoba, M. M. Eds.; AOCS Press: Champaign, IL, 1997; 216–233.

Márquez-Ruiz, G.; Dobarganes, M. C. High-Performance Size-Exclusion Chromatography for Lipid Analysis in Organic Media. Lipid Analysis and Lipidomics: New Techniques and Applications; Mossoba, M. M.; Kramer, J. K. G.; Brenna, J. T.; McDonald, R. E. Eds.; AOCS Press: Champaign, IL, 2006; 205–238.

Massart, D. L.; Vandeginste, B. G. M.; Buydens, L. M. C.; De Jong, A.; Lewi, P. J.; Smeyers-Verbeke, J. Handbook of Chemometrics and Qualimetrics, Part A; Elsevier: Amsterdam, 1998.

Massart, D .L.; Vandeginste, B. G. M.; Deming, S. M.; Michotte, Y.; Kaufmann, L. Chemometrics: a Textbook; Elsevier: Amsterdam, 2003.
Massey, L. K. Permeability Properties of Plastics and Elastomers: A Guide to Packaging and Barrier Materials; William Andrew Publishing: Norwich, NY, 2003; 7.
McDonald, R. E.; Mossoba, M. M., Eds. New techniques and applications in lipid analysis; AOCS Press: Champaign, IL, 1997.
McLaren, D. G.; Miller, P. L.; Lassman, M. E.; Castro-Perez, J. M.; Hubbard, B. K.; Roddy, T. P. An ultraperformance liquid chromatography method for the normal-phase separation of lipids. Anal. Biochem. 2011, 414, 266–272.
McMurray, J. Organic Chemistry; Brooks/Cole: Pacific Grove, CA, 2000.
McNichol, J.; MacDougall, K.; Melanson, J.; McGinn, P. Suitability of soxhlet extraction to quantify microalgal fatty acids as determined by comparison with *in situ* transesterification. Lipids 2012, 47, 195–207.
Melton, S.; Moyers, R.; Playford, C. Lipids extracted from soy products by different procedures. J. Am. Oil Chem. Soc. 1979, 56, 489–493.
Mihaljevic, B.; Katusin-Razem, B.; Razem, D. The re-evaluation of the ferric thiocyanate assay for the lipid hydroperoxides with special consideration of the mechanistic aspects of the response. Free Radic. Biol. Med. 1996, 21 (1), 53–63.
Mildner-Szkudlarz, S.; Jelen, H. H. The potential of different techniques for volatile compounds analysis coupled with PCA for the detection of the adulteration of olive oil with hazelnut oil. Food Chem. 2008, 110, 751–761.
Mildner-Szkudlarz, S.; Jelen, H. H.; Zawirska-Wojtasiak, R.; Wasowicz, E. Application of headspace—Solid phase microextraction and multivariate analysis for plant oils differentiation. Food Chem. 2003, 83, 515–522.
Miyazawa, T. Determination of phospholipid hydroperoxides in human blood plasma by a chemiluminescence-HPLC assay. Free Radic. Biol. Med. 1989, 7, 209–218.
Mobaraki, N.; Hemmateenejad, B. Structural characterization of carbonyl compounds by IR spectroscopy and chemometrics data analysis. Chemometr. Intell. Lab. 2011, 109, 171–177.
Montarini, L.; King, J. W.; List, G. R.; Rennick, K. A. Selective extraction of phospholipid mixtures by supercritical carbon dioxide and cosolvents. J. Food Sci. 1996, 61, 1230–1233.
Moreau, R.; Powell, M.; Singh, V. Pressurized liquid extraction of polar and nonpolar lipids in corn and oats with hexane, methylene chloride, isopropanol, and ethanol. J. Am. Oil Chem. Soc. 2003, 80, 1063–1067.
Moreau, R. A.; Asmann, P. T.; Norman, H. A. Analysis of major classes of plant lipids by high-performance liquid chromatography with flame ionization detection. Phytochem. 1990, 29, 2461–2466.
Moya Moreno, M. C. M.; Mendoza Olivares, D.; Amezquita Lopez, F. J.; Peris Martonez, V.; Bosch Reig, F. Study of the formation of carbonyl compounds in edible oils and fats by 1H-NMR and FTIR. J. Molec. Struct. 1999, 482–483, 557–561.
Nadkarni, D. V.; Sayre, L. M. Structural definition of early lysine and histidine adduction chemistry of 4-hydroxynonenal. Chem. Res. Toxicol. 1995, 8, 284–291.
Nagawade, R. R.; Shinde, D. B. Zirconyl(IV) chloride-catalyzed reaction of indoles: An expeditious synthesis of bis(indolyl) methanes. Acta Chim. Slov. 2006, 53, 210–213.
Nair, V.; Offerman, R. J.; Turner, G. A.; Pryor, A. N.; Baenziger, N. C. Fluorescent 1,4-dihydropyridines—The malondialdehyde connection. Tetrahedron 1988, 44, 2793–2803.

Nakamura, T.; Maeda, H. A simple assay for lipid hydroperoxides based on triphenylphosphine oxidation and high-performance liquid chromatography. Lipids 1991, 26, 765–768.

Navas, J. A.; Tres, A.; Codony, R.; Boatella, J.; Bou, R.; Guardiola, F. Modified ferrous oxidation-xylenol orange method to determine lipid hydroperoxides in fried snacks. Eur. J. Lipid Sci. Technol. 2004, 106, 688–696.

Nazir, H.; Yildiz, M.; Yilmaz, H.; Tahir, M. N.; Ülkü, D. Intramolecular hydrogen bonding and tautomerism in Schiff bases. Structure of N-(2-pyridil)-2-oxo-1-naphthylidenemethylamine. J. Molec. Struct. 2000, 524, 241–250.

Newman, J. W.; Hammock, B. D. Optimizing thiol derivatizing reagent for the mass spectral analysis of disubstituted epoxy fatty acids. J. Chromatography A 2001, 925, 223–240.

Nikolić, N. C.; Cakić, S. M.; Novaković, S. M.; Cvetković, M. D.; Stanković, M. Z. Effect of extraction techniques on yield and composition of soybean oil. Macedonian J. Chem. Chem. Eng. 2009, 28, 173–179.

Nordback, J.; Lundberg, E.; Christie, W. W. Separation of lipid classes from marine particulate material by HPLC on a polyvinyl alcohol-bonded stationary phase using dual-channel evaporative light-scattering detection. Marine Chem. 1998, 60, 165–175.

Nourooz-Zadeh, J.; Tajaddini-Sarmadi, J.; Wolff, S. P. Measurement of hydroperoxides in edible oils using the ferrous oxidation in xylenol orange assay. J. Agric. Food Chem. 1995, 43, 17–21.

Okitsu, K.; Suzuki, T.; Takenaka, N.; Bandow, H.; Nishimura, R.; Maeda, Y. Acoustic multibubble cavitation in water: A new aspect of the effect of a rare gas atmosphere on bubble temperature and its relevance to sonochemistry. J. Phys. Chem. B 2006, 110, 20081–20084.

Orengo, I. F.; Black, H. S.; Kettler, A. H.; Wolf, J. E., Jr. Influence of dietary menhaden oil upon carcinogenesis and various cutaneous responses to ultraviolet radiation. Photochem. Photobiol. 1989, 49, 71–77.

Orr, R. J.; Williams, H. L. Kinetics of the reactions between iron(II) and hydroperoxides based upon cumene and cyclohexane. J. Phys. Chem. 1953, 57, 925–931.

Osawa, C.; Gonçalves, L.; Ragazzi, S. The use of fast methodologies (kits) in evaluating deep-frying oils. J. Am. Oil Chem. Soc. 2007, 84, 893–897.

Osawa, C. C.; Goncalves, L. A. G.; Ragazzi, S.. Determination of hydroperoxides in oils and fats using kits. J. Sci. Food Agric. 2007, 87, 1659–1666.

Otto, M. Chemometrics Statistics and Computer Application in Analytical Chemistry; Wiley-VCH: Weinheim, DE, 1999.

Patterson, W. A. Infrared absorption bands characteristic of the oxirane ring. Anal. Chem. 1954, 26, 823–835.

Pawliszyn, J. Quantitative Aspects of SPME. Applications of Solid Phase Microextraction; Pawliszyn, J. Ed.; RSC Publishing: London, 1999; 3–21.

Peng, T. Evaluation of oxygen bomb methodology for investigating thermal vs oxidative reactions in frying oils. M.S. Thesis, Rutgers University, New Brunswick, NJ, 2010.

Pérez-Serradilla, J.; Ortiz, M.; Sarabia, L.; de Castro, M. Focused microwave-assisted Soxhlet extraction of acorn oil for determination of the fatty acid profile by GC–MS. Comparison with conventional and standard methods. Anal. Bioanal. Chem. 2007, 388, 451–462.

Peter, S. Supercritical Fractionation of Lipids. Supercritical Fluid Technology in Oil and Lipid Chemistry; King, J. W.; List, G. R., Eds.; AOCS Press: Champaign, IL, 1996; 65–100.

Petersen, A. D.; Kleeberg, K. K.; Jahreis, G.; Busch-Stockfish, M.; Fritsche, J. Comparison of analytical and sensory lipid oxidation parameters in conventional and high-oleic rapeseed oil. Eur. J. Lipid Sci. Technol. 2012, 114, 1193–1203.

Pierce, C. Y.; Barr, J. R.; Cody, R. B.; Massung, R. F.; Woolfitt, A. R.; Moura, H.; Thompson, H. A.; Fernandez, F. M. Ambient generation of fatty acid methyl ester ions from bacterial whole cells by direct analysis in real time (DART) mass spectrometry. Chem. Commun. 2007, 807–809.

Pierce Chemical Co.. Quantitative peroxide assay kit (lipid). http://www.piercenet.com/instructions/2160526.pdf (accessed October 1, 2012).

Pomerantsev, A. L.; Rodionova, O. Y. Chemometric view on "comprehensive chemometrics." Chemometr. Intell. Lab. 2010, 103, 19–24.

Ponne, C. T.; Moller, A. C.; Tijskens, L. M. M.; Bartels, P. V.; Meijer, M. M. T. Influence of microwave and steam heating on lipase activity and microstructure of rapeseed (Brassica napus). J. Agric. Food Chem. 1996, 44, 2818–2824.

Povolo, M.; Contarini, G. Comparison of solid-phase microextraction and purge-and-trap methods for the analysis of the volatile fraction of butter. J. Chromatogr. A 2003, 985, 117–125.

Pramanik, M.; Mendon, S. K.; Rawlins, J. W. Determination of epoxy equivalent weight of glycidyl ether based epoxides via near infrared spectroscopy. Polymer Testing 2012, 31, 716–721.

Price, G. J.; Duck, F. A.; Digby, M.; Holland, W.; Berryman, T. Measurement of radical production as a result of cavitation in medical ultrasound fields. Ultrason. Sonochem. 1997, 4, 165–171.

Przybylski, R.; Eskin, N. A. M. Methods to Measure Compounds and the Flavor Significance of Volatile Compounds. Methods to Assess Quality and Stability of Oils and Fat-Containing Foods; Warner, K.; Eskin, N. A. M., Eds.; AOCS Press: Champaign, IL, 1995; 107–133.

Qin, H. Thermal desorption studies of corn oil at frying temperatures: Thermal scission vs autoxidation. M.S. Thesis, Rutgers University, New Brunswick, NJ, 2011.

Raynie, D. E. Modern extraction techniques. Anal. Chem. 2006, 78, 3997–4004.

Redemann, C. E. Peroxides in isopropanol. J. Am. Chem. Soc. 1942, 64, 3049–3050.

Refsgaard, H. H. F.; Tsai, L.; Stadtman, E. R. Modifications of proteins by polyunsaturated fatty acid peroxidation products. Proc. Nat. Acad. Sci. USA 2000, 97, 611–616.

Reis, M. M.; Araujo, P. H. H.; Sayer, C.; Guidici, R. Spectroscopic on-line monitoring of reactions in dispersed medium: Chemometric challenges. Anal. Chim. Acta 2007, 595, 257–265.

Richards, M. P.; Feng, Y. Choosing the proper organic solvent for the determination of lipid hydroperoxides by the ferric thiocyanate assay. Anal. Biochem. 2000, 278, 232–235.

Riesz, P.; Kondo, T. Free radical formation induced by ultrasound and its biological implications. Free Radic. Biol. Med. 1992, 13, 247–270.

Riesz, P.; Kondo, T.; Krishna, C. M. Free radical formation by ultrasound in aqueous solutions. A spin trapping study. Free Radical Res. 1990, 10, 27–35.

Rockenbauer, A.; Tudos, F. Solvent effects of ESR spectra of neutral nitroxide free radicals: I. Localized solvent-radical adducts in mixtures. Adv. Mol. Relax Int. Pr. 1979, 15, 207–212.

Roggo, Y.; Chalus, P.; Maurer, L.; Lema-Martinez, C.; Edmond, A. L.; Jent, N. A review of near infrared spectroscopy and chemometrics in pharmaceutical technologies. J. Pharmaceut. Biomed. Anal. 2007, 44, 683–700.

Rohman, A.; Man, Y. B. C. The chemometrics approach applied to FTIR spectral data for the analysis of rice bran oil in extra virgin olive oil. Chemometr. Intell. Lab. 2012, 110, 129–134.

Rollie, M. E.; Patonay, G.; Warner, I. M. Deoxygenation of solutions and its analytical applications. Ind. Eng. Chem. Res. 1987, 26, 1–6.

Sahasrabudhe, M.; Smallbone, B. Comparative evaluation of solvent extraction methods for the determination of neutral and polar lipids in beef. J. Am. Oil Chem. Soc. 1983, 60, 801–805.

Sayre, L. M.; Arora, P. K.; Iyer, R. S.; Salomon, R. G. Pyrrole formation from 4-hydroxynonenal and primary amines. Chem. Res. Toxicol. 1993, 6, 19–22.

Schaich, K. M. Lipid Oxidation in Fats and Oils: An Integrated Biew. Bailey's Industrial Fats and Oils, 6th ed.; Shahidi, F., Ed.; John Wiley: New York, 2005; 2681–2767.

Schaich, K. M. Cooxidations of Oxidizing Lipids: Reactions with Proteins. Lipid Oxidation Pathways; Kamal-Eldin, A.; Min, D. B., Eds.; CRC Press, Boca Raton, FL, 2008; Vol. 2, 183–274.

Schaich, K. M.; Karel, M. Free radicals in lysozyme reacted with peroxidizing methyl linoleate. J. Food Sci. 1975, 40, 456–459.

Schaich, K. M.; Fisher, C.; King, R. T. Formation and Reactivity of Free Radicals in Curcuminoids: An EPR Study. Food Phytochemicals for Cancer Prevention II: Teas, Spices, and Gerbs; Ho, C. T.; Huang, M. T.; Rosen, R. T., Eds.; American Chemical Society: Washington, D.C., 1993; 204–221.

Schieberle, P.; Tsoukalas, B.; Grosch, W. Decomposition of linoleic acid hydroperoxides by radicals. Z. Lebensm. Unters. Forsch. 1979, 168, 448–456.

Sedman, J.; van de Voort, F. R.; Ismail, A. A. Application of Fourier Transform Infrared Spectroscopy in Edible-Oil Analysis. New Techniques and Applications in Lipid Analysis; McDonald, R. E.; Mossoba, M. M., Eds.; AOCS Press: Champaign, IL, 1997; 283–324.

Seppanen, C. M.; Csallany, A. S. Simultaneous determination of lipophilic aldehydes by high-performance liquid chromatography in vegetable oil. J. Am. Oil Chem. Soc. 2001, 78, 1253–1260.

Shahidi, F. Indicators for Evaluation of Lipid Oxidation and Off-Flavor Development in Food. Food Flavors: Formation, Analysis and Packaging Influences; Contis, E. T., Ho, C.-T.; Mussinan, C. J.; Parliment, T. H.; Shahidi, F.; Spanier, A. M., Eds.; Elsevier: Amsterdam, 1998; 55–68.

Shantha, N. C.; Decker, E. A. Rapid, iron-based spectrophotometric methods for determination of peroxide values of food lipids. J. AOAC Intl. 1994, 77, 421–424.

Shen, J.; Shao, X. A comparison of accelerated solvent extraction, Soxhlet extraction, and ultrasonic-assisted extraction for analysis of terpenoids and sterols in tobacco. Anal. Bioanal. Chem. 2005, 383, 1003–1008.

Shirono, K.; Morimatsu, T.; Takemura, F. Gas solubilities (CO_2, O_2, Ar, N_2, H_2, and He) in liquid chlorinated methanes. J. Chem. Eng. Data 2008, 53, 1867–1871.

Sigma-Aldrich. Solvent stabilizer systems. http://www.sigmaaldrich.com/chemistry/solvents/learning-center/stabilizer-systems.html (accessed August 8, 2012).

Singleton, J.; Stikeleather, L. A solvent extractor system for the rapid extraction of lipids and trace bioactive micronutrients in oilseeds. J. Am. Oil Chem. Soc. 1999, 76, 1461–1466.

Sjovall, O.; Kuksis, A.; Kallio, H. Reversed-phase high-performance liquid chromatographic separation of tert.-butyl hydroperoxide oxidation products of unsaturated triacylglycerols. J. Chromatogr. A 2001, 905, 119–132.

Somaglino, L.; Bouchoux, G.; Mestas, J.-L.; Lafon, C. Validation of an acoustic cavitation dose with hydroxyl radical production generated by inertial cavitation in pulsed mode: Application to *in vitro* drug release from liposomes. Ultrason. Sonochem. 2011, 18, 577–588.

Steltzer, E. Evaluation of chemical assays for determining hydroperoxides levels in oxidized lipids, M.S. Thesis, Rutgers University, New Brunswick, NJ, 2012.

Strange, D. E.; Schaich, K. M. Extraction of lipids from extruded corn-soy blends. J. Food Lipids 2000, 7, 217–224.

Stringer, M. D.; Goergoeg, P. G.; Freeman, A.; Kakkar, V. V. Lipid peroxides and atherosclerosis. BMJ 1989, 298, 281–284.

Suyama, K.; Adachi, A. Origin of alkyl-substituted pyridines in food flavor: Formation of the pyridines from the reaction of alkanals with amino acids. J. Agric. Food Chem. 1980, 28, 546–549.

Suyama, K.; Adachi, S. Reaction of alkanals and amino acids or primary amines. Synthesis of 1,2,3,5- and 1,3,4,5-substituted quarternary pyridinium salts. J. Org. Chem. 1979, 44, 1417–1420.

Suzuki, Y. J.; Forman, H. J.; Sevanian, A. Oxidants as stimulators of signal transduction. Free Radic. Biol. Med. 1997, 22, 269–285.

Talpur, M. Y.; Sherazi, S. T. H.; Mahesar, S. A.; Bhutto, A. A. A simplified UV spectrometric method for determination of peroxide value in thermally oxidized canola oil. Talanta 2010, 80, 1823–1826.

Tappel, A. L.; Tappel, A. A.; Fraga, C. G. Application of simulation modeling to lipid peroxidation processes. Free Radic. Biol. Med. 1989, 7, 361–368.

Temelli, F. Extraction of triglycerides and phospholipids from canola with supercritical carbon dioxide and ethanol. J. Food Sci. 1992, 57, 440–442, 457.

Toschi, T. G.; Bendini, A.; Ricci, A.; Lercker, G. Pressurized solvent extraction of total lipids in poultry meat. Food Chem. 2003, 83, 551–555.

Townshend, A.; Wheatley, R. A. Determination of carbonyl compounds by the oxidative chemiluminescence of 2,4-dinitrophenylhydrazine. Analyst 1998, 123, 1047–1051.

Turk, E. Phosgene from chloroform. Chem. Eng. News 1998, 76, 6.

Turner, C. Overview of Modern Extraction Techniques for Food and Agricultural Samples. Modern Extraction Techniques; American Chemical Society: Washington, D.C., 2006; 3–19.

Uchida, K.; Sakai, K.; Itakura, K.; Osawa, T.; Toyokuni, S. Protein moification by lipid peroxidation products: Formation of malondialdehyde-derived Ne-(2-propenal)lysine in proteins. Archiv. Biochem. Biophys. 1997, 346, 45-52.

Uchida, K.; Stadtman, E. R. Modification of histidine residues in proteins by reaction with 4-hydroxynonenal. Proc. Nat. Acad. Sci. USA 1992, 89, 4544–4548.

Vaclavik, L.; Cajka, T.; Hrbek, V.; Hajslova, J. Ambient mass spectrometry employing direct analysis in real time (DART) ion source for olive oil quality and authenticity assessment. Anal. Chim. Acta 2009, 645, 56–63.

Vaclavik, L.; Hrbek, V.; Cajka, T.; Rohlik, B.-A.; Pipek, P.; Hajslova, J. Authentication of animal fats using direct analysis in real time (DART) ionization-mass spectrometry and chemometric tools. J. Agric. Food Chem. 2011, 59, 5919–5926.

Van de Vort, F. R.; Ismael, A. A.; Sedman, J.; Dubois, J.; Nicodermo, T. The determination of peroxide value by Fourier transform infrared spectroscopy. J. Am. Oil Chem. Soc. 1994, 71, 921–926.

van de Waterbeemd, H.; Carter, R. E.; Grassy, G.; Kubinyi, H.; Martin, Y. C.; Tute, M. S.; Willett, P. Glossary of terms used in computational drug design (IUPAC Recommendations 1997). Pure Appl. Chem. 1997, 69, 1137–1152.

Van Duuren, B. L. Carcinogenicity of epoxides, lactones, and peroxy compounds. 3. Biological activity and chemical reactivity. J. Med. Chem. 1966, 9, 77–79.

Vernon, H. M. The solubility of air in fats, and its relation to Caisson disease. Proc. Royal Soc. London 1907, LXXIX, 366–371.

Vock, E. H.; Lutz, W. K.; Ilinskaya, O.; Vamvakas, S. Discrimination between genotoxicity and cytotoxicity for the induction of DNA double-strand breaks in cells treated with aldehydes and diepoxides. Mutat. Res. 1999, 441, 85–93.

von der Hude, W.; Carstensen, S.; Gürtler, R.; Obe, G. Structure-activity relationships of epoxides: Induction of sister-chromatid exchanges in V79 cells by enantiomeric epoxides. Mutat. Res.-Gen. Tox. 1992, 278, 289–297.

Wang, C.; Chen, L.; Rakesh, B.; Qin, Y.; Lv, R. Technologies for extracting lipids from oleaginous microorganisms for biodiesel production. Front. Energy 2012, 6, 266–274.

White, P. J. Conjugated Diene, Anisidine Value, and Carbonyl Value Analyses. Methods to Assess Quality and Stability of Oils and Fat-Containing Foods; Warner, K., Eskin, N. A. M., Eds.; AOCS Press: Champaign, IL, 1995; 159–178.

Wold, S. Chemometrics, why, what and where to next? J. Pharmaceut. Biomed. Anal. 1991, 9, 589–596.

Wold, S. Chemometrics; what do we mean with it, and what do we want from it? Chemometr. Intell. Lab. 1995, 30, 109–115.

Wold, S.; Sjostrom, M. Chemometrics, present and future success. Chemometr. Intell. Lab. 1998, 44, 3–14.

Wolff, S. P. Ferrous ion oxidation in presence of ferric ion indicator xylenol orange for measurement of hydroperoxides. Methods Enzymol. 1994, 233, 182–189.

Wong, J. W.; Yeo, H. C. H.; Shibamoto, T. Determination of malonaldehyde and formaldehyde formed from fatty acid ethyl esters upon microwave and thermal heating. J. Agric. Food Chem. 1991, 39, 2260–2262.

Wu, Z.; Ahmad, R.; Armstrong, D. A. Formation of lipoamide anion radicals by hydroxyl, formate and alcohol radicals at pH 6–9. Radiat. Phys. Chem. 1977, 23, 251–257.

Xie, Jia and Schaich, K. M. 2012. Looking beyond hydrogen abstraction in lipid oxidation: Evidence of alternate pathways in non-volatile products. American Oil Chemists' Society National Meeting, Long Beach, CA, May, 2012.

XploSafe. XPell CE indicating pellets for peroxides prevention in organic solvents. http://www.xplosafe.com/products/xpell-ce (accessed September 8, 2012).

Yamada, K.; Terao, J.; Matsushita, S. Electrochemical detection of phospholipid hydroperoxides in reverse-phase high performance liquid chromatography. Lipids 1987, 22, 125–128.

Yang, G. C. Detection of lipid hydroperoxides by high-performance liquid chromatography coupled with post-column reaction. Trends in Food Science Technol. 1992, 3, 15–18.

Yang, H.; Irudayaraj, J.; Paradkar, M. M. Discriminant analysis of edible oils and fats by FTIR, FT-NIR and FT-Raman spectroscopy. Food Chem. 2005, 93, 25–32.

Yao, L.; Schaich, K. M. Factors affecting yields and extract stability in accelerated solvent extraction of lipids from dry extruded products. Rutgers University, New Brunswick, NJ, manuscript in preparation, 2012a.

Yildiz, G.; Wehling, R.; Cuppett, S. Monitoring PV in corn and soybean oils by NIR spectroscopy. J. Am. Oil Chem. Soc. 2002, 79, 1085–1089.

Yildiz, G.; Wehling, R. L.; Cuppett, S. L. Comparison of four analytical methods for the determination of peroxide value in oxidized soybean oils. J. Am. Oil Chem. Soc. 2003, 80, 103–107.

Yildiz, M.; Kiliç; Hökelek, T. Intramolecular hydrogen bonding and tautomerism in Schiff bases. Part I. Structure of 1,8-di[N-2-oxyphenyl-salicylidene]-3,6-dioxaoctane. J. Molec. Struct. 1998, 441, 1–10.

Young, C. A.; Vogt, R. R.; Nieuwland, J. A. Colorimetric estimation of peroxides in unsaturated compounds. Ind. Eng. Chem. 1936, 8, 198–199.

Yu, A.; Crawford, E.; Tice, J.; Musselman, B.; Wu, J.-T. Bioanalysis without sample cleanup or chromatography: The evaluation and initial implementation of direct analysis in real time ionization mass spectrometry for the quantification of drugs in biological matrices. Anal. Chem. 2009, 81, 193-202.

Yu, X.; van de Voort, F. R.; Sedman, J. Determination of peroxide value of edible oils by FTIR spectroscopy with the use of the spectral reconstitution technique. Talanta 2007, 74, 241–246.

Zamora, R.; Hidalgo, F. J. Modification of lysine amino groups by the lipid peroxidation product 4,5(E)-epoxy-2(E)-heptenal. Lipids 1994, 29, 243–249.

Zamora, R.; Hidalgo, F. J. Linoleic acid oxidation in the presence of amino compounds produces pyrroles by carbonylamine reactions. Biochim. Biophys. Acta 1995, 1258, 319–327.

Zamora, R.; Hidalgo, F. J. 2-Alkylpyrrole formation from 4,5-epoxy-2-alkenals. Chem. Res. Toxicol. 2005, 18, 342–348.

Zamora, R.; Alaiz, M.; Hidalgo, F. J. Modification of histidine residues by 4,5-epoxy-2-alkenals. Chem. Res. Toxicol. 1999, 12, 654–660.

Zamora, R.; Alaiz, M.; Hidalgo, F. J. Contribution of pyrrole formation and polymerization to the nonenzymatic browning produced by amino-carbonyl reactions. J. Agric. Food Chem. 2000, 48, 3152–3158.

Zamora, R.; Gallardo, E.; Hidalgo, F. J. Amine degradation by 4,5-epoxy-2-decenal in model systems. J. Agric. Food Chem. 2006, 54, 2398–2404.

Zamora, R.; Gallardo, E.; Hidalgo, F. J. Strecker degradation of phenylalanine initiated by 2,4-decadienal or methyl 13-oxooctadeca-9,11-dienoate in model systems. J. Agric. Food Chem. 2007, 55, 1308–1314.

Zamora, R.; Gallardo, E.; Navarro, J. L.; Hidalgo, F. J. Strecker-type degradation of phenylalanine by methyl 9,10-epoxy-13-oxo-11-octadecenoate and methyl 12,13-epoxy-9-oxo-11-octadecenoate. J. Agric. Food Chem. 2005, 53, 4583–4588.

Zamora, R.; Navarro, J. L.; Hidalgo, F. J. Determination of lysine modification product, e-N-pyrrolylnorleucine in hydrolyzed proteins and trout muscle microsomes by micellar electrokinetic capillary chromatography. Lipids 1995, 30, 477–483.

Zamora, R.; Rios, J. J.; Hidalgo, F. J. Formation of volatile pyrrole products from epoxyalkenal/protein reactions. J. Sci. Food Agric. 1994, 66, 543–546.

Zhang, W.-H.; Liu, J.; Xu, G.; Yuan, Q.; Sayre, L. M. Model studies on protein side chain modification by 4-oxo-2-nonenal. Chem. Res. Toxicol. 2003, 16, 512–523.

CHAPTER

3

Oxidation in Different Food Matrices

How Physical Structure Impacts Lipid Oxidation in Oil-in-Water Emulsions and Bulk Oils

Bingcan Chen,[1] D. Julian McClements,[1] and Eric A. Decker[1]

Introduction of Lipid Oxidation in Colloidal Food Systems

Lipid oxidation is one of the major causes of quality deterioration in natural and processed foods. Oxidative deterioration is a large economic concern in the food industry because it affects many quality characteristics such as flavor (rancidity), color, texture, and the nutritive value of foods. In addition, it produces potentially toxic compounds (Halliwell et al., 1995; Frankel, 2005; Decker & McClements, 2008; Decker et al., 2010a, 2010b). Thus, lipid oxidation is one of the major processes that limit the shelf life of foods. The oxidative instability of polyunsaturated fatty acids often restricts their functional use as nutritionally beneficial lipids in food.

Lipid oxidation is a general term used to describe a complex sequence of chemical interactions between unsaturated fatty acyl groups in lipids with active oxygen species (Frankel, 2005; Decker & McClements, 2008; Min & Boff, 2008). The unsaturated fatty acids on triacylglycerols and phospholipids have low volatility and do not directly contribute to the aroma of foods. However, these fatty acids will decompose during lipid oxidation to form small, volatile molecules that produce the off-aromas associated with oxidative rancidity. These volatile compounds are detrimental to food quality, except in the case of food products such as

[1]Department of Food Science, University of Massachusetts, Amherst, MA 01003, USA.

fried foods, dried cereal, and cheeses, where small amounts of these volatile compounds are important for their flavor profile. The mechanisms of lipid oxidation in a particular food depend on the nature of the reactive species present and their physicochemical environment (Coupland & McClements, 1996a, 1996b; Erickson & Sista, 1997; Waraho et al., 2011). Thus, a thorough understanding of lipid oxidation mechanisms is important in developing practical methods for controlling lipid oxidation in foods.

Food manufacturers can use a variety of methods to increase the oxidative stability of their products. One such method is to reduce polyunsaturated fatty acid concentrations. This can be accomplished by replacing polyunsaturated fatty acids with fats high in saturated fatty acids (for example, tropical oils). However, this practice is contrary to current nutritional recommendations that advocate an increase in dietary unsaturated fatty acids for the purpose of decreasing the risk of coronary heart disease. Another method to decrease rancidity is to use partial hydrogenation of fats and oils to remove the most highly unsaturated fatty acids, which are very susceptible to oxidation (for example, linolenic acid). But, partial hydrogenation leads to the formation of *trans* fatty acids. Several studies have shown that *trans* fatty acids are more atherogenic than saturated fats because they increase low-density lipoprotein and decrease high-density lipoprotein, which is often referred to as "good" cholesterol (Duxbury, 2005). These findings prompted the U.S. Food and Drug Administration (FDA) to require food manufacturers to list *trans* fat content on nutrition facts labels as of January 1, 2006 (Duxbury, 2005). In order to list no *trans* fatty acids on the nutritional label of products such as fried foods, cookies, pastries, and crackers, food manufacturers will not be able to use partially hydrogenated fats in their product formulations.

Besides alteration in fatty acid content, only a limited number of approaches can be used to control lipid oxidation in foods. Exclusion of oxygen from products, while effective, is often not practical during processing and storage. Addition of antioxidants that scavenge free radicals and chelate prooxidative metals are the most common methods used to retard lipid oxidation (Decker et al., 2002; Frankel, 2005). Some of the most effective free radical scavengers and metal chelators are synthetic compounds, which are often perceived negatively by consumers. Therefore, the food industry will have to develop new antioxidant technologies to overcome the challenge of developing consumer-acceptable foods (that is, those with no *trans* fatty acids or synthetic food additives) with nutritionally significant amounts of unsaturated fatty acids.

Several antioxidant technologies have the potential to stabilize foods prone to oxidative rancidity. These technologies include controlling reactive oxygen species and prooxidant metals and reducing the damaging effects of free radicals via free-radical scavengers (Chaiyasit et al., 2005; Alamed et al., 2009; Elias et al., 2008; Sasaki et al., 2010). Unfortunately, very few new antioxidant ingredients have been introduced to the food industry over the past few decades. This means that the approved antioxidant ingredients available to solve oxidative rancidity problems in foods must be used more efficiently. One way to increase the efficiency of antioxidant ingredients is to deliver them to the site of the oxidative reactions or to use them in combination with other technologies that can reduce oxidation rates. However, the mechanisms of lipid oxidation reactions in the food must be thoroughly understood to be able to develop technologies that utilize antioxidant ingredients to their full potential.

Lipid oxidation mechanisms are different in bulk fats than in water-in-oil and oil-in-water emulsions (Frankel et al., 1994; Halliwell et al., 1995; Abdalla & Roozen, 1999; Naz et al., 2005; Waraho et al., 2011; Chen et al., 2011b). Understanding these different oxidation pathways will help to develop antioxidant technologies that work in emulsions. This chapter provides a brief background of the physical properties of food emulsions and the ways in which these properties impact lipid oxidation pathways.

Food Emulsions and Lipid Oxidation

Most lipids in foods exist as dispersion, either lipids dispersed in water (oil-in-water) or water dispersed in lipids (water-in-oil). Conventional emulsions consist of small spherical droplets with one immiscible phase embedded in the other phase (Dickinson & Stainsby, 1982; Dickinson, 1992) and with droplet diameters of the dispersed phase ranging from less than 100 nm to greater than 100 μm (McClements et al., 2007). Emulsions can be conveniently divided into three different regions: the continuous phase, the interfacial region, and the interior of the droplets (McClements & Decker, 2000). The physical stability of emulsions are highly dependent on the composition of these three phases. Emulsions are thermodynamically unfavorable systems that require stabilizers, such as emulsifiers or texture modifiers, to prevent gravitational separation, flocculation, coalescence, and Oswald ripening (Dickinson, 1992; Friberg et al., 2004; McClements, 2005). Conventional oil-in-water emulsions are often made by homogenizing the oil and a solution of a water-soluble emulsifier. Homogenizers can be high-shear mixers, high-pressure homogenizers, colloid mills, ultrasonic homogenizers, and membrane homogenizers. The desired droplet characteristics can be manipulated by alterations in homogenizer type, homogenizer operating conditions, and emulsifier type and concentration (Fig. 3.1). For example, the droplet size of oil-in-water

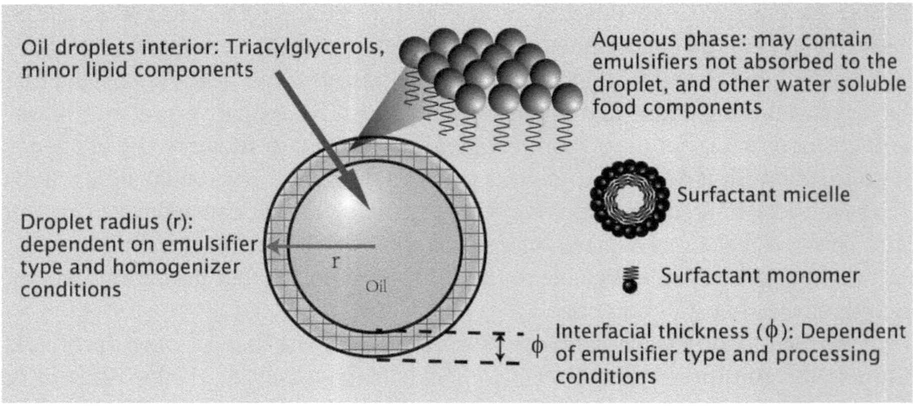

Fig. 3.1. The composition and characteristics of oil-in-water emulsions.

emulsions can be reduced by increasing the homogenization pressure or number of passes through a high-pressure homogenizer. The electrical charge on the droplets depends on the emulsifier type and pH; thus, it can be positive, neutral, or negative (Walstra, 1993, 2003; McClements, 2005). In addition to conventional emulsions, the food industry is interested in novel emulsions, such as nanoemulsions and multiple emulsions.

Nanoemulsions are conventional emulsions whose droplets range from 1 to 800 nm (Bouchemal et al., 2004; Sarker, 2005; Mason et al., 2006). The major advantage of nanoemulsions is that when their particle size is less than 50 nm, they do not scatter light strongly and thus appear clear. Nanoemulsions are metastable systems that can be designed to persist for many months or years if droplet aggregation and Ostwald ripening can be reduced. Nanoemulsions can be formed using high intensity mechanical energy to break up the disperse phase into smaller droplets. Only mechanical devices such as sonicators and high-pressure homogenizers (especially microfluidizers) can form nanoemulsions as they are capable of generating disruptive forces with high-enough intensity. Conversely, low-intensity methods can form nanoemulsions via the spontaneous phase separation of two immiscible liquids, leading to the formation of the dispersion of one liquid in the other liquid. These methods include the phase inversion temperature (PIT), emulsion inversion, and solvent displacement methods. The disadvantage of low-intensity methods is that they require organic solvents and synthetic surfactants or cosurfactants.

Multiple emulsions, such as water-in-oil-in-water emulsions, consist of small water droplets contained within larger oil droplets that are dispersed within an aqueous continuous phase (Garti, 1997; Garti & Bisperink, 1998; Garti & Beni-

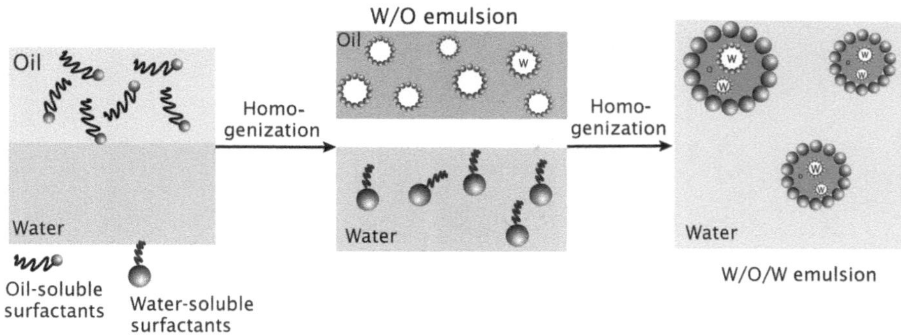

Fig. 3.2. Schematic representation of the formation of water-in-oil-in-water emulsions.

chou, 2004) (Fig. 3.2.). There are two different interfacial layers in this type of emulsion: the water-in-oil layer, surrounding the inner water droplets, and the oil-in-water layer, surrounding the entire oil droplets. Thus, two different types of emulsifier are normally used to stabilize water-in-oil-in-water emulsions: an oil-soluble emulsifier for the inner-water droplets and a water-soluble emulsifier for the oil droplets. Water-in-oil-in-water emulsions are typically created using a two-step procedure. First, a water-in-oil emulsion is produced by homogenizing water, oil, and an oil-soluble emulsifier. Then, a water-in-oil-in-water emulsion is created by homogenizing the water-in-oil emulsion with an aqueous solution that contains a water-soluble emulsifier. These emulsions can be produced using high-shear mixers, high-pressure homogenizers, colloid mills, ultrasonic homogenizers, and membrane homogenizers (McClements, 2005). The homogenization conditions necessary to produce the water-in-oil emulsion are usually more intense than those in the second stage. The size of the water droplets in the water-in-oil emulsion and in the final water-in-oil-in-water emulsion is influenced by varying emulsifier type, emulsifier concentration, and homogenization conditions. Multiple emulsions are not commonly used in foods because they are highly susceptible to destabilization during traditional food processing operations (such as mechanical shearing, thermal processing, chilling, freezing, and dehydration) (McClements et al., 2007). However, research continues in this area because multiple emulsions can have reduced caloric density and could partition hydrophilic labile components in the water phase within the emulsion droplet.

The physical properties of emulsion droplets can have a significant impact on lipid oxidation reactions (Fig. 3.3). Therefore, understanding the most important properties of emulsions droplets is useful for determining their susceptibility to lipid oxidation. The particle size distribution of an emulsion is usually represented

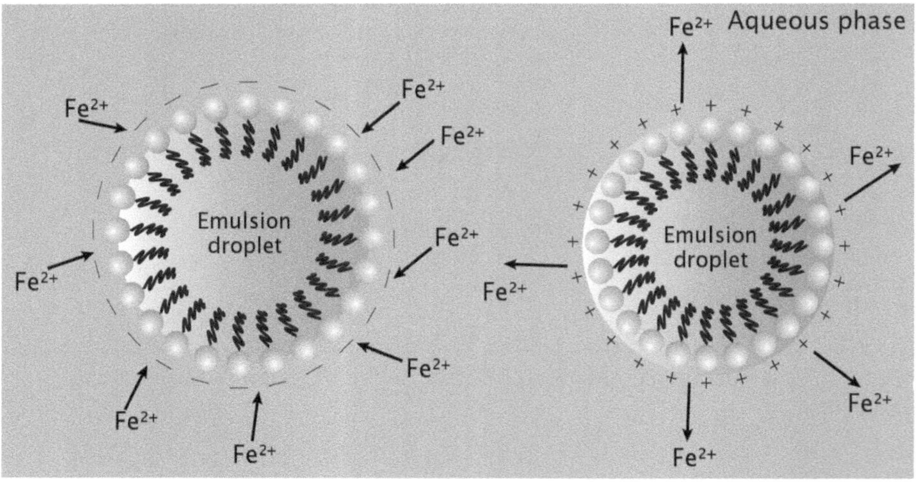

Fig. 3.3. The impact of electrical charge on the oxidative stability of oil-in-water emulsions.

as the particle concentration (that is, the volume or number of droplets present) versus droplet size (that is, its radius or diameter). To determine particle size, most researchers employ light scattering techniques. The electrical properties of the surface of an emulsion droplet (ζ-potential) are conveniently measured by electrophoretic light-scattering techniques (Hunter, 1986; McClements, 2005). The ζ-potential of a droplet depends on the emulsifier type as well as environmental conditions such as ionic strength and pH. Normally, the lipid or water in the emulsion droplet is liquid. However, in some cases the dispersed phase is either partially or fully solidified (Walstra, 2003; Muller & Keck, 2004; Wissing et al., 2004; McClements, 2005). The level of solidified fat in oil-in-water emulsions depends on oil type (that is, solid fat content versus temperature profile), thermal history (that is, temperature versus time profile), the presence of additives (for example, crystal structure modifiers), emulsifier type, and droplet size (Muller et al., 2000; Muller & Keck, 2004; Walstra, 2003).

Mechanisms of Lipid Oxidation in Emulsions

Oxidation chemistry in emulsified oils is different from that of bulk lipids for a number of reasons. First, the aqueous phase can contain prooxidants and antioxidants that are not normally found in bulk oils. Second, the presence of an oil–water interface can allow oil and water-soluble components to interact. Finally, the different phases of the emulsion (lipid phase, aqueous phase, and interfacial region) can allow for the partitioning of antioxidants, prooxidants, and oxidizable substrates in multiple phases (Frankel et al., 1994; McClements & Decker, 2000; Richards et al., 2002). Most oil-in-water emulsions are more prone to oxidation than bulk oils due to the large surface area of the interfacial region, which allows lipids and aqueous-phase components to interact (Chaiyasit et al., 2007).

Transition metals naturally occurring in food emulsions have been implicated as the most important prooxidant in oil-in-water. This is because both metals and lipid hydroperoxides are ubiquitous to food emulsions (Chaiyasit et al., 2007). Water-phase metals interact with surface active lipid hydroperoxides that concentrate at the emulsion–droplet surface (Yoshida & Niki, 1992; Mei et al., 1998a; Mei et al., 1998b; Mancuso et al., 2000; Nuchi et al., 2002; Dimakou et al., 2007). Lipid hydroperoxides react with both reduced and oxidized forms of transition metals to produce peroxyl (LOO•) and alkoxyl (LO•) radicals, respectively. These radicals can oxidize unsaturated lipids (LH) in the interior of the oil droplets. The high energy of the alkoxyl radical allows it to also promote β-scission reactions that decompose fatty acids into the volatile compounds that cause rancidity. The most important transition metal prooxidant is thought to be iron because iron-binding proteins, such as transferrin and lactoferrin, can strongly inhibit lipid oxidation oil-in-water emulsions (Huang et al., 1999; Mancuso et al., 1999). Ferrous (Fe^{2+}) ions have higher solubility and reactivity, making them more effective prooxidants than ferric (Fe^{3+}) ions (Halliwell & Gutteridge, 1990). Controlling the concentrations and/or reactivity of both metals and lipid hydroperoxides is an effective strategy to lipid oxidation rates (Mei et al., 1998a; Nuchi et al., 2002; Katsuda et al., 2008; Wang & Wang, 2008).

Chemical and Physical Properties of Emulsion Droplets That Impact Lipid Oxidation

The chemical and physical properties of the oil–water interface can impact lipid oxidation in oil-in-water emulsions primarily because they impact the ability of metals and lipid hydroperoxides to interact. The lipid droplet and the chemical composition of its interfacial layer are complex because they have numerous components, including emulsifiers, antioxidants, minor lipid components (for example, sterols, triacylglycerol hydrolysis products, and lipid oxidation products), biopolymers, and minerals, of which many are surface-active (McClements, 2005; Chaiyasit et al., 2007; Waraho et al., 2011). The thickness of the interfacial layer ranges from 1 to 40 nm, and emulsifier types can alter its chemical and physical properties. The composition and properties of the emulsion–droplet interface are so complex and dynamic that they can have a large impact on lipid oxidation and antioxidant kinetics.

Surface Area

The size of droplets and, therefore, the surface area of emulsions, varies from less than 0.2 μm in cream liqueurs and soft drinks to 100 μm in salad dressings and mayonnaise. The importance of interactions between metals and hydroperoxides at the interface of emulsion droplets suggests that emulsion droplet surface area is important to lipid oxidation. This proposition has not always been supported because some researchers have found an increase in lipid oxidation rates with increasing surface area (Gohtani et al., 1999; Kargar et al., 2011), while others have reported an increase in the rate of lipid oxidation when the surface area decreased (Hegenauer et al., 1979; Lethuaut et al., 2002; Nakaya et al., 2005; Imai et al., 2008; Lee et al., 2011). Still others found that lipid oxidation rates were independent of surface area (Shimada et al., 1996; Osborn & Akoh, 2004; Dimakou et al., 2007; Paraskevopoulou et al., 2007; Sun & Gunasekaran, 2009).

We do not completely understand why increased surface area does not consistently increase lipid oxidation rates. A proposed mechanism suggests that increases in interfacial area could alter the partitioning of reactants, prooxidants, and antioxidants between the oil, water, and interfacial regions, which could either promote or retard oxidation. Another suggested mechanism is that more mechanical energy required to produce small emulsion droplets might promote oxidation (Nakaya et al., 2005). Overall, the research suggests that emulsion droplet size, and thus interfacial surface area, is not a major factor in lipid oxidation in oil-in-water emulsions. This could be due to the extremely large surface area of emulsions, meaning that surface area never limits reaction rates. In addition, the particle size in most emulsions is the average size of both very small and large droplets. If the small droplets were always the first droplet to oxidize, then small changes in droplet size distribution might not have much effect on oxidation rates.

Emulsion Droplet Charge

The electrical charge of emulsion droplets (Fig. 3.3) has a strong influence on the oxidative stability of oil-in-water emulsions (Mei et al., 1998a; Mancuso et al., 2000; Silvestre et al., 2000; Boon et al., 2009). This is because emulsion droplet interfacial charge will either attract or repulse prooxidant metals (Haahr & Jacobsen, 2008). Droplet charge depends on the type of emulsifiers (that is, cationic, anionic, or neutral), and it can be altered by electrostatic layer-by-layer (LbL) deposition methods, in which charged biopolymers are absorbed onto oppositely charged droplets (Klinkesorn et al., 2005a, 2005b; Djordjevic et al., 2007; Shaw et al., 2007). Several studies have shown that anionic emulsion droplets produced by sodium dodecyl sulfate (SDS) promote lipid oxidation by attracting cationic iron to the droplet surface, while cationic droplets produced by dodecyl trimethyl ammonium bromide (DTAB) retard lipid oxidation by repelling these transition metals away from the surface (Mei et al., 1998a; Mancuso et al., 2000; Silvestre et al., 2000).

Proteins can also alter emulsion droplet surface charge and lipid oxidation rates. Protein-coated emulsion droplets are anionic at pH values greater than their pI and cationic at pH values less than their pI (Djordjevic et al., 2008; Boon et al., 2009). The oxidation of cationic protein–stabilized emulsion droplets at pH 3.0 is less than that of anionic protein–stabilized emulsion droplets at pH 7.0. Oxidative stability of protein-stabilized emulsion varied with protein type, with oxidative stability being in the order sodium caseinate > whey protein isolate > soy protein isolate at pH 3.0 (Hu et al., 2003). The level of cationic charge did not correlate with the oxidative stability of the emulsion droplet, suggesting that droplet interfacial thickness and/or the antioxidant properties of the protein could also be involved in the inhibition of lipid oxidation at pH 3.0. Protein type impacts the oxidative stability of anionic protein–stabilized emulsion droplets as well. Villiere and his coworkers reported that sodium caseinate–stabilized emulsion droplets were more oxidatively stable than a bovine serum albumin–stabilized emulsion at pH 6.5, even though the casein-stabilized emulsions had a higher negative charge (Villiere et al., 2005). When EDTA was added to the emulsion, the oxidation rate was lower in the emulsions stabilized by casein compared to those stabilized by bovine serum albumin, which was attributed to the ability of casein to scavenge free radicals more effectively than bovine serum albumin.

Droplet charge also influences antioxidant activity by impacting antioxidant location and activity via attractive/repulsive electrostatic interactions (Mei et al., 1999). Such an effect could take place if charged antioxidants can concentrate at the surface of an oppositely charged lipid dispersion via electrostatic attraction. Increased antioxidant activity was observed for anionic ascorbic acid with positively charged lipid micelles (Pryor et al., 1993), negatively charged Trolox C with positively charged phospholipids (Barclay & Vinqvist, 1994), and positively charged spermine with negatively charged phospholipids (Kogure et al., 1993). In oil-in-water emulsions, Mei and coworkers (1999) found that the activity of gallic acid decreased in anionic SDS–stabilized emulsion droplets compared to nonionic Brij–stabilized emulsion droplets, presumably due to the electrostatic repulsion of the anionic antioxidants away from the anionic droplets.

Overall, manipulating the charge of emulsion droplets is an effective way to impact lipid oxidation in oil-in-water emulsions. If technologies can be found to make physically stable emulsions that are neutral or positive, then attraction of the cationic prooxidant metals would be less and lipid oxidation reactions could be reduced. Unfortunately, it is very difficult to produce physically stable cationic emulsion droplets at neutral pH values.

Interfacial Thickness

Emulsifiers with different molecular dimensions (for example, their molecular weights, conformations, head-group sizes, or tail-group sizes) can be used to change the interfacial thickness of oil-in-water emulsions. In addition, LbL (Fig. 3.4) deposition methods can increase interfacial thickness by depositing one or more biopolymer layers around emulsion droplets (Klinkesorn et al., 2005a; 2005b; Djordjevic et al., 2007; Shaw et al., 2007). Emulsions with thicker interfacial coatings around droplets can alter lipid oxidation rates by decreasing interactions between lipids and aqueous-phase prooxidants (Silvestre et al., 2000). This concept appeared in experiments where the surfactant head-group size was altered in salmon oil-in-water emulsions using Brij 76 and Brij 700, with Brij 700 having 10 times more polyoxyethylene groups than Brij 76 and therefore creating a thicker interface (Silvestre et al., 2000). Emulsion droplets with thick interfacial layers had slower Fe^{2+}-promoted decomposition of cumene hydroperoxides and slower lipid oxidation rates (Silvestre et al., 2000). Surfactant tail-group size can also alter lipid oxidation rates because emulsion stabilized with Brij-lauryl (12-carbon tail group) oxidizes faster than emulsions stabilized with Brij-stearyl (18-carbon tail group) (Chaiyasit et al., 2000). Surfactant tail-group size played a more minor role in lipid oxidation in oil-in-water emulsions than polar head-group size.

Interfacial Permeability

Lipid oxidation in emulsion droplets could depend on the ability of different aqueous phase or interfacial components to be able to diffuse into the emulsion droplet's core (examples include oxygen, free radicals, and prooxidants). Surfactants can have different packing properties at the oil–water interface, which could im-

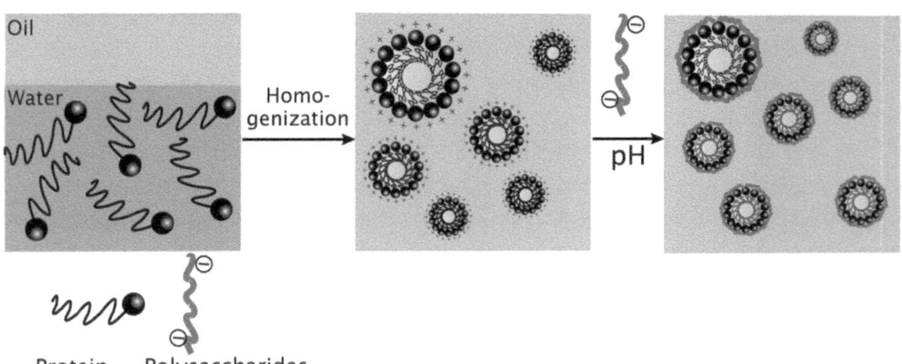

Fig. 3.4. Schematic representation of the formation of layer-by-layer (LbL) emulsions.

pact the diffusion rates and, consequently, oxidation kinetics (Villiere et al., 2005). Kellerby and coworkers examined whether cross-linking casein on the interface of menhaden oil-in-water emulsions could decrease lipid oxidation rates (Kellerby et al., 2006). In their study, transglutaminase treatment of the casein-stabilized emulsion cross-linked the protein at the emulsion–droplet interface so that it could not be displaced by Tween 20. However, the emulsions with cross-linked casein did not show increased oxidative stability when compared to untreated emulsions. Similarly, oil-in-water emulsions stabilized by β-lactoglobulin being crossed by heat-induced disulfide bonds also did not impact the ability of iron to decompose lipid hydroperoxides (Kellerby et al., 2006). The cross-linked proteins at the surface of an emulsion droplet that do not have increased oxidative stability suggests that the cross-linked protein interface is still highly porous, thus allowing aqueous-phase components to react with lipid hydroperoxides.

Interfacial Chemical Composition

Some of the emulsifiers used to stabilize emulsions could have chemical compositions that could alter lipid oxidation reactions by scavenging free radicals, chelating transition metals, interfering with hydroperoxide-transition metal interactions, and/or binding volatile lipid oxidation products (Haahr & Jacobsen, 2008). Rampon and coworkers found that during lipid oxidation, interfacial proteins could form adducts with lipid oxidation products at the emulsion–droplet interface (Rampon et al., 2001). Headspace propanal concentrations have been reported to be lower in protein-stabilized oil-in-water emulsions, again suggesting that lipid-oxidation products could be bound by proteins at the emulsion–droplet interface (Shen et al., 2007). Finally, casein isolated from the interface of oxidized soybean oil-in-water emulsions has been reported to have an increased molecular weight, presumably due to the formation of casein-lipid oxidation adducts (Leaver et al., 1999). These reactions could impact food quality because the formation of lipid-protein adducts will decrease the volatility of oxidation products, which could decrease rancidity (Zhou & Decker, 1999). Although it might be possible to control lipid oxidation by placing chemically reactive emulsifiers at the emulsion droplet surface, very little research has been done in this area, with the exception of studies on surface-active antioxidant compounds, which are discussed in more detail below.

The effectiveness of free radical scavenging antioxidants is dependent on properties such as activation energy, rate constants, and oxidation-reduction potential (Decker & McClements, 2008). However, antioxidative effectiveness of antioxidants is also dramatically dependent on their physical location within an emulsion (Frankel, 2005). The antioxidant polar paradox (Porter et al., 1989; Porter, 1993; Huang et al., 1996; Frankel et al., 1996a, 1996b) has been shown in oil-in-water emulsions by numerous investigators (Frankel et al., 1994; Huang et al., 1996; Frankel et al., 1996a, 1996b; Hopia et al., 1996; Frankel et al., 1997; Huang & Frankel, 1997; Abdalla & Roozen, 1999; Frankel, 1999). This concept states that polar antioxidants are more effective in bulk oil than in oil-in-water emulsions and that nonpolar antioxidants are more effective in oil-in-water emulsions than in bulk oil. Recently, it has been challenged in oil-in-water emulsions using phenolics esterified to fatty acids (Yuji et al., 2007; Laguerre et al., 2009; Laguerre et al., 2010). As an example, the antioxidant activity of rosmarinic acid esterified with butyl (C4) and octyl (C8) alkyl chains was greater than unesterified rosmarinic acid, suggesting that increasing the lipid solubility of rosmarinic acid enhanced its antioxidant activity in oil-in-water emulsions (Panya et al., 2012). Similar results have been observed in liposomes (Panya et al., 2010). However, when rosmarinic acid was esterified with longer-chain fatty acids (C18 and C20), the corresponding antioxidant activity decreased dramatically, even though the antioxidant esterified with C18 and 20 was more nonpolar than with C4 and 8. This phenomenon likely occurs because the highly nonpolar antioxidants partition away from the emulsion droplet surface, where lipid oxidation reactions are most prevalent.

Lipid and Aqueous Phase Components That Impact Lipid Oxidation in Oil-in-Water Emulsions

The chemical and physical properties of the emulsion–droplet interface are very important for lipid oxidation reactions. However, numerous components originate in the emulsion-droplet core, and the continuous phase can also impact lipid oxidation reactions.

Minor Oil Components

Commercial refining of oils utilizes neutralization to remove free fatty acids. Despite this refining step, 0.05–0.70% of free fatty acids remain in the oil (Chen et al., 2011b). Free fatty acids are known to accelerate lipid oxidation in bulk oils (Yoshida, 1993). Recently, Waraho and coworkers (2009) reported that oleic acid, but not methyl oleate (0–5.0% of oil), increased the negative charge of oil-in-water emulsion droplets, along with the formation of lipid hydroperoxides and headspace hexanal. EDTA strongly inhibited free fatty acid–promoted lipid oxidation. Overall, these data suggest that free fatty acids increase the negative charge of emulsion droplets, promoting the attraction of transition metals and thus increasing lipid oxidation rates.

In bulk oils, phospholipids have been reported to both promote and inhibit lipid oxidation. Their prooxidant activity has been attributed to anionic physical structures that can attract prooxidant metals, and their antioxidant activity has been attributed to their ability to chelate metals, decompose hydroperoxides, increase the partitioning of other antioxidants at the oil–water interface, and form complexes with lipid oxidation products that reduce the volatility of the compounds that cause rancidity. When phospholipids are at the interface of the emulsion droplet, they produce an anionic interface and generally accelerate lipid oxidation (Klinkesorn et al., 2005a, 2005b; Shaw et al., 2007). However, when phosphatidylcholine (0.3% of lipid) was added to an emulsion stabilized by Tween 20, the emulsion's ζ potential did not change, suggesting that the higher surface activity of the Tween 20 prevented the anionic phospholipid from accumulating at the droplet's surface. In this emulsion, phosphatidylcholine inhibited lipid oxidation as determined by headspace hexanal (Cardenia et al., 2011).

Continuous-Phase Proteins

In an attempt to replace synthetic food additives, proteins have been considered a natural food antioxidant that is generally recognized as safe (GRAS). When proteins are used to make emulsions, a portion of the protein adheres to the lipid droplet's surface until the droplet becomes saturated, and then the excess protein partitions into the continuous phase (Faraji et al., 2004). Proteins in the continuous phase (such as whey protein isolate, sweet whey, casein, β-lactoglobulin, lactoferrin, and soy protein isolate) can inhibit lipid oxidation in oil-in-water emulsions using several different mechanisms (Faraji et al., 2004; Elias et al., 2006; Elias et al., 2008). For example, the amino acids in continuous phase proteins can act as free radical scavengers. Examples include tyrosine, cysteine, tryptophan (Ostdal et al., 1996), and metal chelators when the pH is greater than the pI of the protein and thus the protein is negatively charged (Shaw et al., 2007).

Faraji and coworkers (2004) reported that nonadsorbed proteins have good antioxidant activity at pH 7 but not at pH 3, supposedly due to the negative charge of the proteins at pH 7.0, which can bind cationic transition metals. Metal chelation by proteins in oil-in-water emulsions can inhibit lipid oxidation by limiting the access of metal to the lipid or decreasing metal reactivity (Elias et al., 2008). The chelating activity of native proteins can be increased by changing their confirmation using thermal processing or enzymatic hydrolysis via the exposure antioxidant amino-acid residues that are usually buried in the protein's interior (Elias et al., 2006; Elias et al., 2008).

The antioxidant activity of various milk fractions can be increased by heating, which exposes free sulfhydryl groups (Taylor & Richardson, 1980). Tong and coworkers later found that continuous-phase, whey-protein fractions (WPF) in salmon oil-in-water emulsions could inhibit lipid oxidation via their cysteine and tyrosine residues. Surface-exposed antioxidant amino acids were found to be preferentially oxidized prior to the emulsified lipids, thereby retarding the rate of lipid oxidation (Tong et al., 2000). When whey proteins were subjected to limited enzymatic hydrolysis, which increased the exposure of antioxidant amino-acid residues, their free-radical scavenging also increased (Elias et al., 2006).

Polysaccharides

Polysaccharides are used as food ingredients in oil-in-water emulsion due to their ability to promote thickening, gelation, and stabilization (Matsumura et al., 2003; Kishk & Al-Sayed, 2007). Some polysaccharides have multiple benefits because they can both physically stabilize emulsions and inhibit lipid oxidation through their ability to scavenge free radicals, bind transition metal, and enhance viscosity (Paraskevopoulou et al., 2007; Chen et al., 2010b). Conversely, some polysaccharide ingredients may contain high levels of transition metal impurities and increase lipid oxidation rates (Katsuda et al., 2008).

Polysaccharides may inhibit lipid oxidation in oil-in-water emulsions because of their ability to increase continuous phase viscosity, which could lower the diffusion rate of oxygen and prooxidant metals and decrease the probability of collision with oil droplets (Shimada et al., 1996). However, most biopolymer solutions have extremely high macroscopic viscosity, but the molecular diffusion of small molecules is not restricted because the microscopic viscosity is not increased in the large space between the polysaccharides chains. For example, pullulan and maltodextrin can greatly increase the viscosity of oil-in-water emulsions but do not greatly retard the lipid oxidation rate, while glycoproteins, gum arabic, and soluble soybean polysaccharides greatly increase viscosity and have good antioxidant activity (Matsumura et al., 2003). Therefore, most polysaccharides that inhibit lipid oxidation do so via interactions with free radicals and metals and not via viscosity enhancement. For example, soluble soybean polysaccharides and gum arabic inhibit lipid oxidation by scavenging free radicals with their antioxidant amino acids on proteins associated with the polysaccharides (Matsumura et al., 2003). Tragacanth gum has been reported to act as a radical chain-breaker due to its ability to donate hydrogen atoms and scavenge free radicals, while xanthan gum can inhibit lipid oxidation via its ability to chelate Fe^{2+} ions with its anionic pyruvate groups (Shimada et al., 1992).

Polysaccharides are commonly used to increase the physical stability of oil-in-water emulsions by increasing viscosity and inhibiting creaming. However, they can also destabilize emulsions by causing depletion flocculation. If polysaccharides are going to be used as antioxidants, this must occur at concentrations that do not promote depletion flocculation. Chen and coworkers used emulsion physical stability as a selection criteria for the optimum concentration of four polysaccachrides in continuous phase, low-methoxyl pectin, high-methoxyl pectin, α-carrageenan sodium alginate (0.1 wt%) that could be used to inhibit lipid oxidation in polyoxyethylene (23) lauryl ether (Brij 35)–stabilized oil-in-water emulsions at neutral pH (Chen et al., 2010b). Low-methoxyl pectin was the most effective polysaccharide at inhibiting lipid oxidation, presumably due to its higher ferrous binding capacity, as none of the polysaccharides were effective free-radical scavengers.

Some polysaccharides may be located at the oil–water interface rather than in the continuous aqueous phase of oil-in-water emulsions if they are surface active. For example, gum arabic, modified starch, and propylene glycol alginate are widely used by the food and beverage industries as functional ingredients to emulsify oils (Paraskevopoulou et al., 2007). Unfortunately, most of the polysaccharide emulsifiers produce anionic emulsion droplets and therefore can increase oxidation rates. Sugars such as pentose, hexose, and reducing disaccharides also have been shown to be prooxidants (Mabrouk & Dugan, 1961; Mabrouk, 1964). This could be because sugars can reduce Fe^{3+} to Fe^{2+}, which is a stronger prooxidant, thus accelerating lipid oxidation (Yamauchi et al., 1984). On the other hand, the sugar alcohols have been shown to exhibit antioxidant activity in safflower oil emulsions (Sims et al., 1979).

Surfactant Micelles

When small molecule surfactants are used to stabilize emulsions, they adsorb onto the emulsion droplet's surface until it is saturated. Then, the remaining surfactant partitions into the continuous phase in a manner similar to that of proteins. The surfactant in the continuous phase will form micelles if surfactant concentration exceeds their critical micelle concentration (CMC). These surfactant micelles can solubilize small molecules and form comicelles with other surface-active compounds, thus altering the distribution of emulsion components between the oil, water, and interfacial regions (Nuchi et al., 2002; Richards et al., 2002). In many cases surfactant micelles inhibit lipid oxidation in oil-in-water emulsions (Richards et al., 2002). Nuchi and coworkers showed that lipid hydroperoxides will partition out of emulsion droplets in the presence of surfactant micelles and thus could inhibit lipid oxidation by preventing the free radicals formed by decomposing hydroperoxides from attacking unsaturated lipids in the droplet's core (Nuchi et al., 2002). Cho and coworkers (2002) found that surfactant micelles could partition iron out of the oil droplets, which could inhibit lipid oxidation. Richards and coworkers showed that Brij micelles in the continuous phase could solubilize antioxidants when the antioxidants moved from emulsion droplet to the continuous phase (Richards et al., 2002). With some antioxidants, this can increase their ability to inhibit lipid oxidation because it favors the partitioning of highly hydrophobic antioxidants at the emulsion–droplet interface (Panya et al., 2012).

Photosensitizers

Lipid oxidation is not a spontaneous reaction because atmospheric oxygen cannot thermodynamically react directly with double bonds due to the spin state restrictions. Ground state atmospheric oxygen is in a triplet state, but when it is converted to a singlet state, it can react directly with double bonds. To promote the formation of singlet oxygen, one can employ photosensitizers, such as riboflavin, hemoglobin, and chlorophyll. Riboflavin has been shown to promote lipid oxidation in oil-in-water emulsions in the presence of light. EDTA, sodium azide, and superoxide dismutase all inhibit riboflavin-promoted lipid oxidation in oil-in-water emulsions, implying that metals, singlet oxygen, and superoxide anion, respectively, are prooxidants. The mechanisms of photosensitized riboflavin-promoted oxidation were proposed to involve the formation of lipid hydroperoxides by singlet oxygen, the enhancement of the prooxidant activity of metals by superoxide anion-promoted metal reduction, and the decomposition of hydroperoxides by reduced metal ions (Lee et al., 2011).

Association Colloids and Lipid Oxidation

Bulk oil traditionally has been considered a homogeneous liquid in which physical structures do not impact chemical reactions. However, edible oils are composed of triacylglycerols and contain polar lipids, such as mono- and diacylglycerols, phospholipids, sterols, tocopherols, and free fatty acids that are not completely removed during refining. In addition to these minor lipid products, bulk oils also contain small amounts of water. Since many of the minor lipid components are surface active—which means that they have an affinity for both nonpolar and polar environments—they tend to form physical structures such as reverse micelles and lamellar structures. These physical structures are known as association colloids. Researchers are just beginning to try to understand how association colloids impact lipid oxidation in bulk oils. Understanding this chemistry could lead to the development of new antioxidative techniques to control rancidity.

Koga and Terao published some of the first work showing that physical structures in bulk oils had an impact on lipid oxidation. Their work showed that phospholipids increased the antioxidant activity of α-tocopherol in stripped oil containing a small amount of water (1%) (Koga & Terao, 1994, 1995). They also reported that phospholipids increased the oxidation of α-tocopherol in the presence of the water-soluble peroxyl radicals, suggesting that the phospholipids increased the concentration of α-tocopherol at the water–oil interface of association colloids. This increase in α-tocopherol degradation rates could also be due to a decrease in the reduction potential of α-tocopherol via partitioning into a more polar environment. This would make tocopherol a more efficient free-radical scavenger when it is at the oil–water interface (Laranjinha, 2001). Koga and Terao (1994) also found that α-tocopherol degradation rates by water-soluble peroxyl radicals decreased as the phospholipid's fatty acid size decreased; thus, the phospholipid lost the ability to form association colloids. Finally, Koga and Terao also found that the activity of α-tocopherol increased when it was conjugated to the polar head group of phosphatidylcholine, presumably due to the increased partitioning of α-tocopherol at the water–oil interface of association colloids.

More recent research by Chaiyasit and coworkers evaluated the impact of cumene hydroperoxides, oleic acid, and phosphatidylcholine on the structure of reverse micelles formed by sodium bis(2-ethylhexyl) sulfosuccinate (AOT). In this study, hydroperoxides and oleic acid increased lipid oxidation rates, while phosphatidylcholine altered reverse micelle size and decreased lipid oxidation rates (Chaiyasit et al., 2007, 2008). Researchers in Russia found that the decomposition of cumene hydroperoxides was accelerated by reverse micelles formed by cationic surfactants in organic media (Trunova et al., 2007; Kasaikina et al., 1999, 2006, 2007, 2008, 2010). This research group also reported that the oxidation of sunflower oil increased in the presence of fatty alcohols such as 1-tetradecanol, 1-octadecanol, and the MAG, 1-monopalmitoylglycerol. They suggested that this increase was due to the ability of these compounds to produce reverse micelles. These reverse micelles could concentrate polar species such as hydroperoxide and peroxyl radicals at the oil–water interface, where they could increase the rate of hydroperoxide decomposition into free radicals that could promote lipid oxidation (Kortenska et al., 2002).

The ability of phosphatidylcholine to form association colloids in stripped oil was determined by measuring the formation of reverse micelles by small-angle x-ray diffraction (Chen et al., 2010a). Phosphatidylcholine with oleic acid (DOPC) was found to form reverse micelles with a critical micelle concentration of approximately 950 μM. However, when phosphatidylcholine with butyric acid (DC_4PC) was added to the stripped oil, no physical structures could be observed. DOPC accelerated lipid oxidation, while DC_4PC had no effect. This suggested that the physical structures formed by DOPC were responsible for increasing lipid oxidation rates because both DOPC and DC_4PC had the same choline head group; thus, the polar head group could not be responsible for the observed difference in lipid oxidation rates.

Phospholipid reverse micelles also can impact the antioxidant mechanisms of free-radical scavengers (Chen et al., 2010a, 2011a). The activity of the

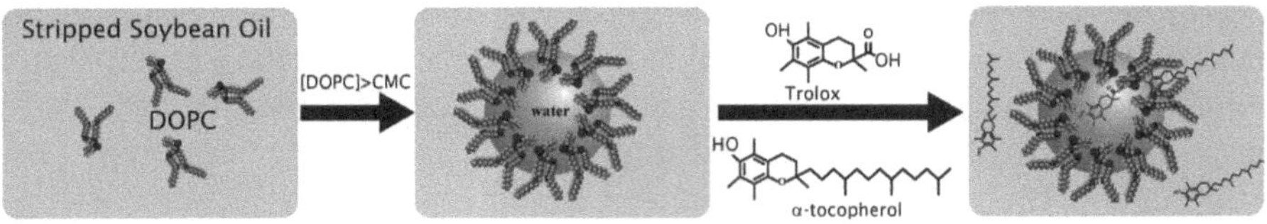

Fig. 3.5. Schematic representation of the location of Trolox and α-tocopherol in stripped soybean oil (SSO) containing 1,2-dioleoyl-sn-glycero-3-phosphocholine (DOPC) reverse micelles.

nonpolar free-radical scavenger α-tocopherol and the polar free-radical scavenger Trolox was compared in the presence and absence of DOPC-reverse micelles. DOPC-reverse micelles increased the activity of both α-tocopherol and Trolox at low concentrations (10 μM), but when antioxidant concentrations increased to 100 μM, their effectiveness decreased. Trolox was more effective than α-tocopherol at inhibiting lipid oxidation; using fluorescence probes, it was determined that Trolox was highly concentrated in the reverse micelles, a factor that could explain its superior antioxidant activity (Fig 3.5).

It is still unclear why the reverse micelles could both increase and decrease antioxidant activity. But the latest results from our group show that ferric (Fe^{3+}) ions can deplete α-tocopherol and Trolox in oil. This reaction in oil with association colloids was not the same as has been observed in oil-in-water emulsions. In oil-in-water emulsions, tocopherol and tocopherol radicals can reduce metal ions to lower-state valences by electron donation (Eqs. 3.1 and 3.2). Our results showed that neither α-tocopheroxyl or Trolox radicals were formed in medium-chain triacylglycerols (MCT) when incubated with either 100 or 600 μM ferric, as determined by EPR and HPLC (Chen et al., 2012). Instead, an unidentified novel product bearing less polarity than α-tocopherol was observed by HPLC upon its reaction with ferric ions. In addition, the reaction rate between ferric ions and antioxidants in MCT was faster when the concentration of antioxidants was less than 60 μM. Thus, this unreported reaction between ferric and α-tocopherol may contribute to the varying role of reverse micelles on increasing and decreasing the activity of antioxidants. Ferrous (Fe^{2+}), on the other hand, did not react with α-tocopherol and Trolox, but it will decompose lipid hydroperoxides and generate alkoxyl free radicals, which further accelerate the propagation step of oil oxidation (Eq. 3.3).

$$Fe^{3+} + AH \rightarrow Fe^{2+} + A\bullet + H^{+} \quad \text{(Eq. 3.1)}$$

$$A\bullet + Fe3+ \rightarrow A+ + Fe^{2+} \quad \text{(Eq. 3.2)}$$

$$Fe^{2+} + LOOH \rightarrow Fe^{3+} + LO\bullet + OH \quad \text{(Eq. 3.3)}$$

The existence of DOPC (1,2-dioleoyl-sn-glycero-3-phosphocholine) reverse micelle decreased the rates of antioxidant consumption by ferric ions (Chen et al., 2012). The same concentration of DC_4PC (1,2-dibutyryl-sn-glycero-3-phosphocholine) that had the same choline group as DOPC but did not form reverse micelles did not decrease ferric ion–promoted tocopherol decomposition. Our findings suggested that the nanostructures formed by phospholipids (that is, DOPC) might be the main factor in determining reactions between antioxidants and iron in bulk oil.

References

Abdalla, A. E.; Roozen, J. P. Effect of plant extracts on the oxidative stability of sunflower oil and emulsion. Food Chem. 1999, 64 (3), 323–329.

Alamed, J.; Chaiyasit, W.; McClements, D. J.; Decker, E. A. Relationships between free radical scavenging and antioxidant activity in foods. J. Agric. Food Chem. 2009, 57 (7), 2969–2976.

Barclay, L. R. C.; Vinqvist, M. R. Membrane peroxidation: Inhibiting effects of water-soluble antioxidants on phospholipids of different charge types. Free Radic. Biol. Med. 1994, 16 (6), 779–788.

Boon, C. S.; McClements, D. J.; Weiss, J.; Decker, E. A. Role of iron and hydroperoxides in the degradation of lycopene in oil-in-water emulsions. J. Agric. Food Chem. 2009, 57 (7), 2993–2998.

Bouchemal, K.; Briançon, S.; Perrier, E.; Fessi, H. Nano-emulsion formulation using spontaneous emulsification: Solvent, oil and surfactant optimisation. International Journal of Pharmaceutics 2004, 280 (1), 241–251.

Cardenia, V.; Waraho, T.; Rodriguez-Estrada, M. T.; McClements, D. J.; Decker, E. A. Antioxidant and prooxidant activity behavior of phospholipids in stripped soybean oil-in-water emulsions. J. Am. Oil Chem. Soc. 2011, 88 (9), 1409–1416.

Chaiyasit, W.; Elias, R. J.; McClements, D. J.; Decker, E. A. Role of physical structures in bulk oils on lipid oxidation. Crit. Rev. Food Sci. Nutr. 2007, 47 (3), 299–317.

Chaiyasit, W.; McClements, D. J.; Decker, E. A. The relationship between the physicochemical properties of antioxidants and their ability to inhibit lipid oxidation in bulk oil and oil-in-water emulsions. J. Agric. Food Chem. 2005, 53 (12), 4982–4988.

Chaiyasit, W.; McClements, D. J.; Weiss, J.; Decker, E. A. Impact of surface-active compounds on physicochemical and oxidative properties of edible oil. J. Agric. Food Chem. 2008, 56 (2), 550–556.

Chaiyasit, W.; Silvestre, M. P. C.; McClements, D. J.; Decker, E. A. Ability of surfactant hydrophobic tail group size to alter lipid oxidation in oil-in-water emulsions. J. Agric. Food Chem. 2000, 48 (8), 3077–3080.

Chen, B. C.; Han, A.; Laguerre, M.; McClements, D. J.; Decker, E. A. Role of reverse micelles on lipid oxidation in bulk oils: Impact of phospholipids on antioxidant activity of alpha-tocopherol and Trolox. Food Funct. 2011a, 2 (6), 302–309.

Chen, B. C.; Han, A.; McClements, D. J.; Decker, E. A. Physical structures in soybean oil and their impact on lipid oxidation. J. Agric. Food Chem. 2010a, 58 (22), 11993–11999.

Chen, B. C.; McClements, D. J.; Decker, E. A. Role of continuous phase anionic polysaccharides on the oxidative stability of menhaden oil-in-water emulsions. J. Agric. Food Chem. 2010b, 58 (6), 3779–3784.

Chen, B. C.; McClements, D. J.; Decker, E. A. Minor components in food oils: A critical review of their roles on lipid oxidation chemistry in bulk oils and emulsions. Crit. Rev. Food Sci. Nutr. 2011b, 51 (10), 901–916.

Chen, B.; Panya, A.; McClements, D. J.; Decker, E. A. New insights into the role of iron in the promotion of lipid oxidation in bulk oils containing reverse micelles. J. Agric. Food Chem. 2012, 60 (13), 3524–3532.

Cho, Y. J.; McClements, D. J.; Decker, E. A. Ability of surfactant micelles to alter the physical location and reactivity of iron in oil-in-water emulsion. J. Agric. Food Chem. 2002, 50 (20), 5704–5710.

Coupland, J. N.; McClements, D. J. Droplet composition affects the rate of oxidation of emulsified ethyl linoleate—Supporting evidence. J. Am. Oil Chem. Soc. 1996a, 73 (9), 1207–1207.

Coupland, J. N.; McClements, D. J. Lipid oxidation in food emulsions. Trends Food Sci. Technol. 1996b, 7 (3), 83–91.

Decker, E. A.; Elias, R. J.; McClements, D. J. Oxidation in Foods and Beverages and Antioxidant Applications; Woodhead Publishing Ltd: Cambridge, UK, 2010a; 1: Understanding Mechanisms of Oxidation and Antioxidant Activity.

Decker, E. A.; Elias, R. J.; McClements, D. J. Oxidation in Foods and Beverages and Antioxidant Applications. Woodhead Publishing Ltd: Cambridge, UK, 2010b; 2: Management in Different Industry Sectors.

Decker, E. A.; McClements, D.; Chaiyasit, W.; Nuchi, C.; Sivestre, M.; Mancuso, J.; Tong, L.; Mei, L. Factors Influencing Free Radical Formation in Food Emulsions: Free Radicals in Health and Food. Food: Chemistry, Nutrition, and Health Effects; Morello, M. J., Shahidi, F., and Ho, C.-T., Eds; American Chemical Society: Washington, DC, 2002; 83–97.

Decker, E. A.; McClements, D. J. Lipids. Fennema's Food Chemistry; 4th edition; CRC Press/Taylor & Francis: New York, 2008; 155–216.

Dickinson, E. An Introduction to Food Colloids; Oxford University Press: Oxford, 1992.

Dickinson, E.; Stainsby, G. Colloids in Food; Applied Science Publishers: London, 1982.

Dimakou, C. P.; Kiokias, S. N.; Tsaprouni, I. V.; Oreopoulou, V. Effect of processing and storage parameters on the oxidative deterioration of oil-in-water emulsions. Food Biophys. 2007, 2 (1), 38–45.

Djordjevic, D.; Cercaci, L.; Alamed, J.; McClements, D. J.; Decker, E. A. Chemical and physical stability of citral and limonene in sodium dodecyl sulfate-chitosan and gum arabic-stabilized oil-in-water emulsions. J. Agric. Food Chem. 2007, 55 (9), 3585–3591.

Djordjevic, D.; Cercaci, L.; Alamed, J.; McClements, D. J.; Decker, E. A. Stability of citral in protein- and gum arabic-stabilized oil-in-water emulsions. Food Chem. 2008, 106 (2), 698–705.

Duxbury, D. Omega-3s offer solutions to *trans* fat substitution problems. Food Technol. 2005, 59 (4), 34–39.

Elias, R. J.; Bridgewater, J. D.; Vachet, R. W.; Waraho, T.; McClements, D. J.; Decker, E. A. Antioxidant mechanisms of enzymatic hydrolysates of beta-lactoglobulin in food lipid dispersions. J. Agric. Food Chem. 2006, 54 (25), 9565–9572.

Elias, R. J.; Kellerby, S. S.; Decker, E. A. Antioxidant activity of proteins and peptides. Crit. Rev. Food Sci. Nutr. 2008, 48 (5), 430–441.

Erickson, M. C.; Sista, R. V. Influence of Microenvironment on Oxidative Susceptibility of Seafood Lipids. Flavor and Lipid Chemistry of Seafoods; Shahidi, F.; Cadwallader, K. R., Eds; American Chemical Society: Washington, DC, 1997; Vol. 674, 175–185.

Faraji, H.; McClements, D. J.; Decker, E. A. Role of continuous phase protein on the oxidative stability of fish oil-in-water emulsions. J. Agric. Food Chem. 2004, 52 (14), 4558–4564.

Frankel, E. N. Food antioxidants and phytochemicals: Present and future perspectives. Fett-Lipid 1999, 101 (12), 450–455.

Frankel, E. N. Lipid Oxidation; The Oily Press: Bridgwater, England, 2005.

Frankel, E. N.; Huang, S. W.; Aeschbach, R. Antioxidant activity of green teas in different lipid systems. J. Am. Oil Chem. Soc. 1997, 74 (10), 1309–1315.

Frankel, E. N.; Huang, S. W.; Aeschbach, R.; Prior, E. Antioxidant activity of a rosemary extract and its constituents, carnosic acid, carnosol, and rosmarinic acid, in bulk oil and oil-in-water emulsion. J. Agric. Food Chem. 1996a, 44 (1), 131–135.

Frankel, E. N.; Huang, S. W.; Kanner, J.; German, J. B. Interfacial phenomena in the evaluation of antioxidants—Bulk oils vs emulsions. J. Agric. Food Chem. 1994, 42 (5), 1054–1059.

Frankel, E. N.; Huang, S. W.; Prior, E.; Aeschbach, R. Evaluation of antioxidant activity of rosemary extracts, carnosol and carnosic acid in bulk vegetable oils and fish oil and their emulsions. J. Sci. Food Agric. 1996b, 72 (2), 201–208.

Friberg, S.; Larsson, K.; Sjöblom, J. Food Emulsions; Marcel Dekker: New York, 2004.

Garti, N. Double emulsions—Scope, limitations and new achievements. Colloid Surface A 1997a, 123, 233–246.

Garti, N. Progress in stabilization and transport phenomena of double emulsions in food applications. Food Science and Technology-Lebensmittel-Wissenschaft & Technologie 1997b, 30 (3), 222–235.

Garti, N.; Benichou, A. Recent Developments in Double Emulsions for Food Applications. In Food emulsions, 4th ed.; Friberg, S. E.; Larsson, K.; Sjoblom, J., Eds.; Marcel Dekker: New York, 2004.

Garti, N.; Bisperink, C. Double emulsions: Progress and applications. Curr. Opin. Colloid In. 1998, 3 (6), 657–667.

Gohtani, S.; Sirendi, M.; Yamamoto, N.; Kajikawa, K.; Yamano, Y. Effect of droplet size on oxidation of docosahexaenoic acid in emulsion system. J. Dispersion Sci. Technol. 1999, 20,(5), 1319–1325.

Haahr, A. M.; Jacobsen, C. Emulsifier type, metal chelation and pH affect oxidative stability of n-3-enriched emulsions. Eur. J. Lipid Sci. Technol. 2008, 110 (10), 949–961.

Halliwell, B.; Gutteridge, J. Role of free radicals and catalytic metal ions in human disease: An overview. Methods in Enzymology 1990, 186, 1–88.

Halliwell, B.; Murcia, M. A.; Chirico, S.; Aruoma, O. I. Free-radicals and antioxidants in food and *in-vivo*—What they do and how they work. Crit. Rev. Food Sci. Nutr. 1995, 35 (1–2), 7–20.

Hegenauer, J.; Saltman, P.; Ludwig, D.; Ripley, L.; Bajo, P. Effect of supplemental iron and lipid oxidation in milk. 1. Comparison of metal complexes in emulsified and homogenized milk. J. Agric. Food Chem. 1979, 27 (4), 860–867.

Hopia, A. I.; Huang, S. W.; Schwarz, K.; German, J. B.; Frankel, E. N. Effect of different lipid systems on antioxidant activity of rosemary constituents carnosol and carnosic acid with and without alpha-tocopherol. J. Agric. Food Chem. 1996, 44 (8), 2030–2036.

Hu, M.; McClements, D. J.; Decker, E. A. Impact of whey protein emulsifiers on the oxidative stability of salmon oil-in-water emulsions. J. Agric. Food Chem. 2003, 51 (5), 1435–1439.

Huang, S. W.; Frankel, E. N. Antioxidant activity of tea catechins in different lipid systems. J. Agric. Food Chem. 1997, 45 (8), 3033–3038.

Huang, S. W.; Hopia, A.; Schwarz, K.; Frankel, E. N.; German, J. B. Antioxidant activity of alpha-tocopherol and Trolox in different lipid substrates: Bulk oils vs oil-in-water emulsions. J. Agric. Food Chem. 1996, 44 (2), 444–452.

Huang, S. W.; Satué-Gracia, M. T.; Frankel, E. N.; German, J. B. Effect of lactoferrin on oxidative stability of corn oil emulsions and liposomes. J. Agric. Food Chem. 1999, 47 (4), 1356–1361.

Hunter, R. J. Foundations of Colloid Science; Oxford University Press: Oxford, 1986; Vol. 1.

Imai, H.; Maeda, T.; Shima, M.; Adachi, S. Oxidation of methyl linoleate in oil-in-water micro- and nanoemulsion systems. J. Am. Oil Chem. Soc. 2008, 85 (9), 809–815.

Kargar, M.; Spyropoulos, F.; Norton, I. T. The effect of interfacial microstructure on the lipid oxidation stability of oil-in-water emulsions. J. Colloid Interface Sci. 2011, 357 (2), 527–533.

Kasaikina, O.; Golyavin, A.; Krugovov, D.; Kartasheva, Z.; Pisarenko, L. Micellar catalysis in the oxidation of lipids. Moscow University Chemistry Bulletin 2010, 65 (3), 206–209.

Kasaikina, O. T.; Kancheva, V. D.; Maximova, T. V.; Kartasheva, Z. S.; Yanishlieva, N. V.; Kondratovich, V. G.; Totseva, I. R. Catalytic effect of amphiphilic components on the lipid oxidation and lipid hydroperoxide decomposition. Oxid. Commun. 2006, 29 (3), 574–584.

Kasaikina, O. T.; Kartasheva, Z. S.; Pisarenko, L. M. Effect of surfactants on liquid-phase oxidation of hydrocarbons and lipids. Russ. J. Gen. Chem. 2008, 78 (8), 1533–1544.

Kasaikina, O. T.; Kortenska, V. D.; Kartasheva, Z. S.; Kuznetsova, G. M.; Maximova, T. V.; Sirota, T. V.; Yanishlieva, N. V. Hydrocarbon and lipid oxidation in micro heterogeneous systems formed by surfactants or nanodispersed Al2O3, SiO2 and TiO2. Colloid Surf. A-Physicochem. Eng. Asp. 1999, 149 (1–3), 29–38.

Katsuda, M. S.; McClements, D.; Miglioranza, L. H. S.; Decker, E. A. Physical and oxidative stability of fish oil-in-water emulsions stabilized with β-lactoglobulin and pectin. J. Agric. Food Chem. 2008, 56 (14), 5926–5931.

Kellerby, S. S.; Gu, Y. S.; McClements, D. J.; Decker, E. A. Lipid oxidation in a menhaden oil-in-water emulsion stabilized by sodium caseinate cross-linked with transglutaminase. J. Agric. Food Chem. 2006, 54 (26), 10222–10227.

Kishk, Y. F. M.; Al-Sayed, H. M. A. Free-radical scavenging and antioxidative activities of some polysaccharides in emulsions. LWT-Food Sci. Technol. 2007, 40 (2), 270–277.

Klinkesorn, U.; Sophanodora, P.; Chinachoti, P.; Decker, E. A.; McClements, D. J. Encapsulation of emulsified tuna oil in two-layered interfacial membranes prepared using electrostatic layer-by-layer deposition. Food Hydrocolloids 2005a, 19 (6), 1044–1053.

Klinkesorn, U.; Sophanodora, P.; Chinachoti, P.; McClements, D. J.; Decker, E. A. Increasing the oxidative stability of liquid and dried tuna oil-in-water emulsions with electrostatic layer-by-layer deposition technology. J. Agric. Food Chem. 2005b, 53 (11), 4561–4566.

Koga, T.; Terao, J. Antioxidant activity of a novel phosphatidyl derivative of vitamin E in lard and its model system. J. Agric. Food Chem. 1994, 42 (6), 1291–1294.

Koga, T.; Terao, J. Phospholipids increase radical-scavenging activity of vitamin E in a bulk oil model system. J. Agric. Food Chem. 1995, 43 (6), 1450–1454.

Kogure, K.; Fukuzawa, K.; Kawano, H.; Terada, H. Spermine accelerates iron-induced lipid peroxidation in mitochondria by modification of membrane surface charge. Free Radic. Biol. Med. 1993, 14 (5), 501–507.

Kortenska, V. D.; Yanishlieva, N. V.; Kasaikina, O. T.; Totzeva, I. T.; Boneva, M. I.; Russina, I. F. Phenol antioxidant efficiency in various lipid substrates containing hydroxy compounds. Eur. J. Lipid Sci. Technol. 2002, 104 (8), 513–519.

Laguerre, M.; Giraldo, L. J. L.; Lecomte, J.; Figueroa-Espinoza, M. C.; Barea, B.; Weiss, J.; Decker, E. A.; Villeneuve, P. Chain length affects antioxidant properties of chlorogenate esters in emulsion: The cutoff theory behind the polar paradox. J. Agric. Food Chem. 2009, 57 (23), 11335–11342.

Laguerre, M. L.; López Giraldo, L. J.; Lecomte, J. R. M.; Figueroa-Espinoza, M.-C.; Baréa, B.; Weiss, J.; Decker, E. A.; Villeneuve, P. Relationship between hydrophobicity and antioxidant ability of "phenolipids" in emulsion: A parabolic effect of the chain length of rosmarinate esters. J. Agric. Food Chem. 2010, 58 (5), 2869–2876.

Laranjinha, J. Redox cycles of caffeic acid with alpha-tocopherol and ascorbate. Methods Enzymol. 2001, 335, 282–295.

Leaver, J.; Law, A. J. R.; Brechany, E. Y. Covalent modification of emulsified beta-casein resulting from lipid peroxidation. J. Colloid Interface Sci. 1999, 210 (1), 207–214.

Lee, S. J.; Choi, S. J.; Li, Y.; Decker, E. A.; McClements, D. J. Protein-stabilized nanoemulsions and emulsions: Comparison of physicochemical stability, lipid oxidation, and lipase digestibility. J. Agric. Food Chem. 2011, 59 (1), 415–427.

Lethuaut, L.; Metro, F.; Genot, C. Effect of droplet size on lipid oxidation rates of oil-in-water emulsions stabilized by protein. J. Am. Oil Chem. Soc. 2002, 79 (5), 425–430.

Mabrouk, A. The kinetics of methyl linoleate emulsion autoxidation in the presence of polyhydroxy compounds. J. Am. Oil Chem. Soc. 1964, 41 (4), 331–334.

Mabrouk, A. F.; Dugan, L. R. Kinetic investigation into glucose-, fructose-, and sucrose-activated autoxidation of methyl linoleate emulsion. J. Am. Oil Chem. Soc. 1961, 38 (12), 692–695.

Mancuso, J. R.; McClements, D. J.; Decker, E. A. The effects of surfactant type, pH, and chelators on the oxidation of salmon oil-in-water emulsions. J. Agric. Food Chem. 1999, 47 (10), 4112–4116.

Mancuso, J. R.; McClements, D. J.; Decker, E. A. Iron-accelerated cumene hydroperoxide decomposition in hexadecane and trilaurin emulsions. J. Agric. Food Chem. 2000, 48 (2), 213–219.

Mason, T. G.; Wilking, J.; Meleson, K.; Chang, C.; Graves, S. Nanoemulsions: Formation, structure, and physical properties. J. of Phys.: Condens. Matter 2006, 18, 635–666.

Matsumura, Y.; Egami, M.; Satake, C.; Maeda, Y.; Takahashi, T.; Nakamura, A.; Mori, T. Inhibitory effects of peptide-bound polysaccharides on lipid oxidation in emulsions. Food Chem. 2003, 83 (1), 107–119.

McClements, D. J. Food Emulsions: Principles, Practices, and Techniques; CRC: Boca Raton, FL, 2005.

McClements, D. J.; Decker, E. A. Lipid oxidation in oil-in-water emulsions: Impact of molecular environment on chemical reactions in heterogeneous food systems. J. Food Sci. 2000, 65 (8), 1270–1282.

McClements, D. J.; Decker, E. A.; Weiss, J. Emulsion-based delivery systems for lipophilioc bioactive components. J. Food Sci. 2007, 72 (8), R109-R124.

Mei, L.; Decker, E. A.; McClements, D. J. Evidence of iron association with emulsion droplets and its impact on lipid oxidation. J. Agric. Food Chem. 1998a, 46 (12), 5072–5077.

Mei, L.; McClements, D. J.; Wu, J.; Decker, E. A. Iron-catalyzed lipid oxidation in emulsion as affected by surfactant, pH and NaCl. Food Chem. 1998b, 61 (3), 307–312.

Mei, L. Y.; McClements, D. J.; Decker. E. A. Lipid oxidation in emulsions as affected by charge status of antioxidants and emulsion droplets. J. Agric. Food Chem. 1999, 47 (6), 2267–2273.
Min, D. B.; Boff, J. M. Lipid oxidation of edible oil. Food Lipids: Chemistry, Nutrition, and Biotechnology; Marcel Dekker: New York, 2008.
Muller, R. H.; Keck, C. M. Challenges and solutions for the delivery of biotech drugs—a review of drug nanocrystal technology and lipid nanoparticles. J. Biotechnol 2004, 113 (1–3), 151–170.
Muller, R. H.; Mader, K.; Gohla, S. Solid lipid nanoparticles (SLN) for controlled drug delivery—a review of the state of the art. Eur. J. Pharm. Biopharm. 2000, 50 (1), 161–177.
Nakaya, K.; Ushio, H.; Matsukawa, S.; Shimizu, M.; Ohshima, T. Effects of droplet size on the oxidative stability of oil-in-water emulsions. Lipids 2005, 40,(5), 501–507.
Naz, S.; Siddiqi, R.; Sheikh, H.; Sayeed, S. A. Deterioration of olive, corn and soybean oils due to air, light, heat and deep-frying. Food Res. Int. 2005, 38, (2) 127–134.
Nuchi, C. D.; Hernandez, P.; McClements, D. J.; Decker, E. A. Ability of lipid hydroperoxides to partition into surfactant micelles and alter lipid oxidation rates in emulsions. J. Agric. Food Chem. 2002, 50 (19), 5445–5449.
Osborn, H. T.; Akoh, C. C. Effect of emulsifier type, droplet size, and oil concentration on lipid oxidation in structured lipid-based oil-in-water emulsions. Food Chem. 2004, 84 (3), 451–456.
Ostdal, H.; Daneshvar, B.; Skibsted, L. H. Reduction of ferrylmyoglobin by beta-lactoglobulin. Free Radic. Res. 1996, 24 (6), 429–438.
Panya, A.; Laguerre, M.; Bayrasy, C.; Lecomte, J.; Villeneuve, P.; McClements, D. J.; Decker, E. A. An investigation of the versatile antioxidant mechanisms of action of rosmarinate alkyl esters in oil-in-water emulsions. J. Agric. Food Chem. 2012, 60 (10), 2692–2700.
Panya, A.; Laguerre, M.; Lecomte, J.; Villeneuve, P.; Weiss, J.; McClements, D. J.; Decker, E. A. Effects of chitosan and rosmarinate esters on the physical and oxidative stability of liposomes. J. Agric. Food Chem. 2010, 58 (9), 5679–5684.
Paraskevopoulou, D.; Boskou, D.; Paraskevopoulou, A. Oxidative stability of olive oil-lemon juice salad dressings stabilized with polysaccharides. Food Chem. 2007, 101 (3) 1197–1204.
Porter, W. L. Paradoxical behavior of antioxidants in food and biological systems. Toxicol. Ind. Health 1993, 9 (1–2), 93–122.
Porter, W. L.; Black, E. D.; Drolet, A. M. Use of polyamide oxidative fluorescence test on lipid emulsions—contrast in relative effectiveness of antioxidants in bulk versus dispersed systems. J. Agric. Food Chem. 1989, 37 (3), 615–624.
Pryor, W. A.; Cornicelli, J. A.; Devall, L. J.; Tait, B.; Trivedi, B. K.; Witiak, D. T.; Wu, M. D. A rapid screening test to determine the antioxidant potencies of natural and synthetic antioxidants. J. Org. Chem. 1993, 58 (13), 3521–3532.
Rampon, V.; Lethuaut, L.; Mouhous-Riou, N.; Genot, C. Interface characterization and aging of bovine serum albumin stabilized oil-in-water emulsions as revealed by front-surface fluorescence. J. Agric. Food Chem. 2001, 49 (8), 4046–4051.
Richards, M. P.; Chaiyasit, W.; McClements, D. J.; Decker, E. A. Ability of surfactant micelles to alter the partitioning of phenolic antioxidants in oil-in-water emulsions. J. Agric. Food Chem. 2002, 50 (5), 1254–1259.
Sarker, D. K. Engineering of nanoemulsions for drug delivery. Current Drug Delivery 2005, 2 (4), 297–310.

Sasaki, K.; Alamed, J.; Weiss, J.; Villeneuve, P.; Giraldo, L. J. L.; Lecomte, J.; Figueroa-Espinoza, M. C.; Decker, E. A. Relationship between the physical properties of chlorogenic acid esters and their ability to inhibit lipid oxidation in oil-in-water emulsions. Food Chem. 2010, 118 (3), 830–835.

Shaw, L. A.; McClements, D. J.; Decker, E. A. Spray-dried multilayered emulsions as a delivery method for omega-3 fatty acids into food systems. J. Agric. Food Chem. 2007, 55 (8), 3112–3119.

Shen, Z.; Rusli, J. K.; Sanguansri, L.; Augustin, M. A. Retention of propanal in protein-stabilised tuna oil-in-water emulsions. Food Chem. 2007, 101 (2), 746–752.

Shimada, K.; Fujikawa, K.; Yahara, K.; Nakamura, T. Antioxidative properties of xanthan on the autoxidation of soybean oil in cyclodextrin emulsion. J. Agric. Food Chem. 1992, 40 (6), 945–948.

Shimada, K.; Okada, H.; Matsuo, K.; Yoshioka, S. Involvement of chelating action and viscosity in the antioxidative effect of xanthan in an oil/water emulsion. Biosci. Biotechnol. Biochem. 1996, 60 (1), 125–127.

Silvestre, M. P. C.; Chaiyasit, W.; Brannan, R. G.; McClements, D. J.; Decker, E. A. Ability of surfactant headgroup size to alter lipid and antioxidant oxidation in oil-in-water emulsions. J. Agric. Food Chem. 2000, 48 (6), 2057–2061.

Sims, R. J.; Fioriti, J. A.; Trumbetas, J. Effect of sugars and sugar alcohols on autoxidation of safflower oil in emulsions. J. Am. Oil Chem. Soc. 1979, 56 (8), 742–745.

Sun, C. H.; Gunasekaran, S. Effects of protein concentration and oil-phase volume fraction on the stability and rheology of menhaden oil-in-water emulsions stabilized by whey protein isolate with xanthan gum. Food Hydrocolloids 2009, 23 (1), 165–174.

Taylor, M. J.; Richardson, T. Antioxidant activity of skim milk: Effect of heat and resultant sulfhydryl groups. J. Dairy Sci. 1980, 63 (11), 1783–1795.

Tong, L. M.; Sasaki, S.; McClements, D. J.; Decker, E. A. Mechanisms of the antioxidant activity of a high molecular weight fraction of whey. J. Agric. Food Chem. 2000, 48 (5), 1473–1478.

Trunova, N. A.; Kartasheva, Z. S.; Maksimova, T. V.; Bogdanova, Y. G.; Kasaikina, O. T. Decomposition of cumene hydroperoxide in the systems of normal and reverse micelles formed by cationic surfactants. Colloid J. 2007, 69 (5), 655–659.

Villiere, A.; Viau, M.; Bronnec, I.; Moreau, N.; Genot, C. Oxidative stability of bovine serum albumin- and sodium caseinate-stabilized emulsions depends on metal availability. J. Agric. Food Chem. 2005, 53 (5), 1514–1520.

Walstra, P. Physical Chemistry of Foods; Marcel Decker: New York, 2003.

Walstra, P. Principles of emulsion formation. Chemical Engineering Science 1993, 48, 333.

Wang, G.; Wang, T. Oxidative stability of egg and soy lecithin as affected by transition metal ions and pH in emulsion. J. Agric. Food Chem. 2008, 56 (23), 11424–11431.

Waraho, T.; Cardenia, V.; Rodriguez-Estrada, M. T.; McClements, D. J.; Decker, E. A. Prooxidant mechanisms of free fatty acids in stripped soybean oil-in-water emulsions. J. Agric. Food Chem. 2009, 57 (15), 7112–7117.

Waraho, T.; McClements, D. J.; Decker, E. A. Mechanisms of lipid oxidation in food dispersions. Trends Food Sci. Technol. 2011, 22 (1), 3–13.

Wissing, S. A.; Kayser, O.; Muller, R. H. Solid lipid nanoparticles for parenteral drug delivery. Adv. Drug Deliver Rev. 2004, 56 (9), 1257–1272.

Yamauchi, R.; Goto, Y.; Kato, K.; Ueno, Y. Prooxidant effect of dihydroxyacetone and reducing sugars on the autoxidation of methyl linoleate in emulsions. Agricultural and Biological Chemistry 1984, 48 (4), 843–848.

Yoshida, H. Influence of fatty acids of different unsaturation in the oxidation of purified vegetable oils during microwave irradiation. J. Sci. Food Agric. 1993, 62 (1), 41–47.

Yoshida, Y.; Niki, E. Oxidation of methyl linoleate in aqueous dispersions induced by copper and iron. Arch. Biochem. Biophys. 1992, 295 (1), 107–114.

Yuji, H.; Weiss, J.; Villeneuve, P.; Lopez Giraldo, L. J.; Figueroa-Espinoza, M. C.; Decker, E. A. Ability of surface-active antioxidants to inhibit lipid oxidation in oil-in-water emulsion. J. Agric. Food Chem. 2007, 55 (26), 11052–11056.

Zhou, S. Y.; Decker, E. A. Ability of amino acids, dipeptides, polyamines, and sulfhydryls to quench hexanal, a saturated aldehydic lipid oxidation product. J. Agric. Food Chem. 1999, 47 (5), 1932–1936.

CHAPTER 4

Substrate and Droplet Size

Important Factors for Understanding Aqueous Lipid Oxidation

Kazuo Miyashita[1] and Masashi Hosokawa[1]

Introduction

The first and rate-limiting step of lipid oxidation is the abstraction of a hydrogen radical (H•) from substrate lipids (LH) to form lipid free radicals (L•). The reaction proceeds thereafter through a free-radical chain reaction (Frankel, 1980, 1998; Porter et al., 1995; Kamal-Eldin et al., 2003). In the propagation stage of the reaction, lipid free radicals react with molecular oxygen to form peroxy radicals (LOO•). Peroxy radicals abstract a hydrogen atom from another unsaturated fatty compound to form a monohydroperoxide (LOOH) and an alkyl radical (L•). The latter reacts with molecular oxygen in a repetition of the first propagation reaction. Subsequently, the initially formed monohydroperoxide may decompose to yield free radicals such as alkoxy (LO•) and hydroxyl (HO•) radicals, which serve as initiators for the above reactions. Therefore, the chain-breaking antioxidants are of considerable practical importance in protecting lipids from oxidative deterioration. These antioxidants inhibit or retard oxidation by interfering with either chain propagation or initiation by readily donating hydrogen atoms to lipid free radicals.

The above mechanism for lipid oxidation has been well studied and confirmed using simple lipid esters, such as methyl esters of polyunsaturated fatty acids (PUFA); however, lipid oxidation and the activity of antioxidants become more complicated in foods and animal tissues, in which lipids are present in complex, multi-component, and heterogeneous systems with various other components in an aqueous medium. Therefore, research on aqueous lipid oxidation is very important if we are to fully understand the factors that affect lipid oxidation in foods

[1]Faculty of Fisheries Sciences, Hokkaido University, 3-1-1 Minato, Hakodate, Hokkaido 041-8611, Japan.

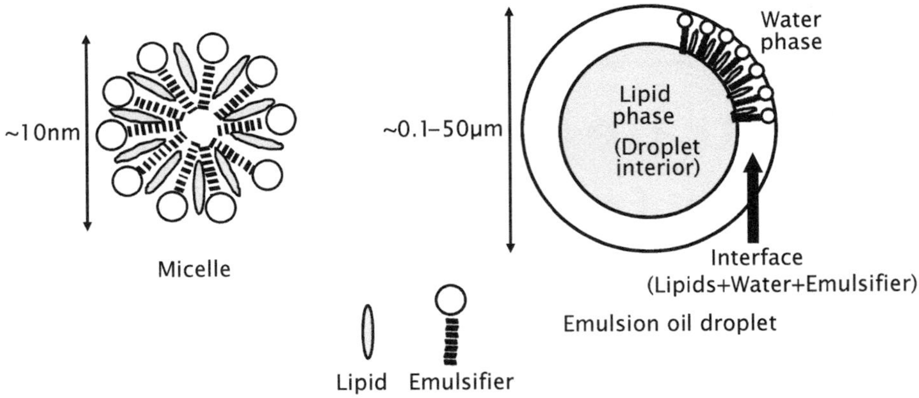

Fig. 4.1. Aqueous structure of micelle and emulsion oil droplets.

and biological systems. Research over past few decades has shown that multiphase lipid oxidation is affected not only by the chemical reactivity of substrate lipids, but also by the physical and/or physicochemical properties of lipid molecules and other related components, such as antioxidants, metals, oxidized molecules, free radicals, and oxygen (Coupland & McClements, 1996; Frankel & Meyer, 2000; McClements & Decker, 2000; Chen et al., 2011; Sun et al., 2011; Waraho et al., 2011).

Oil-in-water emulsions or micelles have been used as models of multiphase lipid oxidation. The various molecules in the micelle and emulsion system are distributed according to their polarity and the surface activity between different phases, which include the oil phase, the water phase, and the interfacial region. Lipid oxidation in such systems is a phenomenon on which the nature of the interface has great influence. The aqueous structure of the interface in an emulsion is basically the same as that of micelles (Fig. 4.1).

Since lipid oxidation generally proceeds from the interface to the interior of the oil droplet in Oil-in-water emulsions, the oxidation of lipids at the interface is an important factor to predict the oxidative stability of lipids in emulsion.

In emulsions, the membranes surrounding the emulsion droplets consist of surface-active substances such as emulsifiers and/or proteins, together with lipids and water. They can sometimes provide a protective barrier for the penetration and diffusion of metals or radicals that initiate lipid oxidation. Therefore, the chemical and physical nature of the emulsifier and proteins is an important factor for the protection of PUFAs against oxidation in the aqueous phase. The different affinities of antioxidants and/or prooxidants to the interface is another important factor. Overall, the major factors affecting lipid oxidation at the interface are:

1. Affinity (partition) of antioxidants and/or prooxidants to the interface.
2. Chemical, physical, and stereochemical susceptibility of interface lipids to oxidation.

The behavior of antioxidants and prooxidants at the interface is related to their polarity and electrical charge. The oxidative stability of lipids at the interface is influenced by the size and concentration of the emulsion droplets, the thickness and the packing degree of interface molecules, and the extent of droplet-droplet interactions. Moreover, substrate lipid class and the degree of unsaturation have been reported to affect the oxidative characteristics at the interface (Hopia et al., 1996; Huang et al., 1996b; Huang and Frankel, 1997; Miyashita, 2002). This chapter mainly focuses on the effect of emulsifier, substrate lipids, and droplet size as a series of factors that affect lipid oxidation at the interface in O/W emulsion.

Affinity of Antioxidants and Prooxidants to the Interface

The affinity (partitioning) of antioxidants and/or prooxidants between the oil phase (oil droplet interior), water phase, and interface is a key factor to predict the susceptibility of lipids to oxidation in an oil-in-water emulsion. For example, the effect of antioxidants and/or prooxidants in an oil-in-water emulsion is easier to observe when their concentration at the interface is higher. Heins et al. (2007) demonstrated that an antioxidant with notable surface-active properties can more easily inhibit lipid oxidation in emulsions because it concentrates at the oil–water interface. Additionally, multiple authors have reported that the partitioning of each antioxidant at the interface is strongly related to its polarity (Schwarz et al., 1996; Huang et al., 1996a, 1996b; Richards et al., 2002). The oxidative stability of lipids increases with an increasing concentration of antioxidants or decreasing concentration of prooxidants (oxidation inducers) at the interface (Fig. 4.2).

A good example has been found in the relationship between the electrically charged interface and electrically charged antioxidants and/or prooxidants. An electrically charged surface attracts oppositely charged ions in the surrounding aqueous phase. These ions include positively charged metal ions, such as Cu^{+}, Cu^{2+}, Fe^{2+}, and Fe^{3+}, along with negatively charged antioxidants, such as ascorbic acid and Trolox (6-hydroxy-2,5,7,8-tetramethylchroman-2-carboxylic acid). Pryor et al. (1993) stabilized two kinds of linoleic acid micelles—one with a positively charged emulsifier, hexadecyltrimethylammonium bromide (HDTBr), and the other with a negatively charged emulsifier, sodium dodecyl sulfate (SDS). They found that the negatively charged antioxidant, ascorbic acid, was ~1000 fold less

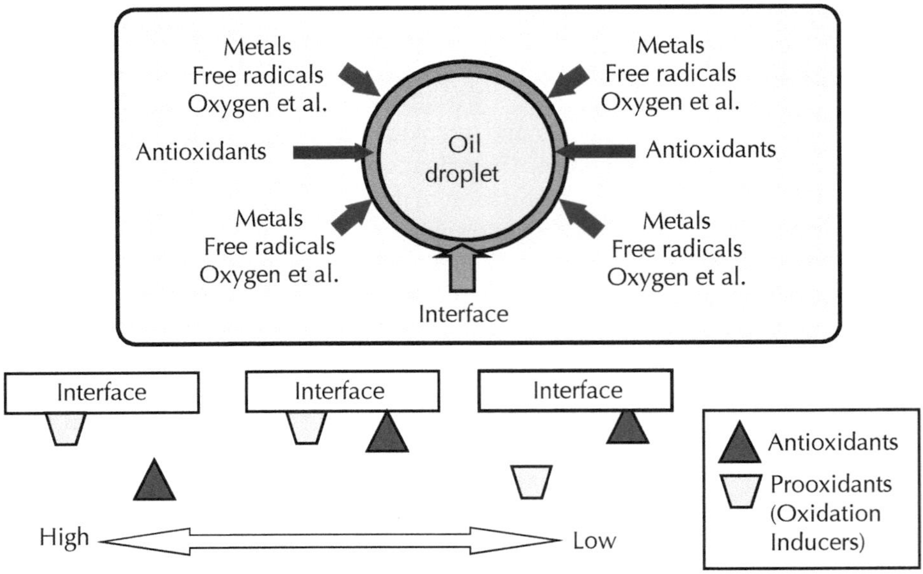

Fig. 4.2. Affinity of antioxidants and/or prooxidants to interface as an important factor affecting the oxidative stability of oil-in-water emulsions.

effective as an antioxidant in the negatively charged micelles than in the positively charged micelles. This occurred because an electrostatic repulsion exists between ascorbic acid and the negatively charged micelles, while there is an electrostatic attraction between ascorbic acid and positively charged micelles. Also negatively charged, Trolox behaves in the same way as ascorbic acid, while the electric status of micelles does not influence the activity of uncharged alpha-tocopherol (Fig. 4.3).

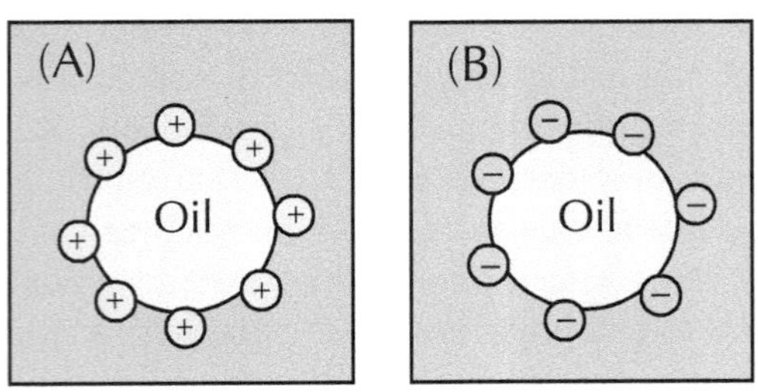

Fig. 4.3. Antioxidant activity of positively charged antioxidants, ascorbic acid and Trolox, and uncharged antioxidant α-tocopherol on the oil droplet with a positively (HDTBr) (A) and negatively (SDS) (B) charged interface.

Relative antioxidant activity
Ascorbic acid: (A)>>(B)
Trolox (A)>(B)
α-Tocopherol: (A)=(B)

Lipid oxidation in food systems is often induced by the hemolytic decomposition of a trace amount of lipid hydroperoxides under the presence of transition metals because it is very difficult or nearly impossible to eliminate traces of metals and lipid hydroperoxides from foods. Due to the positively charged property, the partition of metal ions at the interface of oil-in-water emulsion is greatly affected by its electric charge (Fig. 4.4) (Mei et al., 1998a, 1998b, 1999; Mancuso et al., 1999). When the oil droplet was dispersed with negatively charged (SDS), uncharged (polyoxyethylene 10 lauryl ether; Brij 35), and positively charged (dodecyltrimethylammonium bromide; DTAB) emulsifiers, the oxidative stability was the highest for positively charged droplets, followed by uncharged droplets and negatively charged droplets at pH 6.5. Mei et al. (1998b) observed a more significant effect at a lower pH. Mancuso et al. (1999) found similar results for the oxidation of fish oil-in-water emulsion stabilized by SDS, DTAB, and uncharged Tween 20.

The polar paradox is a well-known theory showing the different activity of an antioxidant in bulk oil and an oil emulsion. This theory depends on the affinity of the antioxidant to the interface between oil and air in the bulk phase and oil and water in an emulsion. The polar paradox was first demonstrated by Porter (1980, 1993) and has been proved experimentally by Frankel's group (Frankel et al., 1994; Frankel & Meyer, 2000). They showed that the hydrophilic (polar) antioxidants are more protective against oxidation in the bulk phase because of their orientation in the air-oil interface, while the lipophylic (nonpolar) antioxidants are less protective because they remain in the oil phase. In contrast, the nonpolar antioxidants are more effective in O/W emulsions than the polar antioxidants because nonpolar antioxidants are oriented at the oil–water interface by their surface activity, while polar antioxidants are dissolved and diluted in the water phase. The difference has been clearly shown by those researchers who used

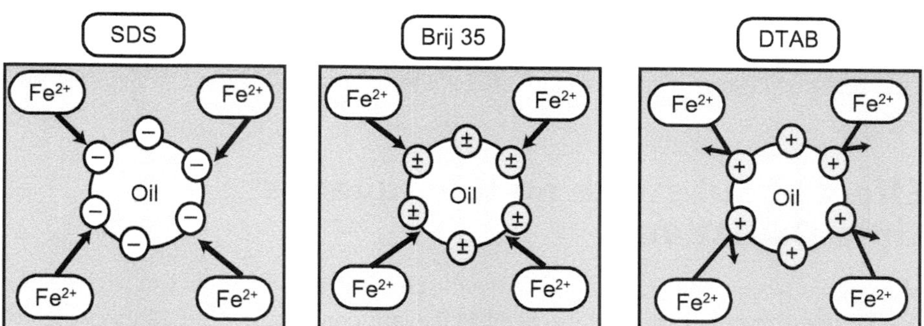

Fig. 4.4. Effect of metal ion on the different electrically charged interface of the oil droplet, prepared with three kinds of emulsifiers: SDS, Brij 35, and DTAB.

polar and nonpolar antioxidant pairs containing the same key structure, such as tocopherol and its polar derivative, Trolox, or ascorbic acid and its nonpolar derivative, ascorbyl palmitate (Frankel et al., 1994; Huang et al., 1994, 1996b).

Although the polarity of antioxidants is an important factor to predict the type of antioxidants that are effective in bulk oils and oil-in-water emulsions, the activity of antioxidants in multiphase systems is not only affected by their polarity. Chaiyasit et al. (2005) analyzed the activities of two sets of antioxidants with different polarities and surface activity but a similar ability to scavenge free radicals—namely, δ-tocopherol (relatively polar) versus α-tocopherol (relatively nonpolar) and 4-hydroxymethyl-2,6-ditertiarybutylphenol (relatively polar) versus butylated hydroxytoluene (BHT) (relatively nonpolar). They found that the nonpolar antioxidants, α-tocopherol and BHT, were more effective in bulk oil than polar antioxidants, δ-tocopherol and 4-hydroxymethyl-2,6-ditertiarybutylphenol, respectively. This relationship did not align with the polar paradox. In oil-in-water emulsions, nonpolar BHT was a more effective antioxidant than polar 4-hydroxymethyl-2,6-ditertiarybutylphenol, while nonpolar α-tocopherol was less effective than polar δ-tocopherol. Other researchers also reported that the antioxidant activity of relatively nonpolar α-tocopherol was lower than that of relatively polar δ-tocopherol (Isnardy et al., 2003; Wagner et al., 2004) and γ-tocoperol (Huang et al., 1996b) in O/W emulsions.

Other studies also indicate that the polarity of antioxidants is not the sole factor that can predict antioxidant activity (Chaiyasit et al., 2005; Di Mattia et al., 2009; Jimenez-Alvarez et al., 2008; Lucas et al., 2010; Sørensen et al., 2011). Medina et al. (2009) analyzed the antioxidant efficiency of hydroxytyrosol fatty acid esters with different alkyl chains in oil-in-water emulsions. They found that hydroxytyrosol octanoate exhibited the highest antioxidant capacity, but the activity of longer alkyl chain esters was lower than that of an octanoate ester, although their polarities were relatively lower than an octanoate ester. A nonlinear relationship between antioxidant effectiveness and the polarity has also been reported in the antioxidant activity of chlorogenic acid esters (Laguerre et al., 2009; Sasaki et al., 2010) and of rosmarinic acid alkyl esters (Laguerre et al., 2010; Panya et al., 2012) in O/W emulsions.

Effect of Substrate on the Aqueous Lipid Oxidation

Although a major lipid class of oil-in-water emulsions in foods is triacylglycerols (TAG), free forms and methyl esters of PUFA are commonly used as model lipid substrates to study lipid oxidation and antioxidant activity (Hopia et al., 1996; Huang et al., 1996b; Nuchi et al., 2002). These lipid substrates have different viscosities and polarities. Fatty acids are amphiphiles to form fatty acid anions that are partially soluble in water and form micelles. In contrast to fatty acids, fatty acid esters and triglycerides are insoluble and are thus dispersed in water as oil-in-water emulsion particles, while esters are solubilized as micelles or microemulsions at lower concentrations. In earlier studies, free fatty acid micelles were used as a model system to asses the effect of different types of antioxidants. Castle and Perkins (1986) found that α-tocopherol was more effective than Trolox in linoleic acid micelles because most of the α-tocopherol molecules were located in the micelles at pH 7, while the distribution of Trolox in the micelles was very low.

Some characteristics of free fatty acids in an aqueous dispersion system are different from those found in TAGs. Specifically, the interaction of antioxidants with ionic free fatty acids and nonpolar TAGs has been reported as significantly different (Hopia et al., 1996; Huang et al., 1996b). Therefore, it may be true that free fatty acids cannot be used as a representative lipid model of a food system, although they are still a useful substrate for obtaining basic scientific information on lipid oxidation in an aqueous system.

The characteristic structure of the docosahexaenoic acid (DHA, 22:6n-3) moiety has been shown in membrane-model liposomes. This was confirmed by a study that used free fatty acids as the substrate. NMR analysis and molecular dynamics simulation of phosphatidylcholine (PC) containing DHA in liposomes and of the free form of DHA in aqueous dispersion indicate that a wide variety of DHA conformation—including back-bended, helical, and angle-iron conformations—occurs in such aqueous systems (Everts & Davis, 2000; Feller et al., 2001; Saiz & Klein, 2001; Huber et al., 2002; Miyashita, 2004). This variety in the DHA chain's conformation packs the lipid chains more loosely (Huster et al. 1997; Olbrich et al., 2000), which causes high water permeability (Saiz & Klein, 2001). Molecular dynamics simulation also indicates the remarkable overlapping of water molecules with double bond regions of the DHA chain. The presence of water molecules near a DHA molecule prevents hydrogen abstraction at the bis-allylic positions and reduces the chain-carrying reaction of lipid oxidation.

Indeed, the DHA molecule in PC liposomes (Miyashita et al., 1994a; Nara et al., 1995, 1997, 1998; Araseki et al., 2002) and in cell membranes (Araseki et al., 2005) was oxidatively stabilized. This stability is the same or higher than that of linoleate in both systems. Previous studies have shown that DHA has an unexpectedly high oxidative stability in animals and humans (Kubo et al., 1998; Ando et al., 2000; Wander & Du, 2000). This may be partly due to the specific conformation of DHA in aqueous biological systems. The protection of cellular EPA against oxidation has also been reported using *E. coli* (Nishida et al., 2006). Analysis of monohydroperoxide (HP) suggested a different oxidation behavior of DHA molecules in aqueous and organic solutions. When oxidizing the DHA of cellular phospholipids, DHA-HPs formed by the abstraction at C-6 (4-HP and 8-HP), C-12 (10-HP and 14-HP), and C-18 (16-HP and 20-HP) were detected, but there were no HPs formed by the hydrogen abstraction at C-9 (7-HP and 11-HP) and C-15 (13-HP and 17-HP) (Kobayashi et al., 2003) (Fig. 4.5). The same tendency was also found in the oxidation of DHA in PC liposomes and in micelles.

Several papers demonstrated the characteristic oxidative stability of the free form of PUFA, such as eicosapentaenoic acid (EPA, 20:5n-3) and DHA in micelles. When the oxidative stabilities of EPA and DHA were compared with those of linoleic acid and α-linolenic acid, EPA and DHA were more stable than linoleic acid and α-linolenic acid (Miyashita et al., 1993, 1994b, 1995). This relationship reverses that found in bulk phase or organic solvents (Miyashita, 2002). The oxidative stability of ethyl linoleate, ethyl linolenate, and ethyl docosahexaenoate in microemulsion is also quite different from that in bulk phase or organic solvent. Ethyl docosahexaenoate was more stable than ethyl linoleate and ethyl α-linolenate in microemulsion oxidized with Fe(II)-ascorbic acid (Hirano et al., 1997).

Although the cause for this observation is not yet clear, it is probably occurred because of the specific stereochemical conformation of substrate EPA and DHA at the interface of micelles and microemulsions (Miyashita, 2004). The conformation and kinetics of a molecule or part of a molecule are reflected in NMR relaxation times, that is, their spin-lattice relaxation (T_1) and spin-spin relaxation (T_2) times. When comparing the proton NMR relaxation times of each signal of linoleic acid and DHA in micelles with those of PUFA in a chloroform solution, Kobayashi et al. (2004) observed two characteristic trends. First, protons in PUFA in micelles had shorter relaxation times than those in a chloroform solution, except for the methylene adjacent to the methyl terminal of DHA. Second, T_2 of protons in micelles was much shorter than T_1, whereas T_1 and T_2 were almost the same in a chloroform solution. The decrease of T_2 in micelles indicates that associations between lipid molecules in micelles are so rigid that molecular motion is more restricted than in non-associated lipid molecules, as found in the chloroform solution. Compared to T_1, the shorter T_2 in micelles means that the aqueous system is closer to a solid state.

Proton T_2 gives more information on the flexibility and mobility of the acyl part of PUFA (Fig. 4.6). There was little difference between the T_2 of corresponding protons on DHA and linoleic acid in a chloroform solution, but the T_2 of methyl protons (a) of DHA in micelles had a much larger value than those of linoleic acid (Kobayashi et al., 2004) (Fig. 4.7). Olefin protons (d) and bisallylic protons (e) showed a tendency similar to that of methyl protons (a). For the proton on the carboxyl terminal (h), linoleic acid had a slightly longer T_2 than DHA. The difference

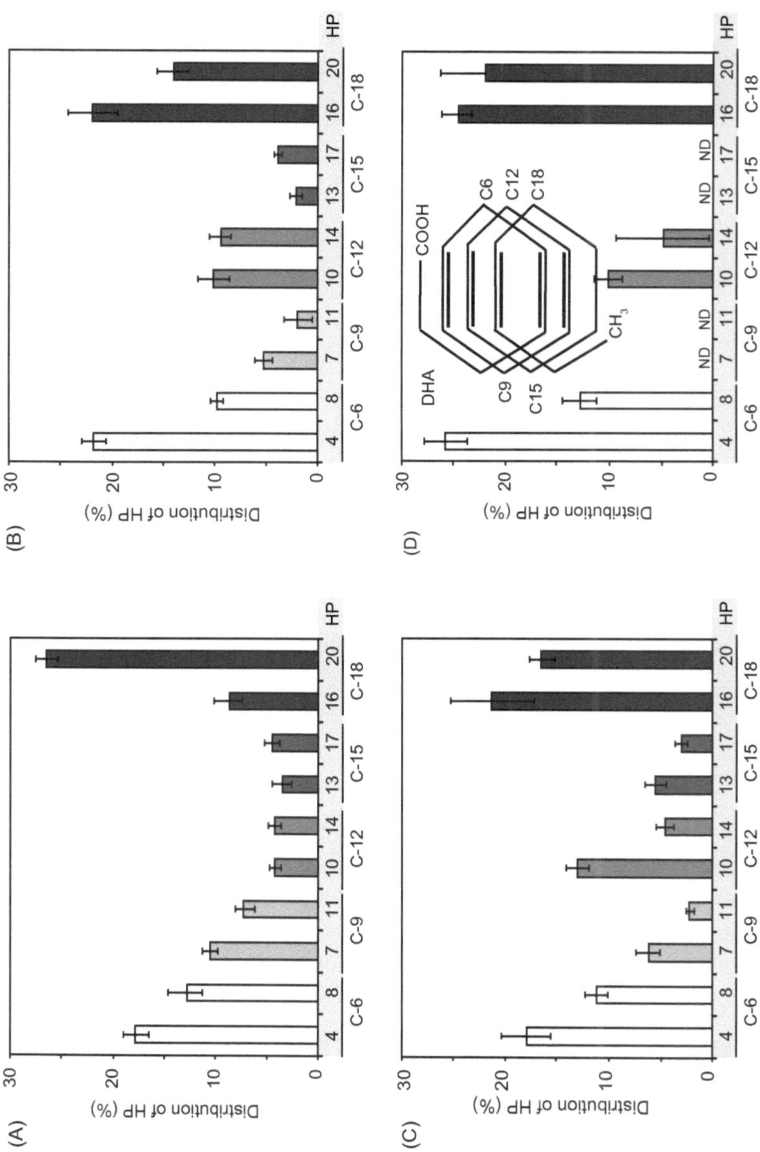

Fig. 4.5. Isomeric distribution of monohydroperoxide (HP) isomers formed in the oxidations of DHA in chloroform (A), micelles (B), liposome (C), and HepG2 cellular lipids (D).

Adapted from Kobayashi, H.; Yoshida, M.; Miyashita, K. Comparative study of the product components of lipid oxidation in aqueous and organic systems. Chem. Phys. Lipids 2003, 126, 111–120.

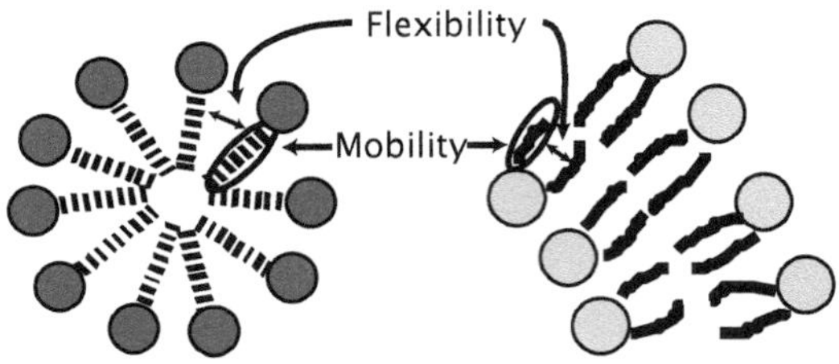

Fig. 4.6. Flexibility and mobility of hydrophobic acyl chains.

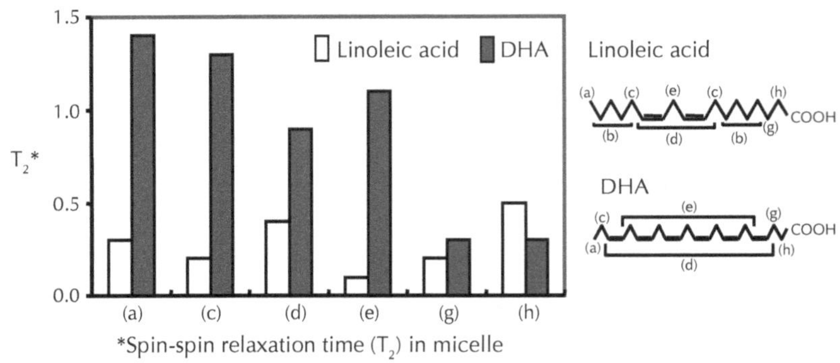

Fig. 4.7. Spin-spin relaxation time (T_2) of DHA and linoleic acid in micelles.

Adapted from Kobayashi, H.; Yoshida, M.; Maeda, I.; Miyashita, K. Proton NMR relaxation times of polyunsaturated fatty acids in chloroform solutions and aqueous micelles. J. Oleo Sci. 2004, 56, 105–108.

in T_2 values suggests that the acyl part of DHA has greater mobility and flexibility of than that of linoleic acid. This flexibility of the DHA molecule allows water molecules to permeate the micelles. The penetration of water molecules to the acyl moieties can inhibit the hydrogen abstraction from bis-allylic positions of DHA.

Droplet Size and Oxidative Stability of Lipids in Oil-in-Water Emulsion

Overall, the chemical properties of interfaces such as electric charge and polarity have been well studied. However, little attention has been paid to the stereochemical structures, interactions, and physical location of interfacial components. The specific stereochemical structure and molecular dynamics of DHA have been found in free fatty acid micelles, in PC liposomes, and in cellular lipids. This characteristic behavior of the DHA molecule may be found in the interfacial region of oil-in-water emulsions.

When the concentration of lipid and emulsifier is the same, the area of interface increases with decreasing droplet size (Fig. 4.8). The opportunity for attack by an oxidation inducer, such as free radicals and metal ions at the interface, increases with the interface's increasing area; therefore, the oxidative stability of lipids in emulsion generally decreases with decreasing droplet size (Lethuaut et al., 2002). However, this relationship may not always be true. If the interface has strong protection against oxidation, as seen in the DHA micelles, the increase in interface area may increase the oxidative stability of lipids. To confirm this assumption, it is necessary to prepare different sizes of monodisperse emulsions with the same substrate and emulsifier concentration. These monodisperse emulsions can be produced using a microchannel (MC) apparatus (Kobayashi et al., 2001; Sugiura et al., 2002). MC emulsification enables the formulation of monodisperse emulsions with coefficients that have less than 5% variation from a channel array with a slit-like terrace micro-fabricated on a silicon plate (Kawakatsu et al., 1997). The

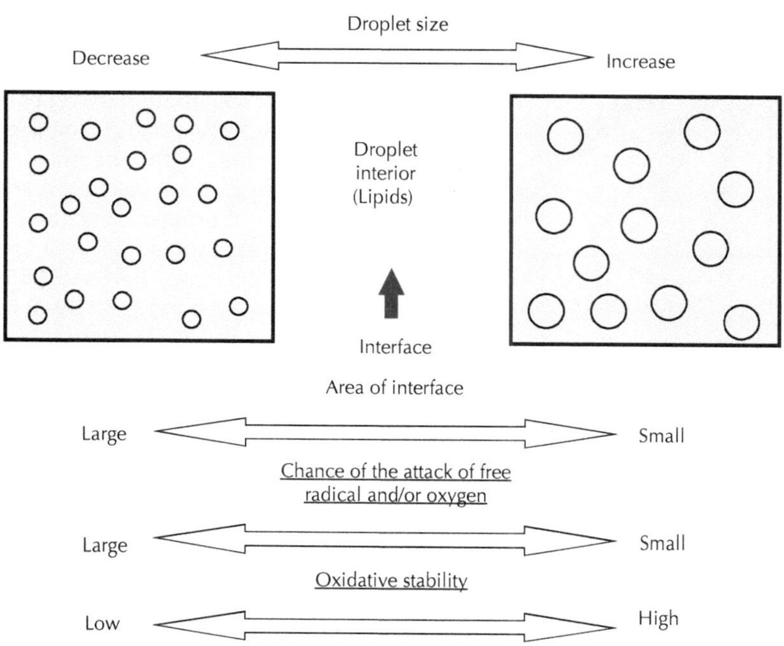

Fig. 4.8. Relationship between the droplet size and oxidative stability of lipids in emulsions.

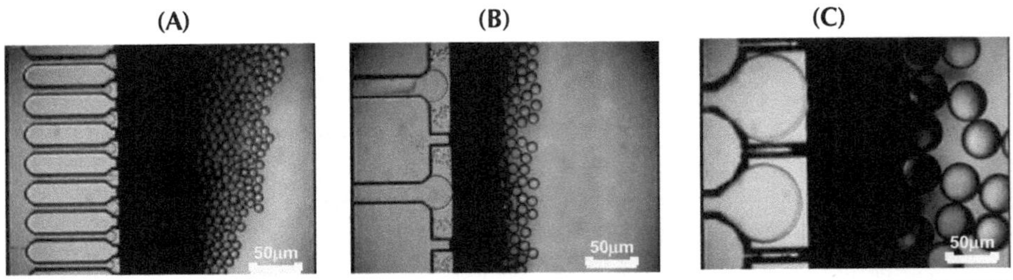

Fig. 4.9. Microscope photographs of microchannel emulsification. (A) channel: MS309; TAG: fish oil TAG; emulsifier: sucrose ester (L1695); (B) channel: MS307; TAG: soybean oil TAG; emulsifier: polyglycerol ester (L-7D); (C) channel: MS337-7; TAG: fish oil TAG; emulsifier: polyglycerol ester (L-7D).

Adapted from Azuma, G.; Kimura, N.; Hosokawa, M.; Miyashita, K. Effect of droplet size on the oxidative stability of soybean oil TAG and fish oil TAG in oil-in-water emulsion. J. Oleo Sci. 2009, 58, 329–338.

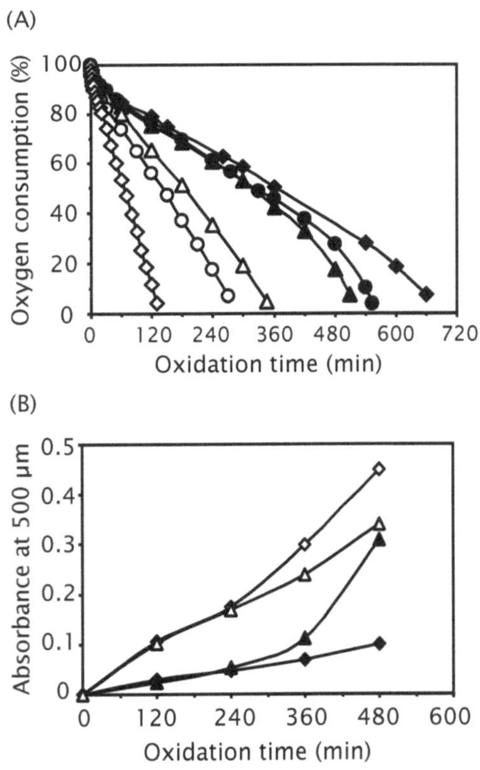

Fig. 4.10. Oxidative stability of soybean oil TAG and fish oil TAG dispersed with polyglycerol ester (L7D) in emulsion. Oxidative stability was measured by oxygen consumption (A) and peroxide formation (B). Emulsion was prepared with different microchannels to make a different diameter of droplet. Soybean oil TAG: solid triangle (6.428+0.065 μm), solid circle (10.469+0.065 μm), and solid diamond (37.499+0.066 μm). Fish oil TAG: open triangle (6.915+0.067 μm), open circle (10.498+0.065 μm), and open diamond (37.235+0.068 μm).

Adapted from Azuma, G.; Kimura, N.; Hosokawa, M.; Miyashita, K. Effect of droplet size on the oxidative stability of soybean oil TAG and fish oil TAG in oil-in-water emulsion. J. Oleo Sci. 2009, 58, 329–338.

resultant droplet size, ranging between 3 and 90 micro μm, is primarily controlled by MC geometry (Fig. 4.9). The droplet size is driven by interfacial tension and the dominant force on a micrometer scale. Droplet formation requires no mechanical stress at very low energy input (Sugiura et al., 2001).

Using the monodispersed emulsion of soybean oil TAG rich in linoleic acid (53.0%) and fish oil TAG rich in DHA (37.3%), Azuma et al. (2009) found that the stability of fish oil TAG in the emulsions increased with a decreasing droplet size, while the reverse result was obtained from the oxidation of soybean oil TAG (Fig. 4.10). The result found in fish oil TAG goes against the general understanding that the oxidative stability of lipids in an emulsion decreases with decreasing droplet size. The characteristic oxidative behavior of fish oil TAG found in Fig. 4.10 may be due to its interface, which is highly protective against oxidation (Fig. 4.11).

When a small amount of fish oil TAG (0.5 wt%) with relatively large amount of emulsifier (0.2 wt%) was sonicated, a small droplet with a diameter around 1 μm could be prepared (Azuma et al., 2009). In this emulsion, the oxidative stability of fish oil TAG was slighly higher than that of soybean oil TAG. The higher oxidative stability of fish oil TAG compared with that of soybean oil was also found in other kinds of microemulsion with low substrate concentration and high emulsifier levels (Miyashita, 2002). The typical micellar size is around 10 μm, while that of emulsion droplet size varies from 0.1 to 50 μm. As described previously, the micellar structure of DHA protects against oxidation. However, the increase in an emulsion's droplet interior decreases the oxidative stability of DHA because DHA is very easily oxidized in the droplet interior. Thus, the stability of DHA in the emulsion decreases with increasing droplet interior. On the contrary, linoleic acid is oxidatively much more stable than DHA in the droplet interior, and the relative oxidative stability of linoleic acid to DHA within an emulsion increases when the droplet interior increases.

Effect of Emulsifier on the Oxidative Stability of Lipids in Oil-in-Water Emulsions

Emulsifiers are one of the major factors affecting lipid oxidation at the interface in oil-in-water emulsions because the interfacial thickness and permeability of antioxidants and/or prooxidants depend upon the emulsifiers' chemical, physical, physicochemical, and stereochemical properties. Emulsifiers in oil-in-water emulsions are generally thought to be oriented on the interface with alkyl residues solubilized in the oil phase and charged or polar groups oriented in the aqueous phase (Fig. 4.1). The thickness of the interfacial layer is usually in the range of a

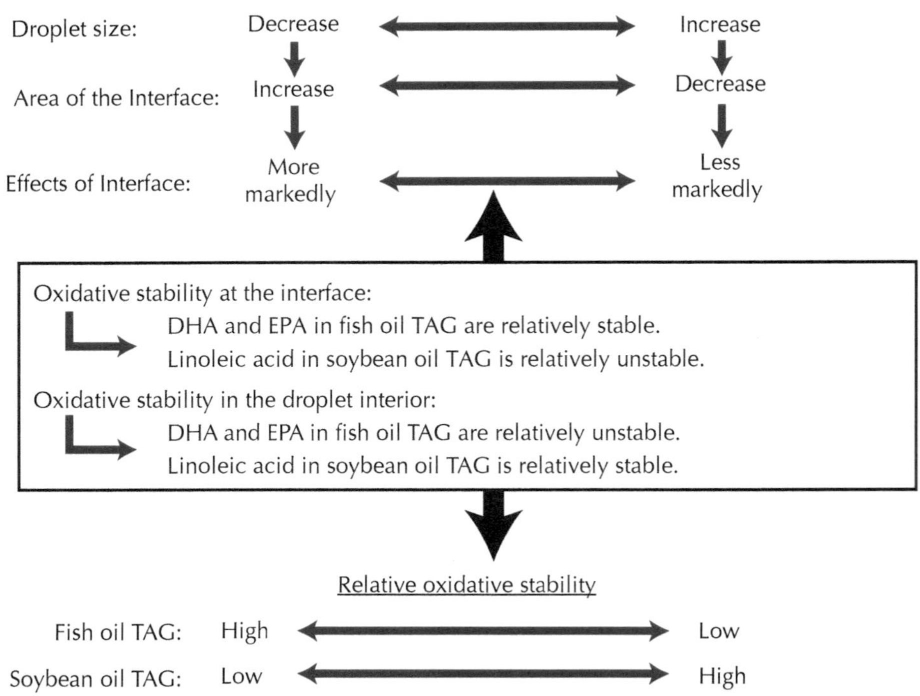

Fig. 4.11. Relationship between droplet size, interface area, and oxidative stability of fish oil TAG and soybean oil TAG.

few nanometers, but it can be modified depending on the type of surface-active materials in the emulsion and the size and conformation of the emulsifier's head and tail group (Waraho et al., 2011). The packing density of emulsifier-emulsifier and/or emulsifier-substrate lipids at the interface has a strong impact on the diffusion of oxygen, free radicals, and other prooxidants through the interfacial layer. High density and strong packing formation form a protective barrier to the penetration and diffusion of oxidation inducers that accelerate lipid oxidation in the droplets. Altering the effectiveness of the packing of the emulsifier-emulsifier and emulsifier-lipid molecules in the droplet membrane is one of the best methods to protect the lipid against oxidation. The higher that surfactant concentrations are, the tighter the packing of the emulsifier at the interface will be; therefore, it is probably easiest to strengthen the interface barrier by increasing the emulsifier's concentration (Coupland & McClements, 1996). Jacobsen et al. (2001) have shown that higher emulsifier concentrations in mayonnaise led to lower lipid peroxide values and less oxidative flavor deterioration. Fomuso et al. (2002) also reported that, in the emulsions with the same droplet size, emulsion with a low emulsifier level (0.25%) showed significantly higher oxidation levels than emulsion with high emulsifier level (1%).

Several types of emulsifiers are known to form a protective interfacial barrier to oxidation. Emulsifiers with large molecular dimensions, especially those with a larger polar head group, appear to form a protective barrier at the interface. Chaiyasit et al. (2000) found that lipids in oil-in-water emulsion dispersed with polyoxyethylene 10 lauryl ether (Brij-lauryl) were more rapidly oxidized than those with polyoxyethylene 10 stearyl ether (Brij-stearyl). When polyoxyethylene 10 stearyl ether (Brij 76) and polyoxyethylene 100 stearyl ether (Brij 700) were used as emulsifiers, the rate of lipid oxidation was lower in Brij 700 stabilized emulsions than those stabilized with Brij 76 (Silvestre et al., 2000).

The hydrophile-lipophile balance (HLB) and acyl size of emulsifiers have been reported to influence the oxidative stability of TAG in oil-in-water emulsions (Kubouchi et al., 2002). The stability increased as the HLB of different kinds of sucrose esters also increased, although these emulsifiers had no effect on the oxidative stability of TAG in the bulk phase. When the HLB of sucrose ester was the same, the oxidative stability increased with increasing acyl chain length of TAG. The effect of HLB or fatty acyl composition may also be explained by the different packing of emulsifiers at the interface. Polyglycerol esters also affected the TAG oxidation in emulsions, but the difference in the oxidative stability of TAG dispersed with polyglycerol ester cannot be explained only with HLB or acyl composition. The wide variation of polymerization degree and presence of unreacted polyglycerol are other factors that could explain the effect of polyglycerol esters on the oxidative stability of lipids in oil-in-water emulsions.

Many studies have shown the antioxidant activity of food proteins in emulsions. Proteins are one of the most important natural food emulsifiers because they are surface active and are effective in preventing the coalescence of oil-in-water emulsions. Protein-stabilized emulsions are complicated because protein molecules have various conformations at the interface that make a physical barrier against prooxidants; in addition, some proteins have the ability to chelate transition metals and to scavenge free radical and reactive oxygen species. Moreover, an interface formed with proteins has an electrical charge at pH values above or below the isoelectric point (pI) of the proteins. Scavenging activity has been mainly considered by aromatic (e.g., tyrosine, tryptophan, and phenylalanine) and sulfur-containing (free cysteine or methionine) amino acids (Tong et al., 2000; Diaz et al., 2003; Faraji et al., 2004; Park et al., 2012). The high oxidative stability of lipids has been reported in emulsions prepared with whey protein (Hu et al., 2003; Osborn & Akoh, 2004; Lee et al., 2011), casein (Hirose et al., 2003; Faraji et al., 2004; Kellerby et al., 2006a; Haahr & Jacobsen, 2008; Mora-Gutierrez et al., 2010; Richards et al., 2011; Dong et al.,

2012; Matalanis et al., 2012), β-lactoglobulin (Kellerby et al., 2006b; Elias et al., 2007), and soy protein (Huang et al., 2011; Park et al., 2012). Transferrin, phosvitin, and lactoferrin have specific iron-binding sites to reduce the oxidation of oil-in-water emulsions containing transition metals (Huang et al., 1999; Mancuso et al., 1999; Jacobsen & Timm, 2001; Nielsen et al., 2004).

Because the radical scavenging and metal chelating activity of proteins is mainly due to the amino-acid side chains, the activity of globular proteins can be increased by hydrolysis during thermal processing or enzymatic treatment (Taylor & Richardson, 1980; Tong et al., 2000; Elias et al., 2007). Many research groups have studied the antioxidant activity of proteins at either the droplet interface or in the continuous phase of emulsions.

Conclusion

Global research has proven that cardiovascular risk lessens if individuals take omega-3 PUFA, especially EPA and DHA. Therefore, food manufacturers are interested in adding these PUFAs to their products and developing effective and nutritionally acceptable techniques to mitigate the oxidative deterioration of EPA and DHA. Oxidative stability of PUFA generally increases by decreasing temperature, avoiding light exposure, inactivating prooxidants, minimizing oxygen concentration, and using antioxidants. The encapsulation of PUFA in solid material can protect them against oxidation. However, food application of the solid particle is limited, and the solid materials for encapsulation are sometimes not favorable to consumers.

Oil-in-water emulsions are one of the most common forms of lipids in foods, found in milk, salad dressing, mayonnaise, sauces, soups, beverages, and infant formula. Therefore, it is important to find a more effective method to prevent the oxidation of omega-3 PUFA in oil-in-water emulsions. The polar paradox is a useful theory to explain the activity of antioxidants in multiphase systems from a general perspective, but the activity of antioxidants cannot be explained only by the overall polarity or electrical charge of the interface.

Interactions at the molecular level between interfacial components have important effects on lipid oxidation in oil-in-water emulsions. Future studies may focus on the interactions of substrate TAG, emulsifiers, antioxidants, and other surface active components—namely, free fatty acids, monoacylglycerols, diacylglycerols, sterols, and lipid oxidation products—in the interfacial region. oil-in-water emulsions stabilized by food proteins and/or phospholipids with natural antioxidants will be particularly interesting as these surface-active compounds and natural antioxidants are known to have healthy benefits on a variety of biological activities.

References

Ando, K.; Nagata, K.; Yoshida, R.; Kikugawa, K.; Suzuki, M. Effect of n-3 polyunsaturated fatty acid supplementation on lipid peroxidation of rat organs. Lipids 2000, 35, 401–407.

Araseki, M.; Kobayashi, H.; Hosokawa, M.; Miyashita, K. Lipid peroxidation of a human hepatoma cell line (HepG2) after incorporation of linoleic acid, arachidonic acid, and docosahexaenoic acid. Biosci. Biotechnol. Biochem. 2005, 69, 483–490.

Araseki, M.; Yamamoto, K.; Miyashita, K. Oxidative stability of polyunsaturated fatty acid in phosphatidylcholine liposomes. Biosci. Biotechnol. Biochem. 2002, 66, 2573–2577.

Azuma, G.; Kimura, N.; Hosokawa, M.; Miyashita, K. Effect of droplet size on the oxidative stability of soybean oil TAG and fish oil TAG in oil-in-water emulsion. J. Oleo Sci. 2009, 58, 329–338.

Castle, L.; Perkins, M. J. Inhibition kinetics of chain-breaking phenolic antioxidants in SDS micelles. Evidence that intermicellar diffusion rates may be rate-limiting for hydrophobic inhibitors such as α-tocopherol. J. Am. Chem. Soc. 1986, 108, 6381–6382.

Chaiyasit, W.; McClements, D.; Decker, E. The relationship between the physicochemical properties of antioxidants and their ability to inhibit lipid oxidation in bulk oil and oil-in-water emulsions. J. Agric. Food Chem. 2005, 53, 4982–4988.

Chaiyasit, W.; Silvestre, M.; McClements, D.; Decker, E. Ability of surfactant tail group size to alter lipid oxidation in oil-in-water emulsions. J. Agric. Food Chem. 2000, 48, 3077–3080.

Chen, B.; McClements, D. J.; Decker, E. A. Minor components in food oils: A critical review of their roles on lipid oxidation chemistry in bulk oils and emulsions. Crit. Rev. Food Sci. Nutr. 2011, 51, 901–916.

Coupland, J. N.; McClements, D. J. Lipid oxidation in food emulsions. Trends Food Sci. Tech. 1996, 7, 83–91.

Diaz, M.; Dunn, C. M.; McClements, D. J.; Decker, E. A. Use of caseinophosphopeptides as natural antioxidants in oil-in-water emulsions. J. Agric. Food Chem. 2003, 51, 2365–2370.

Di Mattia, C. D.; Sacchetti, G.; Mastrocola, D.; Pittia, P. Effect of phenolic antioxidants on the dispersion state and chemical stability of olive oil O/W emulsions. Food Res. Int. 2009, 42, 1163–1170.

Dong, S.; Wei, B.; Chen, B.; McClements, D. J.; Decker, E. A. Chemical and antioxidant properties of casein peptide and its glucose maillard reaction products in fish oil-in-water emulsions. J. Agric. Food Chem. 2011, 59, 13311–13317.

Elias, R. J.; McClements, D. J.; Decker, E. A. Impact of thermal processing on the antioxidant mechanisms of continuous phase β-lactoglobulin in oil-in-water emulsions. Food Chem. 2007, 104, 1402–1409.

Everts, S.; Davis, J. H., ^{1}H and ^{13}C NMR of multilamellar dispersions of polyunsaturated (22:6) phospholipids. Biophys. J. 2000, 79, 885–897.

Faraji, H.; McClements, D. J.; Decker, E. A. Role of continuous phase protein on the oxidative stability of fish oil-in-water emulsions. J. Agric. Food Chem. 2004, 52, 4558–4564.

Feller, S. E.; Garwrish, K.; MacKerell, A. D., Jr. Polyunsaturated fatty acids in lipid bilayers intrinsic and environmental contributions to their unique physical properties, J. Am. Chem. Soc. 2001, 124, 318–326.

Fomuso, L. B.; Corredig, M.; Akoh, C. Effect of emulsifier on oxidation properties of fish oil-based structured lipid emulsions. J. Agric. Food Chem. 2002, 50, 2957–2961.

Frankel, E. N. Lipid oxidation. Prog. Lipid Res. 1980, 19, 1–22.

Frankel, E. N. Lipid Oxidation; The Oily Press: Dundee, Scotland, 1998; 13-77.

Frankel, E. N.; Huang, S.-W.; Kanner, J.; German, B. Interfacial phenomena in the evaluation of antioxidants: Bulk oils vs emulsions. J. Agric. Food Chem. 1994, 42, 1054–1059.

Frankel, E. N.; Meyer, A. S. The problems of using one-dimensional methods to evaluate multifunctional food and biological antioxidants. J. Sci. Food Agric. 2000, 80, 1925–1941.

Haahr, A.; Jacobsen, C. (2008). Emulsifier type, metal chelation and pH affect oxidative stability of n-3-enriched emulsions. Eur. J. Lipid Sci. Tech. 2008, 110, 949–961.

Heins, A.; McPhail, D.; Sokolowski, T.; Stöckmann, H.; Schwarz, K. The location of phenolic antioxidants and radicals at interfaces determines their activity. Lipids 2007, 42 (6), 573–582.

Hirano, S.; Miyashita, K.; Ota, T.; Nishikawa, M.; Maruyama, K.; Nakayama, S. Aqueous oxidation of ethyl linoleate, ethyl linolenate, and ethyl docosahexaenoate. Biosci. Biotechnol. Biochem. 1997, 61, 281–285.

Hirose, A.; Sato, N.; Miyashita, K. Effects of alkylamines and PC on the oxidative stability of soybean oil TAG in milk casein emulsions. J. Am. Oil Chem. Soc. 2003, 80, 431–435.

Hopia, A. I.; Huang, S.-W.; Schwarz, K.; German, B.; Frankel, E. N. Effect of different lipid systems on antioxidant activity of rosemary constituents carnosol and carnosic acid with and without α-tocopherol. J. Agric. Food Chem. 1996, 44, 2030–2036.

Hu, M.; McClements, D. J.; Decker, E. A. (2003). Impact of whey protein emulsifiers on the oxidative stability of salmon oil-in-water emulsions. J. Agric. Food Chem. 2003, 51, 1435–1439.

Huang, S.-W.; Frankel, E. N. Antioxidant activity of tea catechins in different lipid systems. J. Agric. Food Chem. 1997, 45, 3033–3038.

Huang, S.-W.; Frankel, E. N.; Kanner, J.; German, B. Antioxidant activity of α- and γ-tocopherols in bulk oils and in oil-in-water emulsions. J. Agric. Food Chem. 1994, 42, 2108–2114.

Huang, S.-W.; Frankel, E. N.; Schwarz, K.; German, B. Effect of pH on antioxidant activity of α-tocopherol and trolox in oil-in-water emulsions. J. Agric. Food Chem. 1996a, 44, 2496–2502.

Huang, S.-W.; Hopia, A.; Schwarz, K.; Frankel, E. N.; Kanner, J.; German, B. Antioxidant activity of α-tocopherol and Trolox in different lipid substrates: Bulk oils vs oil-in-water emulsions. J. Agric. Food Chem. 1996b, 44, 444–452.

Huang, P.-H.; Lu, H.-T.; Wang, Y.-T.; Wu, M.-C. Antioxidant activity and emulsion-stabilizing effect of pectic enzyme treated pectin in soy protein isolate-stabilized oil/water emulsion. J. Agric. Food Chem. 2011, 59, 9623–9628.

Huang, S.-W.; Satué-Gracia, M. T.; Frankel, E. N.; German, J. B. Effect of lactoferrin on oxidative stability of corn oil emulsions and liposomes. J. Agric. Food Chem. 1999, 47, 1356–1361.

Huber, T.; Rajamoorthi, K.; Kurze, V. F.; Beyer, K.; Brown, M. F. Structure of docosahexaenoic acid-containing phospholipid bilayers as studied by ^{2}H NMR and molecular dynamics simulations. J. Am. Chem. Soc. 2002, 124, 298–309.

Huster, D.; Jin, A. J.; Arnold, K.; Gawrisch, K. Water permeability of polyunsaturated lipid membranes measured by ^{17}O NMR. Biophys. J. 1997, 73, 855–864.

Isnardy, B.; Wagner, K. H.; Elmadfa, I. Effects of alpha, gamma, and delta-tocopherols on the autoxidation of purified rapeseed oil triacylglycerols in a system containing low oxygen. J. Agric. Food Chem. 2003, 51, 7775–7780.

Jacobsen, C.; Hartvigsen, K.; Thomsen, M. K.; Hansen, L. F.; Lund, P.; Skibsted, L. H.; Holmer, G.; Adler-Nissen, J.; Meyer, A. S. Lipid oxidation in fish oil enriched mayonnaise: Calcium disodium ethylenediaminetetraacetate, but not gallic acid, strongly inhibited oxidative deterioration. 2001, J. Agric. Food Chem. 2001, 49, 1009–1019.
Jacobsen, C.; Timm, M. Oxidation in fish oil enriched mayonnaise: Ascorbic acid and low pH increase oxidative deterioration. J. Agric. Food Chem. 2001, 49, 3947–3956.
Jimenez-Alvarez, D.; Giuffrida, F.; Golay, P. A.; Cotting, C.; Lardeau, A.; Kely, B. J. Antioxidant activity of oregano, parsley, and olive mill wastewaters in bulk oils and oil-in-water emulsions enriched in fish oil. J. Agric. Food Chem. 2008, 56, 7151–7159.
Kamal-Eldin, A.; Mäkinen, M.; Lampi, A.-M. In Lipid Oxidation Pathways; Kamal-Eldin, A., Ed.; AOCS Press: Champaign, IL, 2003; 1–36.
Kawakatsu, T.; Kikuchi, T.; Nakajima, M. Regular-sized cell creation in microchannel emulsification by visual microprocessing method. J. Am. Oil Chem. Soc. 1997, 74, 317–321.
Kellerby, S. S.; Gu, Y. S.; McClements, D. J; Decker, E. A. Lipid oxidation in a menhaden oil-in-water emulsion stabilized by sodium caseinate cross-linked with transglutaminase. J. Agric. Food Chem. 2006a, 54, 10222–10227.
Kellerby, S. S.; McClements, D. J.; Decker, E. A. Role of proteins in oil-in-water emulsions on the stability of lipid hydroperoxides. J. Agric. Food Chem. 2006b, 54, 7879–7884.
Kobayashi, H.; Yoshida, M.; Maeda, I.; Miyashita, K. Proton NMR relaxation times of polyunsaturated fatty acids in chloroform solutions and aqueous micelles. J. Oleo Sci. 2004, 56, 105–108.
Kobayashi, H.; Yoshida, M.; Miyashita, K. Comparative study of the product components of lipid oxidation in aqueous and organic systems. Chem. Phys. Lipids 2003, 126, 111–120.
Kobayashi, I.; Nakajima, M.; Nabetani, Y.; Kikuchi, A.; Shohno, A.; Satoh, K. Preparation of monodisperse micron-scale microspheres by microchannel emulsification. J. Am. Oil Chem. Soc. 2001, 78, 797–802.
Kubo, K.; Saito, M.; Tadokoro, T.; Maekawa, A. Dietary docosahexaenoic acid does not promote lipid peroxidation in rat tissue to the extent expected from peroxidizability index of the lipids. Biosci. Biotechnol. Biochem. 1998, 62, 1698–1706.
Kubouchi, H.; Kai, H.; Miyashita, K.; Matsuda, K. Effects of emulsifiers on the oxidative stability of soybean oil TAG in emulsions. J. Am. Oil Chem. Soc. 2002, 79, 567–570.
Laguerre, M.; López Giraldo, L. J.; Lecomte, J.; Figueroa-Espinoza, M.-C.; Baréa, B.; Weiss, J.; Decker, E. A.; Villeneuve, P. Chain length affects antioxidant properties of chlorogenate esters in emulsion: The cutoff theory behind the polar paradox. J. Agric. Food Chem. 2009, 57, 11335–11342.
Laguerre, M.; López Giraldo, L. J.; Lecomte, J.; Figueroa-Espinoza, M.-C.; Baréa, B.; Weiss, J.; Decker, E. A.; Villeneuve, P. Relationship between hydrophobicity and antioxidant ability of "phenolipids" in emulsion: A parabolic effect of the chain length of rosmarinate esters. J. Agric. Food Chem. 2010, 58, 2869–2876.
Lee, S. J.; Choi, S. J.; Li, Y.; Decker, E. A.; McClements, D. J. Protein-stabilized nanoemulsions and emulsions: Comparison of physicochemical stability, lipid oxidation, and lipase digestibility. J. Agric. Food Chem. 2011, 59, 415–427.
Lethuaut, L.; Métro, F.; Genot, C. Effect of droplet size on lipid oxidation rates of oil-in-water emulsions stabilized by protein. J. Am. Oil. Chem. Soc. 2002, 79, 425–430.

Lucas, R.; Comelles, F.; Alcántara, D.; Maldonado, O.; Curcuroze, M.; Parra, J. L.; Morales, J. C. Surface-active properties of lipophilic antioxidants tyrosol and hydroxytyrosol fatty acid esters: A potential explanation for the nonlinear hypothesis of the antioxidant activity in oil-in-water emulsions. J. Agric. Food Chem. 2010, 58, 8021–8026.

Mancuso, J. R.; McClements, D. J.; Decker, E. A. The effects of surfactant type, pH, and chelators on the oxidation of salmon oil-in-water emulsions. J. Agric. Food Chem. 1999, 47, 4112–4116.

Matalanis, A.; Decker, E. A.; McClements, D. J. Inhibition of lipid oxidation by encapsulation of emulsion droplets within hydrogel microspheres. Food Chem. 2012, 132, 766–772.

McClements, D. J.; Decker, E. A. Lipid oxidation in oil-in-water emulsions: Impact of molecular environment on chemical reactions in heterogeneous food systems. J. Food Sci. 2000, 65, 1270–1282.

Medina, I.; Lois, S.; Alcantara, D.; Lucas, R.; Morales, J. C. Effect of lipophilization of hydroxytyrosol on its antioxidant activity in fish oils and fish oil-in-water emulsions. J. Agric. Food Chem. 2009, 57, 9773–9779.

Mei, L.; McClements, D. J.; Decker, E. A. Evidence of iron association with emulsion droplets and its impact on lipid oxidation. J. Agric. Food Chem. 1998a, 46, 5072–5077.

Mei, L.; McClements, D. J.; Decker, E. A. Lipid oxidation in emulsions as affected by charge status of antioxidants and emulsion droplets. J. Agric. Food Chem. 1999, 47, 2267–2273.

Mei, L.; McClements, D. J.; Wu, J.; Decker, E. A. Iron-catalyzed lipid oxidation in emulsion as affected by surfactant, pH and NaCl. Food Chem. 1998b, 61, 307–312.

Miyashita, K. Polyunsaturated lipids in aqueous systems do not follow our preconceptions of oxidative stability. Lipid Tech. Newsletter 2002, 8, 35–41.

Miyashita, K. Effects of flexibility and permeability of polyunsaturated fatty acid molecules on their oxidative stability in aqueous systems. Lipid Tech. Newsletter 2004, 16, 197–202.

Miyashita, K.; Azuma, G.; Ota, T. Oxidative stability of geometric and positional isomers of unsaturated fatty acids in aqueous solution. J. Jpn. Oil Chem. Soc. 1995, 44, 425–430.

Miyashita, K.; Nara, E.; Ota, T. Oxidative stability of polyunsaturated fatty acids in an aqueous solution. Biosci. Biotechnol. Biochem. 1993, 57, 1638–1640.

Miyashita, K.; Nara, E.; Ota, T. Comparative study on the oxidative stability of phosphatidylcholines from salmon egg and soybean in an aqueous solution. Biosci. Biotechnol. Biochem. 1994a, 58, 1772–1775.

Miyashita, K.; Tateda, N.; Ota, T. Oxidative stability of free fatty acid mixtures from soybean, linseed, and sardine oils in an aqueous solution. Fisheries Sci. 1994b, 60, 315–318.

Mora-Gutierrez, A.; Attaie, R.; Farrell, H.M., Jr. (2010). Lipid oxidation in algae oil-in-water emulsions stabilized by bovine and caprine caseins. J. Agric. Food Chem. 2010, 58, 5131–5139.

Nara, E.; Miyashita, K.; Ota, T. Oxidative stability of PC containing linoleate and docosahexaenoate in an aqueous solution with or without chicken egg albumin. Biosci. Biotechnol. Biochem. 1995, 59, 2319–2320.

Nara, E.; Miyashita, K.; Ota, T. Oxidative stability of liposomes prepared from soybean PC, chicken egg PC, and salmon egg PC. Biosci. Biotechnol. Biochem. 1997, 61, 1736–1738.

Nara, E.; Miyashita, K.; Ota, T.; Nadachi, Y. The oxidative stabilities of polyunsaturated fatty acids in salmon egg phosphatidylcholine liposomes. Fisheries Sci. 1998, 64, 282–286.

Nielsen, N. S.; Petersen, A.; Meyer, A. S.; Timm-Heinrich, M.; Jacobsen, C. Effects of lactoferrin, phytic acid, and EDTA on oxidation in two food emulsions enriched with long-chain polyunsaturated fatty acids. J. Agric. Food Chem. 2004, 52, 7690–7699.

Nishida, T.; Orikasa, Y.; Ito, Y.; Yu, R.; Yamada, A.; Watanabe, K.; Okuyama, H. *Eschericha coli* engineered to produce eicosapentaenoic acid becomes resistant against oxidative damaeges. FEBS Lett. 2006, 580, 2731–2735.

Nuchi, C. D.; Hernandez, P.; McClements, D. J.; Decker, E. A. Ability of lipid hydroperoxides to partition into surfactant micelles and alter lipid oxidation rates in emulsions. J. Agricul. Food Chem. 2002, 50, 5445–5449.

Olbrich, K.; Rawicz, W.; Needham, D.; Evans, E. Water permeability and mechanical strength of polyunsaturated lipid bilayers. Biophys. J. 2000, 79, 321–327.

Osborn, H. T.; Akoh, C. C. Effect of emulsifier type, droplet size, and oil concentration on lipid oxidation in structured lipid-based oil-in-water emulsions. Food Chem. 2004, 84, 451–456.

Panya, A.; Laguerre, M.; Bayrasy, C.; Lecomte, J.; Villeneuve, P.; McClements, D. J.; Decker, E. A. An investigation of the versatile antioxidant mechanisms of action of rosmarinate alkyl esters in oil-in-water emulsions. J. Agric. Food Chem. 2012, 60, 2692–2700.

Park, E. Y.; Nakamura, Y.; Sato, K.; Matsumura, Y. Effects of amino acids and peptide on lipid oxidation in emulsion systems. J. Am. Oil Chem. Soc. 2012, 89, 477–484.

Porter, N. A.; Galdwell, S. E.; Mills, K. A. Mechanisms of free radical oxidation of unsaturated lipids. Lipids 1995, 30, 277–290.

Porter, W. L. Autoxidation in Food and Biological Systems; Simic, M. G.; Karel, M. Eds.; Plenum Press: New York, 1980; 295–365.

Porter, W. L. Paradoxical behavior of antioxidants in food and biological systems. Toxicol. Ind. Health 1993, 9, 93–122.

Pryor, W. A.; Cornicelli, J. A.; Devall, L. J.; Tait, B.; Trivedi, B. K.; Witiak, D. T.; Wu, M. A rapid screening test to determine the antioxidant potencies of natural and synthetic antioxidants. J. Org. Chem. 1993, 58, 3521–3532.

Richards, A.; Golding, M.; Wijesundera, C.; Lundin, L. The influence of secondary emulsifiers on lipid oxidation within sodium caseinate-stabilized oil-in-water emulsions. J. Am. Oil Chem. Soc. 2011, 88, 65–73.

Richards, M. P.; Chaiyasit, W.; McClements, D. J.; Decker, E. A. Ability of surfactant micelles to alter the partitioning of phenolic antioxidants in oil-in-water emulsions. J. Agric. Food Chem. 2002, 50, 1254–1259.

Saiz, L.; Klein, M. L. Structural properties of a highly polyunsaturated lipid bilayer from molecular dynamics simulations. Biophys. J. 2001, 81, 204–216.

Sasaki, K.; Alamed, J.; Weiss, J.; Villeneuve, P.; López Giraldo, L. J.; Lecomte, J.; Figueroa-Espinoza, M.-C.; Decker, E. A. Relationship between the physical properties of chlorogenic acid esters and their ability to inhibit lipid oxidation in oil-in-water emulsions. Food Chem. 2010, 118, 830–835.

Schwarz, K.; Frankel, E. N.; German, B. Partition behaviour of antioxidative phenolic compounds in heterophasic systems. Lipid Fett. 1996, 98, 115–121.

Silvestre, M.; Chaiyasit, W.; Brannan, R.; McClements, D.; Decker, E. Ability of surfactant head group size to alter lipid and antioxidant oxidation in oil-in-water emulsions. J. Agric. Food Chem. 2000, 48, 2057–2061.

Sørensen, A. D. M.; Nielsen, N. S.; Decker, E. A.; Let, M. B.; Xu, X.; Jacobsen, C. The efficacy of compounds with different polarities as antioxidants in emulsions with omega-3 lipids. J. Am. Oil Chem. Soc. 2011, 88, 489–502.

Sugiura, S.; Nakajima, M.; Iwamoto, S.; Seki, M. Interfacial tension driven monodispersed droplet formation from microfabricated channel array. Langmuir 2001, 17, 5562–5566.

Sugiura, S.; Nakajima, M.; Seki, M. Preparation of monodispersed emulsions with large droplets in microchannel emulsification. J. Am. Oil Chem. Soc. 2002, 79, 515–519.

Sun, Y.-E.; Wang, W.-D.; Chen, H.-W.; Li, C. Autoxidation of unsaturated lipids in food emulsion. Crit. Rev. Food Sci. Nutr. 2011, 51, 453–466.

Taylor, M. J.; Richardson, T. Antioxidant activity of skim milk: Effect of heat and resultant sulfhydryl groups. J. Dairy Sci. 1980, 63, 1783–1795.

Tong, L. M.; Sasaki, S.; McClements, D. J.; Decker, E. A. Mechanisms of the antioxidant activity of a high molecular weight fraction of whey. J. Agric. Food Chem. 2000, 48, 1473–1478.

Wagner, K. H.; Isnardy, B.; Elmadfa, I. γ- and Δ-tocopherols are more effective than α-tocopherol on the autoxidation of a 10% rapeseed oil triacylglycerol-in-water emulsion with and without a radical initiator. Eur. J. Lipid Sci. Technol. 2004, 106, 44–51.

Wander, R. C.; Du, S.-H. Oxidation of plasma proteins is not increased after supplementation with eicosapentaenoic and docosahexaenoic acids. Am. J. Clin. Nutr. 2000, 72, 731–737.

Waraho, T.; McClements, D. J.; Decker, E. A. Mechanisms of lipid oxidation in food dispersions. Trends Food Sci. Tech. 2011, 22, 3–13.

CHAPTER

5

The Role of the Interfacial Layer and Emulsifying Proteins in the Oxidation in Oil-in-Water Emulsions

Claude Genot[1], Claire Berton[2], and Marie-Hélène Ropers[1]

Introduction

According to food safety agencies, Western consumers need to adjust their lipid intake and opt for polyunsaturated fats—more specifically, those containing n-3 fatty acids. However, these fats tend to oxidize during technological operations, storage, and digestion. The resulting products of lipid oxidation lead to the development of off-flavors that decrease the food acceptability of the products. They also could contribute to the development of pathologies, such as chronic inflammation, metabolic syndrome, and colon cancer (Awada et al., 2012; Guéraud et al., 2010; Kanazawa et al., 2002; Pillon et al., 2009; Zarrouki et al., 2007).

In many enriched-food formulations, polyunsaturated lipids are organized in the form of oil droplets that are dispersed in a more or less hydrated matrix, so-called oil-in-water emulsions. Hydrated matrices, such as dairy products, meat products, dressings, and soups, are also viewed as oil-in-water emulsions. Oil-in-water emulsions can be delivery intermediates used during the formulation process to incorporate healthy oils into foods more easily than the original oils and fats. Additionally, they are the basis for many cosmetic, pharmaceutical, and medical applications, as for instance parenteral emulsions. Consequently, model

[1]INRA, UR1268 Biopolymères Interactions Assemblages, BP 71627 F-44316 Nantes cedex 3 France; [2]Department of Food Science, The Pennsylvania State University, University Park, Pennsylvania 16802.

oil-in-water emulsions have been widely used in recent years to better understand the mechanisms that control their physical and chemical stabilities.

In this chapter, we will first introduce recent findings that address the role of the interfacial layer and emulsifiers in oxidation. Then, we present recent results of our laboratory in this area. We have focused on emulsions stabilized by monolayers of food-grade emulsifiers, and we propose new concepts for the role of proteins and lateral interface homogeneity in the development of oxidation.

Emulsions: Definition and Key Concepts

Food matrices are complex, multiphase systems, a number of which are oil-in-water emulsions. Based on the work of Dickinson (1992) and McClements (2005), this short overview of emulsions aims to describe the main properties of emulsions.

Oil-in-water emulsions are made up of two immiscible liquids: the first, oil, which is dispersed in water, the second, as droplets. These droplets constitute the dispersed phase, while the aqueous phase around them constitutes the continuous phase. The area between the dispersed and continuous phase is the interface. Often characterized by an average diameter, droplet size distribution is one of the main characteristics of oil-in-water emulsions. It determines the number of droplets and their total surface area; the smaller the droplets are, the higher their number and the developed surface for the same amount of oil. Food emulsions display a large range of droplet sizes, from less than 0.1 µm to 20 µm or more.

Oil-in-water emulsions are thermodynamically unstable systems; thus, oil droplets have to be stabilized physically to avoid separation of the oil and aqueous phases. Texture modifiers and emulsifiers can accomplish this goal. Texture modifiers stabilize the emulsions by increasing the viscosity of the continuous phase, which limits the creaming of oil droplets and their eventual flocculation and coalescence. Emulsifiers adsorb at the oil–water interface during homogenization, which decreases the surface tension between the oil and the water phases and the total free energy of the system. Emulsifiers can also prevent the flocculation and coalescence of oil droplets via steric or electrostatic repulsions. Food-grade emulsifiers include high molecular weight emulsifiers (mainly proteins) and low molecular weight emulsifiers (surfactants and polar lipids, for example). They can be used alone, mixed together, or associated with polysaccharides to improve the stability of the emulsions. Additionally, proteins can be used in their native form or partially denatured and aggregated by high-pressure or thermal treatments to improve their adsorption at interfaces.

The composition and properties of interfaces depend on the nature of the adsorbed emulsifiers and on the interactions between the adsorbed molecules. Accordingly, interfaces exhibit various physicochemical characteristics, such as surface activity, surface load, thickness, surface coverage, electric charge, lateral and transversal topography (homogeneity), hydration, lateral mobility, interfacial rheological properties, and permeability. Interfaces are also dynamic systems, which means that they can be affected by the emulsifiers' abilities to change conformation, desorb, or undergo further (secondary) adsorption. All of these properties may evolve with time and when environmental conditions change. For instance, if the amount of unadsorbed emulsifiers evolves, it will modify the physical and chemical characteristics of the system (Euston & Mayhill, 2001; Dalgleish et al., 2002; Bongard et al., 2009). Chemical reactions, such as oxidation, will also modify the composition and properties of the interface layer. First, triacylglycerol hydroperoxides exhibit surface-active properties and tend to migrate toward the interface (Nuchi et al., 2002). Second, interfacial proteins are also subject to oxidative modifications (Leaver et al., 1999; Rampon et al., 2001) that, in turn, may modify the interface properties. These modifications of the interface layer(s) affect the rheological properties and the stability of the emulsions.

The amount of emulsifier forming the interfacial layer, also defined as surface load, depends on the emulsifier concentration and emulsification conditions. Typical surface loads encountered for protein–stabilized interfaces range from 1.5 to 4.5 mg m^{-2}, depending on the protein type and the pH (Atkinson et al., 1995; Bos & van Vliet, 2001). However, only the hydrophobic parts of the polypeptidic chain are actually in direct contact with the oil–water interface and the other protein segments protrude into the aqueous phase. Surface load is usually lower for surfactant–stabilized interfaces, between 1–2 mg m^{-2} (Bos & van Vliet, 2001).

For monolayer- and/or single emulsifier–covered interfaces, the interface thickness varies from 0.5–1 nm for the surfactant-stabilized interface to 1–15 nm for protein emulsifiers (Dalgleish, 1993; Fang & Dalgleish, 1993; Atkinson et al., 1995; Dickinson, 2009; Singh, 2011). Due to internal constraints and the arrangement of the polypeptidic chains on the droplet's surface, only 30–40% of the interface is actually covered with proteins. Proteins form immobile and viscoelastic interfacial films that exhibit non-Newtonian behavior. They are also susceptible to structural and conformational rearrangements after adsorption, which implies that the surface load, the thickness of the protein–stabilized interfaces, and their physical properties may evolve (Dickinson, 1992; Tcholakova et al., 2008). By contrast, surfactant molecules, especially polyoxyethylene sorbitan esters, such as Tween 20, are more tightly packed at the interface than proteins (Wilde et al., 2004; Grigoriev & Miller, 2009). They cover nearly 100% of the interface (Dickinson, 1992) and form fluid interfaces with a measurable coefficient of lateral diffusion (Wilde et al., 2004).

The electrostatic charge and the interface thickness conferred by emulsifiers directly impact steric and electrostatic repulsions, which prevent emulsions from flocculation and coalescence. Interfacial proteins generate both steric and electrostatic repulsions; therefore, they can form a physical barrier to droplet coalescence (Dickinson, 1992; Wilde et al., 2004; Tcholakova et al., 2008; Grigoriev & Miller, 2009). Depending on the protein type, a variety of repulsive interactions may occur. When dissolved in pure water, bovine serum albumin (BSA) generates long-range steric repulsions, whereas β-casein (BCN) exerts mainly electrostatic repulsions. Finally, β-lactoglobulin (BLG) exhibits intermediate behavior, with electrostatic repulsions at large distances and steric repulsions at short distances between the nearby interfaces (Dimitrova & Leal-Calderon, 1999; Dimitrova et al., 2001; Dimitrova et al., 2004). In some cases, proteins may even favor the physical destabilization of emulsions through either droplet–droplet bridging, which leads to flocculation, or depletion flocculation (Dickinson, 1992; Dimitrova & Leal-Calderon, 1999; Tcholakova et al., 2008; Grigoriev & Miller, 2009). Surfactants are generally considered to be less efficient than proteins in avoiding the breakage of the interfacial film and droplet coalescence (Dickinson, 1992; Wilde et al., 2004). However, ionic surfactants induce electrostatic repulsions and nonionic surfactants, such as Tween 20, can exert steric repulsions because of their polyoxyethylene chains. Tween 20–stabilized interfaces have also been shown to fit well with a model of electrostatic repulsion (Dimitrova & Leal-Calderon, 1999; Dimitrova et al., 2001), which the interface ionic environment may explain. In fact, in pH 6.7 and 3.0 Tween 20 stabilized–emulsions (80 mM NaCl), the oil droplets have a slightly negative charge (Berton et al., 2011b).

The amount of emulsifiers used to stabilize oil-in-water emulsions is generally much higher than the amount required to cover the entire interfacial surface. As a consequence, a substantial fraction of water-soluble emulsifiers remains in the aqueous phase of oil-in-water emulsions. Because emulsifiers possess different surface properties, competitive adsorption phenomena occur when mixtures of emulsifiers are used (Courthaudon et al., 1991; Dickinson & Tanai, 1992; Mackie et al., 2000; van Aken, 2003; Kotsmar et al., 2008; Morris & Gunning, 2008; Day et al., 2010). Thus, the nature and concentration of molecules present at the interface do not necessarily reflect the global composition of emulsifiers in the emulsions. The distribution of proteins and surfactants at the oil–water interfaces is determined by the competitive adsorption between the emulsifiers and by the nature of the protein–surfactant interactions, both at the interface and in the bulk phase. The subsequent composition of the interface has a pronounced impact on the rheological properties of the interfaces (Maldonado-Valderrama & Rodríguez Patino, 2010) and on their physical stabilities. In addition, unadsorbed emulsifiers—either as monomers, aggregates, or micelles—affect the physical and chemical stability of emulsions (Casanova & Dickinson, 1998; Donnelly et al., 1998; Ponginebbi et al., 1999; Dickinson et al., 2003; Hu et al., 2003; Faraji et al.,

2004; Villière, 2005; Dauphas et al., 2008; Ye, 2008; Cheng et al., 2010). This is why it is important to quantify the partition of the emulsifiers within the different emulsion phases. We have developed a quick and easy methodology to determine this partition (Berton et al., 2011a), which we present later in this chapter.

Oxidation of Emulsions: Main Findings

The same general laws and reaction schemes for oxidation in bulk oils and fats govern lipid oxidation in emulsions (Chapter 1 of this book). However, research has identified several factors that intervene in the development of lipid oxidation in emulsions due to their multiphase and multicomponent, molecule-molecule and colloidal interactions, and the partition and dynamics of the involved reacting entities. Several reviews based on the extensive work performed by different research teams over the past 15 years synthesize our knowledge about the influence of these factors (McClements & Decker, 2000; Genot et al., 2003; Frankel, 2005; Jacobsen et al., 2008; Waraho et al., 2011). These reviews present concepts founded in experimental results and logical reasoning, along with the shared assumption that oxidation in emulsions is an interfacial phenomenon (McClements & Decker, 2000; Genot et al., 2003; Frankel, 2005; Jacobsen et al., 2008). According to this paradigm, metal ions or free radicals present in the aqueous phase initiate lipid oxidation in oil-in-water emulsions at the oil–water interface. Thus, reacting species have to reach and cross the interface before accessing the oxidizable substrate—namely, the unsaturated fatty acids of the oil phase. The oxidative stability of emulsified lipids is therefore assumed to depend greatly on the properties of the interfacial layer. To learn more about oxidizing emulsion, researchers have varied the emulsifier type, the droplet size distribution, the pH, the homogenization procedure, the composition of the aqueous phase, the lipophilicity of antioxidants, etc. They interpreted these data on the basis of the mechanisms that are listed below and schematized in Fig. 5.1.

1. Because of a subsequent increase in the interfacial area, studies have often postulated that a decrease in the mean size of oil droplets favors the prooxidative species' access to the oil phase. Accordingly, emulsified lipids generally oxidized faster than the bulk oils (Lethuaut et al., 2002). However, multiple studies (see, for example, Lethuaut et al., 2002; Villière, 2005; Dimakou et al., 2007; Let et al., 2007; Sorensen et al., 2007; Nakazawa et al., 2008; Azuma et al., 2009) contradict each other about the effect of the size of oil droplets. On the one hand, a smaller droplet size implies a larger interfacial area and thus favors contact between the lipid's dispersed phase and aqueous phase prooxidants. On the other hand, a higher surface area also implies higher amounts of adsorbed species, which may have

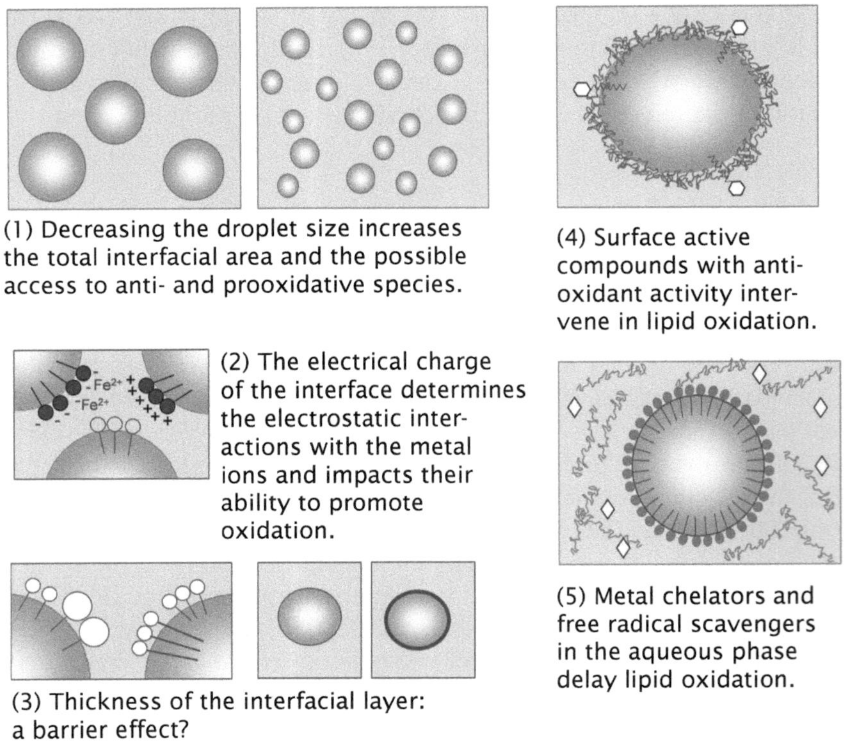

Fig. 5.1. Main interfacial parameters involved in lipid oxidation of oil-in-water emulsions.

some protective effect. Finally, it is noteworthy that the homogenization procedure itself can affect the development of lipid oxidation in emulsions (Let et al., 2007; Horn et al., 2009).

2. The electrical charge of the interfacial layer determines the attractive or repulsive interactions between metal ions and impacts their ability to promote oxidation. According to this model, a positively charged interface that repels cationic metal ions would delay lipid oxidation compared to a negatively charged interface that is thought to favor lipid oxidation (Mei et al., 1998; McClements & Decker, 2000). Such claims maintained that positively charged emulsifying proteins, at pH under their isoelectric point, protected the lipid phase against oxidation (Donnelly et al., 1998; Osborn & Akoh, 2004).

3. A thick interface layer could protect more than a thinner one against oxidation due to its "barrier" effect, which insulates the core of the oil droplets, containing the unsaturated fatty acids, against the aqueous prooxidant species (Silvestre et al., 2000). The authors observed accordingly that when the size of the polar headgroup of Brij (polyoxyethylene glycol dodecyl ether) nonionic surfactants increased, the amount of hexanal produced in emulsions at pH 7 decreased by a third. A longer surfactant hydrophobic tail also decreased oxidation at an acidic pH, whereas it had no effect at a neutral pH. Accordingly, proteins that generate thicker interface layers were often assumed to provide more protection than surfactants and phospholipids. The barrier effect also includes presumed effects of the structure and physical characteristics of the interfaces on reactant diffusion.

4. The presence at the interface of metal chelators and/or free-radical scavengers, such as phenolic compounds or proteins, contributes to the protection afforded by the interface if we assume that specific emulsifier-antioxidant interactions do not hamper the antioxidative effect; in some cases, reverse prooxidative effects have even been evoked (Horn et al., 2009; Oehlke et al., 2011; Medina et al., 2012).

5. Metal chelators—such as EDTA, citric acid, caseins, or xanthan gum—or free-radical scavengers, which include proteins or hydrocolloids, present in the aqueous phase are also capable of retarding oxidative development. Metal chelators prevent the metal redox cycles, form insoluble metal complexes, or provide steric hindrance between metals and sensitive compounds (Choe & Min, 2009); thus, they delay lipid oxidation (Villière et al., 2005). Free-radical scavengers also prevent free radicals from reaching the interface layer.

Additionally, some researchers have assumed that the colloidal structures present in the aqueous phase or inside the oil droplets play a significant role in oxidative development. Aqueous micelles have a protective effect in that they sequester prooxidative species (Nuchi et al., 2002; Cho et al., 2008; Richards et al., 2011; Panya et al., 2012), whereas colloidal structures in the oil phase favor oxidation (Chaiyasit et al., 2008; Chen et al., 2010).

These statements are generally accepted and currently aid in the interpretation of results obtained for various emulsified systems. In that way, protein-stabilized emulsions have repeatedly exhibited better oxidative stability than surfactant-stabilized ones. The researchers had assumed that this resulted from the free radical and chelating properties of the adsorbed and unadsorbed proteins and also from a possible "barrier" effect caused by the protein–covered interface layer. We estimated the concentrations of unadsorbed emulsifiers in these emulsions from the initial amounts of emulsifiers used to prepare the emulsions and the average droplet sizes given by the authors and used the interfacial molecular areas of emulsifiers found in the literature. According to these estimates, the concentrations of unadsorbed emulsifiers in oil-in-water emulsions were generally above 3 g L^{-1} (Fig. 5.2), and the proportions of unadsorbed emulsifiers, were often higher than 70%, and even higher than 90% in some cases. This practice does not permit (i) differentiation between the contributions of adsorbed and unadsorbed emulsifiers, and (ii) separation of the specific effects of the protein accumulated in the aqueous phase from those resulting from the interfacial layers.

To address this question, Faraji et al. (2004) removed unadsorbed proteins from the continuous phase of emulsions stabilized at pH 7 using whey protein isolate through repeated centrifugation and resuspension of the emulsion droplets. They found that the unwashed emulsions were more oxidatively stable than washed emulsions, suggesting that proteins in the continuous phase had antioxidative effects. Accordingly, they observed that when proteins (for example, whey protein isolate, caseinate, or soy protein isolate) were added to the aqueous phase of washed emulsions, oxidation decreased. However, these experiments did not compare the oxidative stability of washed emulsions stabilized by different emulsifiers. Thus, they lack direct evidence for the interfacial layer's effects on the oxidation of emulsified lipids and did not elucidate the mechanisms involved. This is why we decided to address the role of the interfacial layer in emulsion oxidation by formulating emulsions stabilized by a large range of emulsifiers. Their unadsorbed amounts diminished as much as possible.

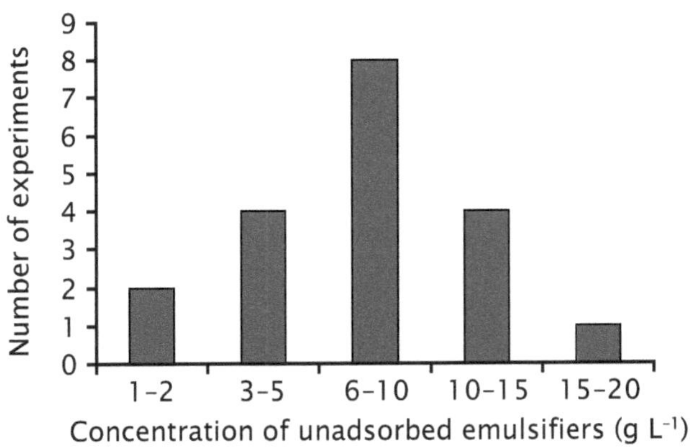

Fig. 5.2. Frequency distribution of the unadsorbed emulsifier concentration in the oil-in-water emulsions used for studying lipid oxidation over the period 2002–2011. The concentration was estimated from the oil droplet size, the oil volume fraction, and the total concentration of the emulsifiers given by authors assuming similar surface load for the same emulsifier in all studies.

Designing Oil-in-Water Emulsions to Study the Role of the Interface Layer on Oxidation

Physically stable emulsions with a controlled structure, controlled composition of the interface, and controlled ratio of adsorbed/unadsorbed emulsifier have been formulated to better assess the role of the interfacial layer in the phenomena of lipid oxidation. We present the scheme of our strategy in Fig. 5.3. We had previously

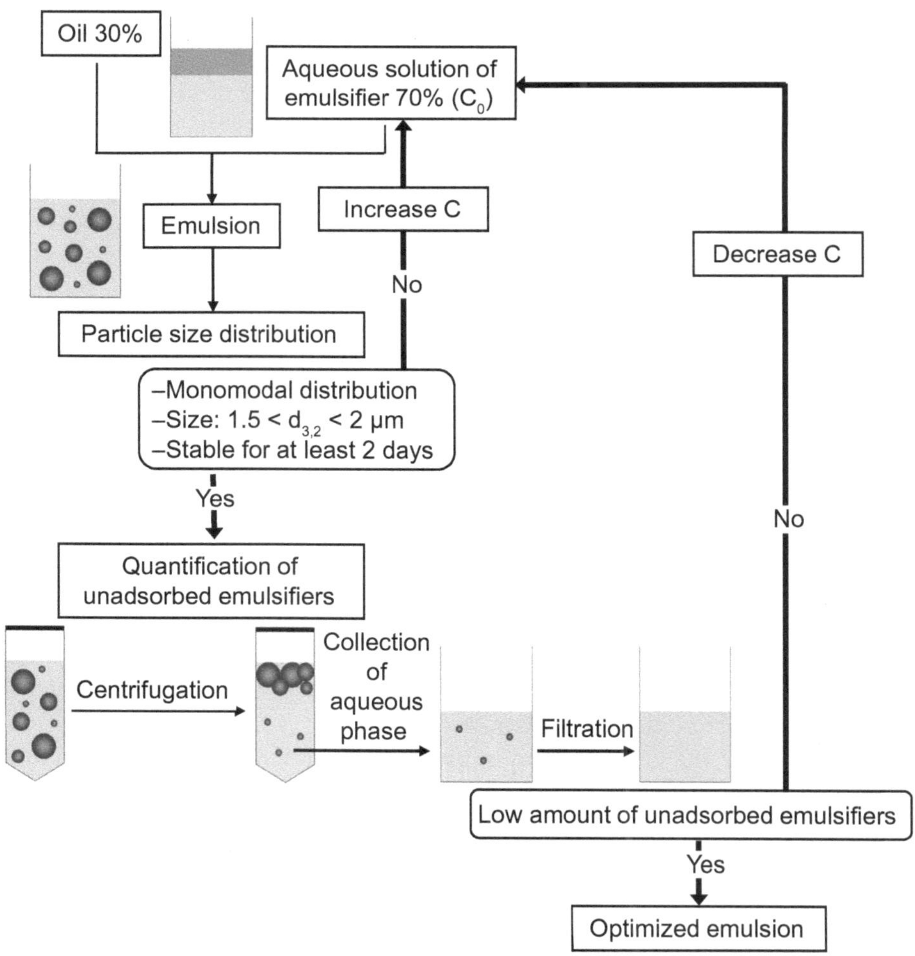

Fig. 5.3. Procedure for optimizing the formulation of emulsions characterized by low amounts of unadsorbed emulsifiers and good physical stability.

tried to wash emulsions as described by Faraji et al., (2004), but the resulting emulsions were physically unstable.

We used stripped rapeseed oil to prepare emulsions with the goal of preventing the antioxidant effect of tocopherols. Chosen emulsifiers included milk proteins like bovine serum albumin (BSA), β-casein (BCN), and β-lactoglobulin (BLG) and surfactants; examples of the latter include Tween 20 [polyoxyethylene (20) sorbitan monolaurate or polysorbate 20] and Tween 80 [polyoxyethylene (20) sorbitan monooleate or polysorbate 80]. They were previously solubilized with an initial concentration C_0 in a buffered aqueous phase that mimics the physicochemical environment of milk and milk products [1,4-piperazinediethanesulfonic acid (PIPES) 10 mM, NaCl 150 mM, pH 6.7]. We prepared oil-in-water emulsions (30/70; w:w) by high-pressure homogenization of a mixture of stripped rapeseed oil and an aqueous solution of one emulsifier. The droplet size was determined using static light scattering. If the distribution of particles was monomodal and the average diameter $d_{3,2}$ was stable and remained between 1 and 2 μm for at least two days, the emulsion passed this step (Fig. 5.3). If the droplet size was not stable over time, we increased the concentration of the initial aqueous solution and repeated the first step. When emulsions passed this stage, the amount of unadsorbed emulsifiers was quantified both in the fresh emulsion and after two days of incubation, because the amount of unadsorbed emulsifiers may evolve during the incubation of emulsions (Euston & Mayhill, 2001; Dalgleish et al., 2002; Bongard et al., 2009; Berton et al., 2011a). We first separated the aqueous phase from the oil droplets covered by the adsorbed emulsifier through centrifugation at a moderate speed (3000 × g) (Rampon et al., 2003). The recovered aqueous phases were slightly turbid, and residual droplets of small diameter (less than 200 nm, on average) were detected by dynamic light scattering. They were thus sequentially filtrated through 0.45 to 0.1 μm cellulose acetate filters (Berton et al., 2011a). Residual droplets represented only about 0.1% of the initial oil mass and less than 1% of the initial interfacial area (Berton et al., 2011a). The unadsorbed emulsifiers were finally quantified on these filtrated aqueous phases using methods specifically adapted and tested for this purpose.

To determine the concentration in unadsorbed proteins, we chose the Lowry method (colorimetric quantification) (Lowry et al., 1951), later modified by Markwell (Markwell et al., 1978), which was compared to the spectrophotometric method (Layne, 1957). When the emulsifier was a protein and the emulsion was freshly prepared, both the quantifications from the implementation of Markwell's method and the UV-spectrophotometric method were successful. After aging the emulsions in oxidizing conditions, the quantification by UV-spectroscopy failed, no matter which protein we used to stabilize the emulsions, because the recovered aqueous phase was yellowish and the spectral background hid the contribution of proteins to the UV spectrum (Berton et al., 2011a). Moreover, as we will see later, proteins changed during aging, and we could not rely on the extinction coefficient of 280 nm because the aromatic amino acids were susceptible to oxidative modifications, which induced spectral changes.

We developed a quick method to quantify surfactants—more specifically, fatty acid esters—in the aqueous phase of emulsions, to avoid numerous extraction steps with organic solvents and the use of sophisticated equipment, as is often proposed. This method is based on the direct acid transesterification of surfactant fatty acid chains. The fatty acid moieties converted into methyl esters are extracted in hexane and quantified with an internal standard using gas chromatography (Berton et al., 2011a). This method applies as well for mixtures of surfactants if they have at least one fatty acid chain.

For protein and surfactant mixtures, quantification using the aforementioned methods also applies: Direct transesterification in the aqueous phase was not affected by the presence of proteins, at least for the three tested proteins—BLG, BCN, and BSA (Berton et al., 2011a).

The emulsions optimized according to this procedure (Fig 5.3) exhibited sufficient physical stability. The concentrations of unadsorbed emulsifiers in the aqueous phase of these emulsions were always inferior to 2.1 g L^{-1}, and more than 60% of the emulsifier was adsorbed at the oil-in-water less than (Berton et al., 2011a, 2011b), values that are comparable to those estimated from the literature (Fig. 5.2). Thus, we optimized our formulations as much as possible to minimize the contributions of unadsorbed emulsifiers while allowing for the physical stability of emulsions (Berton et al., 2011a).

Oxidation of the Lipid Phase When the Interfaces Are Stabilized by Proteins versus Surfactants

We used emulsions, optimized as described above, to investigate the role of the interfacial layer's composition on lipid oxidation in emulsions. The details can be found in previous reports (Berton, 2011; Berton et al., 2011a, 2012b). Chosen to provide different physicochemical properties for the interface layer, emulsifiers included milk proteins—BSA, BCN and BLG—and both nonionic and anionic surfactants, including Tween 20, Tween 80, and Citrem (citric acid ester of monoglycerides). We varied the pH for some emulsifiers above and below the isoelectric point of the proteins to investigate the effects of pH and protein charge. Lipid oxidation was induced by increasing the temperature to 33 °C, or by adding chelated iron [$FeSO_4$/ethylene diamine tetraacetic acid (Fe/EDTA) equimolar mixture; 200 μM], iron-ascorbate redox cycle [$FeCl_3$/sodium ascorbate (Fe/Asc) equimolar mixture; 50 mM], 2,2′-Azobis(2-methylpropionamidine) dihydrochloride (AAPH) (1 mM), or metmyoglobin (MetMb) (1 μM) to the emulsions, which were kept at 25 °C. We chose these incubation conditions to induce the kinetics of oxygen uptake for the Tween 20–stabilized emulsion in the same range.

Lipid oxidation was evaluated in these emulsions according to methods that allowed a characterization of the reaction's different stages: oxygen (O_2) consumption, formation of conjugated dienes (CD), and formation of selected volatile compounds by solid phase micro-extraction (SPME). Finally, the curves were adjusted with a modified-Gompertz equation to calculate the lag period L (h) and the rate of the propagation phase (μ, mmol kg^{-1} oil). In the last step, we used the estimated values to calculate the time ($t_{1/2}$, h) corresponding to the incubation time when half of the maximum oxygen uptake was consumed or half of the maximum CD were produced.

Fig. 5.4 summarizes the results obtained from oxygen uptake measurements for the emulsions stabilized at pH 6.7 using the minimum amount of protein or surfactant emulsifiers and in the different incubation conditions.

The main conclusions of our study appear to contradict the classical schemes evoked above. First, when the oil droplets were covered by the protein emulsifier, they were generally less oxidatively stable than when they were covered with surfactant emulsifier. This trend was clear for the three proteins tested with oxidation induced either by Fe/EDTA or by Fe/Asc (iron redox cycle) and in the other incubation conditions for BLG and BCN. Our results indicate that, generally, in single emulsifier-stabilized emulsions, surfactant-stabilized oil droplets were less oxidizable than protein-stabilized ones. As compared to surfactants, interfacial proteins revealed that they were unable to protect the lipid phase against oxidation. We thereby assumed that surfactants could form a compact and homogeneous barrier to protect the lipid core of oil droplets against oxidation, whereas protein layers would be more heterogeneous and porous. Indeed, at 33 °C, or in the presence of MetMb, the oxidative behaviors of BSA- and Tween 20–stabilized emulsions were very similar and were attributed to specific oxidative interactions between proteins and lipids. Later, we discuss the proteins' possible effects based on their chemical properties.

Second, even if the droplet charge (positive, negative, or charge level) may intervene in the repulsive or attractive interactions of the ions with the interface, our results indicate that it does not govern the oxidative stability of the emulsions.

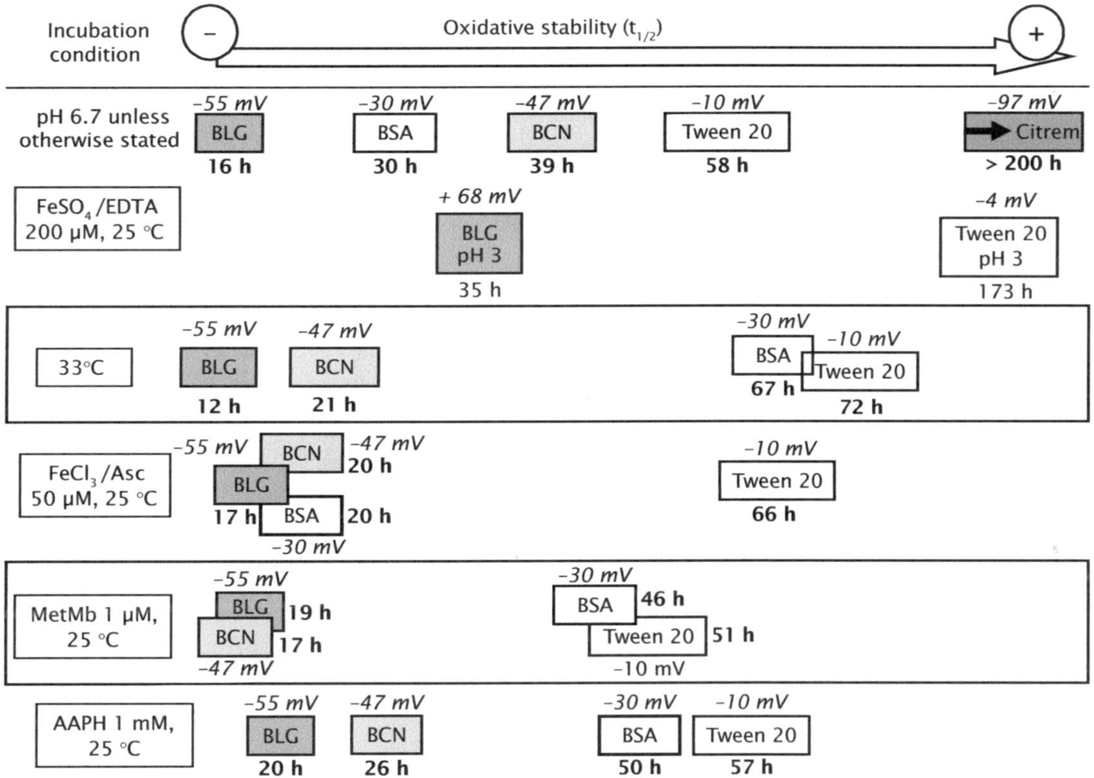

Fig. 5.4. Oxidative stability of oil-in-water emulsions stabilized with minimum concentration of single protein or surfactant. The oxidative stability is expressed as time to reach half of the maximum oxygen consumption ($t_{1/2}$,h). Mean values given near the rectangles symbolize the emulsifier/incubation conditions. Droplet charge is shown in italics.

Original data from Berton et al. (2011b, 2012b).

According to Fig. 5.4, the ranking of the emulsion's oxidative stability did not follow its charge level, especially in the presence of Fe/EDTA, MetMb, or Fe/Asc. We confirmed this finding by replacing the neutral buffer solution with an acidic one to prepare BLG- and Tween 20–stabilized emulsions and oxidation induced by Fe/EDTA. In this condition, the surfactant stabilized droplets, charged both weakly and negatively, remained more oxidatively stable than the positively charged BLG-stabilized droplets. Indeed, the oxidative stability increased in both BLG and Tween 20 pH 3 emulsions, even to a larger extent with Tween 20, as compared to the neutral conditions. This finding demonstrates that better oxidative stability cannot be attributed to the modification of protein and droplet charges. It could result from a decrease in the solubility of the Fe/EDTA complex or from an effect of the phosphate buffer at pH 3, as this buffer is capable of breaking the initial redox cycle through iron binding. Additionally, the Citrem-stabilized emulsion, which exhibited the most negatively charged interface, was the most oxidatively stable; the oxidation markers remained at their lowest level along the kinetics duration. We suspect a specific antioxidant effect of the emulsifier through its citrate moiety, but other mechanisms cannot be excluded. Previous studies have also presented results contradicting the droplet charge hypothesis. For instance, a worse oxidative stability was observed for a BSA-stabilized emulsion prepared at pH 4.0 (initial zeta potential = + 28.2 ± 4.2 mV) than for the same emulsion prepared at pH 6.5 (initial zeta potential = –36.2 ± 1.3 mV) (Villière et al., 2005; Villière & Genot, 2006) (Fig. 5.5).

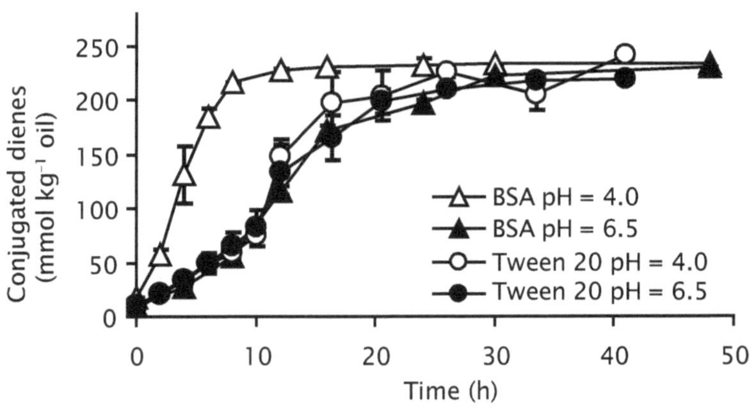

Fig. 5.5. Effect of pH on the formation of conjugated dienes in 30% (w/w) sunflower oil-in-water emulsions stabilized by BSA or Tween 20 (20 g L^{-1}) and stored at 50 °C. Experimental conditions detailed in Villière (2005) and Villière et al., (2005). Unpublished results.

In this example, which was characterized by large amounts of unadsorbed emulsifier, we also found very close kinetics for Tween 20–stabilized emulsions, regardless of their pH. We concluded that neither the pH through the catalytic efficiency of trace metals nor the charge of the protein could explain these results. The effect could be due to the different conformations of the protein according to the pH (Druaux et al., 1995; Burova et al., 2003), which produces different accessibilities and interactions of amino-acid residues between the prooxidant and oxidizing species.

Finally, the results that we summarize in Fig. 5.4 highlight that complex oxidative interactions between the emulsifier and prooxidant species cause oxidation to proceed differently depending on both the emulsifier and the way that oxidation is accelerated or initiated. Different proteins give different rates and/or lag phases and oxidative behavior according to the initiation mechanisms (Berton, 2011; Berton et al., 2011b, 2012b). We investigated protein modifications along with lipid oxidation because emulsifying proteins are modified in oxidizing emulsions (Rampon et al., 2001; Genot et al., 2003), which could interfere with lipid oxidation differently depending on the protein's chemical reactivity.

The Double Roles of Proteins in the Oxidation of Emulsions

To follow simultaneous protein modifications and lipid oxidation in the protein-stabilized emulsions, we measured oxygen uptake in the headspace of the emulsions and protein fluorescence thanks to its direct measurement in front-surface mode. O_2 uptake mainly results from lipid oxidation, while front-surface protein fluorescence is considered a noninvasive and easily accessible marker of protein modification (Genot et al., 2003). In addition, we recorded the front-surface fluorescence spectra of the creamed and aqueous phases recovered after centrifugation of the emulsions at t_0 and after 24 and 48 h of aging to investigate the modifications undergone specifically by either adsorbed or unadsorbed proteins. In the resulting soluble fraction, protein solubility in guanidinium chloride (6 M) and protein carbonyls were measured in these creamed and aqueous phases and in the emulsions themselves. Protein aggregation was also evaluated by SDS-polyacrylamide gel electrophoresis (SDS-PAGE) in denaturing and non-denaturing conditions.

In correspondence with previous results (Rampon et al., 2001), we observed a decrease of the protein fluorescence signal during emulsion incubation. Estimated kinetic parameters showed a good and positive linear relationship between $t_{1/2}$ calculated from oxygen uptake and protein fluorescence measurements, which demonstrated that lipid oxidation and protein modifications are strongly linked (Berton et al., 2012c). However, lag phases and $t_{1/2}$ for oxygen uptake were always greater than lag phases and $t_{1/2}$ calculated from the protein fluorescence decrease, the gap differing according to the emulsion and the incubation conditions (Fig. 5.6). Thus, our results indicate that in the sequence of the reactions, protein modifications probably occurred earlier than lipid oxidation.

The changes in protein fluorescence can result from:

- chemical attack of the aromatic tryptophanyl residues; the fact that in oxidizing emulsions we were not able to directly measure unadsorbed protein from absorbance spectra (see above) gives some argument in favor of a chemical degradation of the residues.
- static or dynamic quenching of the fluorescence due to the formation of various neoformed species in the medium and modifications in the protein conformation.
- changes in the inner filter effect—that is, the direct reabsorption of emitted fluorescence, linked to structural (modification in light scattering properties) and chemical changes (leading to the formation of new chromophores and color modifications) in the systems.

Finally, one should underline that the measured fluorescence signal results from the fluorescence emitted by all of the protein aromatic residues (mainly, tryptophanyl) that can exist in different environments. Measurements of the fluorescence spectra of the creamed and aqueous phase of the emulsions at t_0 after 24 h or 48 h of incubation at 25 °C with $FeSO_4$/EDTA 1/1 M/M 200 μM (Fig. 5.7A) showed that the decrease in fluorescence observed in the total emulsion should be mainly attributed to the protein adsorbed at the interface of the oil droplets, which is mainly represented in the creamed phase of the emulsions.

In a similar way, the adsorbed proteins underwent a great decrease in their solubility, accompanied by carbonyl formation, whereas these phenomena were limited or even absent in the aqueous phase (Fig. 5.7B and C). Only results obtained with β-lactoglobulin-stabilized emulsions are presented here, but roughly similar results were obtained with BSA and β-casein (Berton, 2011; Berton et al., 2012c). We also used SDS-PAGE to investigate the aggregation of proteins from whole emulsions, the aqueous phase, or the creamed phase during incubation of emulsions and the possible involvement of disulfide bonds. In the aqueous phase of BLG-stabilized emulsions, no modifications in the SDS-PAGE profiles occurred during the incubation period. In contrast, for the whole emulsion and creamed phase, the intensity of the band corresponding to the monomers of protein decreased markedly during the incubation period, in both nonreducing and reducing conditions (Fig. 5.7D). At the same time, high–molecular weight

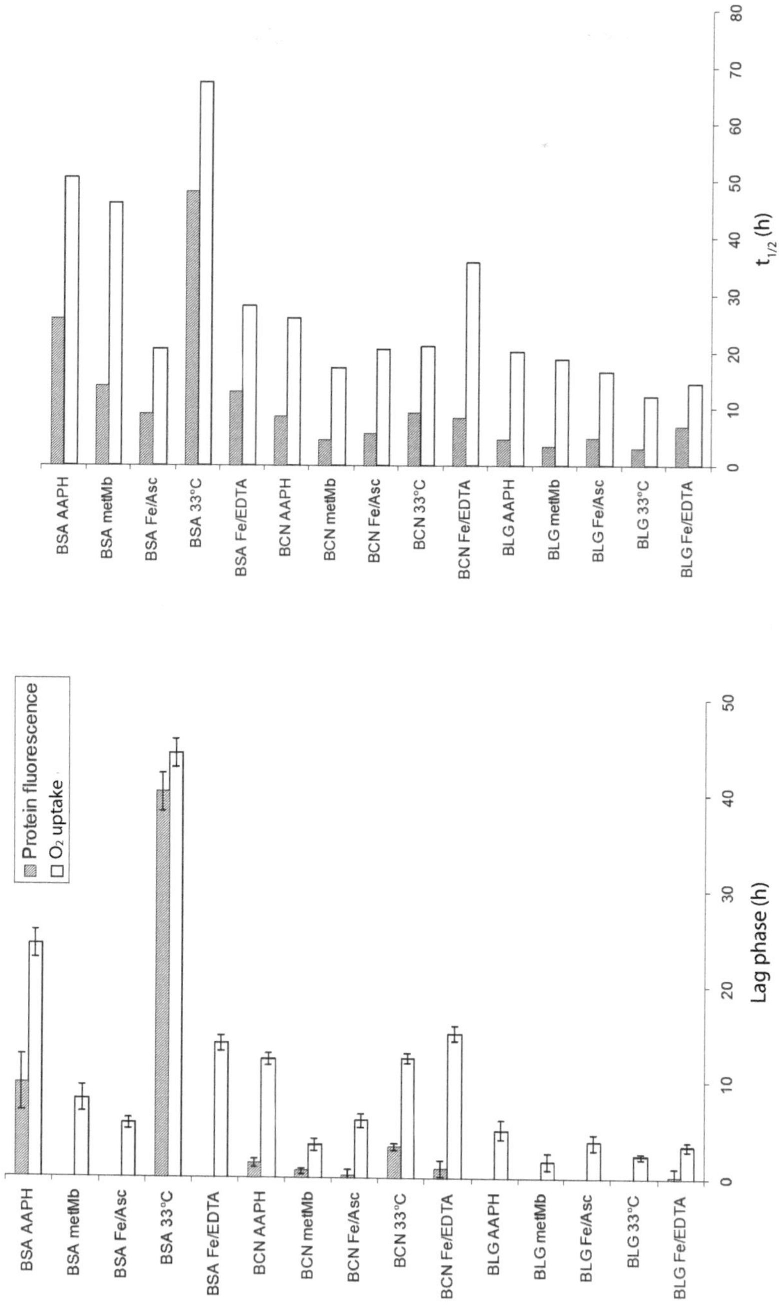

Fig. 5.6. Comparison of lag phase (left) and $t_{1/2}$ (right) for oxygen uptake and protein fluorescence decrease during incubation of emulsions stabilized by different proteins.

Data from Berton (2011) and Berton et al. (2011b, 2012b).

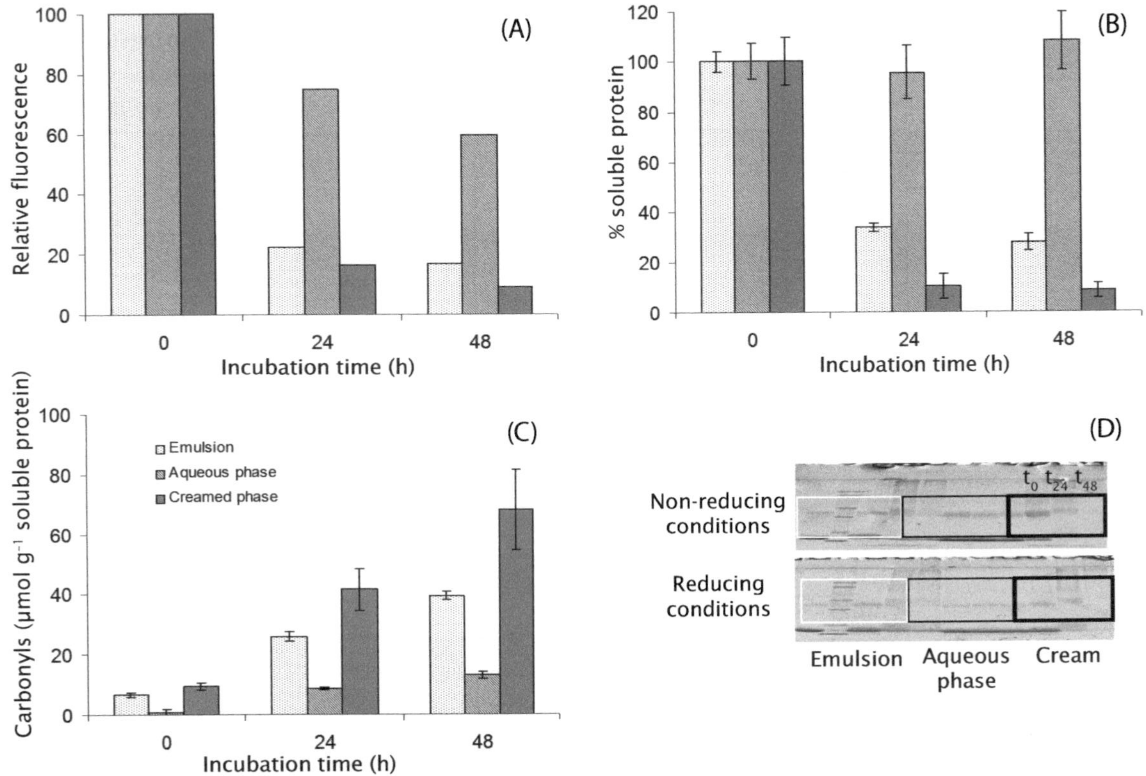

Fig. 5.7. Evolution of β-lactoglobulin in the different phases of emulsions in the presence of $FeSO_4$/EDTA at 25° C: relative protein fluorescence (A), protein solubility in 6M guanidinium chloride (B), carbonyls measured on soluble proteins (C), and SDS-PAGE electrophoresis in reducing and non-reducing conditions (D).

Original data from Berton et al. (2012c).

fractions appeared, which corresponded to protein aggregates ranging from around 30 kDa to more than 100 kDa (not shown). When the electrophoresis of the proteins from the creamed phase was performed in reducing conditions, the intensity of the BLG monomer band was only partially recovered. This demonstrates that protein aggregation was not solely due to the formation of disulfide bonds. BCN also aggregated during emulsion incubation, even though it does not contain any cysteinyl group. This is made possible by the formation of di-tyrosine, discussed widely as a factor involved in the aggregation of proteins in the presence of oxygen-reactive species (Davies et al., 1987; Levine & Stadtman, 2001; Gerrard, 2002; Lund et al., 2011) or radicals arising from oxidizing lipids (Schaich, 2005; Dalsgaard et al., 2010; Dalsgaard et al., 2011). Residues of nucleophilic amino acids of proteins (cysteinyl, histidyl, lysyl, and arginyl residues) can also form covalent adducts with aldehydes arising from lipid oxidation such as hydroxyalkenals (Doorn & Petersen, 2003), alkenals, and even, to a lesser extent, hexanal (Meynier et al., 2004).

To sum up, whatever the chosen marker of protein modification (decrease of protein fluorescence signal, protein carbonylation, loss of solubility, or aggregation), the extent of the modifications in the oxidizing emulsions were much higher for the proteins recovered from the cream than from the aqueous phase. Thus, adsorbed proteins are main targets for protein modifications in oxidizing emulsions.

Finally, we compared the kinetics of O_2 consumption and the decrease of protein fluorescence for the emulsions containing the minimum amount of unadsorbed proteins (BLG or BCN) and for emulsions presenting excess of the proteins in the aqueous phase. Our goal was to better understand the role played by the unadsorbed protein. In the last case, the oxygen uptake and decrease of protein fluorescence were either significantly delayed or slowed down. This implies that unadsorbed proteins can delay the access of prooxidant species to the lipid core of oil droplets. Therefore, we confirm that unadsorbed proteins exert an antioxidant role. This antioxidant effect would be successful during the earliest steps of oxidation, when amounts of free radicals are still too low to propagate the reaction to the lipid phase. This would explain why unadsorbed proteins undergo low damage (that is, unchanged solubility, no aggregation, low levels of protein carbonyls, and weak protein fluorescence decrease) while efficiently delaying lipid oxidation. Once the propagation step is engaged in the lipid phase, larger amounts of reactive species are produced. Then, the reactions at the interface between proteins and oxidizing lipids or with lipid oxidation products become quantitative, inducing protein modifications (for example, loss of solubility, aggregation, formation of protein carbonyl, or total extinction of protein fluorescence). Fig. 5.8. summarizes this view.

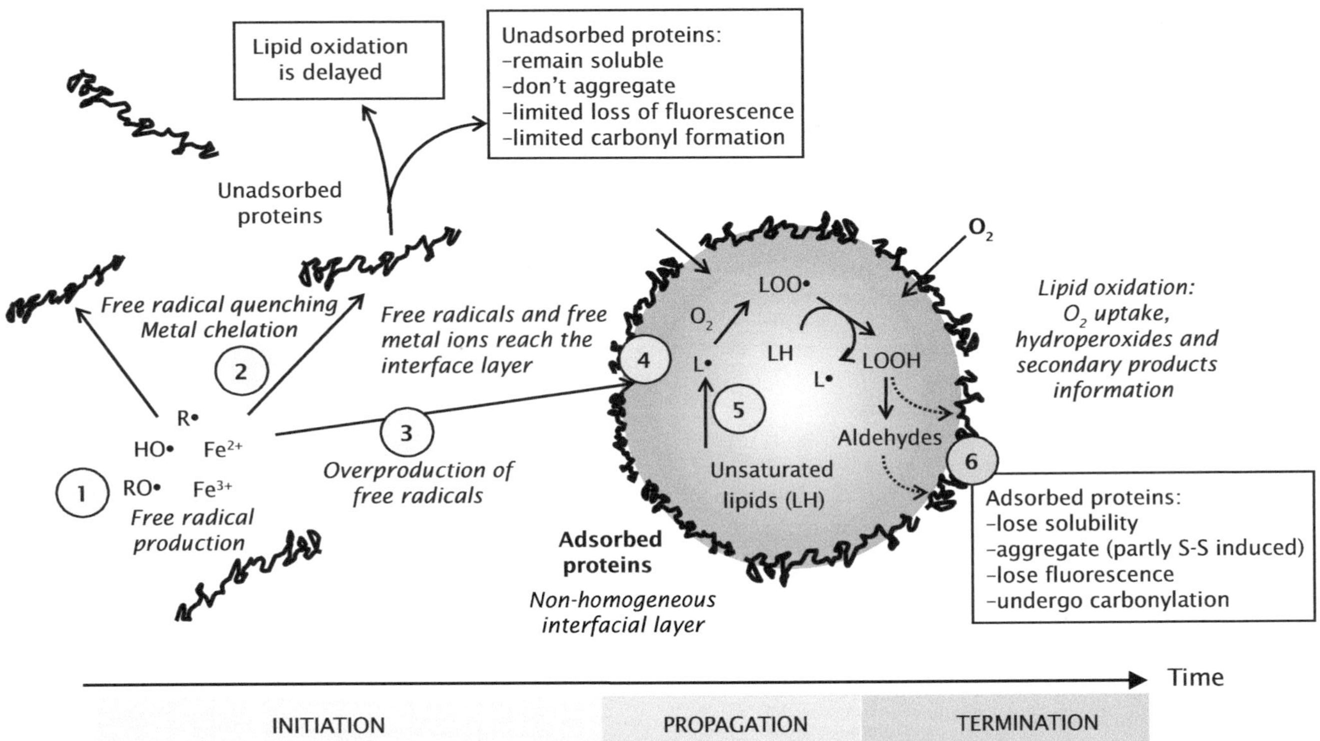

Fig. 5.8. General scheme of the protein-lipid oxidative interactions during oxidation of oil-in-water emulsions.

Homogeneity and Compactness of the Interface Layer: Its Possible Role in Oxidation Development

We have shown that, generally, protein–stabilized interfaces protect emulsified lipids against oxidation less efficiently than surfactant–stabilized interfaces (Berton et al., 2011b, 2012b). These different behaviors have been tentatively attributed to specific characteristics of the surfactant- or protein-covered interfaces. Due to van der Waals' interactions between alkyl chains, Tween 20 molecules are well aligned to form more homogeneous and compact interfacial layers than proteins. The length of the polyoxyethylene (POE) chain constituting the polar head of the surfactant was found to affect oxidative stability in surfactant–stabilized emulsions (Silvestre et al., 2000), indicating that these chains and their interfacial arrangement could intervene in the relative protective effect of Tween 20. Accordingly, the structure of the interfacial layer was recently assumed to intervene in the control of lipid oxidation phenomena (Kargar et al., 2011). To confirm this hypothesis, we measured lipid oxidation in emulsions stabilized by a mixture of Tween 20 and cosurfactants. The chosen cosurfactants [sorbitan monolaurate (Span 20) and monolauroyl glycerol] possessed the same fatty acid tail, but smaller-size polar head groups. Since proteins are widely used as emulsifiers in food products, we attempted to develop strategies to improve their protective effect against lipid oxidation when they are located at the oil–water interface in oil-in-water emulsions. We first expected to overcome the poor protective effect of native BLG against lipid oxidation by increasing the thickness and the coverage homogeneity of the interfacial layer; to do so, we planned to use denatured and partially aggregated BLG as an emulsifier. We also tested phospholipid/BCN or phospholipid/BLG mixed interfaces because phospholipids can exert an antioxidant role in multiphase systems (Fomuso et al., 2002; Frankel, 2005; Haahr & Jacobsen, 2008; Helgason et al., 2009). In any case, we used optimized emulsions in which we minimized the amounts of emulsifiers to keep the contributions of unadsorbed emulsifiers as low as possible while producing physically stable emulsions.

Oxidative Stability of Emulsions Stabilized with Mixtures of Surfactants

Span 20 is a surfactant with the structure of Tween 20, but without the polyoxyethylene chain. Due to its low hydrophilic-lipophilic balance, it cannot stabilize oil-in-water emulsions. However, emulsions prepared with mixtures of Tween 20 and Span 20 are stable. We were aware that partial replacement of Tween 20 by Span 20 molecules would modify two parameters at the same time: the polyoxyethylene chain amount and structure of the interfacial layer recently assumed to intervene in lipid oxidation development (Kargar et al., 2011). Therefore, we produced and incubated emulsions stabilized with Tween 20 and Span 20, as previously described. Monolauryl glycerol (MLG) was also used as a cosurfactant. Each cosurfactant was introduced at a 29.3% molar ratio, which corresponds to a surface coverage of 10% (Berton et al., 2012a). *In fine* less than 11% Tween 20 and 2.5% MLG or Span 20 were present in the aqueous phase of the emulsions. All measured lipid oxidation markers (oxygen uptake, CD formation, and volatile formation) led to corroborating results; that is, lipid oxidation developed more slowly in the single Tween 20–stabilized emulsion than in the surfactant mixture–stabilized ones, with a longer lag phase but with similar oxidation rates (Fig. 5.9).

As mentioned above, the faster lipid oxidation may result from the global decrease of the POE amount per interfacial surface unit, changes in the structure of the mixed interfacial layer, or from both factors. To verify if the cosurfactant actually impacts the in-plane organization of the surfactant film, we analyzed the miscibility of the cosurfactants into Tween 20 layers.

Tween 20/cosurfactant–stabilized interfaces were reconstituted via Langmuir mixed monolayers at the air–water interface. From the isotherms of compression, we extracted the mean molecular areas of the mixtures as a function of the molar

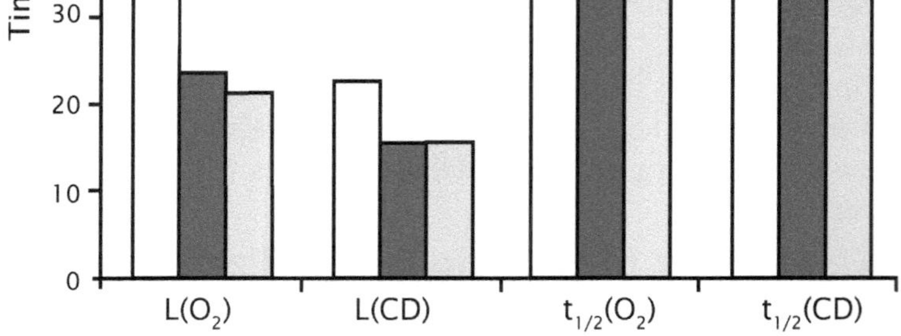

Fig. 5.9. Lag phase (L) and half time ($t_{1/2}$) of oxygen uptake (O_2) and conjugated dienes (CD) formation during the incubation of emulsions stabilized with Tween 20, Tween 20/Span 20 mixture, or Tween 20/MLG mixture at 25 °C in the dark in the presence of $FeSO_4$/EDTA (1/1; M/M; 200 μM).

Data were extracted from adjusted curves (Berton et al., 2012a).

proportion in cosurfactant at the film pressure, which corresponds to the molecular area occupied by Tween 20 at the interfaces of the oil-in-water emulsions. The curves exhibited a positive deviation from the straight line joining the mean molecular areas of pure Tween 20 and pure cosurfactant (Fig. 5.10).

This positive deviation indicates that repulsive interactions occurred between Tween 20 and cosurfactants in the mixed films spread at the air-buffer interface. These repulsive interactions were stronger for Tween 20/MLG mixtures than for Tween 20/Span 20 mixtures because the positive deviation was greater for the

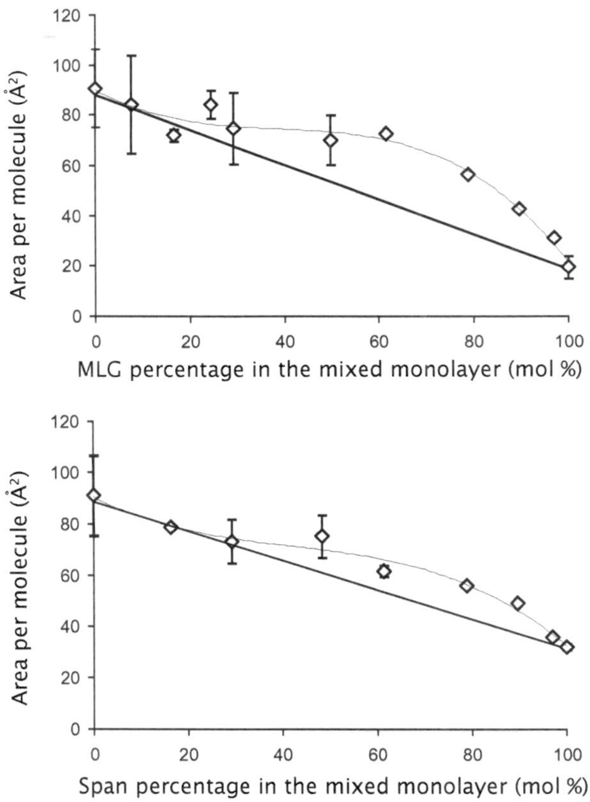

Fig. 5.10. Mean molecular area at a pressure of 16 mN m^{-2} in monolayers of Tween 20 containing variable proportions of monolauroyl glycerol (MLG; top) or Span (bottom). The dotted curve represents the polynomial model adjusting the experimental data ($R^2 > 0.97$). The full lines join the molecular area of both pure surfactants.

former. From these data, one must conclude that Tween 20 and the cosurfactants would form separated domains, leading to an interfacial heterogeneity. Provided that these results can be extrapolated to surfactant behavior and interactions at the oil–water interface, we assume that domain separations induced by repulsive interactions between Span 20 or MLG and Tween 20 led to a lower coverage of the oil droplets at the junctions between the domains. This phase separation at the surface of the oil droplets would make the unsaturated substrate more accessible to the water-soluble prooxidants (such as reactive oxygen species and metal ions) and would favor lipid oxidation in the mixed–surfactant emulsions by decreasing the lag phase (Fig. 5.9).

Oxidative Stability of Emulsions Stabilized with Partially Denatured and Partially Aggregated BLG

Heat treatments are commonly applied to food products at the industrial scale. They lead to partial or total denaturation and aggregation of proteins because the characteristics of the aggregates greatly depend on the applied thermal treatment and the physicochemical environment of the proteins (pH, ionic strength, protein concentration, and so on) (de la Fuente et al., 2002). These aggregates also exhibit interfacial properties that differ from those of the native proteins. For instance, mixtures of aggregates and residual denatured whey proteins were found to exhibit enhanced surface activity, the interface layer forming a weakly dissipative viscoelastic network (Mahmoudi et al., 2010; Mahmoudi et al., 2011). We used this feature to investigate the role of protein denaturation and partial aggregation in lipid oxidation.

A solution of denatured and partially aggregated BLG, obtained by heating a solution of native BLG in ultra-pure water (10 g L^{-1}) at 80 °C for 45 min, was used to prepare emulsions (Berton, 2011). The development of lipid oxidation in this emulsion was very similar to the control BLG-stabilized emulsion, and no decrease of lipid oxidation could be obtained. The partial unfolding of BLG prior to the emulsification did not improve the oxidative stability of the emulsion, as compared to the native BLG-stabilized one. Similarly, the heat denaturation of whey proteins or pure BLG performed post-emulsification did not improve the oxidative stability of emulsions (Kellerby et al., 2006) or even slightly increase the rate of oxidation (Djordjevic et al., 2004). Additionally, the chemical cross-linking of WPI did not decrease the oxygen diffusion coefficient at the oil-in-water interface, as compared to native WPI (Tikekar et al., 2011). In contrast, heat denaturation combined with higher homogenization pressure improved the oxidative stability of fish oil–enriched milk emulsions (Let et al., 2007; Sørensen et al., 2007), and heat-denatured BLG in the aqueous phase of surfactant–stabilized emulsions was found to better protect the oil phase against oxidation, as compared with native BLG (Elias et al., 2007). This was attributed to improved accessibility of radical scavenging amino acid residues (namely, free sulfhydryl groups) to free radicals in the aqueous phase. One may notice that in some of these experiments, large amounts of proteins were used, as concentrations in the aqueous phase were likely high. Possible differences in the composition of the oil-in-water interface and its barrier properties were also evoked.

Oxidative Stability of Emulsions Stabilized with Protein/Phospholipids Interfacial Layers

Phospholipids can exert an antioxidant role in multiphase systems. They are small-sized molecules in comparison to proteins and, at low concentration, could fill out the surface not actually covered by protein residues. Emulsions stabilized by mixed protein/dilauroyl phosphatidylcholine (DLPC) interfacial layers were prepared by replacing 10% of the surface covered by the proteins with DLPC. We did so by taking into account the surface coverage of DLPC. The emulsions exhibited protein concentrations in the aqueous phase that were similar to those of the control emulsions. They were incubated at 25 °C in the presence of Fe/EDTA (200 μM). Contrary to our initial hypothesis, lipid oxidation started earlier in emulsions containing DLPC than in controls (Fig. 5.11).

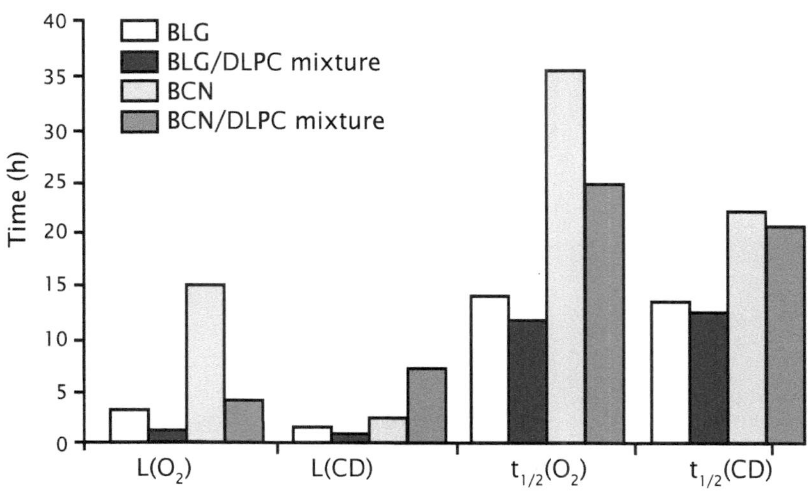

Fig. 5.11. Lag phase (L) and half time ($t_{1/2}$) of oxygen uptake (O_2) and conjugated dienes (CD) formation, during the incubation of emulsions stabilized with BLG, BLG/DLPC mixture, BCN, or BCN/DLPC mixture at 25 °C in the dark with $FeSO_4$/EDTA (1/1 M/M 200 μM).

Data were extracted from adjusted curves (Berton et al., 2012a).

In the BCN/DLPC-stabilized emulsion, the lag phase was markedly shortened: 4.4 h instead of 15 h in the pure BLG-stabilized emulsion. The difference was less marked with BLG (1.4 h in the BLG/DLPC-stabilized emulsion instead of 3.3 h in the pure BLG emulsion). In both BLG/DLPC- and BCN/DLPC-stabilized emulsions, the rates of oxygen uptake remained steady (6.4 against 6.1 mmol O_2 kg^{-1} oil h^{-1} and 3.3 mmol O_2 kg^{-1} oil h^{-1}, respectively). To understand this unexpected behavior, we attempted to reconstruct the films on model interfaces and analyzed their structure using atomic force microscopy. The films were formed at the air-water interfaces, transferred onto a solid substrate according to the Langmuir-Blodgett method, and analyzed under the conditions described in a previous publication (Berton, 2011). The three-dimensional images reported in Fig. 5.12 have dimensions 2 × 2 μm, which are in the same range as the surface of the emulsion droplets (d3,2) between 1.3 and 2.2 μm.

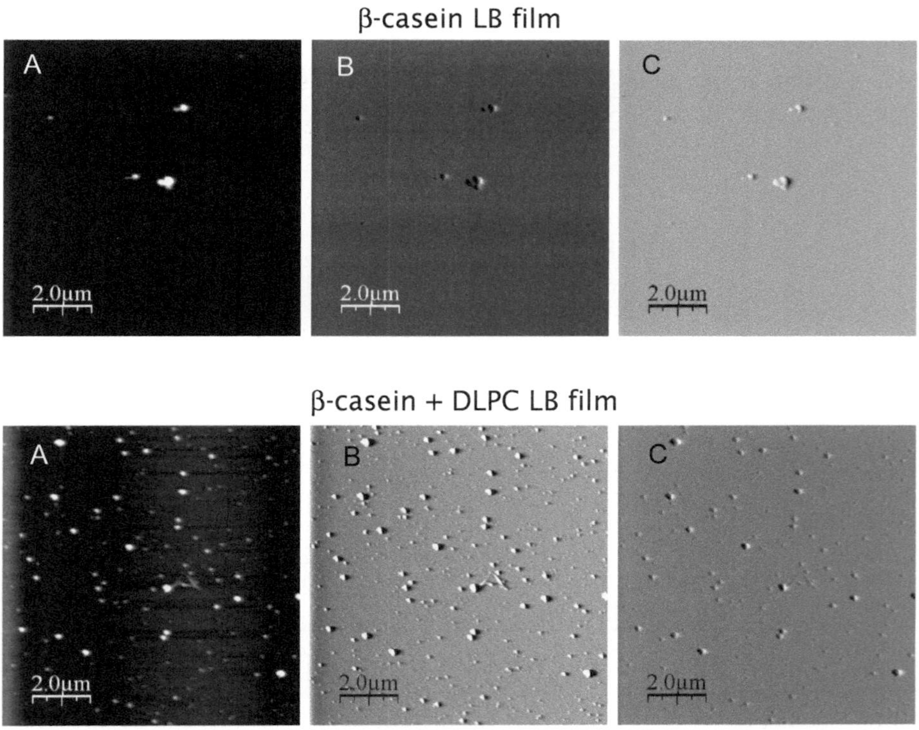

Fig. 5.12. AFM pictures of Langmuir-Blodgett (LB) films of β-casein films (top) or β-casein/DLPC mixture films (bottom): topography images (A), deflection images (B), and phase contrast images (C). In topography images, the brightness is proportional to the height of the sample.

In the case of BCN, the films were very compact and continuous, with a few large particles up to 50 nm in height (Fig. 5.12). When we incorporated DLPC inside the BCN layer, we still observed large particles, and numerous smaller particles of a regular distribution appeared. These particles were rather small (1–2 nm height) and well individualized. Their presence disrupted the homogeneity of the BCN film, as highlighted by the appearance of narrow height peaks in the topography profile (Berton, 2011). BLG formed continuous films containing numerous small particles of around 1 to 2 nm in height and less frequent but larger particles up to 10 nm in height. When DLPC was incorporated into the BLG film, the texture of the BLG/DLPC film was slightly more granular and organized as a network of grains (1–2 nm height). These grains were wider than the small particles found in the single BLG film. Thus, for both protein layers (BLG or BCN), the incorporation of DLPC seemed to induce some heterogeneity in the Langmuir-Blodgett film, despite the antioxidant activity reported in many studies. As the development of lipid oxidation was faster in protein/DLPC-stabilized emulsions and the appearance of protein/DLPC films was more grainy and not homogeneous, we assume that the homogeneity of interfacial films is a parameter that intervenes in the control of lipid oxidation. Further investigation must be conducted to analyze the heterogeneity of interfacial films at the oil–water interface before confirming this hypothesis.

Conclusion

Several models are currently used to interpret the results obtained on various emulsified systems. However, careful examination of the literature allows us to conclude that if these attractive explanations fit most of the results obtained in some simple model emulsions, they often contradict results obtained under different conditions.

Thus, we paid attention here to the formulation of emulsions. The unadsorbed amount of emulsifiers was minimized, while ensuring the physical stability of the emulsions. One main conclusion of our work was that in single emulsifier–stabilized emulsions, interfacial proteins were unable to protect the lipid phase against oxidation and that surfactant-stabilized oil droplets were less oxidizable than protein-stabilized ones. Thermal treatment did not improve the properties. We thereby assumed that surfactants could organize as a compact and homogenous barrier protecting the lipid core of oil droplets against oxidation, whereas protein layers would be more heterogeneous and porous. Therefore, the surfactants could provide some protection of the dispersed oil phase through a physical barrier effect. Our results also show unambiguously that the droplet charge is not a key parameter in controlling lipid oxidation, especially in the presence of protein emulsifiers. Interestingly, the structure of the interfacial layer seems to be another parameter that affects lipid oxidation. We also found that if the adsorbed proteins do not protect the lipid phase from oxidation, they are the main targets for oxidative attack, which causes their serious modifications (for example, loss of fluorescence and solubility, carbonyl formation, and thiol-dependent and not-thiol-dependent aggregation). In contrast, the unadsorbed proteins are able to significantly delay the oxidative modifications of the emulsions, probably by interacting with the prooxidant species in the aqueous phase, but they are poorly modified.

Further work is now needed to investigate more deeply the protein's oxidative modifications and to better understand its protective role in the aqueous phase. Also, the link between the surface homogeneity of the adsorbed layer and its capability to protect the lipid phase could be explored. Finally, one should underline that the knowledge models detailed in this chapter consider separately the effects of different factors on emulsion oxidation. They do not take into account the factors' hierarchy and the multiple interactions within these factors (first, second, and higher order, depending on the system's complexity). This makes it very difficult to actually predict the oxidative behavior of real food formulations, and the unexpected development of oxidation could be observed in real situations. A more integrated view of oxidation in complex systems is now to be built.

Acknowledgment: INRA and region Pays de la Loire founded the Ph.D. grant of CB. Cédric Gaillard, Dominique Guibert, Quentin Pottiez (short-stay trainee) Lucie Ribourg, Véronique Solé, Michelle Viau, and Nadège Wicker (short-stay trainee) are gratefully acknowledged for their participation in the experimental work. Anne Meynier is also recognized for her frequent and valuable scientific exchanges.

References

Atkinson, P. J.; Dickinson, E.; Horne, D. S.; Richardson, R. M. Neutron reflectivity of adsorbed β-casein and β-lactoglobulin at the air/water interface. J. Chem. Soc., Faraday Trans. 1995, 91, 2847–2854.

Awada M.; Soulage, C. O.; Meynier, A.; Debard, C.; Plaisancié, P.; Benoit, B.; Picard, G.; Loizon, E.; Chauvin, M.-A.; Estienne, M. et al. Dietary oxidized *n*-3 PUFA induce oxidative stress and inflammation: Role of intestinal absorption of 4-HHE and reactivity in intestinal cells. J. Lipid Research, 2012, DOI:10.1194/jlr.M026179 doi:10.1194/jlr.M026179.

Azuma, G.; Kimura, N.; Hosokawa, M.; Miyashita, K. Effect of droplet size on the oxidative stability of soybean oil TAG and fish oil TAG in oil-in-water emulsion. J. Oleo Sci. 2009, 58, 329–338.

Berton, C. Construction raisonnée d'interfaces pour protéger les lipides émulsionnés contre l'oxydation; Université de Nantes, 2011.
Berton, C.; Genot, C.; Guibert, D.; Ropers, M.-H. Effect of lateral heterogeneity in mixed surfactant–stabilized interfaces on the oxidation of unsaturated lipids in oil-in-water emulsions. J. Colloid Interface Sci. 2012a, 377, 244–250.
Berton, C.; Genot, C.; Ropers, M.-H. Quantification of unadsorbed protein and surfactant emulsifiers in oil-in-water emulsions. J. Colloid Interface Sci. 2011a, 354, 739–748.
Berton, C.; Ropers, M.-H.; Bertrand, D.; Viau, M.; Genot, C. Oxidative stability of oil-in-water emulsions stabilised with protein or surfactant emulsifiers in various oxidation conditions. Food Chem. 2012b, 131, 1360–1369.
Berton, C.; Ropers, M.-H.; Guibert, D.; Solé, V.; Genot, C. Modifications of interfacial proteins in oil-in-water emulsions prior to and during lipid oxidation. J. Agric. Food Chem. 2012c, 60 (35), 8659–8671.
Berton, C.; Ropers, M.-H.; Viau, M.; Genot, C. Contribution of the interfacial layer to the protection of emulsified lipids against oxidation. J. Agric. Food Chem. 2011b, 59, 5052–5061.
Bongard, S.; Meynier, A.; Riaublanc, A.; Genot, C. Protein-stabilized oil-in-water emulsions and low-fat dairy stirred gels. J. Food Nutr. Res. 2009, 48, 42–50.
Bos, M. A.; van Vliet, T. Interfacial rheological properties of adsorbed protein layers and surfactants: A review. Adv. Colloid Interface Sci. 2001, 91, 437–471.
Burova, T. V.; Grinberg, N. V.; Grinberg, V. Y.; Tolstoguzov, V. B. Binding of odorants to individual proteins and their mixtures. Effects of protein denaturation and association. A plasticized globule state. Colloids Surf., A 2003, 213, 235–244.
Casanova, H.; Dickinson, E. Rheology and flocculation of oil-in-water emulsions made with mixtures of α (s1)-casein plus β-casein. J. Colloid Interface Sci. 1998, 207, 82–89.
Chaiyasit, W.; McClements, D. J.; Weiss, J.; Decker, E. A. Impact of surface-active compounds on physicochemical and oxidative properties of edible oil. J. Agric. Food Chem. 2008, 56, 550–556.
Chen, B.; Han, A.; McClements, D. J.; Decker, E. A. Physical structures in soybean oil and their impact on lipid oxidation. J. Agric. Food Chem. 2010, 58, 11993–11999.
Cheng, Y.; Xiong Y. L.; Chen, J. Antioxidant and emulsifying properties of potato protein hydrolysate in soybean oil-in-water emulsions. Food Chem. 2010, 120, 101–108.
Cho, Y. H.; Kim, S.; Bae, E. K.; Mok, C. K.; Park, J. Formulation of a cosurfactant-free O/W microemulsion using nonionic surfactant mixtures. J. Food Sci. 2008, 73, E115–E121.
Choe, E.; Min, D. B. Mechanisms of antioxidants in the oxidation of foods. Compr. Rev. Food. Sci. Food Saf. 2009, 8, 345–358.
Courthaudon, J. L.; Dickinson, E.; Dalgleish, D. G. Competitive adsorption of β-casein and nonionic surfactants in oil-in-water emulsions. J. Colloid Interface Sci. 1991, 145, 390–395.
Dalgleish, D. G. The sizes and conformations of the proteins in adsorbed layers of individual caseins on latices and in oil-in-water emulsions. Colloids Surf., B 1993, 1, 1–8.
Dalgleish, D. G.; Goffm, H. D.; Brun, J. M.; Luan, B. B. Exchange reactions between whey proteins and caseins in heated soya oil-in-water emulsion systems—overall aspects of the reaction. Food Hydrocolloids 2002, 16, 303–311.
Dalsgaard, T. K.; Nielsen, J. H.; Brown, B. E.; Stadler, N.; Davies, M. J. Dityrosine, 3,4-dihydroxyphenylalanine (DOPA), and radical formation from tyrosine residues on milk proteins with

globular and flexible structures as a result of riboflavin-mediated photo-oxidation. J. Agric. Food Chem. 2011, 59, 7939–7947.
Dalsgaard, T. K.; Sorensen, J.; Bakman, M.; Vognsen, L.; Nebel, C.; Albrechtsen, R.; Nielsen, J. H. Light-induced protein and lipid oxidation in cheese: Dependence on fat content and packaging conditions. Dairy Sci. Technol. 2010, 90, 565–577.
Dauphas, S.; Amestoy, M.; Llamas, G.; Anton, M.; Riaublanc, A. Modification of the interactions between β-casein stabilised oil droplets with calcium addition and temperature changing. Food Hydrocolloids 2008, 22, 231–238.
Davies, K. J.; Delsignore M. E.; Lin, S. W. Protein damage and degradation by oxygen radicals. II. Modification of amino acids. J. Biol. Chem. 1987, 262, 9902–9907.
Day, J. P. R.; Pudney, P. D. A.; Bain, C. D. Ellipsometric study of the displacement of milk proteins from the oil–water interface by the non-ionic surfactant C10E8. Phys. Chem. Chem. Phys. 2010, 12, 4590–4599.
de la Fuente, M. A.; Singh, H.; Hemar, Y. Recent advances in the characterisation of heat-induced aggregates and intermediates of whey proteins. Trends Food Sci. Technol. 2002, 13, 262–274.
Dickinson, E. An Introduction to Food Colloids; Oxford University Press: Oxford, UK, 1992.
Dickinson, E. Hydrocolloids as emulsifiers and emulsion stabilizers. Food Hydrocolloids 2009, 23, 1473–1482.
Dickinson, E.; Radford, S. J.; Golding, M. Stability and rheology of emulsions containing sodium caseinate: Combined effects of ionic calcium and non-ionic surfactant. Food Hydrocolloids 2003, 17, 211–220.
Dickinson, E.; Tanai, S. Protein displacement from the emulsion droplet surface by oil-soluble and water-soluble surfactants. J. Agric. Food Chem. 1992, 40, 179–183.
Dimakou, C. P.; Kiokias, S. N.; Tsaprouni, I. V.; Oreopoulou, V. Effect of processing and storage parameters on the oxidative deterioration of oil-in-water emulsions. Food Biophys. 2007, 2, 38–45.
Dimitrova, T. D.; Leal-Calderon, F. Forces between emulsion droplets stabilized with Tween 20 and proteins. Langmuir 1999, 15, 8813–8821.
Dimitrova, T. D.; Leal-Calderon, F.; Gurkov, T. D.; Campbell, B. Disjoining pressure vs thickness isotherms of thin emulsion films stabilized by proteins. Langmuir 2001, 17, 8069–8077.
Dimitrova, T. D.; Leal-Calderon, F.; Gurkov, T. D.; Campbell, B. Surface forces in model oil-in-water emulsions stabilized by proteins. Adv. Colloid Interface Sci. 2004, 108–109, 73–86.
Djordjevic, D.; McClements D. J.; Decker, E. A. Oxidative stability of whey protein-stabilized oil-in-water emulsions at pH 3: Potential omega-3 fatty acid delivery systems (Part B). J. Food Sci. 2004, 69, C356–C362.
Donnelly, J. L.; Decker, E. A.; McClements D. J. Iron-catalyzed oxidation of menhaden oil as affected by emulsifiers. J. Food Sci. 1998, 63, 997–1000.
Doorn, J. A.; Petersen, D. R. Covalent adduction of nucleophilic amino acids by 4-hydroxynonenal and 4-oxononenal. Chem.-Biol. Interact. 2003, 143–144, 93–100.
Druaux, C.; Lubbers, S.; Charpentier, C.; Voilley, A. Effects of physicochemical parameters of a model wine on the binding of gamma-decalactone on bovine serum albumin. Food Chem. 1995, 53, 203–207.

Elias, R. J.; McClements D. J.; Decker, E. A. Impact of thermal processing on the antioxidant mechanisms of continuous phase beta-lactoglobulin in oil-in-water emulsions. Food Chem. 2007, 104, 1402–1409.

Euston, S. R.; Mayhill, P. G. Time- and temperature-dependent changes in the adsorbed layer on caseinate-stabilized emulsion droplets. Food Res. Int. 2001, 34, 369–376.

Fang, Y.; Dalgleish, D.G. Casein adsorption on the surfaces of oil-in-water emulsions modified by lecithin. Colloids Surf., B 1993, 1, 357–364.

Faraji, H.; McClements D. J.; Decker, E. A. Role of continuous phase protein on the oxidative stability of fish oil-in-water emulsions. J. Agric. Food Chem. 2004, 52, 4558–4564.

Fomuso, L. B.; Corredig, M.; Akoh, C. C. Effect of emulsifier on oxidation properties of fish oil-based structured lipid emulsions. J. Agric. Food Chem. 2002, 50, 2957–2961.

Frankel, E. N. Lipid Oxidation, Second Edition, The Oily Press: Bridgwater, England, 2005.

Genot, C.; Meynier, A.; Riaublanc, A. Lipid oxidation in emulsions. Lipid Oxidation Pathways; Kamal-Eldin, A., Ed.; AOCS Press: Champaign, IL, 2003; 190–244.

Gerrard, J. A. Protein-protein crosslinking in food: Methods, consequences, applications. Trends Food Sci. Technol. 2002, 13, 391–399.

Grigoriev, D. O.; Miller, R. Mono- and multilayer covered drops as carriers. Curr. Opin. Colloid Interface Sci. 2009, 14, 48–59.

Guéraud, F.; Atalay, M.; Bresgen, N.; Cipak, A.; Eckl, P. M.; Huc, L.; Jouanin, I.; Siems, W.; Ushida, K. Chemistry and biochemistry of lipid peroxidation products. Free Radic. Res. 2010, 44, 1098–1124.

Haahr, A. M.; Jacobsen, C. Emulsifier type, metal chelation and pH affect oxidative stability of n-3-enriched emulsions. Eur. J. Lipid Sci. Technol. 2008, 110, 949–961.

Helgason, T.; Awad, T.; Kristbergsson, K.; Decker, E. A.; McClements, D. J.; Weiss, J. Impact of surfactant properties on oxidative stability of β-carotene encapsulated within solid lipid nanoparticles. J. Agric. Food Chem. 2009, 57, 8033–8040.

Horn, A. F., Nielsen, N. S.; Jacobsen, C. Additions of caffeic acid, ascorbyl palmitate or Δ-tocopherol to fish oil-enriched energy bars affect lipid oxidation differently. Food Chem. 2009, 112, 412–420.

Hu, M.; McClements, D. J.; Decker, E. A. Impact of whey protein emulsifiers on the oxidative stability of salmon oil-in-water emulsions. J. Agric. Food Chem. 2003, 51, 1435–1439.

Jacobsen, C.; Let, M. B.; Nielsen, N. S.; Meyer, A. S. Antioxidant strategies for preventing oxidative flavour deterioration of foods enriched with n-3 polyunsaturated lipids: A comparative evaluation. Trends Food Sci. Technol. 2008, 19, 76–93.

Kanazawa, A.; Sawa, T.; Akaike, T.; Maeda, H. Dietary lipid peroxidation products and DNA damage in colon carcinogenesis. Eur. J. Lipid Sci. Technol. 2002, 104, 439–447.

Kargar, M.; Spyropoulos, R.; Norton, I. T. The effect of interfacial microstructure on the lipid oxidation stability of oil-in-water emulsions. J. Colloid Interface Sci. 2011, 357, 527–533.

Kellerby, S. S.; Gu, Y. S.; McClements, D. J.; Decker, E. A. Lipid oxidation in a menhaden oil-in-water emulsion stabilized by sodium caseinate cross-linked with transglutaminase. J. Agric. Food Chem. 2006, 54, 10222–10227.

Kotsmar, C.; Grigoriev, D. O.; Xu, F.; Aksenenko, E. V.; Fainerman, V. B.; Leser, M. E.; Millert, R. Equilibrium of adsorption of mixed milk protein/surfactant solutions at the water/air interface. Langmuir 2008, 24, 13977–13984.

Layne, E. Spectrophotometric and turbidimetric methods for measuring proteins. Methods Enzymol. 1957, 447–454.

Leaver, J.; Law, A. J. R.; Brechany, E. Y. Covalent modification of emulsified beta-casein resulting from lipid peroxidation. J. Colloid Interface Sci. 1999, 210, 207–214.

Let, M. B.; Jacobsen, C.; Sørenson, A. D. M.; Meyer, A. S. Homogenization conditions affect the oxidative stability of fish oil enriched milk emulsions: Lipid oxidation. J. Agric. Food Chem. 2007, 55, 1773–1780.

Lethuaut, L.; Metro, F.; Genot, C. Effect of droplet size on lipid oxidation rates of oil-in-water emulsions stabilized by protein. J. Am. Chem. Oil Soc. 2002, 79, 425–430.

Levine, R. L.; Stadtman, E. R. Oxidative modification of proteins during aging. Exp. Gerontol. 2001, 36, 1495–1502.

Lowry, O. H.; Rosebrough, N. J.; Farr, A. L.; Randall, R. J. Protein measurement with the folin phenol reagent. J. Biol. Chem. 1951, 193, 265–275.

Lund, M. N.; Heinonen, M.; Baron, C. P.; Estevez, M. Protein oxidation in muscle foods: A review. Mol. Nutr. Food Res. 2011, 55, 83–95.

Mackie, A. R.; Gunning, A. P.; Wilde, P. J.; Morris, V. J. Orogenic displacement of protein from the oil/water interface. Langmuir 2000, 16, 2242–2247.

Mahmoudi, N.; Axelos, M. A. V.; Riaublanc, A. Interfacial properties of fractal and spherical whey protein aggregates. Soft Matter 2011, 7, 7643–7654.

Mahmoudi, N.; Gaillard, C.; Boue, F.; Axelos, M. A. V.; Riaublanc, A. Self-similar assemblies of globular whey proteins at the air-water interface: Effect of the structure. J. Colloid Interface Sci. 2010, 345, 54–63.

Maldonado-Valderrama, J.; Rogriguez Patino, J. M. Interfacial rheology of protein-surfactant mixtures. Cur. Op. Colloid & Interf. Sci. 2010, 15 271–282.

Markwell, M. A. K.; Haas, S. M.; Bieber, L. L.; Tolbert, N. E. A modification of the lowry procedure to simplify protein determination in membrane and lipoprotein samples. Anal. Biochem. 1978, 87, 206–210.

McClements, D. J. Food Emulsions: Principles, Practices, and Techniques; CRC Press: Boca Raton, FL, 2005.

McClements, D. J.; Decker, E. A. Lipid oxidation in oil-in-water emulsions: Impact of molecular environment on chemical reactions in heterogeneous food systems. J. Food Sci. 2000, 65, 1270–1282.

Medina, I.; Undeland, I.; Larsson, K.; Storro, I.; Rustad, T.; Jacobsen, C.; Kristinova, V.; Gallardo, J. M. Activity of caffeic acid in different fish lipid matrices: A review. Food Chem. 2012, 131, 730–740.

Mei, L. Y.; Decker, E. A.; McClements, D. J. Evidence of iron association with emulsion droplets and its impact on lipid oxidation. J. Agric. Food Chem. 1998, 46, 5072–5077.

Meynier, A.; Rampon, V.; Dalgalarrondo, M.; Genot, C. Hexanal and t-2-hexenal form covalent bonds with whey proteins and sodium caseinate in aqueous solution. Int. Dairy J. 2004, 14, 681–690.

Morris, V. J.; Gunning, A. P. Microscopy, microstructure and displacement of proteins from interfaces: Implications for food quality and digestion. Soft Matter 2008, 4, 943–951.

Nakazawa, R.; Shima, M.; Adachi, S. Effect of oil-droplet size on the oxidation of microencapsulated methyl linoleate. J. Oleo Sci. 2008, 57, 225–232.

Nuchi, C. D.; Hernandez, P.; McClements, D. J.; Decker, E. A. Ability of lipid hydroperoxides to partition into surfactant micelles and alter lipid oxidation rates in emulsions. J. Agric. Food Chem. 2002, 50, 5445–5449.

Oehlke, K.; Heins, A.; Stockmann, H.; Sonnichsen, F.; Schwarz, K. New insights into the antioxidant activity of trolox in O/W emulsions. Food Chem. 2011, 124, 781–787.

Osborn, H. T.; Akoh, C. C. Effect of emulsifier type, droplet size, and oil concentration on lipid oxidation in structured lipid-based oil-in-water emulsions. Food Chem. 2004, 84, 451–456.

Panya, A.; Laguerre, M; Bayrasy, C.; Lecomte, J.; Villeneuve, P.; McClements, D. J.; Decker, E. A. An investigation of the versatile antioxidant mechanisms of action of rosmarinate alkyl esters in oil-in-water emulsions. J. Agric. Food Chem. 2012, 60, 2692–2700.

Pillon, N.; Zarrouki, B.; Lagarde, M.; Soulage, C. Alteration of intracellular signaling pathways of insulin in muscle and fat cells by products of lipid peroxidation 4-hydroxy-2-nonenal (4-HNE) and 4-hydroxy-hexenal (4-HHE). Diabetes Metab. 2009, 35, A46–A47.

Ponginebbi, L.; Nawar, W. W.; Chinachoti, P. Oxidation of linoleic acid in emulsions: Effect of substrate, emulsifier, and sugar concentration. J. Am. Oil Chem. Soc. 1999, 76, 131–138.

Rampon, V.; Genot, C.; Riaublanc, A.; Anton, M.; Axelos, A. V.; McClements, D. J. Front-face fluorescence spectroscopy study of globular proteins in emulsions: Displacement of BSA by a nonionic surfactant. J. Agric. Food Chem. 2003, 51, 2482–2489.

Rampon, V.; Lethuaut, L.; Mouhous-Riou, N.; Genot, C. Interface characterization and aging of bovine serum albumin stabilized oil-in-water emulsions as revealed by front-surface fluorescence. J. Agric. Food Chem. 2001, 49, 4046–4051.

Richards, A.; Golding, M.; Wijesundera, C.; Lundin, L. The influence of secondary emulsifiers on lipid oxidation within sodium caseinate-stabilized oil-in-water emulsions. J. Am. Chem. Oil Soc. 2011, 88, 65–73.

Schaich, K. Lipid Oxidation: Theoretical Aspects. Bailey's Industrial Oil and Fat Products; Shahidi, F., Ed.; John Wiley & Sons: Hoboken, NJ, 2005.

Silvestre, M. P. C.; Chaiyasit, W.; Brannan, R. G.; McClements, D. J.; Decker, E. A. Ability of surfactant headgroup size to alter lipid and antioxidant oxidation in oil-in-water emulsions. J. Agric. Food Chem. 2000, 48, 2057–2061.

Singh, H. Aspects of milk-protein-stabilised emulsions. Food Hydrocolloids 2011, 25, 1938–1944.

Sørensen, A. D. M.; Baron, C. P.; Let, M. B.; Bruggemann, D. A.; Pedersen, L. R. L.; Jacobsen, C. Homogenization conditions affect the oxidative stability of fish oil enriched milk emulsions: Oxidation linked to changes in protein composition at the oil–water interface. J. Agric. Food Chem. 2007, 55, 1781–1789.

Tcholakova, S.; Denkov, N. D.; Lips, A. Comparison of solid particles, globular proteins and surfactants as emulsifiers. Phys. Chem. Chem. Phys. 2008, 10, 1608–1627.

Tikekar, R. V.; Johnson, A.; Nitin, N. Real-time measurement of oxygen transport across an oil-water emulsion interface. J. Food Eng. 2011, 103, 14–20.

van Aken, G. A. Competitive adsorption of protein and surfactants in highly concentrated emulsions: Effect on coalescence mechanisms. Colloids Surf., A 2003, 213, 209–219.

Villière, A. Approche physico-chimique et sensorielle de l'oxydation des lipides dans des émulsions stabilisées par des protéines. Université de Nantes, 2005.

Villière, A.; Genot, C. Approche physico-chimique et sensorielle de l'oxydation des lipides en émulsions. Oléagineux, Corps Gras, Lipides 2006, 13, 152–159.

Villière, A.; Viau, M.; Bronnec, I.; Moreau, N.; Genot, C. Oxidative stability of bovine serum albumin- and sodium caseinate-stabilized emulsions depends on metal availability. J. Agric. Food Chem. 2005, 53, 1514–1520.

Waraho, T.; McClements, D. J.; Decker, E. A. Mechanisms of lipid oxidation in food dispersions. Trends Food Sci. Technol. 2011, 22, 3–13.

Wilde, P.; Mackie, A.; Husband, F.; Gunning, P; Morris, V. Proteins and emulsifiers at liquid interfaces. Adv. Colloid Interface Sci. 2004, 108, 63–71.

Ye, A. Q. Interfacial composition and stability of emulsions made with mixtures of commercial sodium caseinate and whey protein concentrate. Food Chem. 2008, 110, 946–952.

Zarrouki, B.; Soares, A. F.; Guichardant, M.; Lagarde, M.; Geloen, A. The lipid peroxidation end-product 4-HNE induces cox-2 expression through p38MAPK activation in 3T3-L1 adipose cell. FEBS Lett. 2007, 581, 2394–2400.

CHAPTER

6

Oxidative Stability of Enzymatically Processed Oils and Fats

Derya Kahveci[1,2], Zheng Guo[1], Ling-Zhi Cheong[1], Mia Falkeborg[1], Warowan Panpipat[1], and Xuebing Xu[1]

Introduction

Fats and oils with modified functional and nutritional properties are in high demand in the food industry. The physical characteristics of fats and oils—for instance, those related to their melting and crystallization profiles—are important for food technologists to consider because nutritionists and consumers display great interest in their nutritional properties, based on their composition.

The fats and oils industry has extensively used chemical interesterification to alter the physical and/or nutritional properties of fats and oils. The process produces a complete positional randomization of acyl groups in triacylglycerols (TAGs). However, the random end products of chemical interesterification cannot meet the demand for lipids with specific structures and compositions. Besides, some fatty acids (FAs) are heat sensitive, which means that they cannot survive the harsh conditions of chemical interesterification. The enzymatic process, on the other hand, requires mild reaction temperature and pressure, which is vital for handling unsaturated FAs. Other advantages of enzymatic modification, as opposed to the chemical method, include (i) the specificity of enzymes, providing products with specific structures; (ii) less waste production and, thus, less environmental pollution; (iii) the possibility of enzymes with different characteristics,

[1]Department of Engineering, Aarhus University, 8000 Aarhus C, Denmark; [2]Department of Food Engineering, Faculty of Engineering and Architecture, Yeditepe University, 34755 Kayisdagi, Istanbul, Turkey.

obtained from several sources as well as through protein engineering; and (iv) when immobilized, reuse of the same batch of enzyme.

The enzymatic modification of fats and oils has been a popular topic for the past few decades. However, information on the oxidative stability of such products remains limited. This chapter reviews the newest developments regarding the oxidative stability of enzymatically processed fats and oils.

Enzymes for the Modification of Fats and Oils

Enzymatic modification of fats and oils involves biocatalytic cleavage and/or re-formation of carboxylester bonds. Some enzymes with this function exist in nature, such as carboxylesterases (carboxyl ester hydrolase), which include two groups of enzymes—namely, esterases (EC 3.1.1.1) and lipases (EC 3.1.1.3)—differentiated on the basis of their substrate specificity. Esterases hydrolyze solutions of water-soluble, short-acyl-chain esters and are inactive against water-insoluble, long-chain TAGs, which, in turn, are specifically hydrolyzed by lipases. When comparing the structural features of sequence-related esterases and lipases, it's seen that lipases, unlike esterases, display a significant difference in the distribution of hydrophobic amino-acid residues around their active site (Chahinian & Sarda, 2009). Thus, the molecular basis for lipases' attraction to oils and fats is the existence of a particular surface domain in lipases that specifically interacts with lipid–water interfaces and contributes to the transfer of a single substrate molecule from the organized lipid–water interface (supersubstrate) to the catalytic site of the enzyme. Because esterases are not active against water-insoluble TAGs, they are of less interest in the modification of fats and oils; therefore, no further attention will be given to esterases.

As enzymes found throughout nature, lipases occur in mammals, plants, fungi, and bacteria (Malcata et al., 1992) (Table 6.A). These differently sourced lipases may differ in molecular size, stability, catalytic property, or specificity; however, all lipases sequenced to date share sequence homologies, including a significant region, His-X-Y-Gly-Z-Ser-W-Gly or Y-Gly-His-Ser-W-Gly (W, X, Y, and Z denote unspecified amino acid residues) that all enzymes conserve. Another structural feature common among lipases is a lid or flap composed of an amphiphilic helix peptide sequence, which in its closed conformation (in the absence of an interphase or organic solvent) prevents the substrate from accessing the catalytic triad. When the lid opens, it creates a large hydrophobic surface to which the hydrophobic supersubstrate (usually, the oil drop) binds (Schmid & Verger, 1998).

Plant lipases hydrolyze TAGs at much lower rates (usually < 0.5 μmol min^{-1} mg^{-1}) as compared with animal or microbial lipases, and they also occur naturally

Table 6.A. Some Representative Lipases and Their Characteristics

Origin/Sources	pH Optima	Temperature Optima	Specificity	Commercial Producer	Remarks
Of plant					
Carica papaya	9.0	37	*sn*-3, short-medium chain FA	—	Modification of oils and fats, synthesis
Rice bran	7.5–8.0	75–80	Nonspecific (Phospholipase A2 activity)	—	Hydrolysis of oils and fats
Castor bean	4.8	30	Nonspecific, Ricinoleic acid	—	Hydrolysis of castor oil
Of mammal					
Human pancreatic lipase	8.5	37	*sn*-1,3	Sigma-Aldrich	Diagnostic, clinic
Human gastric lipase	5.4	37	*sn*-3 (acid-stable)	—	Diagnostic, clinic
Porcine pancreatic lipase	7.5	37	*sn*-1,3	Sigma-Aldrich	Digestive aid
Of fungal					
Candida rugosa	7.5	37	Nonspecific	Amano, Sigma-Aldrich	Hydrolysis of lipids and organic synthesis
Candida antarctica lipase A	8.0	65	Nonspecific, trans-FA selective, high activity towards secondary alcohols	Novozymes, Codexis	Hydrolysis of lipids by selectively removing trans-FAs, and organic synthesis
Candida antarctica lipase B	7.0	65	*sn*-1,3 specific	Novozymes, Amano	Modification of oils and fats, organic synthesis

(continued)

Table 6.A. *Continued*

Origin/Sources	pH Optima	Temperature Optima	Specificity	Commercial Producer	Remarks
Of fungal *(continued)*					
Geotrichum candidum	6.0	40	*cis*-Δ^9 oleochemistry (unsaturated fatty acids)	Biocatalysts, Amano	Selective hydrolysis of *cis*-Δ^9 unsaturated FAs, synthesis
Thermomyces lanuginosus	8.0	60	*sn*-1,3	Novozymes, Boehringer Manheim	Hydrolysis and synthesis, specifically for interesterification
Rhizomucor miehei	8.0	40	*sn*-1,3	Novozymes	Modification of oils and fats and synthesis
Aspergillus oryzae			*sn*-1,3	Amano	Hydrolysis and synthesis
Penicillium camembertii			*sn*-1,3	Amano	Hydrolysis and synthesis
Rhizopus delemar	5.6	35	*sn*-1,3 (phospholipase A1 activity)	Tanabe Seiyaku	Hydrolysis and synthesis
Rhizopus oryzae	8.5	30	*sn*-1,3 (phospholipase A1 activity)	Amano	Hydrolysis and synthesis
Rhizopus arrhizus	7.5	37	*sn*-1,3 (phospholipase A1 activity)	Sigma-Aldrich	Hydrolysis, diagnostic, and analysis

Of bacteria					
Burkholderia cepacia	7.0	50	Nonspecific	Amano, Sigma-Aldrich	Hydrolysis, diagnostic and analysis
Pseudomonas glumae	9.0	50	Nonspecific	Asahi, Sigma-Aldrich	Hydrolysis and synthesis
Pseudomonas alcaligenes	6–10	40	*sn*-1,3	Danisco	Hydrolysis, detergent additive
Pseudomonas fluorescens	8.0	55	*sn*-1,3	Amano (Lipase AK)	Hydrolysis and synthesis
Chromobacterium viscosum	7.0	45	*sn*-1,3	Asahi, Amano, Sigma-Aldrich	Hydrolysis and synthesis
Pseudomonas mendocina	7–8	50	*sn*-1,3	Danisco	Hydrolysis, detergent additive
Bacillus thermocatenulatus	8–9	60–70	*sn*-1,3 (thermoalkalophilic)	Boehringer Manheim	Hydrolysis and synthesis
Bacillus sp.	6–8	65	Nonspecific, monoglyceride lipase	Asahi	Hydrolysis and synthesis

at lower concentrations. Therefore, plant lipases are of less interest for low-value products like bulk oils and fats on an industrial scale (Mukherjee, 1994). However, some plant lipases have special value because of their substrate specificity and/or selectivity. For example, castor oil cannot cause the hydrolysis of ricinoleic acid (12-hydroxy-9-octadecenoic acid) with the conventional technology of "steam splitting" at high temperature and pressure, due to dehydration and intermolecular esterification yielding estolides of ricinoleic acid. Castor-bean lipase provides an alternative for *in situ* lipolysis of castor to yield ricinoleic acid in the presence of acid. As an emerging and versatile biocatalyst, *Carica papaya* lipase (CPL) is receiving increasing attention (de María et al., 2006). CPL has some unique properties, including its (i) *sn*-3 selectivity toward glycerides; (ii) preference for short-chain FAs; and (iii) "naturally immobilized" form (tightly attached to the dry matter present in the *C. papaya* latex, making it insoluble in water). These unique features of CPL make it of special interest for specific applications in organic synthesis and lipid modification that target regio- and stereo-structured lipids.

On the contrary, few studies have reported practical applications for mammal lipases. The main applications of mammal lipases are as digestive additives, diagnostics, and analytical reagents. Thanks to strict *sn*-1,3 selectivity, porcine pancreatic lipase is widely used as a tool for *sn*-2 FA composition analysis.

However, microbial lipases are enzymes that have potential industrial applications in the modification of fats and oils (Sharma et al., 2011). They have several advantages over lipases from plants and mammals, such as (i) more genetic diversity; (ii) possession of the extremophile property from the beginning of life; (iii) quick and easy genetic evolution; and (iv) easy industrial production. To date, progress in biotechnological engineering has greatly shortened the period necessary for developing a microbial enzyme for a specific task, and the cost is also decreasing to an acceptable level. Therefore, almost all lipases applied at an industrial level are from fungal or bacterial sources, such as Novozyme 435 (*Candida antartica* lipase B), Lipozyme TL IM (*Thermomyces lanuginosus* lipase), and Lipozyme RM IM (*Rhizomucor miehei* lipase) (Table 6.A).

Besides their high catalytic efficiency (that is, the turnover number of substrate molecules/enzyme molecules), enzymes have extraordinary specificity, which is a technological advantage over chemical catalysts. In the case of lipases, specificities include: (i) lipid class specificity (high activity for a specific lipid like monoacylglycerol lipase); (ii) positional specificity (that is, *sn*-1,3 specificity); (iii) stereochemical specificity (such as a selective catalysis *R-/S-* isomer configuration); and (iv) fatty acid specificity (e.g., *Geotrichum candidum* lipase selectively hydrolyzes *cis*-Δ^9 unsaturated FAs) (Table 6.A). These lipase specificities are one of the leading reasons for their exclusive applications (Table 6.A).

Main Reaction Routes for the Modification of Fats and Oils

Fats and oils are modified for several different reasons, which include (i) changing the TAG structure to provide the chemo-physical properties, such as the transition from palm stearin to cocoa butter substitutes; and (ii) incorporating specific FAs or other reactive functional groups for nutritional considerations, such as the production of MLM type (M, medium chain FA; L, long chain FA) lipids and the incorporation of omega-3 polyunsaturated fatty acids (PUFAs) into children's dietary lipid formulations. The term "modification" implies that the process has a starting material. So far, almost all processes have used natural materials, either from plant, microbial, marine, or animal sources. Some processes may involve the direct esterification of glycerol, but the FAs employed are all from natural sources.

FA composition and positional distribution provide the molecular basis for the chemo-physical properties and physiological activities of fats and oils. Depending on their sources, composition and positional distribution of FAs, which are formed from their own biosynthesis pathway and are a result of millions of years' evolution, are specific. Statistical analysis demonstrates that in most fats and oils from plants or other sources, positional distribution is naturally random. For example, in most plant oils, oleic and linoleic acids are preferably located in *sn*-2 position, while in pork fat, palmitic acid is enriched in the *sn*-2 position (Hirayama & Nakae, 1964). To explain the positional distribution patterns of FAs in fats and oils, Litchfield (1972) proposed the *sn*-1,3-random, *sn*-2-random theory, which has been widely accepted. Many nutrition studies have provided evidence that this positional distribution is also linked to the digestion and absorption of lipids in the human body. Therefore, a positional concern for FA distribution will be of great importance when a strategy for modification of fats and oils is designed. Obviously, *sn*-1,3–specific lipases have an advantage over nonspecific chemical catalysts.

Fig. 6.1 presents the most investigated reaction patterns for the modification of oils and fats: interesterification, transesterification, esterification, and glycerolysis. In general, interesterification refers to the acyl-exchange reaction that occurs between two TAG molecules. As depicted in Fig. 6.1, if a chemical catalyst or a nonspecific lipase catalyzes the reaction between ABA and CBC triglycerides, it can produce up to 27 kinds of TAG with different structures (with consideration for stereo-isomers). With a strictly *sn*-1,3–specific lipase as a catalyst, only four different types of TAG will exist in the final reaction mixture. With *sn*-1,3 lipase, the reaction is confined to the glycerides' 1- and 3- positions, and *sn*-2 fatty acid B will not migrate to the 1,3 position, which chemical or nonspecific lipase-based processes cannot reach. Aarhus Karlshamn has industrialized this reaction

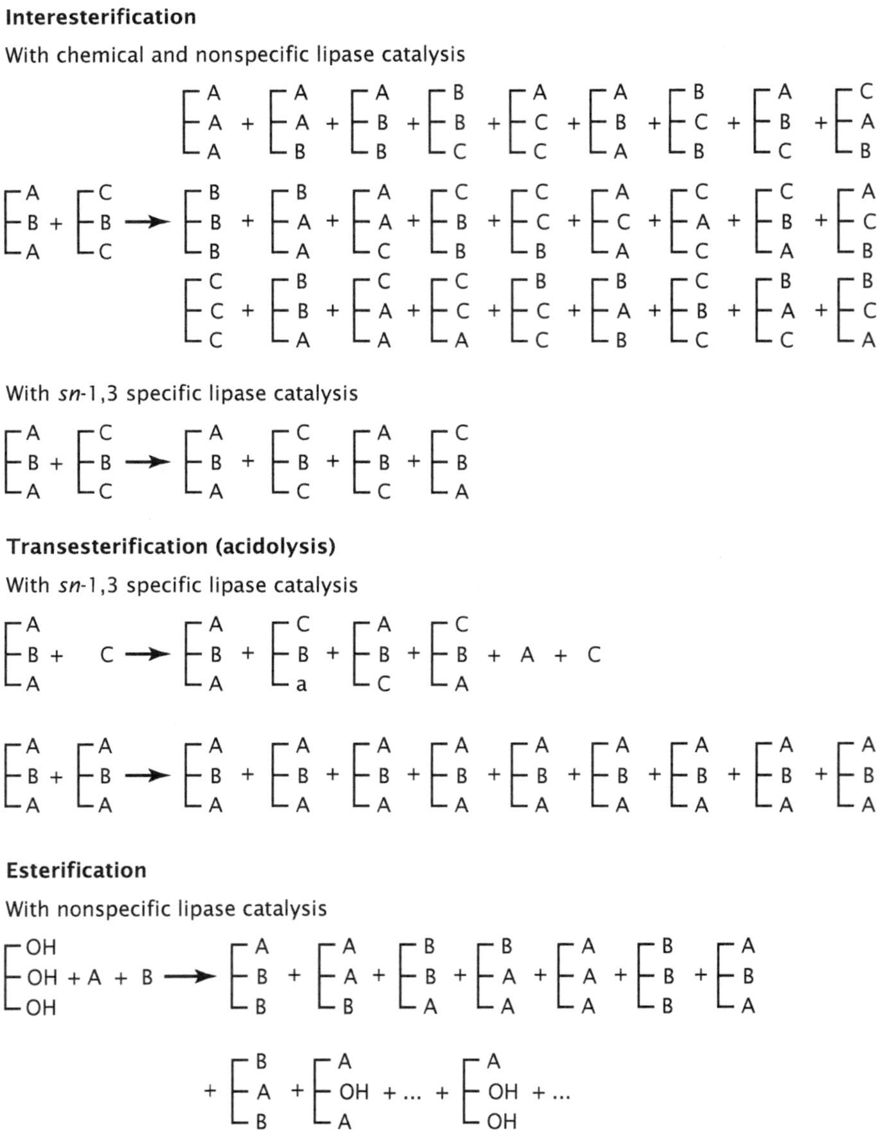

Fig. 6.1. Representative reaction routes for the lipase-catalyzed modification of oils and fats.

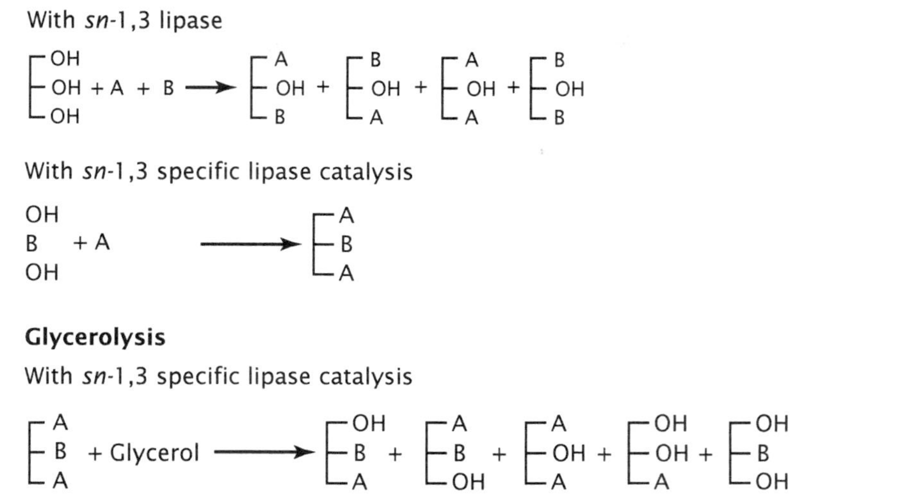

Fig. 6.1. *Continued*

to produce trans-free fats from palm stearin and plant oils employing Lipozyme TL IM (*sn*-1,3 specific lipase) (Walsh, 2007).

Transesterification is an acyl exchange reaction between TAG and FA molecules, also called acidolysis. As Fig. 6.1 shows, the reaction between TAG in the ABA structure and fatty acid C yields TAGs in the CBA and CBC structures and releases fatty acid A. Unilever employs an industrial application for this reaction to produce cocoa-butter equivalents from palm stearin (POP) and stearic acid (S) by Lipozyme RM IM–catalyzed transesterification. The target triglycerides POS and SOS are the main components of cocoa butter, which provide its unique melting profile for chocolate manufacturing.

The chemical or nonspecific lipase-catalyzed esterification of glycerol with FAs produces a mixture of mono-, di-, and triacylglycerols (MAGs, DAGs, and TAGs, respectively) with quite a number of possible structures. Theoretically, the reaction can yield TAGs if water is effectively removed as a by-product. However, if a *sn*-1,3–specific lipase is used, the resulting product will be 1,3-DAG (Fig. 6.1). An example for the industrial application of this reaction is the production of DAG oils (Enova, claimed as healthy oil) by Archer Daniels Midland, where hydrolyzed FAs from plant oils are used as materials for *sn*-1,3–specific lipase-catalyzed esterification with glycerol. Starting from 2 MAGs, *sn*-1,3–specific lipase-catalyzed esterification could be employed to synthesize ABA-type TAGs.

The transesterification reaction between TAGs and glycerol is called glycerolysis, which is an important industrial process for the production of food emulsifiers, such as MAGs and DAGs (Fig. 6.1). It is generally carried out at a high temperature, which cannot be applied to oils containing unsaturated FAs. Lipase-catalyzed glycerolysis can be performed under mild conditions, and the reaction can be adjusted toward the target product, MAG or DAG, by shifting its equilibrium through medium engineering or by manipulating other parameters (Guo & Xu, 2005; Guo et al., 2009).

Oxidative Stability of Enzymatically Modified Oils and Fats

Structured Lipids

Structured lipids (SLs) are restructured TAGs with special functional or nutritional properties. Because of their nutritional properties, SLs containing medium-chain FAs (M) at the *sn*-1 and *sn*-3 positions and omega-3 PUFAs (L) at the *sn*-2 position are of interest. Such lipids, called MLM-type SLs, combine the beneficial effects of FAs with different chain lengths, and their nutritional value lies in the metabolism route during digestion. Upon consumption, the body TAG is hydrolyzed into 2-MAG and two FAs, mainly by pancreatic lipase but also lingual and gastric lipases. In the case of MLM-type SLs, digestion products will be 2-MAG rich in omega-3 PUFAs and medium-chain free fatty acids (FFA). The latter will have higher plasma clearance, higher oxidation rate, improved nitrogen-sparing action, and fewer tendencies to be deposited in the adipose tissue. Thus, medium-chain FFAs quickly provide energy to patients suffering from malabsorption, short bowel syndrome, burn wounds, and for preterm infants as well. On the other hand, TAG oil, which is rich in medium-chain FAs, lacks essential omega-3 PUFAs. Providing omega-3 PUFAs esterified to the *sn*-2 position better protects these FAs from oxidation (Wijesundera et al., 2008) and provides better absorption of them in the 2-MAG form (Christensen et al., 1995). MLM-type SLs have beneficial effects on immune function, nitrogen balance, and lipid clearance from the bloodstream, and they can be used not only as nutrition for patients with maldigestion and malabsorption of lipids, but also as high–value added nutraceuticals for the elderly (Mu & Porsgaard, 2005).

Due to the required positional distribution of FAs, as well as the high risk of oxidation when omega-3 PUFAs are handled, SLs are produced by means of selective enzymatic reactions under mild conditions. The oxidative stability of SLs, including omega-3 PUFAs, has been a major problem to solve, which Martin et al. (2010) have recently reviewed. Compared to substrates, modified oils generally had lower oxidative stability, even though substrates included more omega-3 PUFAs. Akoh and Moussata (2001) suggested that a combination of chemical and instrumental tests, such as the oxidative stability index used in their study, were necessary to assess the oxidative stability of SLs. Measurement of the peroxide value, conjugated dienes, and thiobarbituric acid value alone may not fully explain the oxidative changes that occur in SLs, as well as in unmodified TAGs. Other factors may contribute to the stability of SLs; examples include temperature, storage and assay conditions, presence, amount or absence of tocopherols and phospholipids, FA types, and downstream processing.

Capric acid was incorporated into single-cell oil rich in docosapentaenoic acid (DPA) and docosahexaenoic acid (DHA) by lipase-catalyzed acidolysis in hexane (Hamam & Shahidi, 2005). Contrary to expectations, the incorporation of capric acid did not increase the oxidative stability of SL; instead, the reaction decreased it. Hamam and Shahidi (2005) linked this observation to the loss or structural alteration of natural antioxidants present in the oil during the preparation process of SL. They proved this hypothesis because a substrate mixture treated in the same manner in the absence of any enzyme still had significantly lower oxidative stability than the unmodified oil. Additionally, the incorporation of capric acid into arachidonic acid (AA)-rich single-cell oil negatively affected the oxidative stability of the resulting SL (Hamam & Shahidi, 2004). Similarly, SLs produced through the acidolysis of fish oil with caprylic acid had lower oxidative stability than non-treated fish oil, as well as the oil produced by chemical interesterification between fish oil and tricaprylin. This result occurred even though fish oil had more omega-3 PUFAs than SLs. The reduced oxidative stability of SLs was linked to long processing periods (Nielsen et al., 2004).

Caprylic acid, oleic acid, and stearic acid were incorporated into DHA-rich algal oil by lipase-catalyzed acidolysis in hexane (Yankah & Akoh, 2000). 1,3-specific lipase from *Rhizomucor miehei* prevented incorporation of these FAs at the *sn*-2 position. The concentration of stearic acid reacted with the oil was less than that of either caprylic and oleic acid (that is, 1 mole per mole of oil compared to 6 and 2 moles, respectively). As a result, the DHA content of stearic acid–enriched oil was the highest. Moreover, the amount of tocopherol in the product was reduced more than it was for the other two oils; the reasons for this difference remain unexplained. Therefore, the oxidative stability of stearic acid–enriched oil was lower than that of capric–acid or oleic–acid enriched oils, but still higher than that of the original DHA-rich oil. Oxidation of SLs, including stearic acid (Yankah & Akoh, 2000) and caprylic acid (Lee & Akoh, 1998), increased rapidly after 24 hours of storage at room temperature. The result was linked to a loss of tocopherols.

Antioxidants were added in several cases for improved stability of SLs. Addition of antioxidants before the reaction did not improve the oxidative stability of SLs (Xu et al., 2005). Adding α-tocopherol after the reaction and purification, on the other hand, significantly improved the oxidative stability of MLM-type SLs

(Lee & Akoh, 1998). Compared to α-tocopherol, *tert*-Butylhydroxyquinone exhibited better antioxidative properties when added to SLs produced by incorporating caprylic acid into fish oil (Akoh & Moussata, 2001). Additionally, adding a combination of lecithin, ascorbyl palmitate, and tocopherols was superior to adding them separately for improved oxidative stability of SL produced by incorporation of caprylic acid into soybean oil (Turan et al., 2007).

Interesterified Lipids

Aside from MLM-type SLs, a number of different SLs have been developed via lipase-catalyzed reactions to provide fats and oils with desired functional and/or nutritional properties.

Omega-3 PUFAs were incorporated into oils with longer-chain FAs to improve their nutritional properties. Such lipids are primarily useful as human milk fat substitutes (HMFS). Human milk's TAG has a unique structure, 60–70% of which is palmitic acid, located at the *sn*-2 position; stearic, oleic, and linoleic acids are preferentially esterified to the *sn*-1,3 positions. Such regiospecificity improves the infant's absorption of fat and calcium while preventing the formation of calcium soaps produced by the interaction of long-chain saturated FAs and calcium (Innis, 1991). To obtain optimal fat and calcium absorption, the FA composition and molecular structure of fats in infant formulas should be the same as those of human milk fat. HMFS have been developed to mimic the fat composition and structure of human milk from materials based on palm oil and lard, tripalmitin, and vegetable oil blends (Quinlan and Chandler, 1992; Yang et al., 2003; Nielsen et al., 2006; Maduko et al., 2008; Giet et al., 2009; Pina-Rodriguez & Akoh, 2009) and butterfat (Sørensen et al., 2010). Long-chain PUFAs—such as α-linolenic acid, linoleic acid, eicosapentaenoic acid (EPA), DHA, and ARA—are of major importance for an infant's development (Wan et al., 2011). Consequently, the SLs for infant formulas should be enriched with PUFAs. As mentioned above, PUFAs are highly susceptible to the oxidation process, which leads to the formation of undesirable products such as free radicals, hydroperoxides, aldehydes, and polymers, which are responsible for the loss of PUFAs' health properties. Food products rich in PUFAs, such as infant formula containing SLs, may deteriorate rapidly during storage and produce off-flavors and odors (Choe & Min, 2006; Martin et al., 2010; Wirkowska et al., 2012). Special attention should be paid to oxidation processes in those infant formulae and baby food products. Table 6.B summarizes knowledge currently available about the oxidative stability of enzymatically produced HMFS, as well as several other SLs.

Most studies have shown that similar to MLM-type SLs, interesterified lipids have lower oxidative stability compared to their unmodified counterparts because

Table 6.B. A Compilation of Recent Research on the Oxidative Stability of Interesterified SLs and HMFS

Substrates	Lipase	Product	Methods Used to Measure Oxidative Status	Oxidative Stability[a]	Reference
Tripalmitin, vegetable oil, or fish oil	*Rhizomucor miehei*	HMFS	Tocopherol content, FFA, peroxide value (PV), *p*-anisidine value (PAV)	↓	Maduko et al., 2008
Amaranth oil, palmitic acid, DHA	*Candida antarctica* lipase B, *Rhizomucor miehei*	HMFS	PV, PAV, tocopherol, and phytosterol content	↓	Pina-Rodriguez & Akoh, 2009
Palmitic acid-enriched amaranth oil, DHA	*Rhizomucor miehei*	HMFS	Oxidative induction time (OIT)	↓	Pina-Rodriguez & Akoh, 2009
Stearidonic acid soybean oil, tripalmitin	*Thermomyces lanuginosus*, *Candida antarctica* lipase B	HMFS	Tocopherol content	↓	Teichert & Akoh, 2011a
Palmitic acid-enriched stearidonic acid soybean oil, γ-linolenic acid, or DHA	*Thermomyces lanuginosus*, *Candida antarctica* lipase B	HMFS	PV, PAV, OIT, tocopherol content	↓	Teichert & Akoh, 2011b
Butterfat, FFA from rapeseed oil and soybean oil	*Rhizomucor miehei*	HMFS	FFA, PV, PAV, volatile compound content, OIT, sensory panel	↓	Sørensen et al., 2010
Lard, FFA from soybean oil	*Rhizomucor miehei*	HMFS	FFA, PV, tocopherol, and volatile compound content	↓	Nielsen et al., 2006
Olive oil, omega-3 PUFAs	*Candida antarctica* lipase B	SL	OIT	↓	Kahveci et al., 2009
Extra-virgin olive oil, conjugated linoleic acid (CLA)	*Rhizomucor miehei*	SL	Tocopherol and phytosterol, content, 2,2-diphenyl-1-picrylhydrazyl radical (DPPH•) quenching test, OIT	↓	Lee et al., 2006
Roasted sesame oil, caprylic acid	*Rhizomucor miehei*	SL	Tocopherol, phytosterol, and volatile compound content, OIT	↓	Kim & Akoh, 2006
Rice-bran oil, caprylic acid	*Rhizomucor miehei*	SL	Tocopherol content	↓	Jennings & Akoh, 2009a

(continued)

Table 6.B. *Continued*

Substrates	Lipase	Product	Methods Used to Measure Oxidative Status	Oxidative Stability[a]	Reference
Soybean oil or sunflower oil, CLA	*Rhizomucor miehei*	SL	PV, PAV, 2-thiobarbituric acid reactive substance value (TBARS)	↓	Lee et al., 2003
Borage oil or evening primrose oil, omega 3 PUFAs	*Pseudomonas sp.*	SL	Conjugated dienes (CD), TBARS, volatile compound content	↓	Senanayake & Shahid, 2002
Anhydrous milk fat, linseed oil	*Thermomyces lanuginosus*	SL	OIT	↓	Aguedo et al., 2008
Anhydrous milk fat, rapeseed and/or linseed oil	*Thermomyces lanuginosus*	SL	PV, PAV, TBARS, volatile compound content	↓	Giet et al., 2009
Extra virgin olive oil, fully hydrogenated palm oil	*Thermomyces lanuginosus*, *Candida antarctica* lipase B, *Rhizomucor miehei*	SL	OIT	↓	Criado et al., 2007
Sesame oil, fully hydrogenated soybean oil	*Thermomyces lanuginosus*	SL	OIT	↓	Lopez-Hernandez et al., 2007
Safflower oil, CLA	*Rhizomucor miehei*	SL	PV, PAV, TBARS	↓	Lee et al., 2004
Soybean oil, perilla oil	*Rhizomucor miehei*	SL	Tocopherol and phytosterol content, OIT	↓	Mitra et al., 2010

[a]Compared to unmodified substrates.

they lose tocopherols during the reaction and the purification of their products. Adding antioxidative compounds improved the stability of these products (Lee et al., 2003, 2004, 2006; Maduko et al., 2008; Giet et al., 2009; Kahveci et al., 2009; Mitra et al., 2010; Teichert & Akoh, 2011b).

Plastic Fats and Fat Emulsions

A frequent application of enzymatically processed oils and fats is the development of plastic fats. Plastic fats usually exhibit distinct melting and crystallization behavior, solid fat content, crystal structure, and polymorphic form in order to impart desirable tenderness, texture, mouthfeel, air incorporation, and shelf life (Ghotra et al., 2002). Thus, numerous studies have investigated the physicochemical properties of enzymatically processed fats to ensure that they function as plastic fats (Rousseau & Marangoni, 1999; Farmani et al., 2007; Lee et al., 2008; Cheong et al., 2009; Adhikari et al., 2011; Zhu et al., 2012). In order to demonstrate the value of enzymatically processed fats to the food industry, they must meet not only the functional but also the required sensory characteristics, in addition to the necessary oxidative and shelf-life stability.

Lately, the industry has focused more on evaluating the oxidative stability of plastic fats (Table 6.C). Different indicators have been used to evaluate the oxidative stability of enzymatically processed plastic fats; these include oxidative induction time (OIT), peroxide value (PV), *p*-anisidine value (PAV), thiobarbituric acid-reactive substances (TBARS) value, analysis of volatile compounds, and sensory properties. The following section explains the oxidative stability of enzymatically processed plastic fats.

Farmani and colleagues (2007, 2009) conducted enzymatic transesterification of different fat blends with characteristics suited for applications such as soft-tub, stick, roll-in, and baker's margarines; cake shortening; and vanaspati fat. Both studies evaluated the oxidative stability of the enzymatically processed plastic fats by measuring the OIT at 120 °C. Enzymatically transesterified blends were found to be less stable to oxidation than unmodified blends. The authors attributed the decrease in oxidative stability to the formation of impurities (such as FFA and partial acylglycerols) and the removal of naturally occurring antioxidants during the enzymatic processing. Indeed, as the impurities were removed and antioxidant (TBHQ) was added (post-treatment), the OIT increased significantly. In fact, enzymatically processed fat blends with lower degrees of saturation had oxidative stability comparable to unmodified blends following the post-treatment.

The aforementioned findings accorded with two subsequent findings by Adhikari and colleagues (2010, 2011), who enzymatically interesterified rice-bran oil and coconut oil with either palm stearin or fully hydrogenated soybean oil to produce *trans*-free margarine and shortening. These two studies measured OIT at

Table 6.C. A Compilation of Recent Research on the Oxidative Stability of Plastic Fats

	Indicators for the Measurement of Oxidative Stability					
Composition of Enzymatically Processed Plastic Fats	**PV**	**PAV**	**Volatile Compounds**	**OIT**	**Findings**	**References**
Fully hydrogenated soybean, rapeseed, and sunflower oils				X (IP_{120})	Enzymatically transesterified blends became less stable to oxidation than initial blends with lower oxidation induction time.	Farmani et al., 2007
Canola oil, palm olein, and fully hydrogenated soybean oil				X (IP_{120})	Enzymatically transesterified blends became less stable to oxidation than initial blends with lower oxidation induction time.	Farmani et al., 2009
Rice-bran oil, palm stearin, and coconut oil				X (IP_{100})	Oxidative stability of enzymatically interesterified fats can be improved by adding the antioxidant catechin.	Adhikari et al., 2010
Rice-bran oil, fully hydrogenated soybean oil, and coconut oil				X (IP_{100})	Oxidative stability of enzymatically interesterified fats can be improved by adding the antioxidant catechin.	Adhikari et al., 2011
Pine nut oil and palm stearin	X	X	X		Oxidative stability of enzymatically interesterified fats can be improved by adding the antioxidants quercetin and ascorbyl palmitate.	Zhu et al., 2012

100 °C. The enzymatically processed plastic fats were less stable to oxidation than those produced by physical blending, but their oxidative stability increased significantly with the addition of catechin.

Zhu and colleagues (2012) adopted a different approach to evaluate the oxidative stability of enzymatically processed plastic fats. By interesterifying pine-nut oil with palm stearin, they aimed to produce soft margarine fat with enhanced nutritional properties, namely for the reduction of cholesterol and body weight. They investigated the primary and secondary oxidation products by measuring PV and PAV values, respectively. Both PV and PAV increased significantly ($P < 0.05$) within 30 days, indicating that both primary and secondary oxidations had occurred. Nevertheless, adding antioxidants (for example, quercetin and ascorbyl palmitate) can effectively retard the increase of PV and PAV. In these cases, leaving the unsaturated polymethylene-interrupted FAs from pine-nut oil unoxidized better retains the product's beneficial properties.

Unlike most plastic fats, which contain 100% fat, margarines and other more complex food lipids are present in the form of water-in-oil emulsions. Oxidation has long been known to occur via a different mechanism in fat emulsions than it does in bulk fat. More factors come into play when evaluating the oxidative stability of fat emulsions, including the type of lipids or SLs, emulsifiers, emulsion droplet size distribution, and characteristics, in addition to interactions between the fatty and aqueous phases (Coupland & McClements, 1996). Thus, it is also important to investigate the oxidative stability of such emulsions rather than just the fat stock. Table 6.D lists recent studies in the oxidative stability of fat emulsions.

Fomuso et al. (2002) and Osborn and Akoh (2004) attempted to investigate factors affecting oxidative stability in SL-based emulsions. They agreed that emulsifier type and concentration play a major role in oxidative stability. Fomuso and colleagues (2002) found that high concentrations of ionic emulsifiers, namely whey protein isolate and lecithin, are effective in retarding oxidation. They create multilayers of surfactants surrounding the lipid droplets in oil-in-water emulsions, thus reducing the access of prooxidants to lipid droplets. Osborn and Akoh (2004) found whey protein isolate to be effective in retarding the oxidation process. Whether oil in water or water-in-oil, emulsion type also contributes significantly to oxidative stability. These studies suggest that prooxidants are located in the aqueous phase; thus, water-in-oil emulsions with less water content are less likely to oxidize. McClements and Decker (2002) postulated that at high oil concentration, unsaturated lipids may move to the interior of oil droplets and reduce their accessibility to prooxidants located in the aqueous phase. A limited number of studies have verified the aforementioned postulate; however, in 2004, Osborn and Akoh reported higher oxidative stability in SL-based emulsions with higher oil concentration.

Table 6.D. A Compilation of Recent Research on the Oxidative Stability of Fat Emulsions

Composition of Enzymatically Processed Fat Emulsions	Type of Emulsions[a]	Indicators for the Measurement of Oxidative Stability					Findings	Reference
		PV	PAV	TBARS	Volatile Compounds	Sensory Evaluation		
Menhaden oil and caprylic acid	O/W	X	X	X			Oxidative stability of enzymatically processed fat emulsions can be improved by emulsifier type and concentration.	Fomuso et al., 2002
Sunflower oil and caprylic acid	O/W	X			X	X	Milk drink based on enzymatically processed lipids is less stable to oxidation. Oxidative stability can be improved by adding EDTA.	Timm-Heinrich et al., 2003
Sunflower oil and caprylic acid	W/O	X			X		Mayonnaise based on enzymatically processed lipids is less stable to oxidation. Oxidative stability can be improved by adding EDTA.	Jacobsen et al., 2003
Canola oil and caprylic acid	O/W	X	X		X		Whey protein and high oil concentration improved oxidative stability of enzymatically processed fat emulsions.	Osborn & Akoh, 2004
Palm stearin and coconut oil	W/O	X			X	X	Margarine produced from enzymatically processed lipids has similar oxidative stability to that of unprocessed lipids.	Hong et al., 2006
Capric, oleic acid, and glycerol	O/W	X			X	X	Salad dressings produced from enzymatically processed lipids have oxidative stability similar to that of unprocessed lipids.	Koh et al., 2008

[a]Oil-in-water, O/W; Water-in-oil, W/O

Several studies have also compared the oxidative stability of unprocessed and enzymatically processed fat emulsions. In some cases, the enzymatically processed fat emulsions—especially those containing higher oil content (50% or above)—demonstrated similar oxidative stability with the addition of antioxidants to that of unprocessed fat emulsions (Zhang et al., 2006; Koh et al. 2008). Zhang and colleagues (2006) found that enzymatic interesterified margarines with tocopherols added have a comparable PV, amount of volatile compounds, and sensory characteristics to those of commercial margarine. Similarly, Koh et al. (2008) found enzymatic synthesized structured lipid-based salad dressings with antioxidants (such as TBHQ or oleoresin sage extracts) added to be as stable as commercial salad dressings to oxidation. Jacobsen et al. (2003) had similar results. Mayonnaise produced from enzymatically processed structured lipid was relatively stable with the addition of antioxidants EDTA and propyl gallate. Nevertheless, it is worth mentioning that without added antioxidants, the mayonnaise produced from enzymatically processed SL was less stable to oxidation than that produced from unprocessed lipids. Jacobsen and colleagues (2003) proposed that oxidation in SL-based emulsions has multiple factors that range from the lipid's structure to processing conditions. Purification processes involving high temperature treatment, such as distillation and deodorization, are sometimes required after an enzymatic reaction. These processes reduce the amount of naturally occurring antioxidants, which render the structured lipid less stable to oxidation.

In short, enzymatic processing coupled with high temperature purification may make enzymatically processed SL products less stable to oxidation. During enzymatic processing, impurities such as FFA and partial acylglycerols may form. Meanwhile, the high temperature purification process may reduce the number of naturally occurring antioxidants, causing the SL to decrease in oxidative stability; however, the addition of antioxidants may enhance oxidative stability. Oxidative stability is further complicated when such SLs are used to form fat emulsions. Other factors—especially emulsifier type and concentration and also oil concentration—contribute significantly to the oxidative stability of fat emulsions.

Functional Foods Containing Structured Lipids

Functional food can be defined as food containing a compound that positively affects human health beyond providing basic nutrition. SLs have been reported as health-beneficial substances. Therefore, SLs could have medical and nutraceutical, as well as functional food, applications. For functional food applications, several SLs have been added to food products, but the literature has little information available about the oxidative stability of SLs in functional foods. Recently, some research articles have reported the oxidative stability of food commodities fortified in SLs. For example, mayonnaise or milk drinks containing SLs were studied for their oxidative stability and effects on patients with malabsorption problems.

Additional studies have investigated the oxidative stability of mayonnaise and milk drinks containing SLs produced from sunflower oil and caprylic acid (Jacobsen et al., 2003; Timm-Heinrich et al., 2003). They compared mayonnaise and milk drinks based on enzymatically produced specific SLs from sunflower oil and caprylic acid with those produced from traditional sunflower oil and chemically randomized lipids, with respect to their oxidative stability during storage. Products made from specific SLs oxidized faster than those based on traditional oil and chemically randomized lipids. The reduced oxidative stability was most likely influenced by the lipid's structure, the lower tocopherol content, and the higher initial levels of lipid hydroperoxides and secondary volatile oxidation compounds in the specific SLs. To overcome such problems, the researchers added antioxidants. These studies compared EDTA with propyl gallate and lactoferrin, and the results showed that EDTA at a concentration of 75 and 240 ppm for mayonnaise and milk drinks, respectively, was the most effective antioxidant to retard lipid oxidation in stored mayonnaise and milk drinks containing SLs produced from sunflower oil and caprylic acid.

Possible Factors Influencing the Oxidative Stability of Enzymatically Modified Oils and Fats

Many factors are important for the oxidative stability of enzymatically modified oils and fats. One that has received the most attention is the loss of non-TAG compounds during purification and further processing steps. In the production of randomized SLs by chemical catalysis, batch deodorization is frequently used at the end of the production to remove hydroperoxides, FFAs, and volatiles. However, we do not recommend batch deodorization for the purification and deodorization of enzymatically modified lipids because these products often contain PUFAs, which are sensitive to the high temperatures employed in the process. Additionally, batch deodorization could cause undesirable acyl migration. Thus, short-path distillation (SPD) is preferred for enzymatically modified lipids because it efficiently removes various volatiles but operates under much milder conditions, such as lower temperatures and shorter processing times, compared to conventional batch deodorization (Xu et al., 2002a; Martin et al., 2010). However, SPD has been found to increase lipids' susceptibility to oxidation, which was mainly linked to a loss of minor components during the short-path distillation process. Minor components are naturally occurring compounds, some of which have antioxidative properties that protect the oil against oxidation. These compounds include tocopherols, phenols, and sterols (Shahidi, 2000). Removing these compounds would render the final oil less stable against oxidation. Xu et al. (2002a) used a chemical process to produce SLs from fish oil and tricaprin and compared SPD and batch deodorization for purification of the product. They found that the oxidative stability, measured as OIT, was shorter for the SLs purified by SPD than for the SLs purified by batch deodorization. This result was primarily attributed to the fact that a higher amount of tocopherol was lost during SPD than during batch deodorization. Similarly, several authors working with the production of enzymatically modified lipids found that the SPD process removed a large amount of minor antioxidative compounds. Jennings and Akoh (2009b) enzymatically modified rice-bran oil to have a higher content of caprylic acid and purified the products through SPD. They found that the tocopherol and tocotrienol content fell by 43.3% in the SL product, compared to their levels in the parent rice-bran oil. These authors then analyzed the waste from SPD and found that it contained significant amounts of tocopherols and tocotrienols, indicating that these compounds were lost during the SPD process. Consequently, the oxidative stability index of the rice-bran oil was found to be significantly higher than the oxidative stability index of the SL product. Lee et al. (2008) produced interesterified solid fat from fully hydrogenated soybean oil, rapeseed oil, and palm stearin in a lipase-catalyzed reaction and purified the product using SPD. They found that, in addition to reducing the FFA content, the distillation process decreased the amount of tocopherol in the purified solid fats. Compared with the noninteresterified blends, the content of α-tocopherol fell by 63–80%, while the content of γ-tocopherol was reduced by 17–32%. As tocopherol is one of the natural antioxidants that can retard or prevent lipid oxidation, the decreased presence of tocopherol resulted in a decrease in the oxidative stability of the interesterifed solid fats. Maduko et al. (2008) produced SLs for infant milk formulation by an enzymatic interesterification of tripalmitin with vegetable-oil blends and fish oil. They found that the interesterification process increased lipids' susceptibility to oxidation; they argued that this was associated with the loss of tocopherols and phospholipids during the SPD process. Akoh and Moussata (2001) also attributed the decrease in oxidative stability to the loss of non-TAG compounds during the SPD of their fish- and canola-based SLs.

Besides the SPD process, other types of processing have been reported to alter the oxidative stability of modified lipid products. One example is the alkaline deacidification process, which involves reacting FFA with an alkaline to form a soap stock and subsequently removing the soap stock from the lipid product. Lee et al. (2006) produced structured lipids from extra-virgin olive oil and CLA via a lipase-catalyzed reaction and purified the product through alkaline deacidification. They found that the content of natural phenolic compounds decreased by 97%, and the content of tocopherols was reduced by 56%. Consequently, they argued that the loss of these minor compounds might be due to the deacidification process. As a consequence, testing the oxidative stability of the parent and modified olive oil showed that the parent oil was more protected against oxidation than the modified product.

In addition to the purification processes, the process of enzymatic modification itself also involves several factors that alter the final modified lipid product's oxidative stability. One of these factors is the water content of the reaction system. Interesterification reaction systems must contain a certain amount of water, and this water content is associated with the amount of FFA formed. With the increased level of FFA and partial glycerides, the oxidative stability will contract more because these compounds have prooxidative properties (Frega et al., 1999; Paradiso et al., 2010). Several authors have shown a connection between the reaction system's water content and the oxidative stability of the final product. Kowalski et al. (2004) enzymatically interesterified beef tallow and rapeseed oil, varying the initial water content in the system from 2 to 10% by adding water. They analyzed the crude post-reaction mixtures by determining the FFA, MAG, DAG, and TAG percentages. The amount of these fractions in the interesterified product strongly depended on the system's water content, and a greater amount of water in the reaction medium resulted in more FFA and partial glycerides. Interesterified fats with high FFA and partial glyceride contents had shorter induction times—that is, reduced oxidative stability—compared to those of their counterparts, which had less of such compounds. Ledóchowska and Wilczynska (1998) produced SLs from low erucic acid rapeseed oil and tallow via enzymatic catalysis. They, too, found that the contents of FFA and partial glycerides in the enzymatically interesterified product were closely related to the water content of the starting reaction mixture: The smaller the amount of water in the reaction system, the smaller the fraction of non-TAG in the interesterified product. These results agree with the findings of Aguedo et al. (2008), who studied the interesterification of anhydrous milkfat with linseed oil, and with those of Costales-Rodríguez et al. (2009), who studied the interesterification of palm stearin and soybean oil.

The final FFA content will naturally depend on further processing steps of the modified lipid product, such as batch deodorization or SPD. Maduko et al. (2008) found that the SPD process successfully removed most of the FFA generated during enzymatic interesterification, giving the final product a FFA content similar to that of the substrate oil. These results agree with those of Lee et al. (2008). Although FFA can be removed from the final lipid product, those present during the acidolysis reaction will also influence the oxidative stability of the final modified lipid product. Hamam and Shahidi (2006) showed that tocopherol present in the oil can be esterified with FFA in the reaction medium, leading to the formation of tocopheryl esters. Tocopheryl esters are not detected as free tocopherols and do not give any stability to the resulting modified oils; furthermore, they can be removed during SPD. Thus, high FFA contents in the reaction system can contribute to the loss of tocopherols during the purification step. A loss of tocopherols will, in turn, reduce the oxidative stability of the final lipid product.

Another factor that influences the oxidative stability of modified lipids is the configuration of the reactor type in which the lipids' enzymatic modification takes place. Most studies on enzymatically modified lipids have been applied for batch production (Xu et al., 2002a; Kowalski et al., 2004; Osorio et al., 2005, 2006; Ronne et al., 2005; Aguedo et al., 2008; Lee et al., 2008, 2006; Maduko et al., 2008; Costales-Rodriguez et al., 2009) and/or continuous or semi-continuous production, either in a packed bed reactor (Akoh & Moussata, 2001; Xu et al., 2002b; Ronne et al., 2005; Osorio et al., 2006; Costales-Rodriguez et al., 2009; Jennings & Akoh, 2009b) or in a fluidized-bed reactor (Osorio et al., 2005). Osorio et al. (2005) interesterified palm stearin with soybean oil batch-wise and in a fluidized-bed reactor. They reported that the level of FFA was higher in lipid materials prepared in a batch reactor compared to those prepared using a continuous fluidized–bed reactor. The FFA content of interesterified fat in a continuous fluidized–bed reactor remained approximately constant at 0.7%–1.2% during the 21-day running period, whereas the content of FFA in the fat obtained batch-wise was 2–6%. Osorio et al. (2005) also found that the formation of oxidation products during the continuous interesterification process could be neglected. Later, Osorio et al. (2006) interesterified different fat blends in a continuous packed–bed reactor and batch-wise. Similar to the experiments described above, which were conducted in a continuous fluidized–bed reactor (Osorio et al., 2005), the FFAs obtained in these studies were considerably lower upon interesterification than those obtained batch-wise, catalyzed by the same biocatalyst under similar reaction conditions. Costales-Rodríguez et al. (2009) modified a blend of palm stearin and soybean oil by an enzymatic interesterification batch-wise and continuously in a laboratory-scale packed-bed reactor. They found that the tocopherol contents were slightly higher, and the contents of partial acylglycerols were slightly lower in materials from the continuous interesterification compared to these compounds in the materials from the interesterifications conducted batch-wise.

Not only the reactor type but also the reactor's settings and configuration can influence the oxidative stability of the final lipid product. Kowalski et al. (2004) found that the increase in partial acylglycerols (hydrolysis by-products) was proportional to the temperature of the reaction. This effect was also shown by Xu et al. (2002b); however, they found that an increase in temperature during the enzymatic interesterification only slightly increased the content of hydrolysis by-products. Furthermore,

Kowalski et al. (2004) found that the increase in hydrolysis by-products was proportional to both the reaction time and the catalyst concentration.

Approaches to Improve Oxidative Stability of Enzymatically Processed Oils and Fats

Lipid oxidation is a multifactorial phenomenon. Thus, its retardation can be a daunting task, requiring an approach that accounts for many considerations. Aside from enzymatic processing conditions, the following section highlights two other strategies that can improve the oxidative stability of enzymatically processed oils and fats—namely, the implementation of a delivery system and the addition of antioxidative compounds.

Most enzymatically processed oils and fats are incorporated into more complex food systems. They can be added to foods in different manners, namely as neat oil, emulsified oil, or encapsulated oil. The delivery system of structured lipids has been found in several occasions to exert different effects on the oxidative stability of enzymatically processed oils and fats (Fomuso et al., 2002; Zhang et al., 2006; Koh et al. 2008). These studies investigated the oxidative stability of oil emulsions with contrasting results (Table 6.D). To date, only limited studies compare the oxidative stability of enzymatically processed oils and fats using different delivery systems. Nevertheless, the comparison of lipids' oxidative stability in various delivery systems can provide a basis for further exploration.

Different food systems require different delivery systems. Nielsen and Jacobsen (2011) found that in fish pate, fish oil delivered in the form of microencapsulated oil was most stable to oxidation, followed by emulsified oil and neat oil. In contrast, neat oil was found to provide the highest oxidative stability as a delivery system in salad dressing and yogurt, as compared to an emulsified system (Let et al., 2007). To enhance the oxidative stability of lipids delivered as emulsions, particular attention should be given to the type and concentration of emulsifiers used. Positively charged emulsifiers were found to repel metal ions and reduce oxidation, as compared to negatively charged emulsifiers (Mei et al., 1999). In addition, a thick emulsion interface consisting of differently charged polymers acted as a shield protecting the oil droplets from prooxidants. This thick and positively charged emulsion interface can be created with electrostatic layer-by-layer deposition technology (Klinkesorn et al., 2005). Sørensen et al. (2007) reported a similar finding. A high concentration of β-lactoglobulin (positively charged) located at the interface of oil–water emulsion (thicker interface) resulted in increased oxidative stability.

Antioxidative compounds are often added to retard oxidation in food systems. Some of the popular antioxidants used in lipid food systems include tocopherols, ascorbic acid, ascorbyl palmitate, EDTA, lactoferin, gallic acid, caffeic acid, and propyl gallate. Spices such as oregano, rosemary, and thyme are also added to lipid-containing food systems (Koh et al., 2008). In addition, peptides were found to have antioxidative properties. Farvin and colleagues (2010a, 2010b) reported metal-chelating and iron-reducing properties of low molecular–weight peptides in fish oil–enriched milk. High molecular–weight peptides, on the other hand, have radical-scavenging properties. All of these peptides were reported to contain hydrophobic amino-acid residues, such as proline, histidine, and valine. Lately, lipophilized phenolics are gaining attention as effective amphiphilic antioxidants in lipid systems (Laguerre et al., 2009; Medina et al., 2009; Lue et al., 2010; Sørensen et al., 2011).

Although not thoroughly investigated in enzymatically processed lipids, antioxidants' efficiency was found to depend on their partitioning in multiphase food systems (Huang et al., 1996; Shahidi & Zhong, 2011). Polar antioxidants are more effective in bulk oil due to their partitioning at the oil–air interface, while amphiphilic antioxidants are more effective in emulsion systems due to their partitioning at the interface. Jacobsen et al. (2008) reported that tocopherols lack antioxidative properties in emulsion systems, such as mayonnaise and salad dressing. They attributed the lack of antioxidative efficacy to the partitioning of tocopherols into the interior oil phase, whereas oxidation takes place at the interface. Thus, amphiphilic antioxidants, mainly located at the interface, may more efficiently retard oxidation in emulsion systems. Several studies have examined the efficacy of lipophilized phenolic compounds for retarding oxidation in oil emulsion systems (Laguerre et al., 2009; Medina et al., 2009; Sørensen et al., 2011). Results from these studies uniformly pointed to chain-length effects in addition to amphiphilicity. Lipophilized phenolic compounds with short- and medium-chain length (less than 10) showed excellent antioxidative properties in emulsion systems. Nevertheless, their antioxidative properties decreased drastically as chain length increased beyond C10. This was explained by the fact that increased chain length leads to an increase in lipophilicity and a reduction in the concentration of critical micelles. Thus, instead of positioning at the interface, long-chain lipophilized phenolic compounds have the tendency to form micelles in the aqueous phase (Sørensen et al., 2011).

In short, it is possible to increase the oxidative stability of enzymatically processed oils and fats. Possible approaches include designing a suitable delivery system and adding antioxidative compounds. When delivering the structured lipid in the form of an emulsion, one can enhance its oxidative stability by designing a positively charged thick interface area. Depending on the delivery system, suitable antioxidative compounds can be added.

Conclusion

Enzymes for lipid modification were first documented in the literature about 30 years ago. Technology has now been applied in the industry for a variety of products. Intense research now exists that focuses on possibilities and strategies for applying enzyme-catalyzed reactions to the manufacture of tailor-made products. When these reactions, which are proven to be highly successful at lab scale in terms of productivity and selectivity, are scaled up to industrial production, a number of concerns affecting the end product have been raised. Among these, the oxidative stability of enzymatically modified products has received much attention.

Many researchers have noted that the reduced oxidative stability of fats and oils produced by enzyme-involving steps presents a disadvantage. As discussed in this chapter, many explanations exist for the low oxidative stability of these products. First, a change in lipid structure affects stability by increasing or decreasing the exposure of unsaturated FAs to oxygen, as well as the oxygen solubility or diffusivity of sensitive FAs. Moreover, structural change may alter the state of metal ions in the system. The multiple steps involved in processing could have another effect on oxidative stability. Particularly for laboratory studies, the enzymatic process has not taken place in a closed, automatic, continuous process. Contact with air and contamination with other materials could lead to stability issues. Therefore, some of these concerns could easily be minimized when scaling this process up to industrial operation. A systematic comparison of the oxidative stability of different products from different processes has not yet been reported in industrial operations. The authors believe that this information would be useful for revealing the source of these products' reduced oxidative stability. A third concern is the loss of minor compounds with antioxidative properties in enzymatically processed oils and fats, which is one of the main factors causing a change in stability. A fourth concern is the enzyme itself, which as a protein has a list of properties that may affect lipid oxidation. Immobilized enzymes with carrier materials that affect the lipid oxidation have also been reported. Furthermore, these materials all have effects on the state of metals in the lipid system. Thus, it can be expected that the production process is a complex system that influences the product's oxidative stability.

Lipid oxidation and antioxidation present challenging issues, as seen from the other chapters in this book. The present chapter is a pure review of available research on the oxidative stability of enzymatically processed fats and oils. Given that research on the topic is limited, we conclude that when it comes to enzymatically modified fats and oils, more data, with proper evaluation of the individual as well as combined effects of reaction conditions, reaction route, and composition of substrates and products, both at laboratory as well as industrial scales, are needed for a thorough evaluation.

References

Adhikari, P.; Shin, J. A.; Lee, J. H.; Hu, J. N.; Zhu, X. M.; Akoh, C. C.; Lee, K. T. Production of *trans*-free margarine stock by enzymatic interesterification of rice bran oil, palm stearin and coconut oil. J. Sci. Food Agric. 2010, 90, 703–711.

Adhikari, P.; Shin, J. A.; Lee, J. H.; Kim, H. R.; Kim, I. H.; Hong, S.T.; Lee, K. T. Crystallization, physicochemical properties and oxidative stability of the interesterified hard fat from rice bran oil, fully hydrogenated soybean oil and coconut oil through lipase-catalyzed reaction. Food Bioprocess Technol. 2011, 5, 2474–2487.

Aguedo, M.; Hanon, E.; Danthine, S.; Paquot, M.; Lognay, G.; Thomas, A.; Vandenbol, M.; Thonart, P.; Wathelet, J. P.; Blecker, C. Enrichment of anhydrous milk fat in polyunsaturated fatty acid residues from linseed and rapeseed oils through enzymatic interesterification. J. Agric. Food Chem. 2008, 56, 1757–1765.

Akoh, C. C.; Moussata, C.O. Characterization and oxidative stability of enzymatically produced fish and canola oil-based structured lipids. J. Am. Oil Chem. Soc. 2001, 78, 25–30.

Chahinian, H.; Sarda, L. Distinction between esterases and lipases: Comparative biochemical properties of sequence-related carboxylesterases. Protein Pept. Lett. 2009, 16 (10), 1149–1161.

Cheong, L. Z.; Tan, C. P.; Long, K.; Yusoff, M. S. A.; Lai, O. M. Physicochemical, textural and viscoelastic properties of palm diacylglycerol bakery margarine during storage. J. Am. Oil Chem. Soc. 2009, 86, 723–731.

Choe, E.; Min, D. Mechanisms factors for edible oil oxidation. Compr. Rev. Food Sci. F. 2006, 5, 169–186.

Christensen, M. S.; Høy, C. E.; Becker, C. C.; Redgrave, T. G. Intestinal-absorption and lymphatic transport of eicosapentaenoic (EPA), docosahexaenoic (DHA), and decanoic acids—Dependence on intramolecular triacylglycerol structure. Am. J. Clin. Nutr. 1995, 61, 56–61.

Costales-Rodríguez, R.; Gibon, V.; Verhe, R.; De Greyt, W. Chemical and enzymatic interesterification of a blend of palm stearin: Soybean oil for low trans-margarine formulation. J. Am. Oil. Chem. Soc. 2009, 86, 681–697.

Coupland, J. N.; McClements, D. J. Lipid oxidation in food emulsions. Trend. Food Sci. Technol. 1996, 7, 83–91.

Criado, M.; Hernández-Martín, E.; López-Hernández, A.; Otero, C. Enzymatic interesterification of extra virgin olive oil with a fully hydrogenated fat: Characterization of the reaction and its products. J. Am. Oil Chem. Soc. 2007, 84, 717–726.

de María, P. D.; Sinisterra, J. V.; Tsai, S.-W.; Alcántara, A. R. *Carica papaya* lipase (CPL): An emerging and versatile biocatalyst. Biotechnol. Adv. 2006, 24, 493–499.

Farmani, J.; Hamedi, M.; Safari, M.; Madadlou, A. Trans-free Iranian vanaspati through enzymatic and chemical transesterification of triple blends of fully hydrogenated soybean, rapeseed and sunflower oils. Food Chem. 2007, 102, 827–833.

Farmani, J.; Safari, M.; Hamedi, M. Trans-free fats through interesterification of canola oil/palm olein or fully hydrogenated soybean oil blends. Eur. J. Lipid Sci. Technol. 2009, 111, 1212–1220.

Farvin, K. H. S.; Baron, C. P.; Nielsen, N. S.; Jacobsen, C. Antioxidant activity of yoghurt peptides: Part 1—*In vitro* assays and evaluation of n-3 enriched milk. Food Chem. 2010a, 123, 1081–1089.

Farvin, K. H. S.; Baron, C. P.; Nielsen, N. S.; Otte, J.; Jacobsen, C. Antioxidant activity of yoghurt peptides: Part 2—Characterisation of peptide fractions. Food Chem. 2010b, 123, 1090–1097.

Fomuso, L.; Corredig, M.; Akoh, C. C. Effect of emulsifier on oxidation properties of fish oil-based structured lipid emulsions. J. Agric. Food Chem. 2002, 50, 2957–2961.

Frega, N.; Mozzon, M.; Lercker, G. Effects of free fatty acids on oxidative stability of vegetable oil. J. Am. Oil. Chem. Soc. 1999, 76, 325–329.

Ghotra, B. S.; Dyal, S.D.; Narine, S. S. Lipid shortening: A review. Food Res. Int. 2002, 35, 1015–1048.

Giet, J.-M.; Aguedo, M.; Danthine, S.; Paquot, M.; Thomas, A.; Vandenbol, M.; Thonart, P.; Wathelet, J.-P.; Blecker, C.; Lognay, G. Enzymatic interesterification of anhydrous milk fat with rapeseed and/or linseed oil: Oxidative stability. J. Agric. Food Chem. 2009, 57, 6787–6794.

Guo, Z.; Kahveci, D.; Özçelik, B.; Xu, X. Improving enzymatic production of diglycerides by engineering binary ionic liquid medium system. New Biotechnol. 2009, 26, 37–43.

Guo, Z.; Xu, X. New opportunity for enzymatic modification of fats and oils with industrial potentials. Org. Biomol. Chem. 2005, 3, 2615–2619.

Hamam, F.; Shahidi, F. Acidolysis reactions lead to esterification of endogenous tocopherols and compromised oxidative stability of modified oils. J. Agric. Food Chem. 2006, 54, 7319–7323.

Hamam, F.; Shahidi, F. Enzymatic acidolysis of an arachidonic acid single-cell oil with capric acid. J. Am. Oil Chem. Soc. 2004, 81, 887–892.

Hamam, F.; Shahidi, F. Enzymatic incorporation of capric acid into a single cell oil rich in docosahexaenoic acid and docosapentaenoic acid and oxidative stability of the resultant structured lipid. Food Chem. 2005, 91, 583–591.

Hirayama, O.; Nakae, T. Glyceride structure and biosynthesis of natural fats. Part II. Positional distribution of fatty acids in the glycerides of animal and vegetable fats. Agr. Biol. Chem. 1964, 28, 201–205.

Hong, Z.; Jacobsen, C.; Pedersen, L. S.; Christensen, M. W.; Nissen, J. A. Storage stability of margarines produced from enzymatically interesterified fats compared to those prepared by conventional methods—chemical properties. Eur. J. Lipid Sci. Technol. 2006, 108, 227–238.

Huang, S.W.; Hopia, A.; Schwarz, K.; Frankel, E. N.; German, J. B. Antioxidant activity of α-tocopherol and trolox in different lipid substrates: Bulk oils vs. oil-in-water emulsions. J. Agric. Food Chem. 1996, 44, 444–452.

Innis, S. M. Essential fatty acids in growth and development. Prog. Lipid Res. 1991, 30, 39–103.

Jacobsen, C.; Let, M. B.; Nielsen, N. S.; Meyer, A. S. Antioxidant strategies for preventing oxidative flavor deterioration of foods enriched with n-3 polyunsaturated lipids: A comparative evaluation. Trends Food Sci. Technol. 2008, 19, 76–93.

Jacobsen, C.; Xu, X.; Nielsen, N. S.; Timm-Heinrich, M. Oxidative stability of mayonnaise containing structured lipids produced from sunflower oil and caprylic acid. Eur. J. Lipid Sci. Technol. 2003, 105, 449–458.

Jennings, B. H.; Akoh, C. C. Characterization of a rice bran oil structured lipid. J. Agric. Food Chem. 2009a, 57, 3346–3350.

Jennings, B. H.; Akoh, C. C. Effectiveness of natural versus synthetic antioxidants in a rice bran oil-based structured lipid. Food Chem. 2009b, 114, 1456–1461.

Kahveci, D.; Can, A.; Özçelik, B. Production of lipase-catalyzed structured lipid from olive oil with omega-3 polyunsaturated fatty acids. Food Sci. Biotechnol. 2009, 18, 79–83.

Kim, B. H.; Akoh, C. C. Characteristics of structured lipid prepared by lipase-catalyzed acidolysis of roasted sesame oil and caprylic acid in a bench-scale continuous packed bed reactor. J. Agric. Food Chem. 2006, 54, 5132–5141.

Klinkesorn, U.; Sophandora, P.; Chinachoti, P.; McClements, D. J.; Decker, E. A. Increasing the oxidative stability of liquid and dried tuna oil-in-water emulsions with electrostatic layer-by-layer deposition technology. J. Agric. Food Chem. 2005, 53, 4561–4566.

Koh, S. P.; Arifin, N.; Tan, C. P.; Yusoff, M. S. A.; Long, K.; Idris, N. A.; Lai, O. M. Rheological properties, oxidative stability and sensory evaluation of enzymatically synthesized medium- and long-chain triacylglycerol-based salad dressings. Eur. J. Lipid Sci. Technol. 2008, 110, 1116–1126.

Kowalski, B.; Tarnowska, K.; Gruczynska, E.; Bekas, W. Chemical and enzymatic interesterification of a beef tallow and rapeseed oil equal-weight blend. Eur. J. Lipid Sci. Technol. 2004, 106, 655–664.

Laguerre, M.; Giraldo, L. J. L.; Lecomte, J.; Figueroa-Espinoza, M. C.; Bruno, B.; Weiss, J.; Decker, E. A.; Villeneuve, P. Chain length affects antioxidant properties of chrologenate esters in emulsion: The cutoff theory behind the polar paradox. J. Agric. Food Chem. 2009, 57, 11335–11342.

Ledóchowska, E.; Wilczynska, E. Comparison of the oxidative stability of chemically and enzymatically interesterified fats. Fett-Lipid 1998, 100, 343–348.

Lee J.; Akoh, C.; Lee, K. Physical properties of trans-free bakery shortening produced by lipase-catalyzed interesterification. J. Am. Oil Chem. Soc. 2008, 85, 1–11.

Lee, J.-H.; Kim, M.R.; Kim, H.-R.; Kim, L.-H.; Lee, K.-T. Characterization of lipase-catalyzed structured lipids from selected vegetable oils with conjugated linoleic acid: Their oxidative stability with rosemary extracts. J. Food Sci. 2003, 68, 1653–1658.

Lee, J.-H.; Lee, K.-T.; Akoh, C. C.; Chung, S. K.; Kim, M. R. Antioxidant evaluation and oxidative stability of structured lipids from extravirgin olive oil and conjugated linoleic acid. J. Agric. Food Chem. 2006, 54, 5416–5421.

Lee, J.-H.; Shin, J-A.; Lee, J.-H.; Lee, K.-T. Production of lipase-catalyzed structured lipids from safflower oil with conjugated linoleic acid and oxidation studies with rosemary extracts. Food Res. Int. 2004, 37, 967–974.

Lee, K.-T.; Akoh, C. C. Characterization of enzymatically synthesized structured lipids containing eicosapentaenoic, docosahexaenoic, and caprylic acids. J. Am. Oil Chem. Soc. 1998, 75, 495–499.

Let, M. B.; Jacobsen, C.; Meyer, A. S. Lipid oxidation in milk, yoghurt and salad dressing enriched with neat fish oil or pre-emulsified fish oil. J. Agric. Food Chem. 2007, 55, 7802–7809.

Litchfield, C. Analysis of Triglycerides; Academic Press: New York, 1972; 233–264.

Lopez-Hernandez, A.; Otero, C.; Hernández-Martín, E.; Garcia, H. S.; Hill Jr., C. G. Interesterification in batch and continuous flow processes of sesame oil and fully hydrogenated fat catalyzed by immobilized lipase. Eur. J. Lipid Sci. Technol. 2007, 109, 1147–1159.

Lue, B. M.; Nielsen, N. S.; Jacobsen, C.; Hellgren, L.; Guo, Z.; Xu, X. Antioxidant properties of modified rutin esters by DPPH, reducing power, iron chelation and low density lipoprotein assays. Food Chem. 2010, 123, 221–230.

Maduko, C. O.; Park, Y. W.; Akoh, C. C. Characterization and oxidative stability of structured lipids: Infant milk fat analog. J. Am. Chem. Soc. 2008, 85, 197–204.

Malcata, F. X.; Reyes, H. R.; Garcia, H. S.; Hill, C. G.; Amundson, C. H. Kinetics and mechanisms of reactions catalyzed by immobilized lipases. Enzyme Microb. Technol. 1992, 14, 426–446.

Martin, D.; Reglereo, G.; Sennñoráns, F. J. Oxidative stability of structured lipids. Eur. Food Res. Technol. 2010, 231, 635–653.

McClements, D. J.; Decker, E. A. Lipid oxidation in oil-in-water emulsions: Impact of molecular environment on chemical reactions in heterogeneous food systems. J. Food Sci. 2000, 65, 1270–1282.

Medina, I.; Lois, S.; Alcantara, D.; Lucas, R.; Morales, J. C. Effect of lipophilization of hydroxytyrosol on its antioxidant activity in fish oils and fish oil-in-water emulsions. J. Agric. Food Chem. 2009, 57, 9773–9779.

Mei, L. Y.; McClements, D. J.; Decker, E. A. Lipid oxidation in emulsions as affected by charge status of antioxidants and emulsion droplets. J. Agric. Food Chem. 1999, 47, 2267–2273.

Mitra, K.; Lee, J.-H.; Lee, K.-T.; Kim, S.-A. Production tactic and physiochemical properties of low ω6/ω3 ratio structured lipid synthesised from perilla and soybean oil. Int. J. Food Sci. Tech. 2010, 45, 1321–1329.

Mu, H.; Porsgaard, T. The metabolism of structured triacylglycerols. Prog. Lipid Res. 2005, 44, 430–448.

Mukherjee, K. D. Plant lipases and their application in lipid biotransformations. Prog. Lipid Res. 1994, 33, 165–174.

Nielsen, N. S.; Jacobsen, C. Retardation of lipid oxidation in fish oil-enriched fish-pate—Combination effects. J. Food Biochem. 2011, DOI: 10.1111/j.1745-4514.2011.00605.x.

Nielsen, N. S.; Xu, X.; Timm-Heinrich, M.; Jacobsen, C. Oxidative stability during storage of structured lipids produced from fish oil and caprylic acid. J. Am. Oil Chem. Soc. 2004, 81, 375–384.

Nielsen, N. S.; Yang, T.; Xu, X.; Jacobsen, C. Production and oxidative stability of a human milk fat substitute produced from lard by enzyme technology in a pilot packed-bed reactor. Food Chem. 2006, 94, 53–60.

Osborn, H. T.; Akoh, C. C. Effect of emulsifier type, droplet size and oil concentration on lipid oxidation in structured lipid-based oil-in-water emulsions. Food Chem. 2004, 84, 451–456.

Osorio, N. M.; da Fonseca, M. M. R.; Ferreira-Dias, S. Operational stability of *Thermomyces lanuginosa* lipase during interesterification of fat in continuous packed-bed reactors. Eur. J. Lipid Sci. Technol. 2006, 108, 545–553.

Osorio, N. M.; Gusmao, J. H.; da Fonseca, M. M.; Ferreira-Dias, S. Lipase-catalysed interesterification of palm stearin with soybean oil in a continuous fluidised-bed reactor. Eur. J. Lipid Sci. Technol. 2005, 107, 455–463.

Paradiso, V. M.; Gomes, T.; Nasti, R.; Caponio, F.; Summo, C. Effects of free fatty acids on the oxidative processes in purified olive oil. Food Res. Int. 2010, 43, 1389–1394.

Pina-Rodriguez, A. M.; Akoh, C. C. Synthesis and characterization of a structured lipid from amaranth oil as a partial fat substitute in milk-based infant formula. J. Agric. Food Chem. 2009, 57, 6748–6756.

Pina-Rodriguez, A. M.; Akoh, C. C. Composition and oxidative stability of a structured lipid from amaranth oil in a milk-based infant formula. Food Chem. 2010, 75, C140–C146.

Quinlan, P. T.; Chandler, I. C. Human milk fat substitutes. Australian Patent: Au-A-10344/92, 1992.

Ronne, T. H.; Yang, T. K.; Mu, H. L.; Jacobsen, C.; Xu, X. B. Enzymatic interesterification of butterfat with rapeseed oil in a continuous packed bed reactor. J. Agric. Food Chem. 2005, 53, 5617–5624.

Rousseau, D.; Marangoni, A. The effects of interesterification on physical and sensory attributes of butterfat and butterfat-canola oil spreads. Food Res. Int. 1999, 31, 381–388.

Schmid, R. D.; Verger, R. Lipases: Interfacial enzymes with attractive applications. Angew. Chem. 1998, 37, 1608–1633.

Senanayake, S. P. J. N.; Shahidi, F. Chemical and stability characteristics of structured lipids from borage (*Borago officinalis* L.) and evening primrose (*Oenothera biennis* L.) oils. J. Food Sci. 2002, 67, 2038–2045.

Shahidi, F. Antioxidants in food and food antioxidants. Nahr.-Food 2000, 44, 158–163.

Shahidi, F.; Zhong, Y. Revisitting the polar paradox theory: A critical overview. J. Agric. Food Chem. 2011, 59 (8), 3499–3504.

Sharma, D.; Sharma, B.; Shukla, A. K. Biotechnological approach of microbial lipase: A review. Biotechnology 2011, 10 (1), 23–40.

Sørensen, A. D. M.; Baron, C. P.; Let, M. B.; Bruggermann, D.; Pedersen, L. R. L.; Jacobsen, C. Homogenization conditions affects the oxidative stability of fish oil enriched milk emulsions: Oxidation linked to changes in the protein composition at the oil-water interface. J. Agric. Food Chem. 2007, 55, 1781–1789.

Sørensen, A. D. M.; Nielsen, N. S.; Yang, Z.; Xu, X.; Jacobsen, C. Lipophilization of dihydrocaffeic acid affects its antioxidative properties in fish oil-enriched emulsions. Eur. J. Lipid Sci. Technol. 2011, DOI: 10.1002/ejlt.201100002.

Sørensen, A. D. M.; Xu, X.; Zhang, L.; Kristensen, J. B.; Jacobsen, C. Human milk fat substitute from butterfat: Production by enzymatic interesterification and evaluation of oxidative stability. J. Am. Oil Chem. Soc. 2010, 87, 185–194.

Teichert, S. A.; Akoh, C. C. Characterization of stearidonic acid soybean oil enriched with palmitic acid produced by solvent-free enzymatic interesterification. J. Agric. Food Chem. 2011a, 59, 9588–9595.

Teichert, S. A.; Akoh, C. C. Modifications of stearidonic acid soybean oil by enzymatic acidolysis for the production of human milk fat analogues. J. Agric. Food Chem. 2011b, 59, 13300–13310.

Timm-Heinrich, M.; Xu, X.; Nielsen, N. S.; Jacobsen, C. Oxidative stability of milk drinks containing structured lipids produced from sunflower oil and caprylic acid. Eur. J. Lipid Sci. Technol. 2003, 105, 459–470.

Turan, S.; Karabulut, I.; Vural, H. Influence of *sn*-1,3-lipase-catalysed interesterification on the oxidative stability of soybean oil-based structured lipids. J. Sci. Food Agric. 2007, 87, 90–97.

Walsh, M. K. Immobilized enzyme technology for food application. Novel enzyme technology for food applications; Rastall, R. Ed.; Woodhead Publishing Limited: Cambridge, England, 2007; pp. 66–84.

Wan, Y.; Bechtel, P. J.; Sathivel, S. Physical and nutritional properties of baby food containing menhaden oil (*Breoortia tyrranus*) and microencapsulated menhaden oil. Food Sci. Technol. 2011, 44, 576–581.

Wijesundera, C.; Ceccato, C.; Watkins, P.; Fagan, P.; Fraser, B.; Thienthong, N.; Perlmutter, P. Docosahexaenoic acid is more stable to oxidation when located at the *sn*-2 position of triacylglycerol compared to *sn*-1(3). J. Am. Oil Chem. Soc. 2008, 85, 543–548.

Wirkowska, M.; Gorska, A.; Brys, J.; Ostrowska-Ligeza, E.; Koczon, P. Oxidative stability and triacylglycerols structure of lipid fraction from cookies for infants. Int. J. Food Sci. Nutr. 2011, 1–7.

Xu, X. B.; Jacobsen, C.; Nielsen, N. S.; Heinrich, M. T.; Zhou, D. Q. Purification and deodorization of structured lipids by short path distillation. Eur. J. Lipid Sci. Technol. 2002a, 104, 745–755.

Xu, X. B.; Porsgaard, T.; Zhang, H.; Adler-Nissen, J.; Hoy, C. E. Production of structured lipids in a packed-bed reactor with *Thermomyces lanuginosa* lipase. J. Am. Oil. Chem. Soc. 2002b, 79, 561–565.

Xu, X.; Timm-Heinrich, M.; Nielsen, N. S.; Porsgaard, T.; Jacobsen, C. Effects of antioxidants on the lipase-catalyzed acidolysis during production of structured lipids. Eur. J. Lipid Sci. Technol. 2005, 107, 464–468.

Yang, T.; Xu, X.; He, C.; Li, L. Lipase-catalysed modification of lard to produce human milk fat substitutes. Food Chem. 2003, 80, 473–481.

Yankah, V. V.; Akoh, C. C. Batch enzymatic synthesis, characterization and oxidative stability of DHA-containing structured lipids. J. Food Lipids 2000, 7, 247–261.

Zhang, H.; Jacobsen, C.; Pedersen, L. S.; Christensen, M. W.; Adler-Nielsen, J. Storage stability of margarines produced from enzymatically interesterified fats compared to those prepared by conventional methods—Chemical properties. J. Am. Oil Chem. Soc. 2006, 108 (3), 227–238.

Zhu, X. M.; Hu, J. N.; Xue, C. L.; Lee, J. H.; Shin, J. A.; Hong, S. T.; Sung, C. K.; Lee, K. T. Physicochemical and oxidative stability of interesterified structured lipid for soft margarine fat containing Δ 5-UPIFAs. Food Chem. 2012, 131(2), 533–540.

CHAPTER

7

The Polar Paradox

How an Imperfect Conceptual Framework Accelerated Our Knowledge of Antioxidant Behavior

Roger Nahas[1] and Donald Berdahl[1]

Introduction

> "At this time organic chemistry can drive one completely crazy. It seems to me like a primeval tropical jungle, full of the most remarkable things, an amazing thicket, without escape or end, into which one would not dare to enter."

Thus wrote Friedrich Wöhler in a famous and often quoted letter to Jöns Jacob Berzelius (Wöhler, 1835), describing the state of confusion then reigning in the discipline of organic chemistry, as seemingly unrelated observations defied comprehension, absent a sufficiently developed theoretical framework. As patterns gradually began to emerge from countless studies, as hypotheses were formulated and theories developed and tested, confusion gave way to greater clarity and the field of organic chemistry became the useful, predictive science that we know today.

Teasing out patterns from volumes of confusing data is one method by which scientists begin to form testable hypotheses. This is what William Porter, a scientist at the U.S. Army Natick Research, Development and Engineering Center, did while sorting through confusing and apparently contradictory data on the performance of various antioxidants in food applications. He recognized a pattern suggesting that, all other factors being equal, polar or hydrophilic antioxidants are relatively more effective in applications characterized by low surface-to-volume (LSV) ratio (for example, bulk oils). In contrast, he observed that nonpolar or lipophilic antioxidants seem relatively more effective in applications character-

[1]Kalsec®, Inc., 3713 West Main Street, Kalamazoo, MI 49005.

ized by high surface-to-volume (HSV) ratios, such as those found in oil-in-water emulsions (Porter, 1993). Polar antioxidants work better in nonpolar matrices, and nonpolar antioxidants work better in polar matrices—hence, the "polar paradox." This pattern has provided a useful framework for further thought and experimentation about the function of antioxidants.

To be sure, the polar paradox is an oversimplification. But its simplicity and apparently counterintuitive nature suggest that underlying features affect antioxidant behavior in complex systems. Porter originally used a coarse screen to sift the literature. The examples he cited comprised a wide range of antioxidant species in various applications using data generated from widely differing antioxidant tests. Porter was aware of the difficulties inherent in assigning "cause" based on this approach. For example, he recognized that the increasing methyl substitution that occurs in the δ-, γ-, β-, α-tocopherol series serves not only to change lipophilicity/hydrophilicity but also influences oxidation/reduction potentials, radical stability, steric properties, and other features that might play a significant role in antioxidant effectiveness (Porter, 1993). For this reason, he pointed out that a homologous series would provide a better system for testing the polar paradox. One of the examples he cited was based on work by Sherwin and Thompson (1976) and compared the effect of increasing the length of an alkyl chain on the antioxidant activity of gallic acid esters. But the data from these studies, as it was presented by the original literature and Porter, is not as convincing as he originally suggested. The experiments, which measured the antioxidant performance of esters in both cottonseed and safflower oil, were run using equivalent weight percentages of gal-

Table 7.A. Performance of a Series of Gallate Esters in Cottonseed and Safflower Oils, Corrected for Concentration

AO Treatment	Wt. %	mMolal Conc.	% Conc. Reduction	AOM	% AOM Reduction
AOM Stability of Cottonseed Oil (Hours to PV 70)					
Propyl Gallate	0.01	0.588	—	19	—
Octyl Gallate	0.01	0.354	40%	11	42%
Dodecyl Gallate	0.01	0.295	50%	10	47%
AOM Stability of Safflower Oil (Hours to PV 70)					
Gallic Acid	0.05	2.939	—	19	—
Propyl Gallate	0.05	2.356	11%	11	20%
Hexyl Gallate	0.05	1.966	19%	10	33%

Sherwin & Thompson (1976).

lic acid ester derivatives (Table 7.A). The experiments would have been better run at equivalent molal levels.

When molecular weight differences are accounted for in the cottonseed oil example, the experimenters added 40% fewer moles of octyl gallate than propyl gallate and observed a corresponding 42% reduction in performance. Likewise, 50% fewer moles of dohexyl gallate were added, relative to propyl gallate, and the researchers observed a 47% reduction in performance. When accounting for differences in the molar amounts of antioxidant added, differences in performance largely disappear. The decrease in molal concentration in the safflower experiment may account for half to two-thirds of the observed effect. Under further scrutiny, these Sherwin and Thompson experiments, cited by Porter to support the polar paradox, do not provide convincing evidence.

Porter's statement of the polar paradox hypothesis did relatively little to explain the behavior of antioxidants in the experimental literature he had scrutinized. The real value came after he proposed it—from the planning and execution of experiments expressly designed to test it. In the 30 years since Porter's observation, these experiments have added greatly to our understanding of the factors that influence antioxidant behavior. They have clearly established that chemical properties profoundly affect antioxidant performance because they influence where an antioxidant resides in a complex system.

In the mid-1990s, Frankel and various colleagues published several papers (Frankel et al., 1994; Huang et al., 1994; Frankel, 1996; Frankel et al., 1996; Huang et al., 1996) on the polar paradox and hypothesized that the effect could be explained on the basis of interfacial phenomena and the affinities of antioxidants for interfaces (for example, oil/air, in the case of bulk oils, and oil/water, in the case of emulsions). These authors made another significant contribution by pointing out the difficulty of comparing data from experiments that used different methods of measuring antioxidant performance. They showed that one could achieve conflicting results depending on whether one used conjugated diene hydroperoxides as the measure of oxidation or secondary oxidation products such as hexanal, for example.

Others began to report experiments designed to examine interfacial effects more closely, measuring factors such as octanol/water partition coefficients (Torres de Pineda et al., 2007), surface tension modulation (Lucas et al., 2010), and the like. Some researchers may have lost sight of the key driving force behind polar paradox chemistry—namely, that observed effects are a consequence of existing interfaces and the location of the antioxidants. For example, experimenters who determine the FRAP or ORAC values of antioxidants in a single-phase system like methanol and find that a given set of antioxidants follows or departs from the polar paradox seem to be missing the point (Mateos et al., 2008).

More recent reports indicate that the lipophilicity effect is not linear (Lucas et al., 2010) and postulate that oxidation in bulk-oil systems is not a simple surface effect. One of the most recent papers on the polar paradox suggests that the effect may depend on concentration, obeying the paradox at one level and behaving contrarily at another level (Zhong & Shahidi, 2012). William Porter's simple enunciation of the polar paradox should receive some credit for the amazing progress that has been made in our understanding of oxidation and antioxidants over the past 30 years, as it provided a framework within which questions could be asked, experimentally probed, and answered. The rest of this chapter will explore such questions, experiments, and hypotheses.

Discussion

Edwin Frankel was among the first to intentionally test the polar paradox by systematically comparing antioxidants and their conformity to the newly identified "polar paradox." One paper (Frankel et al., 1994) compared lipophilic and hydrophilic antioxidants side by side in bulk oil, stripped of tocopherols, and an oil-in-water emulsion model incubated at 60 °C. The authors evaluated oxidation by tracking conjugated diene hydroperoxides and the secondary oxidation product, hexanal. They studied antioxidants ascorbic acid and α-tocopherol and their lipophilic versions, ascorbyl palmitate and Trolox, respectively. The researchers also assessed the potentially synergistic effects that might result from combinations of the antioxidants. The results demonstrated that polar antioxidants outperformed their nonpolar counterparts in bulk oil, while the opposite was found to be true in the oil-in-water emulsion. The polar paradox effect was proposed to be due to the interfacial position of the antioxidants (for example, at the air–oil interface, in the case of polar antioxidants in oil, and at the oil–water interface, in the case of nonpolar antioxidants). This proposal provided the basis for a widely acknowledged explanation of the polar paradox. However, it lacked a quantitative measure of lipophilicity or hydrophilicity. Stated another way, it did not provide the exact requirements that would allow an antioxidant to position itself at the oil–water interface. It was not clear at what point it might become too lipophilic/nonpolar, such that it no longer stays at the interface but migrates deep into the oil droplet, where it might not provide much oxidative protection. In the same paper (Frankel et al, 1994), the authors noted that while mixtures of α-tocopherol and ascorbic acid were more active in bulk oils, the combination of α-tocopherol and ascorbyl palmitate outperformed the former. They based their explanation on the synergism that resulted from the application of multiple antioxidants, but the experimental design failed to fully separate the effect of polarity on antioxidant activity from the effect of antioxidant functionality. Further, the explanation does not provide insights on how functionality can contribute to the explanation of some of the paradoxical behavior of antioxidants. Frankel (1996) confirmed this theory by expanding the range of test antioxidants to include rosemary extract components: rosmarinic acid, carnosic acid, and carnosol (see Fig. 7.1). The order of polarity affected the antioxidant activity differently in bulk oil than it did in emulsions. Hence, the order of reactivity in bulk oil was: rosmarinic acid ≈ carnosic acid > carnosol (by hydroperoxides formation) and rosmarinic acid > carnosic acid > carnosol (by hexanal formation). There is a general trend in which the polarity dictates reactivity, perhaps qualitatively. However, a deeper examination of % hydroperoxides inhibition or % hexanal formation vs. log P (the partition coefficient between water and octanol) of carnosic acid, rosmarinic acid, and carnosol shows no apparent quantitative relationship between polarity and antioxidant activity (see Table 7.B).

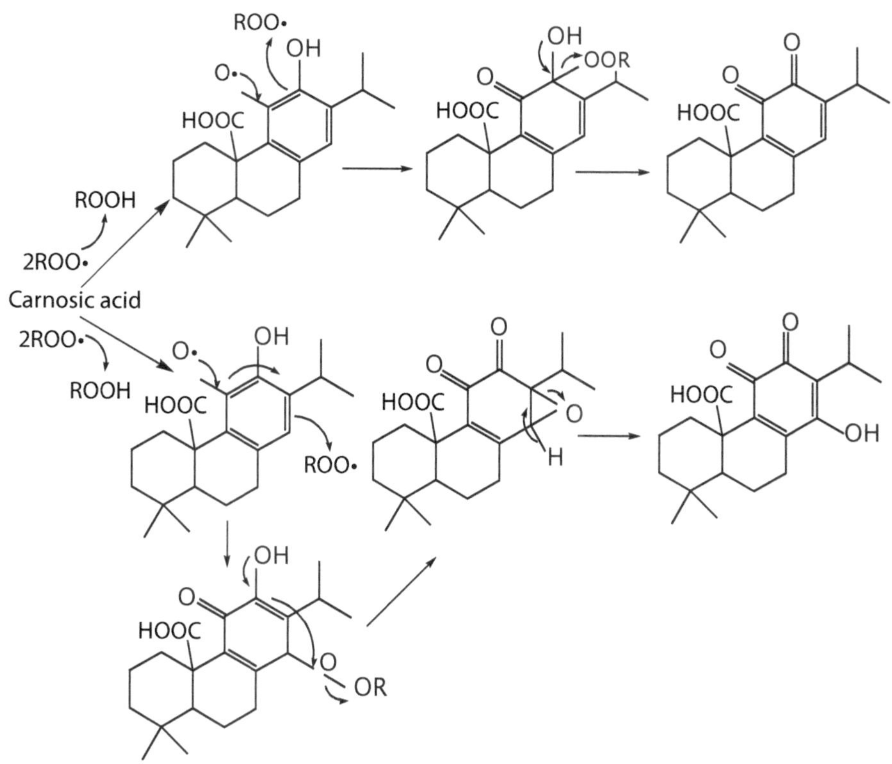

Fig. 7.1. The cascade of chemistry that regenerates antioxidants in the oxidation of carnosic acid.

Table 7.B. Correlation of Polarity with Primary and Secondary Oxidation Markers

	LogP[1]	**Inhibition of Hydroperoxides**	**Formation of Hexanal**
Carnosol	4.58	6.30%	−20% (decreased)
Carnosic Acid	5.13	70%	29%
Rosmarinic Acid	2.07	68%	78%

[1]Log P values were calculated from ChemDraw®.

The contradictory nature of antioxidant behavior might be attributed to many complex factors that could be antioxidant specific, such as functionality (that is, radical scavenging vs. metal chelation) and the optimal polarity of antioxidants in any given system. One can add another layer of complexity when the antioxidant possesses a unique capability or mode of action, such as in the case of carnosic acid. Carnosic acid has super-stoichiometric radical-scavenging capabilities based on its ability to form sequential intermediates that also act as antioxidants as it is being oxidized. Fig. 7.2 depicts plausible rearrangement mechanisms (Masuda et al., 2001).

The food matrix itself can complicate and certainly confuse any prediction of antioxidant behavior based on the polarity of the antioxidant in question. This fact is highlighted in another paper by Edwin Frankel (Huang et al., 1996), in which α-tocopherol and Trolox were evaluated in three different lipid systems (corn oil triglycerides, methyl linoleate and linoleic acid) in bulk form and in Tween 20–based emulsions. Linoleic acid formed mixed micelles with Tween 20, which changed the migratory properties of the antioxidant and made Trolox more effective than tocopherols in these emulsions.

Both the way that oxidation is initiated and evaluated influence whether the data generated will support or contradict the polar paradox. Huang and colleagues (1996) concluded that linoleic acid–mixed micelles affect antioxidant performance as polar antioxidants drawn into water come into proximity with linoleic acid and exert greater antioxidant protection. Cuvelier and colleagues (2000) found that the results differ depending on how oxidation is initiated. They differentiated between the ability to slow down the slope of conjugated-diene formation (proportional to the kinetics rate) and the ability to reduce the maximum level of primary oxidation products generated. The ranking of antioxidants tested was different based on these two inhibition parameters, and acted in accordance with the polar paradox. The authors did not find a close correlation between the metal-chelation potential of a series of antioxidants and their activity. These antioxidants included α-tocopherol, quercetin, butylated hydroxytoluene, butylated hydroxy-

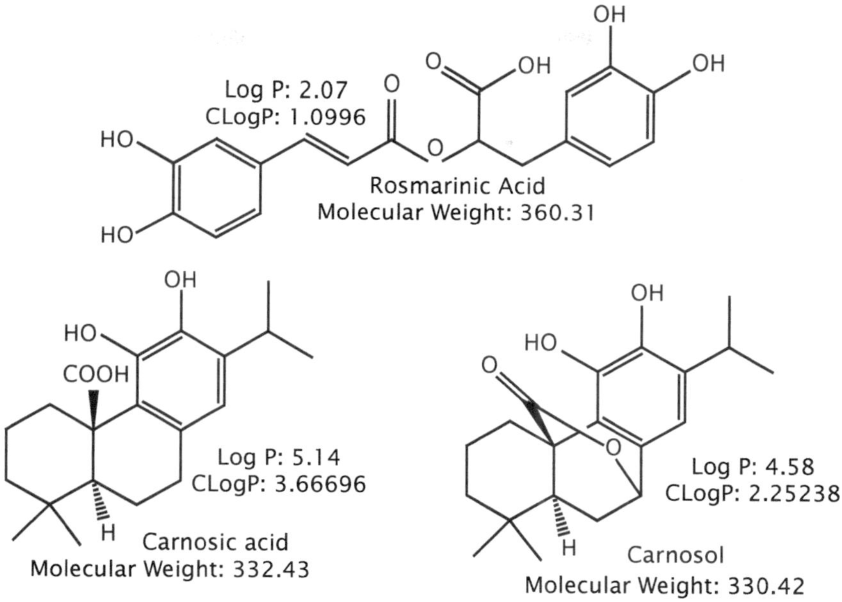

Fig. 7.2. Structures and polarity data for rosmarinic acid, carnosic acid, and carnosol.

anisole, isoquercetin, eriodictyol, isoeugenol, rosmarinic acid, DL-catechin, eugenol, (-)-epicatechin, Trolox, caffeic acid, naringenin, guaiacol, ferulic acid, and p-coumaric acid. As we show in the following section, however, these conclusions might not apply to the amazingly effective chelating antioxidant ethylenediamine-tetraacetic acid (EDTA) and its salts.

Calcium disodium ethylenediaminetetraacetic acid is the food-grade, FDA-approved, metal-sequestering version of EDTA. It is a potent antioxidant widely used in the United States to stabilize mayonnaise and salad dressings; however, its use is more restricted in Europe and Japan (Berdahl et al., 2010). It can slow down the loss of color and flavors and prevent transition metal ions from initiating the Fenton reaction that generates hydroxyl radicals in the initiation stage of the lipid peroxidation cycle. Hence, it serves as a preventative antioxidant. Because of its low solubility in oil, EDTA's usefulness is mainly limited to emulsions. Its ineffectiveness in bulk oil is undoubtedly due in part to its limited solubility, but more factors may be involved. Another reason may be that bulk oils and transition metals have a limited ability to initiate oxidation because their water activity is not high enough for optimal metal solubility and mobility. Zama and colleagues (1979) found that Cu^{++} accelerated the oxidation at the highest water activity (a_w) of the fish oil used in their study. The authors reported an antioxidant effect of EDTA at this highest a_w, but it was still minor compared to the activity of α-tocopherol and butylated hydroxyanisole in bulk oil. Simplistic "polar-paradox thinking" would predict that EDTA (polar) would be more effective than α-tocopherol and butylated hydroxyanisole, but in this case, other factors are clearly more important. For instance, Nielsen and colleagues (2004) showed that EDTA effectively inhibited oxidation in mayonnaise, using a mayonnaise with 16% fish oil and 64% rapeseed oil. The authors even went one step further, concluding that not all metal chelators are suitable for preventing metal-catalyzed oxidation in emulsions; some turn out to be ineffective, and some even act as prooxidants. These predictions were later confirmed by work done in 2% salmon oil, 0.2% Brij 35, pH 7 emulsion models, with and without EDTA. Oxidation assessment was done by measuring thiobarbituric acid reactive species (TBARS) and lipid hydroperoxides (Alamed, 2006). The antioxidative effect of EDTA was described as "dramatic," and the addition of calcium ions largely diminished the effect of EDTA, another clue showing that the antioxidative effect is due to the sequestration of trace-metal ions.

Although research on EDTA was not intended to either prove or disprove the polar paradox, it did shed light on the following key points:

- EDTA clearly demonstrates that metal chelators appear not to be bound by this explanation.
- "Not all chelators are created equal": EDTA is unique in its ability to retain functionality under acidic conditions. It can bind iron(III) with a high affinity, which helps to prevent recirculating the iron between iron(II) and iron(III).

In recent years, a series of papers have described experiments designed specifically to explain the polar paradox, with the added intent of generalizing and building upon it.

William Porter clearly thought of polarity and surface activity as drivers of "where" antioxidants "located" themselves, which means that later experimenters had to examine the effect of surface activity and polarity in a more quantitative manner (Chaiyasit et al., 2005). In this case, polarity was measured by observing the partitioning behavior between hexadecane (a relatively non-oxidizable lipid) and water, and the surface activity was assessed by measuring the ability of antioxidants to alter interfacial tension, as measured by the Du Noüy ring method. Two pairs of structurally related antioxidants were carefully selected (see Fig. 7.3), and their antioxidant activity was assessed in bulk menhaden oil and in a 5% menhaden oil-in-water emulsion. Surprisingly, the polarity and surface activity did not fully explain antioxidant effectiveness, showing that there is more to be explored and understood.

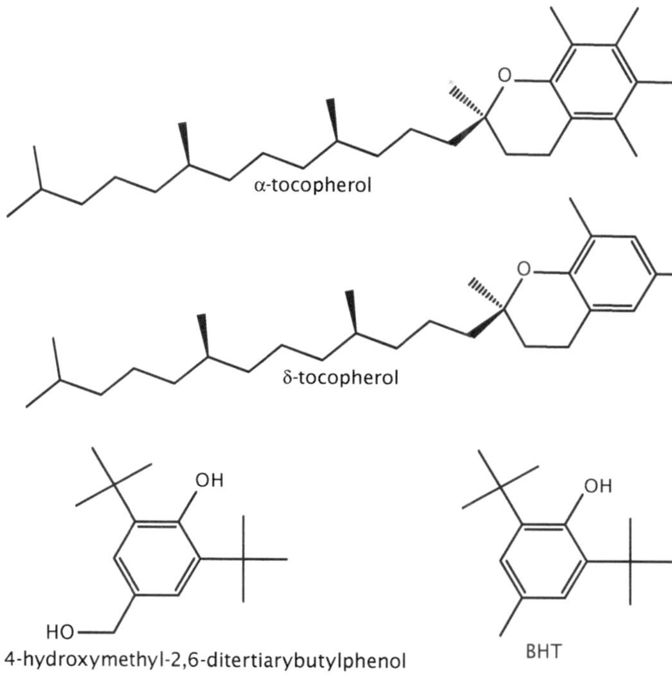

Fig. 7.3. Antioxidants studied in Chaiyasit (2005).

As part of a more generalized structure–activity relationship study, Torres de Pinedo and colleagues. (2007) examined the relative effectiveness of two series of four analogous phenolic-based derivatives in stabilizing refined olive oil using the Rancimat method. One homologous series contained the diphenolic compounds shown in Table 7.C and included hydroxytyrosol, a compound found naturally in olives. The second homologous series contained the methoxy-substituted compounds in Table 7.C. The analysis of results is complicated by a number of factors:

- The Rancimat experiments were conducted based on equivalent weight rather than equivalent concentration, so the differences described in the paper were not as large as they would have been if equivalent concentrations of the compounds were used. When corrected for concentration, the normalized effectiveness values of the compounds are as shown in Table 7.C.
- The olive oil used in the Rancimat experiments was refined, not stripped, and as such, it contained an unknown amount of native antioxidants, such as tocopherols or other phenolics, which would have complicated the interpretation of results.
- The Rancimat tests were run at a very high temperature, 120 °C.

The results of the Rancimat experiments, corrected for concentration differences (assuming that concentration effects are linear) and normalized for "effectiveness," are shown in Table 7.C. As seen here, both series show slightly better performance as the alkyl chain grows longer. This finding does not agree with the predictions that one would make based on the polar paradox, where the more polar antioxidant should be more effective in a bulk-oil system. Given the low molecular weight of these compounds and the high temperatures at which the Rancimat experiment was run, it is possible that these results could be explained by evaporative loss from the oil during the test. If the more volatile, lower molecular-weight compounds were more readily stripped from the oil, one might see the observed relative performance.

Table 7.C. Homologous Series of Phenolic Compounds

Structure	MW (g/mol)	Normalized AO Effectiveness
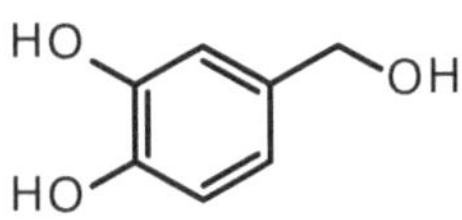	140.14	1.5
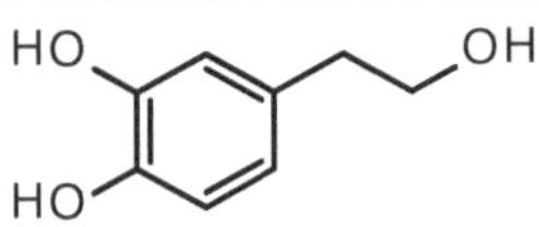	154.16	1.7
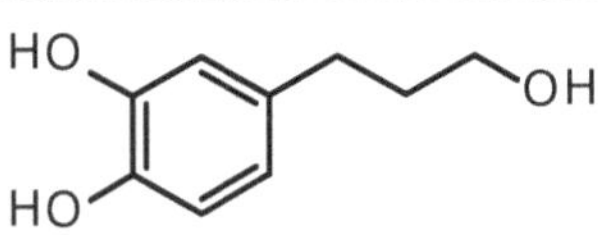	168.19	2.0
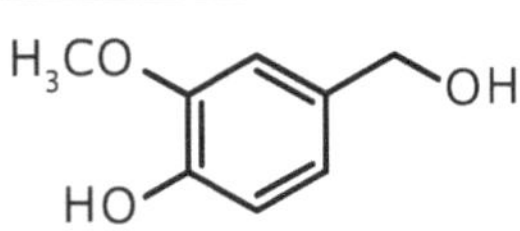	154.16	1.0
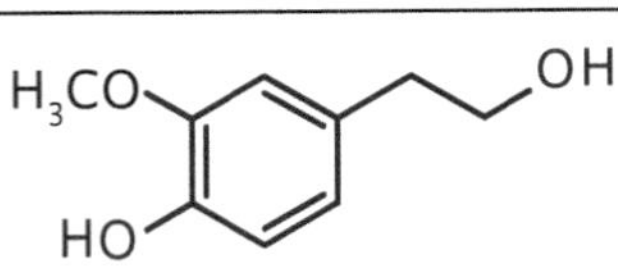	168.19	1.1
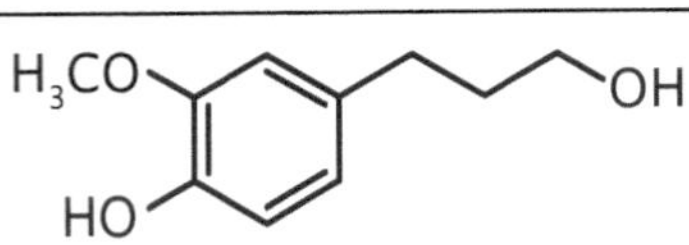	182.21	1.2

Although some of the compounds studied exhibited *ortho*-diphenolic structures known to be capable of coordinating transition-metal ions, the potential effect of metal chelation by these antioxidants was not evaluated.

Another study evaluated the structure–activity relationships of a series of tyrosol lipophilic ester derivatives that were synthesized through an enzymatic chemoselective process using lipase (Mateos et al., 2008). The homologous derivatives were evaluated using the Rancimat method, the Ferric Reducing Antioxidant Power (FRAP) test, and 2,2'-azino bis(3-ethylbenzothiazoline-6-sulfonic acid) diammonium salt (ABTS), which measures free-radical scavenging ability. In this case, the olive oil was stripped of antioxidants prior to its use in the Rancimat test, and antioxidants were added at the same concentration, not weight. The tests were run at a much lower temperature (80°C) than the tests described above, minimizing the chances for physical or chemical loss. Compared to their polar counterparts hydroxytyrosol and tyrosol, the lipophilic hydroxytyrosol esters and tyrosol esters were weaker antioxidants in the lipid matrix, according to the Rancimat method, but there did not seem to be a significant effect of polarity (within the esters) or acyl side length (see Fig. 7.4). The authors concluded that the performance of esters vs. hydroxytyrosol or tyrosol in bulk oils generally agreed with the polar paradox. Regarding the FRAP and TEAC test results, things were more complicated. The behavior of hydroxytyrosol and its derivatives appeared to accord with the polar paradox, whereas the behavior of tyrosol and its derivatives did not.

The same group published another study using a series of alkyl hydroxytyrosyl ethers (Fig. 7.5). They obtained similar results in that the ethers were somewhat less active than their parent structure, hydroxytyrosol, in bulk oil and more active in polar media (organic solvents, for example). Similarly, these researchers did not observe a notable dependency of antioxidant activity in bulk oil on the alkyl chain length (methyl through octadecyl) (Pereira-Caro et al., 2009).

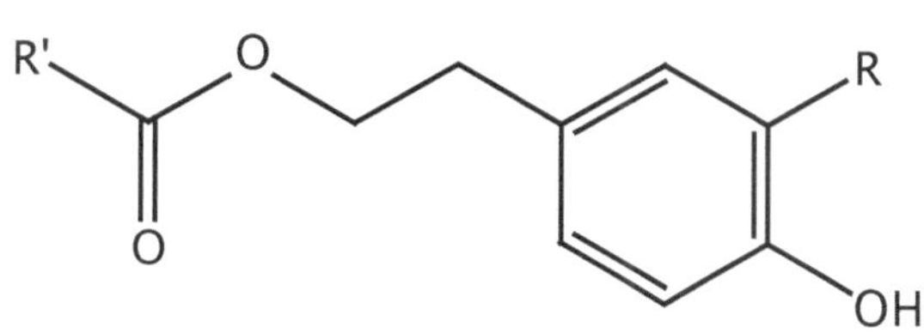

R'=CH_3, C_3H_7, $C_{11}H_{23}$, $C_{15}H_{31}$, $C_{17}H_{35}$, $C_{17}H_{33}$, $C_{17}H_{31}$

R=H, OH

Fig. 7.4. Structure of hydroxytyrosol esters.

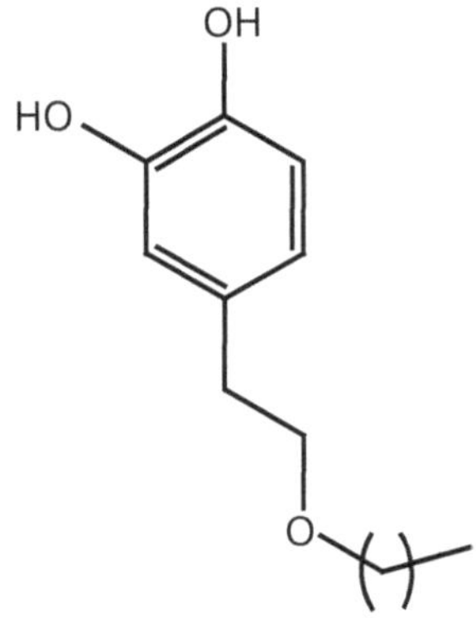

Fig. 7.5. Structure of alkyl hydroxytyrosyl ethers.

The results of these last three studies require comment on the use of tests like FRAP and TEAC. The most general expression of the polar paradox relates to the effectiveness of antioxidants in bulk lipids and in oil-in-water emulsions; therefore, we question the relevance of ranking performance in the FRAP test (in an aqueous solution or methanol) in relation to the polar-paradox question. Similarly, we wonder how a ranking of free-radical screening with ABTS (in water/methanol) or DPPH (also run generally in methanol or ethanol) can provide helpful information on antioxidant performance in radically different environments. Such tests rank the relative effectiveness of these antioxidants under test conditions—not those present in a bulk lipid or an oil-in-water emulsion. Therefore, they are not useful in discussions of the polar paradox, in the opinion of these authors.

The polar paradox is tied to interfaces and surfaces, whether they are in oil-in-water emulsions or in assemblies of more polar constituents in bulk oils. The polar paradox goes to the very nature of the problem: We need to better understand the molecular-scale environment in which these oxidation processes occur and this environment's effect on the location, orientation, and chemical expression of the antioxidants themselves. It comes down to the "ecosystem," if you like—how all of the dynamically changing components in these systems, including the antioxidants, interact in space and time. Tests like the FRAP and TEAC are useful but misapplied in a discussion of the polar paradox.

The Rancimat test is convenient and valuable in certain contexts, but it is our opinion that the polar paradox is better studied under real-world conditions, as opposed to those fostered by accelerated tests.

In many cases, the polar paradox could not explain the effect of antioxidants, which led to the suggestion of a more generalized theorem known as the cut-off effect (Laguerre et al., 2009). A chemo-enzymatic esterification of chlorogenic acid yielded the methyl, butyl, octyl, dodecyl, hexadecyl, octadecyl, and eicosyl corresponding esters, which were evaluated in a solution of 5 mg of stripped Tung oil in 25mL of pH 7.2 buffer using a test that measures conjugated trienes. The esters were also compared by measuring their partition in a 2% sunflower oil-in-water emulsion, at varying concentrations of the emulsifier (Brij 35). The results highlight the importance of the micellization process, where the effect of emulsifier concentration can drastically change the partitioning of the antioxidant. The critical micelle concentration (CMC) is the amount of surfactant above which the surfactant leaves the interface and detaches into the aqueous phase, forming micelles. Laguerre and colleagues postulated in this study that exceeding the CMC alters antioxidant partitioning, especially among the dodecyl esters of the series. This will affect the non-linear behavior observed below the CMC, where a cut-off effect was observed due to an optimal partitioning of esters with a chain length of 8–16 carbons, followed by a less effective partitioning of esters with a chain length of more than 16 carbons. We wonder whether these conclusions can be extended to other antioxidant classes (outside the chlorogenic ester series) and emulsion systems (a system with higher oil concentration), or a system with a different emulsifier (cationic vs. anionic vs. neutral). The following publications might answer a few of these questions.

The same group (Laguerre, 2010) published an article on the effect of the same alkyl chain variation in rosmarinate esters. They noticed a similar increase in antioxidant activity accompanying the increase of carbon chain length, with optimal activity for the octyl chain (methyl, butyl, octyl). Further extension of the carbon chain (dodecyl, hexadecyl, octadecyl, and eicosyl) led to a decrease in activity. The fact that the optimal carbon chain length for maximum antioxidant effect differed between the two studies (that is, C_8 for the chlorogenate esters vs. C_{12} for the rosmarinate esters) shows that the effect is very sensitive to the model employed. However, the trends were very similar and identically support idea of a cut-off effect, a more refined explanation of the polar paradox.

A similar study assessed the C_4, C_8, and C_{12} esters of chlorogenic acid in a 20% menhaden oil–in-water emulsion (Sasaki, 2010). They measured the antioxidants' surface activity using interfacial tensiometry, antioxidant partitioning by separating the aqueous phase through centrifugation, and lipid oxidation with gas chromatography. Interestingly, esterification increased both lipophilicity and surface activity; yet, it failed to improve the lipid oxidation inhibition effect of the parent compound, chlorogenic acid. The authors attributed this effect to the possibility that chlorogenic acid exerts a metal-chelating effect on prooxidant metals or regenerates antioxidants within the emulsion droplet. Although the polar paradox was initially designed for radical scavengers, it is difficult to separate the role of an antioxidant's properties (such as metal chelation or quenching) in any given study. Antioxidants with multiple functionalities can potentially function in multiple ways, and although radical scavenging is often quantified, metal chelation is generally not.

Another attempt to explain the polar paradox and the nonlinear dependency of antioxidant effectiveness in emulsions relied on the surfactant effectiveness of the antioxidants themselves (Lucas, 2010). The "phenolipids" used consisted of both tyrosol and hydroxytyrosol fatty esters. The authors observed a strong correlation between the right hydrophilic–lipophilic balance (HLB) and inhibition of lipid oxidation. The best antioxidant in the series was also the one with the most efficient surfactant property (hydroxytyrosol octanoate) (see Fig. 7.6).

Another complexity in studying lipid oxidation in bulk oils stems from the presence of minor oil components like metals, phospholipids, tocopherols, sterols, carbohydrates, carotenoids, chlorophyll, and free fatty acids. While minor components like tocopherols and carotenoids are beneficial for the oxidative quality of oils, others like trace metal ions and chlorophyll (a photosynthesizer) are pro-oxidants. Since the edible-oil refining process is not totally efficient, these minor components may exist at various levels in consumed oils. Hence, some residual components influence lipid oxidation by altering an oil's physical properties (Chaiyasit et al., 2007). These authors draw attention to the fact that bulk oils have been studied as homogeneous systems, while, in fact, they are heterogeneous. The presence of minor oil components that are amphiphilic—such as mono- and diacylglycerols, phospholipids, and polar fatty acid oxidation breakdown products create conditions that foster the formation of association colloids, such as reverse micelles and lamellar structures (Fig. 7.7).

This phenomenon has implications for explaining why polar antioxidants can be more effective than lipophilic antioxidants because of surface activity at the

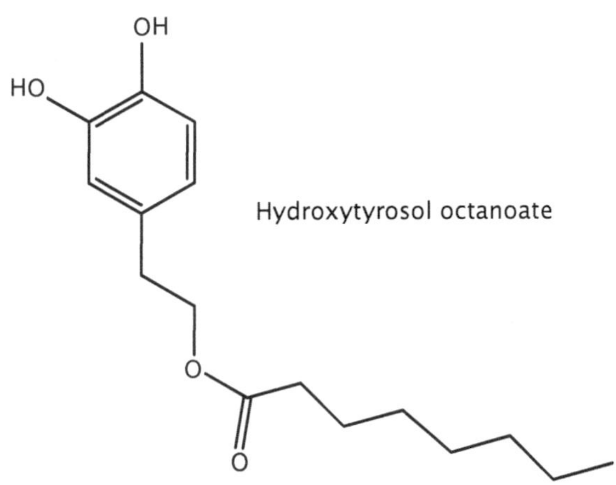

Fig. 7.6. Hydroxytyrosol octanoate.

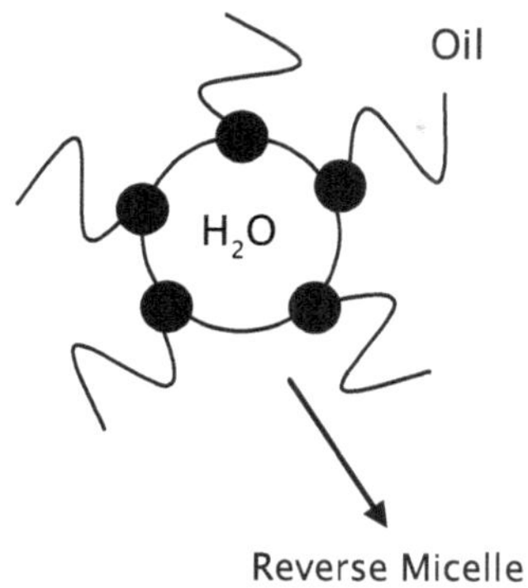

Fig. 7.7. Formation of reverse micelles in bulk oils.

reverse micelle interface with the oil. Consequently, a new explanation arises for polar-paradox behavior in bulk oils, as earlier research suggested that polar antioxidants were more effective in bulk oils because of their position at the oil–air interface (that is, at the surface).

Chen at al. (2011) published a study that tested the activity of α-tocopherol and Trolox in stripped soybean oil using 1,2-dioleyl-sn-*glycero*-3-phosphocholine to form reverse micelles. The formation of reverse micelles had an overall prooxidant effect; however, it improved the activity of both α-tocopherol and Trolox at low concentrations but decreased the activity of higher antioxidant concentrations. The authors postulated that Trolox performed better than α-tocopherol due to its different physical location in the reverse micelle (that is, one is better positioned than the other).

Some recent studies of the polar paradox have examined the effect of concentration on four pairs of polar and nonpolar antioxidants in bulk oil (Shahidi & Zhong, 2011; Zhong & Shahidi, 2012). In these works, application of the polar paradox depends on the systems' concentration of antioxidants. The authors conclude that the polar paradox only operates at certain ranges of concentration.

Conclusion

The polar paradox will be problematic if we expect a generalized explanation to fit all situations. Recent efforts have tried to explain the relationship between polarity and antioxidant behavior, especially in emulsions. These efforts to better understand the effect of minor oil components and various emulsion constituents (like surfactants and iron) and to utilize new series of antioxidants have led to better explanations of paradoxical behavior in system models. However, it remains difficult to predict antioxidant performance and behavior without considering the effect of many other factors, such as:

- *Nature and amount of prooxidants, such as metal ions, and possible presence of photosensitizers (when applicable):* Antioxidants react differently to these elements based on their antioxidant functionality, their concentration, and the concentration and location of prooxidants.
- *Functionality of antioxidants:* Differently functional antioxidants (that is, radical scavengers vs. metal chelators) react and behave differently. Metal chelators can be completely water soluble in an emulsion, distant from the oil–water interface yet exert remarkable antioxidant activity (EDTA). The effect of polarity depends on functionality. A water-soluble chelator like EDTA can outperform any radical scavenger, even those that can be placed at the emulsion's oil–water interface.
- *The emulsion:* The emulsion's ingredients can have a profound effect on antioxidant performance. Whether cationic, anionic, or amphiphilic, the emulsifier type and the antioxidant concentration can impact or even reverse the antioxidant's effects, thus confounding any conclusions drawn on the effect of polarity alone.

The value of the polar paradox lies not in its ability to explain antioxidant performance but in the questions that it generates in the minds of researchers as they seek to explain the complex phenomenon of oxidation in foods. William Porter himself (1993) described the value in this way:

> "The citing of paradoxical behavior is more than a matter of fact. The identification of a paradox often changes a mere recital of facts into a fascinating demonstration of biphasic or conflicting behavior of compounds which sharpens interest and fixes that behavior in mind. The identification also serves as a reminder of the upper and lower limits of action or usefulness, i.e., a little may be good but more may not, or vice versa."

References

Alamed, J.; McClements, D. J.; Decker, E. A. Influence of heat processing and calcium ions on the ability of EDTA to inhibit lipid oxidation in oil-in-water emulsions containing omega-3 fatty acid. Food Chem. 2006, 95, 585–590.

Berdahl, D. R.; Nahas, R I.; Barren, J. P. Synthetic and Natural Antioxidant Additives in Food Stabilization: Current Applications and Future Research," in Oxidation in Foods and Beverages and Antioxidant Applications; Decker, E. A.; Elias, R. J; Julian McClements, D., Eds.; Woodhead Publishing: Oxford, 2010; Vol. 1, 272–313.

Chaiyasit, W.; Elias, R. J.; McClements, D. J.; Decker, E. A. Role of physical structures in bulk oils on lipid oxidation. Crit. Rev. Food Sci. 2007, 47, 299–317.

Chaiyasit, W.; McClements, D. J.; Decker, E. A. The relationship between the physicochemical properties of antioxidants and their ability to inhibit lipid oxidation in bulk oil and oil-in-water emulsions. J. Agric. Food Chem. 2005, 53, 4982–4988.

Chen, B.; Han, A.; Laguerre, M.; McClements, D. J.; Decker, E. A. Role of reverse micelles on lipid oxidation in bulk oils: Impact of phospholipids on antioxidant activity of α-tocopherol and Trolox. Food Funct. 2011, 2, 302–309.

Cuvelier, M.-E.; Bondet, V.; Berset, C. Behavior of phenolic antioxidants in a partitioned medium: Structure—activity relationship. J. Am. Oil Chem. Soc. 2000, 77 (8), 819–823.

Frankel, E. N. Antioxidants in lipid foods and their impact on food quality. Food Chem. 1996, 57 (1), 51–55.

Frankel, E. N.; Huang, S.-W.; Kanner, J.; German, J. B. Interfacial phenomena in the evaluation of antioxidants: Bulk oils vs emulsions. J. Agric. Food Chem. 1994, 42, 1054–1059.

Frankel, E. N.; Huang, S.-W.; Aeschbach, R.; Prior, E. Antioxidant activity of a rosemary extract and its constituents carnosic acid, carnosol, and rosmarinic acid, in bulk oil and oil-in-water emulsion. J. Agric. Food Chem. 1996, 44, 131–135.

Huang, S.-W.; Frankel, E. N.; German, J. B. Antioxidant activity of α- and γ-tocopherols in bulk oils and in oil-in-water emulsions. J. Agric. Food Chem. 1994, 42, 2108–2114.

Huang, S.-W.; Hopia, A.; Schwarz, K.; Frankel, E. N.; German, J. B. Antioxidant activity of α-tocopherol and Trolox in different lipid substrates: Bulk oils vs oil-in-water emulsions. J. Agric. Food Chem. 1996, 44, 444–452.

Laguerre, M.; Giraldo, L. J. L.; Lecomte, J.; Figueroa-Espinoza, M. C.; Barea, B.; Weiss, J.; Decker, E. A.; Villeneuve, P. Chain length affects antioxidant properties of chlorogenate esters in emulsion: The cutoff theory behind the polar paradox. J. Agric. Food Chem. 2009, 57, 11335–11342.

Laguerre, M.; Giraldo, L. J. L.; Lecomte, J.; Figueroa-Espinoza, M. C.; Barea, B.; Weiss, J.; Decker, E. A.; Villeneuve, P. Relationship between hydrophobicity and antioxidant ability of "phenolipids" in emulsion: A parabolic effect on the chain length of rosmarinate esters. J. Agric. Food Chem. 2010, 58, 2869–2876.

Lucas, R.; Comelles, F.; Alcantara, D.; Maldonado, O. S.; Curcuroze, M.; Parra, J. L.; Morales, J. C. Surface-active properties of lipophilic antioxidants tyrosol and hydroxytyrosol fatty acid esters: A potential explanation for the nonlinear hypothesis of the antioxidant activity in oil-in-water emulsions. J. Agric. Food Chem. 2010, 58, 8021–8026.

Masuda, T.; Inaba, Y.; Takeda, Y. Antioxidant mechanism of carnosic acid: Structural identification of two oxidation products. J. Agric. Food Chem. 2001, 49, 5560–5565.

Mateos, R.; Trujillo, M.; Pereira-Caro, G.; Madrona, A.; Cert, A.; Espartero, J. L. New lipophilic tyrosyl esters. Comparative antioxidant evaluation with hydroxytyrosyl esters. Agric. Food Chem. 2008, 56, 10960–10966.

Nielsen, N. S.; Petersen, A.; Meyer, A. S.; Timm-Heinrich, M.; Jacobsen, C. Effects of lactoferrin, phytic acid, and EDTA on oxidation in two food emulsions enriched with long-chain polyunsaturated fatty acids. J. Agric. Food Chem. 2004, 52, 7690–7699.

Pereira-Caro, G.; Madrona, A.; Bravo, L.; Espartero, J. L.; Cert, A.; Mateos, R. Antioxidant activity evaluation of alkyl hydroxytyrosyl ethers, a new class of hydroxytyrosol derivatives. Food Chem. 2009, 115, 86–91.

Porter, W. L. Paradoxical behavior of antioxidants in food and biological systems. Toxicol. Ind. Health 1993, 9 (1–2), 93–122.

Sasaki, K.; Alamed, J.; Weiss, J.; Villeneuve, P.; Giraldo, L. J. L.; Lecomte, J.; Figueroa-Espinoza, M.-C.; Decker, E. A. Relationship between the physical properties of chlorogenic acid esters and their ability to inhibit lipid oxidation in oil-in-water emulsions. Food Chem. 2010, 118, 830–835.

Shahidi, F.; Zhong, Y. Revisiting the polar paradox theory: A critical overview. J. Agric. Food Chem. 2011, 59, 3499–3504.

Sherwin, E. R.; Thompson, J. R. Tertiary-butylhydroquinone—An antioxidant for fats and oils and fat-containing foods. Food Technol. 1967, 21, 106–110.

Torres de Pinedo, A.; Peñalver, P.; Morales, J. C. Synthesis and evaluation of new phenolic-based antioxidants: Structure activity relationship. Food Chem. 2007, 103, 55–61.

Wöhler, Friedrich; Letter to Jöns Jacob Berzelius, January 1835. http://www.chem.yale.edu/~chem125/125/history99/4RadicalsTypes/onLiebig.html

Zama, K.; Takama, K.; Mizushima, Y. Effect of metal salts and antioxidants on the oxidation of fish lipids during storage under the conditions of low and intermediate moistures. J. Food Process. Pres. 1979, 3, 249–257.

Zhong, Y.; Shahidi, F. Antioxidant behavior in bulk oil: Limitations of polar paradox theory. J. Agric. Food Chem. 2012, 60, 4–6.

CHAPTER

8

Role of Hydrophobicity on Antioxidant Activity in Lipid Dispersions

From the Polar Paradox to the Cut-Off Theory

Mickaël Laguerre[1], Ann-Dorit Moltke Sørensen[2], Christelle Bayrasy[1], Jérôme Lecomte[1], Charlotte Jacobsen[2], Eric Andrew Decker[3], and Pierre Villeneuve[1]

Introduction

Lipid dispersions, such as emulsions, micelles, and membrane systems, are ever-present inside and outside of our bodies. Life as we know it is based on compartmentation through phospholipid and archeolipid membranes. In recent years, the use of oil-in-water emulsions, liposomes, and micelles to entrap, protect, and deliver unstable components (drugs, antimicrobials, flavors, antioxidants, nutritional lipids, etc.) has been extensively investigated in the food industry as well as in pharmacology. In cosmetics, emulsions and liposomes are undoubtedly the most widespread skin-care formulations. Oil-in-water emulsions are also important for food products. Many of them exist either partly or wholly as emulsions, or they have existed at some point in an emulsified state (McClements, 2008). Milk is the first food emulsion that we encounter. Even nonemulsified lipid-containing foods form lipid dispersions in the digestive tract thanks to the combined action of shearing forces and bile salts.

---

[1]CIRAD, UMR IATE, Montpellier F-34398, France; [2]National Food Institute, Technical University of Denmark, DK-2800 Kgs. Lyngby, Denmark; [3]Department of Food Science, University of Massachusetts, Amherst, MA 01003, USA.

One of the main problems for practical applications is that lipid dispersions are highly susceptible to lipid oxidation due to their large surface area. Emulsions and membranes contain unsaturated lipids, which are prone to oxidation induced by transition metals, light, temperature, oxygen, etc. Thus, the oxidative stability of such systems depends greatly on the chemical composition of the lipid phase. Apart from noticeable exceptions (Miyashita et al., 1993, 1994), the more unsaturated a dispersion is, the more prone it is to lipid oxidation. Other factors can also influence the rate of lipid oxidation in lipid dispersions; they include aqueous phase pH and ionic composition, surface charge, chemical composition, permeability and thickness of the interfacial layer or the lipid bilayer, along with lipid droplet and membrane characteristics, including particle size, concentration, viscosity, and physical state (Laguerre et al., 2007; Waraho et al., 2011).

Besides altering the sensory (rancidity) and nutritional quality (loss of vitamins and essential fatty acids) of food or cosmetic products, the ultimate oxidation of lipids into highly reactive and toxic compounds (for example, malondialdehyde, 4-hydroxynonenal) is a real danger for consumers. This is especially important for foods that have become more oxidatively liable in recent years. To avoid *trans* fatty acids (which are more atherogenic than saturated fats), food manufacturers are unable to use partially hydrogenated fats in their product formulations. Likewise, manufacturers have improved the nutritional profile of foods by adding more unsaturated fatty acids, such as *n*-3 fatty acids, which make food more prone to oxidation. To complicate matters further, the number of antioxidants available to food or cosmetic manufacturers to control oxidative rancidity is limited and the approval of new antioxidants is difficult due to economic barriers in obtaining government approval for new additives (Chayasit et al., 2007). Thus, industries are more concerned now than ever about lipid oxidation. Typically, there are two different and complementary strategies to inhibit oxidation in dispersed lipid systems. The first is lipid dispersion engineering to control the composition, thickness, or charge of the interfacial layer that separates the encapsulated lipids from the surrounding aqueous phase. This strategy is also used to fine-tune the size, composition, and physical state of lipid phases. Since oxidation cannot be totally controlled by the manipulation of interface or lipid phase properties, a second strategy incorporates antioxidants to further protect against rancidity (Waraho et al., 2011).

In lipid dispersions, the presence of distinct phases (that is, lipids, interface, and water) results in the partitioning of antioxidants according to their polarity. For maximum effectiveness, antioxidants should be localized in the microenvironment where free radicals or transition metals meet oxidizable substrates. The ideal location is assumed to be at the interfacial region. An efficient antioxidant should also be able to move freely and quickly to the site of action. Although often overlooked by scientists working in this field, the diffusibility of antioxidants toward free radicals, transition metals, and oxidizable substrates is of prime importance.

This chapter addresses the versatile role of hydrophobicity on antioxidant capacity in lipid dispersions. This crucial parameter affects both the location and the diffusivity of antioxidants in dispersed lipid systems, which, in turn, impacts their antioxidant capacity. Until recently, the polar paradox was the only theory available to explain how hydrophobicity impacts antioxidant capacity in oil-in-water emulsions, micelles, liposomes, and membranes. However, one cannot but notice the great number of discrepancies and contradictions to the polar paradox when scrutinizing data accumulated for more than 20 years in this field. Since 2009, we have been developing an alternative hypothesis—the cut-off theory—which accounts for the polar paradox's limitations (Laguerre et al., 2009, 2010, 2011, 2013).

In this chapter, we present an overview of the latest experimental and theoretical advances in the field. First, we re-examine the polar paradox assumption wherein fewer polar antioxidants tend to be more effective than their polar counterparts in lipid dispersions. After a brief review of the experimental data supporting our cut-off theory, we address the physicochemistry behind antioxidant action by developing two new, complementary approaches based on location and diffusion. While the polar paradox is founded exclusively on the prominent influence of the antioxidant location (especially through Frankel's interfacial concept), we will see that the cut-off theory tries to account for both the impact of diffusion and location of the antioxidant on its activity.

How Does Hydrophobicity Impact Antioxidant Capacity?

Numerous in vitro studies performed in heterogeneous systems have shown that the antioxidant capacity of phenolic compounds is governed not only by intrinsic properties such as O-H bond dissociation energy (BDE) or steric hindrance but also by physicochemical phenomena such as partitioning in the different phases, diffusion, self-aggregation, and association of antioxidants with amphiphiles (for example, surfactants and phospholipids), which are, in turn, mainly governed by the hydrophobicity of the antioxidant. However, we do not yet understand the influence of hydrophobicity on the activity of phenolic, especially in complex systems such as lipid dispersions.

The Polar Paradox Theory: The Linear Trend

The first investigation of the hydrophobicity influence on antioxidant activity in lipid dispersions took place in the 1950s. Before 1955, a large body of data for the relative effectiveness of antioxidants was generated, and very often it has been extrapolated uncritically from bulk lipids to emulsions, micelles, and membranes in whole tissue foods. Later, however, Chipault et al. (1956), Uri (1958, 1961), Simpson and Uri (1956), and Abbot and Waite (1962) began to comment on the unexpected contrast in relative effectiveness between the same antioxidant used in bulk oil versus emulsion. In 1974, Scott et al. found that decarboxylation or esterification of Trolox (a water-soluble derivative of α-tocopherol) produced increased antioxidant activity in emulsion. Cort (1974, 1982) demonstrated that ascorbic acid was much less effective than ascorbyl palmitate in the hemoglobin–safflower oil emulsion test. Even though these results gave occasional hints, no one tried to summarize these empirical observations to draw a general model before the 1980s. Porter (1980) and Porter et al. (1989) were the first to propose a rule to explain how hydrophobicity impacts antioxidant capacity in lipid-based systems. According to their so-called polar paradox, nonpolar antioxidants tend to be more effective at counteracting lipid oxidation than their polar homologues in oil-in-water emulsions, liposomes, and biological tissues. The reverse is supposed to be true in bulk oils. In lipid dispersions, the concept of Porter's polar paradox is the less polar it is, the more antioxidant.

On Which Data Does the Polar Paradox Rely?

Challenge to the polar paradox has been brought by Porter and colleagues themselves (Porter et al., 1989). Their milestone article is considered to be the first experimental demonstration of the polar paradox in lipid dispersions. However, the data reported in this article do not follow a linear trend. They drew conclusions from this study based on 16 antioxidants. Ten of these were not directly structurally related, such as BHT, caffeic acid, and quercetin. As we recently pointed out, the analysis of different antioxidants introduces many new variables besides polarity, such as number and location of OH groups, double bonds, side chain substitutions, etc. The concern is that these new variables can impact the reactivity of antioxidants toward free-radical and transient metals through a change of hydrogen transfer tendencies (BDE) or a modification of the kinetics and stability of antioxidant radicals. This makes it difficult, if not impossible, to investigate the sole effect of polarity in structurally unrelated antioxidants (Laguerre et al., 2009, 2013). Shahidi and Zhong (2011) came to a similar conclusion, stating that it is inappropriate to attribute the difference in efficiency of ascorbic acid and BHT solely to their polarity. In the article by Porter et al. (1989), the only structurally related antioxidants are a homologous series of gallate alkyl esters, but when we look at these compounds, they do not appear to follow the linear trend proposed by the polar paradox. As Fig. 8.1 shows, the first section of the relationship between hydrophobicity and chain length is linear, but after the propyl chain attains a threshold, the antioxidant activity decreases for an octyl chain and then increases

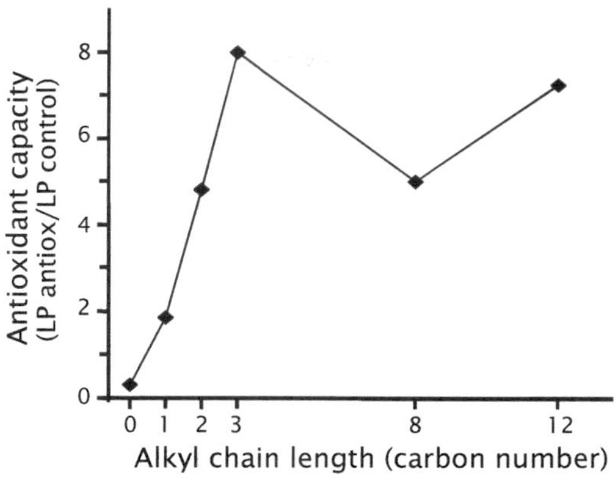

Fig. 8.1. Relationship between antioxidant capacity in lipid dispersion and alkyl chain length of gallate alkyl esters.

Adapted from data reported by Porter et al. (1989) and supposedly supporting a linear trend.

LP: lag phase.

slighty with a dodecyl chain. Consequently, the polar paradox can already be ruled out as consistently predicting the antioxidant capacity beyond three carbon atoms in the side chain of gallate alkyl esters.

It would be particularly unfair, however, not to acknowledge that the authors observed and commented on this deviation from the linear trend. They noted that "the relationship is not rigidly linear, but the trend is there." We agree with the first point (factual constatation) but not with the second one (subjective interpretation). Porter et al. did not write "fewer polar antioxidants are more active in lipid dispersion" but rather that "they tend to be more active." They have obviously added this nuance to account for some compounds that did not follow the linear trend, and therefore, the polar paradox. Then, the crucial question is: Are these compounds (that do not follow the trend) minority cases that could be neglected as experimental artefacts, or majority cases that should be taken into account? If such discrepancies are rare, then recalcitrant nonpolar antioxidants that do not follow the polar paradox are the exception to the rule. Otherwise, if such cases are frequently encountered, we have to look beyond the polar paradox to find a new rationale that accounts more accurately for the antioxidant capacity in lipid dispersions. The following section will present a brief overview of the many studies that contradict the polar paradox in emulsified and membrane systems.

Experimental Contradictions to the Polar Paradox in Model Lipid Dispersions

Stöckmann et al. (2000) tested the polar paradox in a stripped corn oil-in-water emulsion stabilized by different surfactants using the same homologous series of gallic acid and its alkyl esters. Measuring lipid hydroperoxides by the thiocyanate-ferric ion color reaction, they found different orders of antioxidant effectiveness according to the emulsifier used. In a lecithin-stabilized emulsion, the order was approximately propyl gallate ~ butyl gallate > octyl gallate ~ ethyl gallate ~ methyl gallate >> gallic acid. Thus, apart from octyl gallate, the polar paradox seems to be more or less confirmed, as the most effective antioxidants were more or less the more hydrophobic molecules. On the contrary, in both sodium dodecyl sulphate (SDS) and Brij 58–stabilized emulsions, results were not consistent with the polar paradox. Indeed, the order with SDS was methyl gallate ~ ethyl gallate > propyl gallate >> butyl gallate >> octyl gallate ~ gallic acid, while with Brij 58 it was methyl gallate >> octyl gallate ~ butyl gallate > ethyl gallate > propyl gallate >> gallic acid. Stöckmann et al. (2000) concluded that the polar paradox may be limited to emulsions containing emulsifiers with properties similar to phospholipids, a finding that considerably narrowed the applicability of the polar paradox in emulsion. We noted with interest the authors' statement that "there was a nonlinear relationship between decreasing polarity of the gallates and antioxidant activity in SDS, Brij 58, and PHLP (partially hydrolyzed soybean lecithin) emulsions...."

In another study contradicting the polar paradox in emulsions, Yuji et al. (2007) observed that although butyl or dodecyl esters of p-hydroxyphenylacetic acid (HPA) concentrated at lipid/water interfaces more than free HPA, the latter antioxidant was more effective at inhibiting the oxidation of Brij 35–stabilized, non-stripped, Menhaden oil-in-water emulsions. One can also cite the work of Sasaki et al. (2010), who demonstrated in the same emulsified system that phenolic antioxidants conjugated with hydrocarbon chains are more highly associated with lipid emulsions droplets, but these changes in physical properties did not increase antioxidant activity. Sørensen et al. (2012b) also achieved interesting results, which showed that oleyl dihydrocaffeate was a less efficient antioxidant than octyl ester in fish oil-in-water emulsion. Finally, experiments with rutin and lipophilized rutin (rutin laurate and rutin palmitate) as antioxidants in oil-in-water emulsion support the fact that highly lipophilic polyphenols display low antioxidant capacity because the esters were consistently less effective than rutin (Lue et al., 2009).

Experimental data challenging the polar paradox can also be found in membrane systems. For instance, Kagan et al. (1990c) showed that α-tocopherol homologs differing in the lengths of their hydrocarbon side chains (α-Cn) manifest strikingly different antioxidant potencies in membranes. The antioxidant activity of tocopherol homologs during (Fe^{2+}-ascorbate)- or (Fe^{2+}-NADPH)-induced lipid peroxidation in rat liver microsomes increased in the order α-tocopherol (α-C16) < α-C11 < α-C6 < α-C1. These results follow a trend that is exactly contrary to the polar paradox trend since the less polar a compound is, the less antioxidant. This result has been confirmed in liver microsomes, mitochondria, and brain synaptosomes (Kagan et al., 1990a).

Experimental Contradictions to the Polar Paradox in Dispersed Food Lipid Systems

Recently, Sørensen et al. (2012a) evaluated the antioxidative effects of dihydrocaffeic acid and its octyl and oleyl esters, as well as rutin, rutin laurate, and rutin palmitate in fish oil–enriched milk. For both types of compounds, the medium-chain ester was a better antioxidant than the long-chain esters and the nonlipophilized phenolics. This confirmed the findings from Sørensen et al. (2012b) mentioned previously, that octyl dihydrocaffeate was a more active antioxidant than oleyl dihydrocaffeate in model oil-in-water emulsion. These results contradict the polar paradox hypothesis because octyl dihydrocaffeate and rutin laurate were more efficient antioxidants than oleyl dihydrocaffeate and rutin palmitate, respectively. Another challenge to the polar paradox came from the study of two pairs of homologue antioxidants (ascorbic acid/ascorbyl palmitate and gallic acid/propyl gallate) in fish oil–enriched mayonnaise (Jacobsen et al., 1999, 2001, 2008). In these reports, all tested compounds acted as prooxidants. Thus, the antioxidant activity did not improve with increased lipophilicity. The authors concluded that the polar paradox hypothesis was too simple to predict antioxidant efficacy in this complex food emulsion in which lipid oxidation is catalyzed by iron stemming from the egg yolk used as an emulsifier. Indeed, emulsified food systems like milk and mayonnaise are more complex compared to model oil-in-water emulsions. According to the authors, many different molecules can interact and influence the antioxidant activity, and this may be one of the reasons why some results do not support the polar paradox hypothesis.

The Cut-Off Theory: The Nonlinear Trend

Before 2009, no theory could have been considered an alternative to the polar paradox theory in emulsified and liposomal systems. To our knowledge, the cut-off theory was the first to contradict the polar paradox in lipid dispersions and to propose a new scenario wherein the polar paradox became a mere linear portion of a much broader nonlinear phenomenon. In this section, we will present a panorama of the data on which this alternative theory is based.

Cut-Off Effect in Oil-in-Water Emulsions

In 2009, using a wide range of chlorogenate esters in a stripped tung oil-in-water emulsion stabilized with Brij 35, we showed that the antioxidant capacity increased as the alkyl chain was lengthened until a threshold in hydrophobicity was reached (Laguerre et al., 2009). Beyond this threshold (which later received the name critical chain length), the antioxidant capacity suddenly collapsed

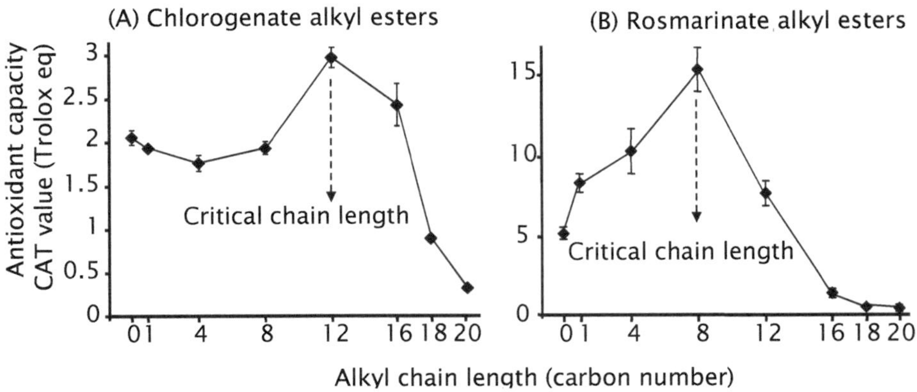

Fig. 8.2. Influence of the alkyl chain length of chlorogenate (A) and rosmarinate (B) alkyl esters on antioxidant activity in stripped tung oil-in-water emulsion.

(A) After Laguerre et al., 2009; (B) After Laguerre et al., 2010.

(Fig. 8.2A). Long-chain alkyl esters of chlorogenic acid displayed near-zero antioxidant capacity, which totally contradicted the polar paradox. Based on these unexpected results, we proposed a new model called the cut-off theory to account for this nonlinear relationship between hydrophobicity and antioxidant capacity. It can be defined for a homologous series of hydrocarbon chain surface-active compounds as the tendency for physical-chemical activity (such as that of antioxidants) to increase stepwise (or stay stable) with increasing chain length up to a critical point, beyond which this activity suddenly collapses.

Since then, this nonlinear trend has been confirmed with other compounds and/or other systems. Using rosmarinate alkyl esters, we showed the same cut-off influence of the alkyl chain length on antioxidant capacity in the same emulsified system (Laguerre et al., 2010). In this work, the cut-off was even more evident, with an almost-perfect bell curve (Fig. 8.2B). We recently found two other cases where phenolipid antioxidants behave in a manner proposed by the cut-off theory. In a stripped soybean oil-in-water emulsion stabilized with Tween 20, butyl, octyl, and dodecyl, rosmarinates were much better antioxidants than rosmarinic acid and its octadecyl and eicosyl esters (Panya et al., 2012). Lee et al. (2013) have also investigated the effect that alkyl chain length has on rosmarinate esters in riboflavin-photosensitized oil-in-water emulsions stabilized with Tween 20. On the basis of lipid hydroperoxides, hexanal, and *t*-2-heptenal concentrations, the order of antioxidant effectiveness was found to be: butyl ~ octyl > dodecyl > nonlipophilized rosmarinic acid > octadecyl > eicosyl > control. Again, medium-chain esters (butyl and octyl rosmarinates) were the more active antioxidants of the series. Apart from our group's findings, Medina et al. (2009) also published evidence that supports the cut-off theory. In a nonstripped cod liver oil-in-water emulsion stabilized with lecithin, they found that the antioxidant capacity of hydroxytyrosol fatty esters increases as the chain lengthens up to eight carbons. The lauryl ester of hydroxytyrosol displayed equivalent or less antioxidant capacity than the octyl ester, which contradicted the polar paradox while supporting the cut-off theory.

Cut-Off Effect in Membrane Systems

Apart from oil-in-water emulsions, a cut-off effect has also been observed in other lipid dispersions, such as liposomes. Kikuzaki et al. (2002) found that alkyl chain length of two homologous series of phenolipids (lipophilized phenolics) had a strong nonlinear influence on their antioxidant capacity in egg yolk phosphatidylcholine liposomes. According to these results, Kikuzaki et al. improved the antioxidant capacity after lengthening the carbon chain to a critical chain length of 6–8 carbons atoms for ferulate alkyl esters and ~12 carbon atoms for gallate alkyl esters. Beyond this hydrophobicity threshold, the antioxidant activity dramatically decreased in a manner identical to that already encountered in an oil-in-water emulsion. One can also cite the work of Kagan et al. (1990b), who measured the efficiency of ubiquinols of varying isoprenoid chain length (from Q0 to Q10) in preventing (Fe^{2+}-ascorbate)-dependent or (F^{2+}-NADPH)-dependent lipid peroxidation in rat liver microsomes and brain synaptosomes and mitochondria. While in a homogenous solution, the radical scavenging activity of ubiquinol homologues did not depend on the length of their isoprenoid chain; however, ubiquinols with short isoprenoid chains (Q1-Q4) were much more potent than the longer chain homologues (Q5-Q10) at inhibiting lipid peroxidation in membranes. It was found that the antioxidant efficiency of ubiquinols decreases in the order Q1 (6 carbons in the aliphatic chain) > Q2 (10 carbons) > Q3 (14 carbons) > Q4 (18 carbons) > Q5 (22 carbons) > Q6 (26 carbons) > Q7 (30 carbons) > Q8 (34 carbons) > Q9 (38 carbons) > Q10 (42 carbons). The antioxidant efficiency of Q0 is markedly less than that of other homologues.

Likewise, Takahashi et al. (1992) reported that distearate esters of ascorbic acid were less efficient to counteract lipid oxidation mediated by AAPH than monostearate of ascorbic acid when dealing with soybean phosphatidylcholine liposomes. Within the ascorbate diester series, dicaprilate (C8) was more effective than dipalmitate (C16), which was, in turn, more effective than distearate (C18). This pioneering work showed that the more hydrophobic ester is not always the more antioxidant in lipid dispersion. Both the number and the length of aliphatic chains exert a negative impact on the ability of ascorbic acid to counteract lipid oxidation. More recently, we also observed that antioxidant capacity of rosmarinate alkyl esters suddenly collapsed after a critical hydrophobicity in multilayer

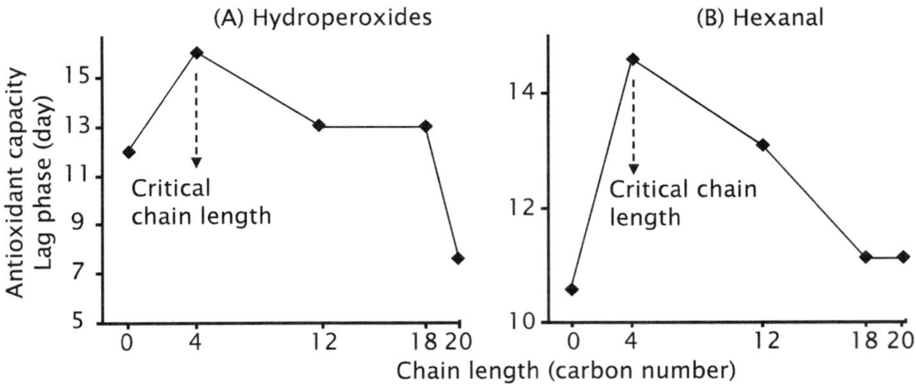

Fig. 8.3. Influence of the alkyl chain length of rosmarinate alkyl esters on antioxidant capacity in chitosan-coated liposomes. Antioxidant activity is expressed as lag phase exit (day) for lipid hydroperoxides (A) and hexanal (B).

Adapted from Panya et al. (2010).

liposomal systems (Panya et al., 2010). This collapse is highly reminiscent of a cut-off effect. It has been observed in chitosan-coated liposomes that the order of antioxidant effectiveness toward lipid oxidation was: butyl rosmarinate > free rosmarinic acid > dodecyl rosmarinate ~ octadecyl rosmarinate > eicosyl rosmarinate (Fig. 8.3). Note that, unfortunately, octyl ester (corresponding to the critical chain length in oil-in-water emulsions) has not been assessed in this study. And yet, better activity for butyl ester, compared to free rosmarinic acid, and poor antioxidant values for long-chain esters (octadecyl, eicosyl rosmarinates) are highly reminiscent of a cut-off effect. Another case that contradicts the polar paradox has recently been suggested by An et al. (2011) using phosphatidylcholine liposomes and daidzein-7-*O*-alkoxyethers as phenolipid antioxidants. AAPH-induced oxidation of phosphatidylcholine at 43 °C monitored at 234 nm through the formation of conjugated dienes showed that (i) nonlipophilized daidzein and its butyl derivative were prooxidant and, above all, that (ii) dodecyl derivative was a better antioxidant than both octyl and hexadecyl derivatives. The fact that the antioxidant capacity of daidzein-7-*O*-alkoxyethers decreased beyond a critical chain length of 12 carbon atoms clearly supported our assumption that the cut-off effect is a general phenomenon occurring in lipid dispersions.

Cut-Off Effect in Living Cells

What is truly fascinating about the cut-off phenomenon is that it does not seem to be restricted to nonliving systems. Recent results have demonstrated that this nonlinear effect is also encountered when examining antioxidant capacity in cultured cells of animal origin. Using a fluorescent molecule to probe the level of reactive oxygen species in p43 overexpressed human dermal fibroblasts, we found a cut-off phenomenon for chlorogenate (Fig. 8.4) (Laguerre et al., 2011) and rosmarinate (not yet published) alkyl esters with a critical chain length of 12 and 8–10 carbon atoms, respectively. In L6 rat muscles cells, Tofani et al. (2010) also reported a collapse of the antioxidant activity of hydroxytyrosol alkyl esters after a dodecyl chain. In all cases, longer alkyl chain length rendered the corresponding phenolipids almost nonreactive toward reactive oxygen species. We hope that these preliminary results will generate a host of research projects to confirm whether cut-off effect dictates (among other factors) the antioxidant capacity in emulsions, liposomes, and, above all, in living cells. It is also worth mentioning that the cut-off phenomenon is not limited to the sole phenolic family as Kimura et al. (2003) have reported such an effect with acyl ascorbates. In intestinal epithelial cells oxidized by $FeSO_4$, they used the fluorescent probe BODIPY 581/591 C11 to find that dodecyl and tetradecyl ascorbate were more antioxidant than ascorbic acid and its decyl, hexadecyl, and octadecyl esters.

Even more interesting is the fact that this behavior has also been observed for activities other than antioxidant properties. Apart from our results, more than 50 studies have been published so far reporting a cut-off effect for a homologous series

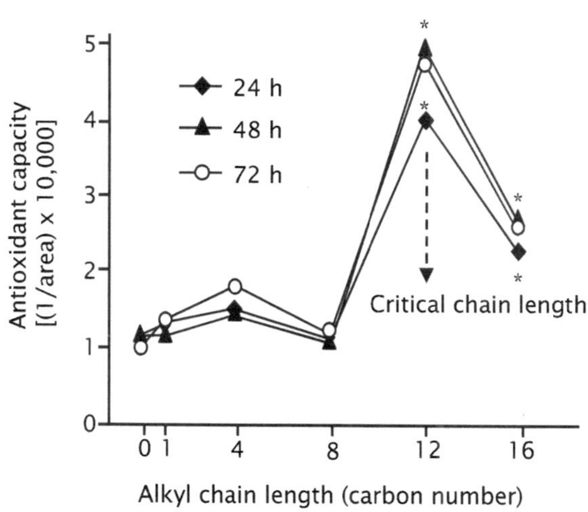

Fig. 8.4. Influence of the alkyl chain length of chlorogenate alkyl esters on antioxidant activity in p43 overexpressing human dermal fibroblasts.

After Laguerre et al. (2011).

of molecules as different as antimicrobials, anaesthetics, antiproliferatives, antivirals, cytotoxics, eye irritants, alcohol intoxication inducers, spermicidal agents, immunosuppressive drugs, antiphotosynthetics, phytotoxics, corneal drugs in rabbits, drugs in mice or in cultured cells, inhibitors of enzymes and receptors, membrane-perturbing (or -destroying) surfactants, environmental pollutants, hydrogen-bond breakers, and finally, protein-binding agents and enzyme inhibitors (for review, see Laguerre et al., 2013). With such a large number of experimental validations, we are now inclined to consider the cut-off effect as a universal phenomenon occurring in various systems, regardless of whether they are living.

The Physicochemistry behind the Cut-Off Effect

If confirmed, the cut-off effect would significantly change our understanding of lipid oxidation and antioxidant chemistry. Indeed, we now have a more global picture of the role of hydrophobicity on antioxidant capacity, especially for highly lipophilic compounds. However, researchers must develop an underlying theoretical framework to relate this large body of empirical observations, with the goal of fully describing both the phenomenology and the mechanisms of antioxidant action in lipid dispersions. This section attempts to account for the cut-off effect, and it is by far the most speculative part of the chapter.

In a recent review (Laguerre et al., 2013), we developed three putative mechanisms of action to account for the cut-off phenomenon: the "reduced mobility," the "internalization," and the "self-aggregation" hypotheses. Briefly, the reduced mobility hypothesis suggested that the mobility of an antioxidant decreases as its alkyl chain lengthens, consequently decreasing its ability to move toward numerous oxidation sites. The *internalization hypothesis* makes the assumption that increasing the hydrocarbon chain from medium to long chains could drive the antioxidant away from the interface (where oxidation is supposed to primarily occur) into the lipid core of emulsions and micelles or into the membrane interior of liposomes and cells, where an antioxidant would be a poor antioxidant. Finally, the self-aggregation hypothesis proposed that beyond the critical chain length, the antioxidant capacity collapse occurs because of antioxidant self-aggregation and the fact that long-chain antioxidants mainly exist as colloidal aggregates. This self-aggregation could have two drawbacks that would have negative impact on antioxidant activity. First, self-aggregation removes the antioxidant from the interface where oxidation is most prevalent. Second, micellization makes long-chain antioxidants bulkier than free molecules, which obviously makes them less mobile toward the oxidizable substrate, free radicals, and transition metals. The last point shows that the self-aggregation and the reduced mobility hypotheses are not in opposition, but rather are complementary. In a sense, the reduced mobility is a consequence of self-aggregation. To better grasp the dependence of the different hypotheses, we used this chapter to fuse them into two new approaches based exclusively on location and diffusion.

Before entering in more details, hereafter we are only taking into account cases in which the grafting of lipid moiety did not supposedly affect the intrinsic antioxidant activity of the molecule. That is why we mainly focused our study on homologous series of antioxidants whose lipid moiety could not induce any mesomer effect with the antioxidant head (mainly, the phenolic ring). In theory, we cannot avoid some scenarios in which the grafted chain may still affect reactivity through its coiling, which may shield the phenolic groups responsible for the antioxidant reactivity. Another scenario could be the modification of the intra- and intermolecular H-bond network between water and phenolic groups through self-aggregation. For the sake of clarity, we will not discuss such theoretical possibilities.

The Location-Based Approach: Predominant Effect of Antioxidant Location

The location-based approach implies that antioxidant location primarily determines antioxidant activity. Presently, most of the research done in the field takes this approach. As an attempt to predict antioxidant activity, the location-based approach may help to elucidate the cut-off mechanism of action. In such a case, the cut-off effect would be mainly due to the effect of hydrophobicity on location, which, in turn, would modulate antioxidant activity. Beyond a critical chain length, the antioxidant would be improperly localized toward reactive centers such as free radicals, transition metals, and, above all, oxidizable substrates. Such a "bad" location for long-chain antioxidants would lead to a collapse in their antioxidant activity. This consequence contradicts the polar paradox theory, which assumes that nonpolar antioxidants are better localized than their polar counterparts.

Hypothesis One (Dogma): Long-Chain Antioxidants Are Localized at the Interface

Predicting antioxidant efficiency in lipid dispersions is complex because they contain different chemical environments in different regions. For instance, oil-in-water emulsions have three distinct phases—oil, interface, and water. To our knowledge, Frankel et al. (1994) were the first to show the importance that an antioxidant has at the interface between water and lipids for efficiently reducing oxidant species like free radicals. From a historical perspective, they developed the first location-based approach. To explain Porter's polar paradox, Frankel's interfacial concept makes two hypotheses. For the sake of clarity and to further explain our point, we will consider each hypothesis independently.

(i) The first hypothesis states that increasing hydrophobicity (all other factors being equal) would lead to more partitioning into the interface. Thus, polar antioxidants such as Trolox would be mainly localized in the water phase, while nonpolar homologues such as α-tocopherol would be mainly concentrated at the interface.

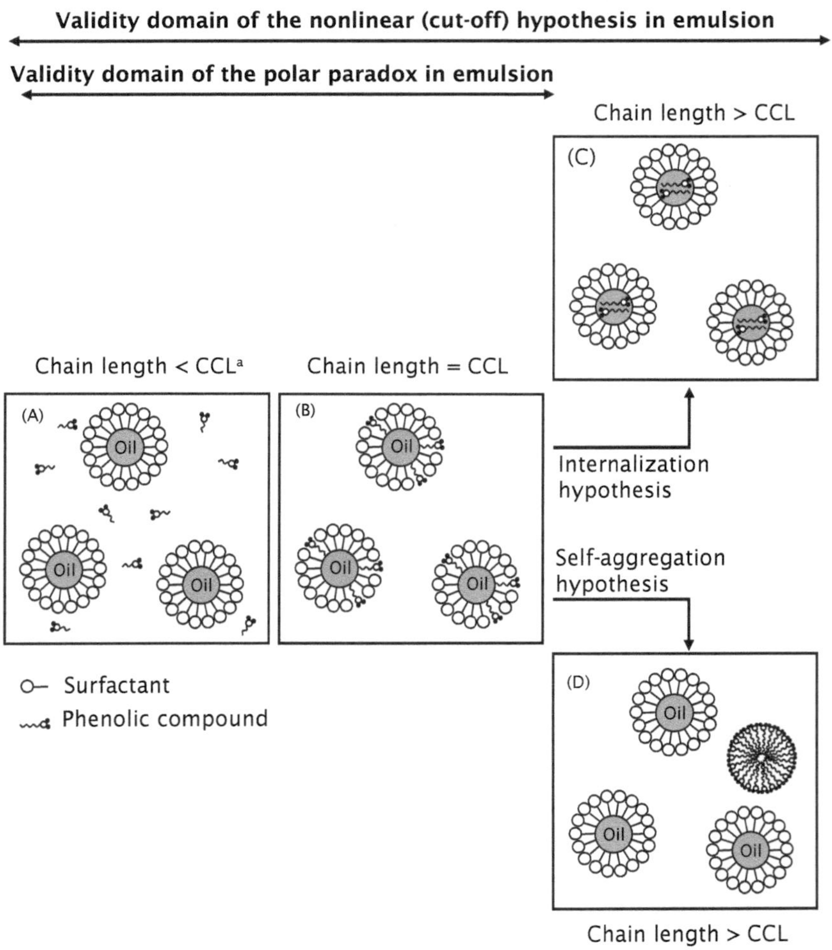

Fig. 8.5. Putative scheme of the distribution of antioxidant in emulsified system (location-based approach). CCL : Critical chain length.

(ii) The second hypothesis of Frankel's interfacial concept is that the interfacial region is a crucial site where the initiation of lipid oxidation is supposed to occur. This said, an interfacial location would enhance antioxidant properties compared to an aqueous location, where antioxidants are thought to be less effective.

To be clear, we do not question the second hypothesis because, to our knowledge, no one has ever produced experimental data contradicting this scenario. On the contrary, we think that the first assumption (the less polar, the more accumulated at the interface) is not always valid, especially for long-chain antioxidants.

From a thermodynamic point of view, entropy drives the transfer of an antioxidant from the aqueous phase to the interface. When increasing the lipophilicity of an antioxidant, the entropy penalty resulting from the mismatch between the hydrophobic part of the antioxidant and the surrounding water molecules becomes more and more important. This entropic penalty will overwhelm the energetic contribution (ΔH), allowing antioxidant solubilization in the water phase. To minimize the entropic penalty, the system favors the transfer of antioxidants from the aqueous phase to the interface.

If this reasoning holds for simple binary systems containing water and a surfactant, what about more complicated systems, such as lipid dispersions (oil-in-water emulsions and liposomes)? What if, beyond a critical chain length, long-chain antioxidants are less concentrated in the lipid–water interface than medium-chain antioxidants for reasons that are not yet fully understood? This would contradict the first hypothesis of Frankel's interfacial concept. Yet, recent results obtained from a broad array of partitioning and tensiometry measurements suggest this surprising conclusion: If the length of the aliphatic chain is too short or too long, antioxidants like phenolipids may not position themselves at the proper location (the interface of lipid particles), and antioxidant effectiveness will consequently decrease.

The two next sections propose the location(s) of long-chain antioxidants, if they are not located at the interface (Fig. 8.5).

Hypothesis Two: Long-Chain Antioxidants Are Localized in the Lipid Phase

In a recent review article (Laguerre et al., 2013), we coin the new concept of the "internalization hypothesis." According to our assumptions, increasing the hydrocarbon chain from medium to long could drive the antioxidant away from the interface into the emulsion droplet core, where a long-chain antioxidant would be a poor antioxidant. Numerous authors have suggested this putative effect. Among others, Medina et al. (2009) used the internalization hypothesis to explain why the octyl ester of hydroxytyrosol exhibits higher antioxidant capacity than the dodecyl ester in fish oil-in-water emulsion. We also put forward this hypothesis to try to rationalize the sudden collapse of rosmarinate esters' antioxidant capacity with long chains in stripped tung oil-in-water emulsions (Laguerre et al. 2010; Panya et al., 2012). Far from solely applicable to emulsified media, the internalization hypothesis may also be applied to micellar and membrane systems, as we will show in the following section.

Thermodynamic Aspects

From a theoretical standpoint, the entropic penalty resulting from the mismatch between the antioxidant hydrophobic tail and the surrounding water molecules actually moves hydrophobic antioxidants from the aqueous phase to the interface. This is at least true for nonionic surfactants for which ionic interactions are impossible. In this case, however, we argue that the interface (composed of nonionic surfactants) is still rich in water molecules intercalating between the hydrophilic groups. Therefore, it might be more thermodynamically favorable for long-chain antioxidants to migrate into the lipid's interior (in the case of micelles or emulsions) or deeper into the lipid membrane (in the case of liposomes, lipoprotein particles, or cells) where virtually no water is present. Consequently, the entropic penalty of the antioxidant hydrophobic chain would tend to zero. But, lipophilized antioxidants also contain a hydrophilic part (that is, a phenolic part) that needs to be stabilized by intermolecular H-bonds, either with water, polyoxyethylene groups of Tweens or Brij, or phosphate groups of phospholipids (energetic contribution, ΔH).

(i) If localized within the membrane or the lipid phase's interior, lipophilized antioxidants will have a low entropic penalty, corresponding to their hydrophobic tail, but a high entropic penalty, resulting from the mismatch between their phenolic head and the lipid chains (unless they form reverse micelles, as described later).

(ii) If localized within the aqueous phase, we suggest that the opposite would take place.

This reasoning suggests two observations. First, the system tends to minimize the sum of these two entropic penalties, which leaves a balance. With a very hydrophobic antioxidant initially located at the interface, the entropic penalty of its hydrocarbon chain overwhelms that of its phenolic head. From a thermodynamical standpoint, the transfer of a long-chain antioxidant from the interface to the lipid phase's interior (for emulsions or micelles) or the membrane (for liposomes, lipoprotein particles, or cells) is favorable. If too hydrophilic, an antioxidant would instead undergo the reverse phenomenon, which would transfer it in the aqueous phase. Our second observation is that an intermediate situation may exist with medium-chain antioxidants. A thermodynamic trade-off, wherein the two entropic penalties of a medium-chain antioxidant are conciliated in the interfacial surfactant monolayer of emulsions and micelles or in the outer layer of membranes, should exist. We propose three different locations, depending on the chain length of the lipophilized antioxidant: an aqueous location for nonlipophilized or short-chain antioxidants, an interfacial location for medium-chain antioxidants, and a lipid core location for long-chain antioxidants. The latter assumption corresponds *stricto sensu* to the recently advanced internalization hypothesis (Laguerre et al., 2013).

Theory versus Experimental Data in Emulsions

Some of our partitioning data show that this hypothesis cannot always explain the cut-off effect of chlorogenate alkyl esters because long-chain chlorogenates (octadecyl and eicosyl) were equally or even more concentrated (depending on the emulsifier's concentration) in the aqueous phase of an emulsion than medium-chain esters (Laguerre et al., 2009) (Fig. 8.6).

On the contrary, results from rosmarinate alkyl esters strongly support the internalization hypothesis by showing that eicosyl rosmarinate (20 carbons) is mainly localized in the oil droplet, while butyl and dodecyl rosmarinates are preferentially localized at the interface of a stripped soybean oil-in-water emulsion (Panya et al., 2012). Adding an emulsifier in excess (that is, more than is necessary to saturate the emulsion droplet's surface) leads to the formation of micelles, which can solubilize eicosyl rosmarinate out of the emulsion droplet's core and carry it to both the aqueous phase and the surfactant interface. This is evident

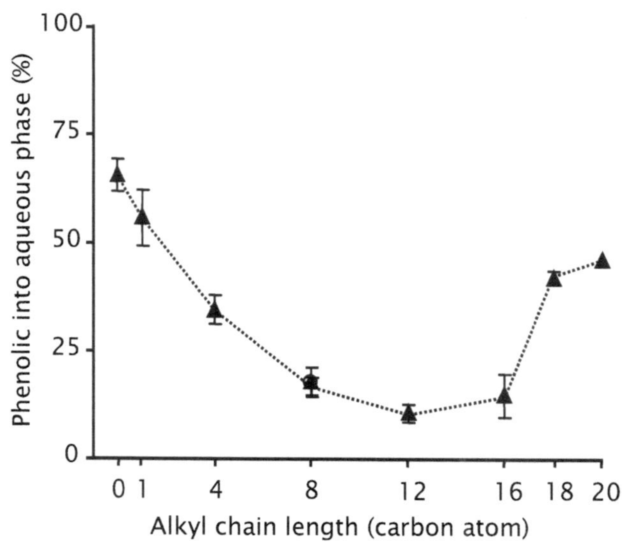

Fig. 8.6. Alkyl chain length effect on the partition behavior of chlorogenic acid and its alkyl esters in a mixture of sunflower oil and phosphate buffer solution at pH 7.2 with 17 μM Brij 35 used as emulsifier.

After Laguerre et al. (2009).

from front-face fluorescence, spectrophotometric analysis [which traps an unreacted arenediazonium ion by *N*-(1-naphtyl)ethylenediamine], and a centrifuge/HPLC partitioning study. The addition of excess surfactant increases the antioxidant capacity of eicosyl rosmarinate in a dose-dependent manner but only slightly decreases that of butyl and dodecyl rosmarinates. Consequently, the cut-off effect observed for medium-chain esters (such as butyl, octyl, and dodecyl esters) in oil-in-water emulsion progressively disappears with the addition of excess surfactant, which allows long-chain esters to move from the oily core into the aqueous phase and, above all, to the interface. Here, they will be more effective at counteracting lipid oxidation (Panya et al., 2012).

Theory versus Experimental Data in Micelles

In micellar systems, Heins et al. (2007) demonstrated with gallic acid and its alkyl esters (methyl, ethyl, propyl, butyl, and octyl esters) that the longer the alkyl chain, the deeper the polar head (phenolic part) would be located in the micelle, especially for SDS and Brij 58. According to these authors, the location of gallate alkyl esters and free radicals determines their antioxidant activity (a location-based approach). As expected, gallate alkyl esters with the deepest internalization within micelles were found to have the lowest stoichiometry for reducing the water-soluble Fremy's radical in SDS and Brij 58 micelles. Even with a lipid-soluble free radical such as galvinoxyl radical, octyl gallate ester showed lower free-radical scavenging capacity than its short-chain homologues in Brij 58 and cetyltrimethylammonium bromide. The deeper that the phenolic is in the micelle, the less antioxidant the phenolic is.

Theory versus Experimental Data in Membranes

Using α-tocopherol homologues with various hydrocarbon chains (1, 6, 11, and 16 carbon atoms for tocopherol itself), Kagan and Quinn (1988) have shown that as the hydrophobic character becomes more dominant, the chromanol nucleus (polar head) has an increasing tendency to leave the aqueous interfacial region of the bilayer and to occupy a position within the membrane's hydrocarbon interior. Kleszczynska et al. (2000) obtained a similar result with a homologous series of pirolidinium bromides and chlorides in an erythrocyte membrane. Using anisotropy fluorescence with two probes (1,6-diphenyl-1,3,5-hexatriene and [1-(4-trimethylammoniumphenyl)-6-phenyl-1,3,5-hexatriene *p*-toluenesulfonate]), they clearly showed a linear relationship between the alkyl chain length of pirolidinium salts and their internalization in the membrane's alkyl chain region. An et al. (2011) have performed another experimental proof in soybean phosphatidylcholine liposomes, showing that the longer the alkyl evidence chain, the deeper it is in the membrane. In this study, the liposomal incorporation of daidzein (D) and its 7-n-alkoxy derivatives (butyl, D4; octyl, D8; dodecyl, D12; and hexadecyl, D16) was monitored by fluorescence of 8-anilino-1-naphthalenesulfonic acid (ANS) for bilayer surface rigidity. The addition of daidzein and its alkyl derivatives to liposomes enhanced the ANS fluorescence, showing that all phenolipids were incorporated in the membrane. The concentration of D and derivatives in the proximity of the membrane surface follows the order D > D4 > D8 > D12 > D16 (Fig. 8.7).

Hypothesis Three: Long-Chain Antioxidants Are Localized in the Aqueous Phase

In 2009, we found that in sunflower oil–in-water emulsion, the partitioning of chlorogenate alkyl esters into the water phase decreases as the chain length increases, up to a dodecyl chain (corresponding to the critical chain length observed for antioxidant capacity), beyond which the corresponding esters (hexadecyl, octadecyl, eicosyl esters) unexpectedly accumulate into the water phase (Fig. 8.6) (Laguerre et al., 2009). Although the curves depicted in Figs. 8.6 and 8.2a are not perfectly symmetrical, the partitioning behavior of chlorogenic acid and its alkyl esters shows the same threshold for antioxidant capacity as for a dodecyl chain length. This result suggests that the better antioxidants are those with the lowest concentration in the water phase, which agrees somewhat with the second hypothesis of Frankel's interfacial concept (discussed earlier). In contrast, the fact that long-chain antioxidants are more concentrated in the aqueous phase than medium-chain ones contradicts the first hypothesis of Frankel's interfacial concept (the less polar the antioxidant, the more concentrated it is at the interface).

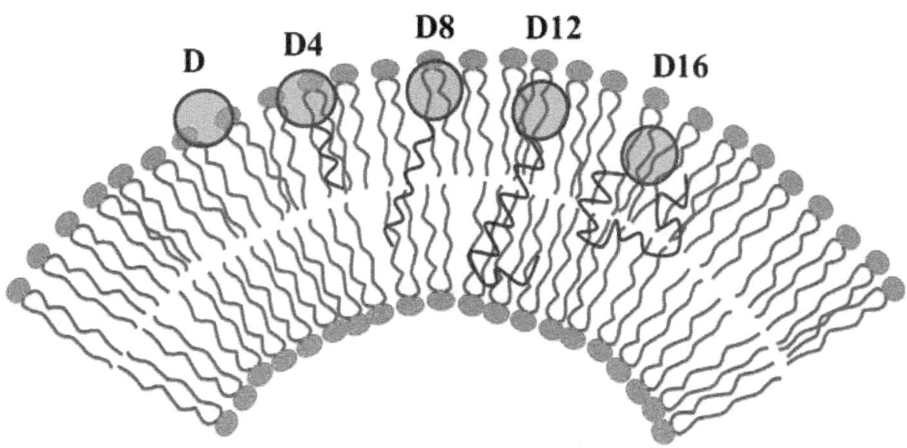

Fig. 8.7. Model of daidzein and its 7-n-alkoxy derivative incorporation into liposome lipid/water interface (location-based approach).

After An et al. (2011).

At low emulsifier concentration (that is, below the emulsifier's CMC), the only physicochemical process that can accumulate oil-soluble antioxidants in water is a self-aggregation of the antioxidant. From this result, we made the assumption (Laguerre et al. 2009), later referred to as the "self-aggregation hypothesis" (Laguerre et al. 2013), that beyond the critical chain length, antioxidant capacity collapses due to the antioxidant's self-aggregation and the fact that long-chain antioxidants mainly exist as aggregates in the water phase of an oil-in-water emulsion (Figure 8.5). We assumed that this tendency for longer chains to form aggregates becomes greater than the tendency to move toward the interfacial membrane; thus, antioxidant concentration at the site of oxidation decreases. Consequently, the micellization process could have two major drawbacks for the long-chain antioxidants' ability to counteract lipid oxidation. First, micellization leads to the removal of the antioxidant from the interface where oxidation is most prevalent (a location-based approach). Second, micellization makes long-chain antioxidants much bulkier than free molecules, which obviously makes them less mobile toward the oxidizable substrate, free radicals, and transition metals (a diffusion-based approach). We will discuss the last point on mobility/diffusion later. In this section, we exclusively focus on the hypothetical effect of antioxidant self-aggregation on its location.

Evidence for Self-Aggregation of Long-Chain Antioxidants

Scientists working on antioxidants, such as tocopherols, tocotrienols, phenolipids and long-chain ascorbates, often overlook the fact that these molecules have a long aliphatic chain and an adequate polar head that produce separate hydrophilic and hydrophobic portions. This structure could allow these amphiphilic molecules to self-assemble into micelles, lamellar structures, and other association colloids.

Among natural phenolipids, resorcinolic lipids have the most precisely defined amphiphilic properties. They form stable monomolecular layers at the air–water interface (Kozubek, 1989, 1995), where the dihydroxybenzene rings are oriented perpendicular to the surface of the subphase (Kato et al., 1990). According to Stasiuk and Kozubek (2010), resorcinolic lipids have very low values for critical micelle concentration. The CMCs determined for different homologues in neutral pH through the solubilization of 1,6-diphenyl-1,3,5-hexatriene were in the range of 4.5–8.5 μM, depending on the length and saturation of the hydrocarbon chains. The CMCs obtained for long-chain homologue phenolipids by measuring the surface pressure were lower (0.5–2.6 μM) (Gulati et al., 1964). It has also been reported that under alkaline conditions, cardol and methylcardol form liposomal structures alone as well as when mixed with cholesterol, fatty acids, or phosphatidylethanolamine (Stasiuk & Kozubek, 2010). Moreover, these authors reported that the lipid aggregates of 3-pentadecylphenol in aqueous solution have a micellar character. Kagan et al. (1990a) reported that α-tocopherol homologues existed in the lipid bilayer not only in a monomeric form but also in clusters, thus decreasing the efficiency of radical scavenging. To our knowledge, this is the first report that the self-aggregation of lipophilic antioxidants may lower their antioxidant capacity. The second important point is that the self-aggregation observed by Kagan et al. occurred in the membrane, probably as reverse micelle-like aggregates.

For synthetic phenolipids, Lucas et al. (2010) have recently determined the CMC of hydroxytyrosol alkyl esters by measuring surface tension. They were found to have very low CMCs for long-chains esters—namely, 2, 3.5, 5.5 μM for hexadecyl, tetradecyl, and dodecyl esters, respectively. As expected, decreasing the chain length drastically increases the CMC—90, 380, 1200, 1500, and 3000 μM for decyl, octyl, hexyl, butyl, and ethyl esters, respectively (Lucas et al., 2010). Other phenolipids—such as tyrosol alkyl esters (Lucas et al., 2010), gallates, glucosyl gallates, and glucuronosyl gallates (Maldonado et al., 2011)—have been recently found to self-assemble into micelles. In addition, they showed that γCMC (that is, the surface tension at the CMC) of hydroxytyrosol alkyl esters suddenly increases after reaching critical chain length for an alkyl chain of 8–10 carbon atoms (Lucas et al., 2010). Interestingly, this finding corresponds to the critical chain length (octyl chain) already found by the same team regarding the antioxidant properties of hydroxytyrosol alkyl esters measured in fish oil-in-water emulsion (Medina et al., 2009). This sudden increase in γCMC after attaining a threshold in hydrophobicity has also been observed for alkyl gallates, glucosyl alkyl gallates, and glucuronosyl alkyl gallates (Maldonado et al., 2011), although their antioxidant properties in emulsion have not yet been measured. Consequently, long-chain phenolipids are no more localized at the air–water interface; otherwise, γCMC should have continued to decrease. Again, this suggests that long-chain phenolipids self-aggregate and return to the water phase as micelles or other associated colloids. Finally, Bayrasy et al. (not yet published) recently used the dynamic light scattering technique to demonstrate that rosmarinate alkyl esters formed microscopic self-assembled aggregates (~200–1000 nm) at concentrations well above the CMC. Taken together, these data tend to support the fact that long-chain antioxidants self-assemble into various aggregate types and morphologies depending on the geometry of the antioxidant (Israelachvili's packing parameter), environmental conditions (especially ionic strength, pH, and temperature), and, above all, antioxidant concentration. A given set of conditions and concentrations logically corresponds to a given aggregate.

In parallel to phenolipids, other long-chain antioxidants have been shown to undergo self-aggregation in water. There are plenty of scientific reports clearly showing that ascorbic acid alkyl derivatives can easily produce supramolecular aggregates, such as monolayers (Capuzzi et al., 1996), micelles (Wen et al., 1998; Lo

Nostro et al., 2000; Palma et al., 2002a, 2002b), liposomes (called "aspasomes," Gopinath et al., 2004), and even nanotubes, in particular cases (Ambrosi et al., 2006). Importantly, the shape and the structure of the various supramolecular aggregates are not fixed. As for phenolipids, the aggregates formed by ascorbic acid esters constantly evolve under the complex influence of their chemical geometry and environment. According to Palma et al. (2002a), aqueous dispersions of 6-*O*-ascorbic acid alkanoates form coagels at room temperature and low water content (10% w/v). These hydrated crystalline phases transform into micellar solutions or gel phases, depending on the length of the hydrophobic chain in the surfactant. When heated, the shorter ascorbyl alkanoates ($n \leq 10$) exhibit a coagel-to-micelle phase transition. By contrast, the longer homologues ($n \geq 11$) exhibit a coagel-to-gel transition. The temperature, ΔH, and ΔS of the transitions depend on the length of the aliphatic side chain. The different phase behavior is related to the interplay of the competing molecular interactions that involve the amphiphiles' hydrophobic chains, their headgroups, and the solvent. In another case, nanotubes have been found to form from vitamin C–based bola-amphiphile (bolaform)—that is, 1,12-diascorbyl dodecanedioate—with ascorbic acid units as the polar headgroups (Ambrosi et al., 2006). Once dispersed in water above 0.5% w/w, this bola-amphiphile forms hollow nanotubes, as revealed by cryo-TEM experiments. These nanostructures transform into clear micellar solutions when heated.

Theory versus Experimental Data: Where Do These Aggregates Partition?
Studying the cut-off effect in emulsified systems also challenges the self-aggregation hypothesis. Indeed, some authors, including us, have not always found long-chain phenolipids in the aqueous phase of an oil-in-water emulsion. However, theoretically, if long-chain phenolipids are self-assembled into micelles (the postulated basis for discrimination), one would expect that these compounds are localized in the aqueous phase. Sørensen et al. (2012b) did not detect octyl or oleyl dihydrocaffeate esters in the aqueous phase of an oil-in-water emulsion. The same result was observed when studying the partitioning behavior of rosmarinic acid fatty esters. For the lowest concentration of emulsifier (17 μM Brij 35), no ester with a chain equal to or longer than 8 carbons was detected in the aqueous phase of a sunflower oil-in-water emulsion. This result suggests that although the self-aggregation hypothesis explained the cut-off effect for chlorogenate esters, it failed to explain the cut-off effect of the chain length of rosmarinate esters (Laguerre et al., 2010).

Yet, a recent re-examination of the partitioning methodology by Sørensen et al. (2012b) pointed out that the absence of long-chain phenolipids in the water phase may be due to a limitation of the applied method (based on the filter) rather than a failure of the self-aggregation hypothesis. Indeed, in their partitioning study on emulsion, they separated the aqueous phase from the discontinuous phase (that is, surfactant plus oil) using ultracentrifugal filter devices (with a regenerated cellulose 3,000 MW cut-off). Micelles or aggregates that may form could potentially be retained in the tube filter and not measured in the aqueous phase collected downstream. Although the partitioning methodology used by Laguerre et al. (2010) did not involve an ultracentrifugal filter, the supernatant (aqueous phase) obtained after the phase separation was filtered through a 0.22 μm-filter prior to HPLC analysis. Again, this is a reasonable explanation of the absence of observed long-chain phenolipids in the water phase. At that time, research had neglected the phenolipids' self-assembly properties well above micellization. As mentioned earlier, we now know that phenolipids, such as long-chain rosmarinate alkyl esters, are able to self-assemble in much bigger objects than micelles with a size ranging between 500 and 1000 nm (not yet published). These large aggregates did not pass through the filter; thus, partitioning studies using either ultracentrifugal filter phase separation or filtration prior to HPLC may obtain a false negative for the concentration of long-chain antioxidant in the aqueous phase. Consequently, further partitioning studies in this field should avoid a filtration step. A suitable and promising alternative is the method based on trapping an unreacted arenediazonium ion (16-ArN_2^+) using *N*-(1-naphtyl)ethylenediamine to form an azo dye that absorbs at 572 nm (Sánchez-Paz et al., 2008). Because 16-ArN_2^+ (localized at the interface) reacts with an antioxidant, the detection of unreacted 16-ArN_2^+ using spectrophotometry allows us to evaluate the partitioning of antioxidants at the interface (Panya et al., 2012).

Unexpected results on the role of surfactant concentration in antioxidant capacity suggest another challenge to the self-aggregation hypothesis, as it was initially formulated. Above a certain concentration, the surfactant added to an oil-in-water emulsion saturates the interface and starts to self-assemble into micelles in the aqueous phase. These micelles may act as carriers for antioxidants, transferring them from the interface to the aqueous phase (Richards et al., 2002). Apart from nonlipophilized chlorogenic and rosmarinic acids, all phenolics have a clear tendency to accumulate in the aqueous phase with an increase in the surfactant concentration. However, if long-chain phenolipids were already self-assembled in the aqueous phase, as proposed by the self-aggregation hypothesis, how do we explain that an excess of surfactant leads to an increase of the long-chain antioxidant content in the aqueous phase? This has been repeatedly observed for both chlorogenate and rosmarinate esters in various oil-in-water emulsion systems (Laguerre et al., 2009, 2010; Panya et al., 2012). The reason for this inconsistency is not yet clear. Likely, the situation described in the initial formulation of the self-aggregation hypothesis is oversimplified and needs sophistication, as we suggested from the beginning (Laguerre et al., 2009).

First, we should keep in mind that we are talking about statistical distribution of many antioxidant molecules in a dynamic system that evolves constantly through material exchange between different phases, collision and fusion of lipid and surfactant phases, etc. Moreover, antioxidants are far from having a 100% specificity for a given phase. For the sake of clarity, we simplified the partitioning by hypothesizing, for instance, that long-chain antioxidants are localized in the aqueous phase. Of course, it is more accurate to say that a significant proportion of self-assembled long-chain antioxidants are localized in the water phase, while non-negligible proportions of these antioxidants (self-assembled or not) remain in other parts of the system, such as the interface and oil.[4] In other words, the problem is quantitative and not just qualitative. As for surfactants, equilibrium exists between their self-aggregated and free forms. Within the self-aggregated forms, it is conceivable that reverse micelles of long-chain antioxidants form in the oil droplet interior of an emulsion. These putative reverse micelles will parallel the micelles of long-chain antioxidants in the aqueous phase and may exchange materials. Obviously, this is a pure speculation that needs to be tested. In this case, it is conceivable that an excess of surfactant leads to an increase of the long-chain antioxidant content in the aqueous phase.

Finally, if we have no doubt that long-chain antioxidants can self-assemble, it is much more difficult to conclude the distribution pattern of such aggregates in lipid dispersions. Further experiments will have to clarify this important question.

The Diffusion-Based Approach: Predominant Effect of Antioxidant Diffusivity

The diffusion-based approach implies that antioxidant activity is primarily affected by the diffusivity of antioxidants. Although leading research teams such as Etsuo Niki's and Valerian E. Kagan's groups use the diffusion/mobility concept, the diffusion-based approach is much less common today than the location-based one when trying to account for antioxidant capacity. We will provide hypothetical examples that show how this approach may benefit the field when trying to explain the cut-off effect. Accordingly, the cut-off effect would be due to an effect of hydrophobicity on the antioxidant's molecular diffusivity. In turn, antioxidant diffusivity modulates antioxidant activity. Beyond a critical chain length, an antioxidant would move too slowly toward reactive centers (examples include metal transition, free radicals, and, above all, oxidizable substrates), which would lead to a collapse in the antioxidant's activity. Here, we postulate that such reduced diffusivity may be due to a rise in hydrophobic interactions between long-chain antioxidants and lipid phases (hypothesis one) and/or the self-aggregation tendency of these long-chain antioxidants (hypothesis two).

[4]This remark is valid for all location-based theories, such as Frankel's interfacial concept as well as the internalization and self-aggregation hypotheses.

Hypothesis One: Hydrophobic Interactions of Long-Chain Antioxidants Reduce Their Diffusivity

In a recent review article (Laguerre et al., 2013), we coined a new concept, the reduced mobility hypothesis, with the idea that the mobility of the lipophilic antioxidant decreases as its alkyl chain lengthens, consequently decreasing its ability to move toward numerous oxidation sites. Increased hydrophobic interactions between the alkyl chains of antioxidants and their hydrophobic environment (that is, their protein and surfactant hydrophobic part, the hydrophobic surface and core of the oil droplet, and the phospholipid membrane) may lower the diffusivity of long-chain antioxidants to the reaction centers, that is, oxidizable substrates, free radicals, and transition metals. Let us consider first that an antioxidant needs to be mobile in the lipid–water interfaces to meet free radicals and, second, that each time an antioxidant interacts with molecules through hydrophobic interaction, it is called hydrophobic contact. The reduced mobility hypothesis suggests that the frequency of hydrophobic contact will increase for long-chain antioxidants, compared to medium chains, as long-chain antioxidants are more strongly bound to the molecular environment by hydrophobic interactions. Thus, they have a lower degree of freedom. This may result in reduced antioxidant activity.

The Reduced Mobility Hypothesis: An Old Idea That Makes Its Way

The putative effect of aliphatic chain length on the mobility of antioxidants in lipid dispersions has been proposed for decades now. In 1985, Niki et al. suggested that the intermembrane mobility of α-tocopherol containing a phytyl chain is reduced in liposomal systems compared to 2,2,5,7,8-pentamethyl-6-chromanol, in which a methyl group has replaced the phytyl chain. According to the authors, the phytyl chain would enhance the retainment of α-tocopherol in liposomes; hence, it would suppress the ability to transfer between liposomal membranes. One year later, Castle and Perkins (1986) pointed out the importance of the intermicellar diffusion rate in the antioxidant capacity of a hydrophobic antioxidant such as α-tocopherol in SDS micelles. In rat liver microsomes, along with brain synaptosomes and mitochondria, Kagan et al. (1990b) showed that when ubiquinol homologues have longer isoprenoid chains, they also have the fewer antioxidants. Among other explanations (which include partitioning behavior and nonuniform distribution through clusterization), the differences in the antioxidant potency of ubiquinols in membranes were suggested to result from differences in intramembrane mobility. The longer the alkyl chain, the less mobile the antioxidant. Likewise, Takahashi et al. (1992) concluded that the intra- and intermembrane mobilities of ascorbic acid alkyl esters in liposomal systems decreased with the increasing number and length of fatty-acid chains. With gallate alkyl esters, Stöckmann et al. (2000) postulated that increased hydrophobic interactions (for longer alkyl chain) may lower the diffusion of gallates into an SDS-enriched environment, resulting in reduced antioxidant activity. In 2010, we used the same interpretation to explain why long-chain rosmarinate alkyl esters exhibited much lower antioxidant capacity than their short- and medium-chain homologues in a stripped tung oil-in-water emulsion stabilized with Brij 35 (Laguerre et al., 2010) and chitosan-coated liposomes (Panya et al., 2010). The same year, Tofani et al. (2010) came to a similar conclusion to explain the cut-off effect of hydroxytyrosol alkyl esters on antioxidant capacity in L6 rat muscle cells. They assumed that at a certain level of lipophilicity, the easy diffusion of esters into the cells could be balanced (for example, in hydroxytyrosol decanoate) or even made unproductive (in hydroxytyrosol dodecanoate or octadecanoate) by trapping them in the plasma membrane. This latter effect could be caused by the higher affinity of long acyl chains to phospholipids or the hydrophobic proteins inside the bilayer.

Theory versus Experimental Data in Micelles

The hypothesis that alkyl chain length may lower mobility in lipid phases and in the vicinity of lipid–water interfaces has been fully documented. In micellar solutions of conventional surfactants, either ionic, nonionic or zwitterionic, research has consistently found that the entry of a surfactant into its micelles is nearly controlled by diffusion, with values of the entry rate constant k^+ in the 10^8–10^9 M^{-1} s^{-1} range for surfactants with an alkyl chain containing between 14 and 18 carbon atoms, depending on the surfactant (Zana, 1996, 2005). The value of k^+ decreases very slowly as the surfactant chain length increases. On the contrary, the rate constant k^- for the exit of a surfactant from its micelles decreases exponentially when the carbon number of the surfactant alkyl chain increases by a factor close to 3 per additional methylene group (Zana, 1996, 2005). This factor corresponds to an increment of about 1.1 kT per methylene group in the free energy of surfactant transfer from a micelle to the aqueous phase. To summarize, the discrimination between medium- and long-chain surfactants only occurs for the exit of a surfactant from its micelle, not from the entry.

Theory versus Experimental Data in Membranes

Apart from micelles, the effect of chain length on mobility within and between synthetic or biomembranes has been well studied (Niki et al., 1988; Gotoh et al., 1996). For example, the short-chain α-tocopherol homologue C_1 exhibited high transbilayer mobility, whereas the long-chain homologue C_{16} (α-tocopherol) underwent no transbilayer migration within minutes (Kagan et al., 1990b). The absence of the so-called "flip-flop" (that is, transbilayer migration) for α-tocopherol has already been demonstrated by the same group (Tyurin et al., 1987). Indeed, the interaction of α-tocopherol with liposomes leads to its incorporation only into the outer monolayer, in which it remains without penetrating the inner monolayer. Likewise, some researchers have reported that 5-hexadecanoylaminofluorescein stays predominantly in the outer membrane leaflet of epithelia and does not pass through tight junctions at 20 °C, whereas the dodecanoyl derivative can "flip-flop" to the inner leaflet and may also pass through tight junctions (Dragsten et al., 1981; Tournier et al., 1989).

Using the two extremes (ubiquinol-1 and ubiquinol-10) of the homologous series employed by Kagan et al. (1990b), Niki (1997) brilliantly demonstrated that the side-chain of the lipophilic antioxidant is important for its incorporation and retainment in biomembranes and lipoproteins because it can drastically slow down intermembrane mobility. Niki used a mixture of two liposomes: soybean phosphatidylcholine (PC) and dimyristoyl phosphatidylcholine liposomes. The antioxidant activities of ubiquinol-10 and ubiquinol-1 against the oxidation of soybean PC liposomes are shown in Fig. 8.8. Only soybean PC liposomes were oxidized to produce PC hydroperoxide. Ubiquinol-10 incorporated into the soybean PC membranes efficiently suppressed oxidation, while ubiquinol-10 incorporated into dimyristoyl PC liposomal membranes did not exert substantial antioxidant

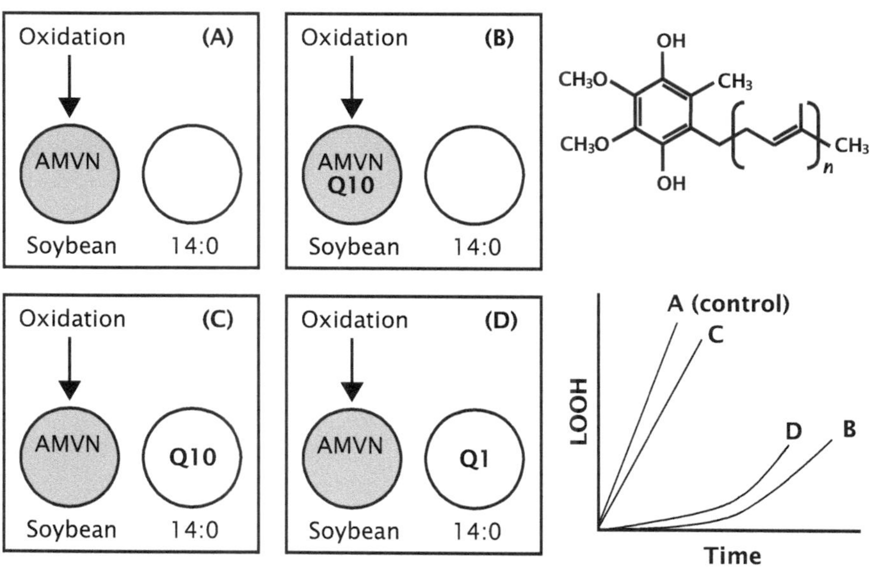

Fig. 8.8. Impact of the isoprenoid chain length of ubiquinol (10 μM) on its antioxidant capacity and intermembrane mobility in a mixture of two types of multilayer liposomes (soybean PC or dimyristoyl PC) at 37 °C in air. A lipophilic free radical initiator, 2,2'-azobis (2,4-dimethylvaleronitrile) (AMVN, 1 mM) inducing lipid oxidation is only added to soybean PC liposomes. Q1 (ubiquinol with $n = 1$); Q10 (ubiquinol with $n = 10$).

Adapted from Niki (1997).

activity against the oxidation of soybean PC liposomes. However, ubiquinol-1 in the dimyristoyl PC liposomes could inhibit the oxidation of soybean PC liposomes as efficiently as ubiquinol-10 present in the soybean PC liposomes (Fig. 8.8). Since the two types of liposome were mixed together, these results clearly show that the mobility of ubiquinol-10 between the membranes (intermembrane mobility) is almost totally restricted, whereas ubiquinol-1 moves readily between them. Thus, experiments have demonstrated the impact of the chain length on the inter-membrane mobility of antioxidants.

Hypothesis Two: Self-Aggregation of Long-Chain Antioxidants Reduces Their Diffusibility

The rise of hydrophobic interactions between long-chain antioxidants and hydrophobic materials is not the only physicochemical method to cause these antioxidants to move less than medium-chain ones. In 2009, we pointed out that a self-aggregation of long-chain antioxidants, such as chlorogenate alkyl esters, may cause a major drawback for their antioxidant capacity by reducing their mobility toward crucial reaction centers in aqueous phase (Laguerre et al., 2009). Self-aggregation is not only supposed to affect the location of antioxidants (based on a location-based approach), it is also expected to strongly impact their mobility/diffusivity in water (a diffusion-based approach). In an aqueous environment, self-aggregation (starting well below CMC) likely decreases the mobility of long-chain antioxidants by making them too bulky: The more aggregated they are, the less mobile they are in aqueous media. To be clear, individual antioxidants (that is, "monomers") are more mobile than "dimers," which, in turn, are more mobile than micelles, and so on. What is interesting is that this rule may introduce a way to discriminate medium- and long-chain antioxidants. It is logical to consider that long-chain antioxidants are less diffusible in the water phase than their medium-chain counterparts because long-chain antioxidants would be in more aggregated forms at the same molecular concentration. Considering that self-aggregation of long-chain antioxidants is well documented, such discrimination on the basis of their diffusivity in water might be responsible for the cut-off effect in lipid dispersions. In contrast, as we will address later, the situation becomes more complicated in a lipid environment as aggregate size may come into play.

Theory versus Experimental Data

Today, it is well recognized that in a model system of surfactants, where molecules occur in either a free (monomeric) or bound (aggregated) form, the latter is much less mobile than the former. Above the CMC, the self-diffusion coefficient (of micelles) declines drastically when micelles are formed in the solution, as Söderman et al. (2004) have reported.

Is that to say that the decrease of self-diffusion is an all-or-nothing behavior occurring suddenly at the CMC or rather that it is an ongoing decline that may begin when even two surfactants (or antioxidants) associate themselves to generate the so-called "dimers"? We strongly believe in the second option. Omitting high overall concentration, when intermicellar interactions (such as collision and aggregation) complicate the situation, it should be noted that a marked decrease in the self-diffusion coefficient was experimentally observed below the CMC (Lindman et al., 1982; Griffiths et al., 1997; Annuziata et al., 1999; Siderius et al., 2002; Zuev et al., 2011). According to Zuev et al. (2007), this phenomenon may be caused by the premicellar association of surfactant molecules, especially under the form of "dimers" (which are noncovalently bound). These authors measured the self-diffusion coefficient of SDS at different ionic strengths and SDS concentration using NMR. They observed a collapse of self-diffusivity after reaching a certain concentration. According to Zuev et al. (2007), this critical concentration corresponds to the formation of "SDS dimers," which are less mobile than SDS "monomers."

Diffusivity of Antioxidant Aggregates in Lipid Phases

It is important to point out that although self-aggregation (such as micellization) would decrease diffusivity of long-chain antioxidants in an aqueous environment, it is much more difficult to speculate in the case of a hydrophobic medium. Indeed, would not the fusion/fission mechanism between lipid phases accelerate rather than stop the diffusion of supramolecular aggregates of long-chain antioxidants in lipid phases? Future contributions will have to balance, on one hand, the fusogenicity of aggregates for lipid phases (that is, their ability to fuse with lipid objects), and, on the other hand, the loss of self-diffusivity in water for these aggregates. Finally, aggregate size is another structural characteristic that may come into play. Aggregates larger than a theoretically critical aggregate size may penetrate very slowly through liposomal or biological membranes. There are many investigations on nanotechnology that show such a size-dependent internalization of aggregates/particles (Rejman et al., 2004; Grosse et al., 2005; von Gersdorff et al., 2006; Wang et al., 2010; Busch et al., 2011).

Conclusion

Until recently, it was widely accepted in the field of lipid oxidation and antioxidant chemistry that nonpolar antioxidants tend to be more active in lipid dispersions than their polar counterparts, as postulated by the polar paradox. However, when scrutinizing the data, one cannot help but notice a significant number of contradictions to this theory. In this context, we have proposed an alternative hypothesis that relies on a nonlinear relationship between antioxidant activity and hydrophobicity. Indeed, we have reported that in oil-in-water emulsions, liposomes, and human dermal fibroblasts, the activity of a homologous series of antioxidants increases as the alkyl chain is lengthened, with a threshold for a medium chain (typically, 8–12 carbon atoms). Further chain extension leads to a collapse of this activity. This nonlinear trend, also referred to as the cut-off effect, has been observed with various homologous series (such as chlorogenate, rosmarinate, hydroxytyrosolate, ferulate, gallate, and ascorbate alkyl esters, along with daidzein-7-*O*-alkoxyethers) by independent research groups in numerous compartmentalized systems, regardless of whether they are living or not.

In this chapter, we dichotomized the effect of location and diffusion on antioxidant activity. A location-based approach implies that hydrophobicity dictates the location of antioxidants, which, in turn, primarily determines their antioxidant activity. On the contrary, the diffusion-based approach implies that antioxidant activity is primarily affected by diffusivity, which, in turn, is impacted by the respective hydrophobicity of antioxidants. Currently, the diffusion-based approach is much less common than the location-based approach when accounting for antioxidant capacity. While the theoretical background of the polar paradox is mainly based on location, we tried to combine both approaches to explain the cut-off effect.

The major drawback of listing the possible cut-off mechanisms of action is to set up an artificial opposition between location and diffusion. We separated these two concepts for the sake of clarity, even though they are intricate and interdependent. The frontier between them is fuzzy, if not invisible. In compartmentalized systems, a difference in the diffusion coefficient may be accompanied by a difference in location. An antioxidant with poor plasmic membrane diffusibility stays in the extracellular medium and cannot be located in the cytoplasm of a living cell. In the case of ubiquinol, a difference in isoprenoid-side chain length results in a difference in intermembrane diffusibility (Fig. 8.8), which necessarily leads to a difference in location. To be clear, location and diffusion are two facets of the same concept, which is exclusively based on intermolecular interactions. Long-chain antioxidants are hydrophobic and tend to minimize the entropic penalty caused by contact with water. It is precisely this fundamental property that may cause long-chain antioxidants to self-assemble, to migrate into lipid phases, and to have a reduced self-diffusion coefficient in membranes, association colloids, or oil droplets. Thus, hydrophobicity appears to be the common cause that impacts both location and diffusivity, which, in turn, impact antioxidant properties.

Future hypotheses to predict antioxidant activity in foods and biological tissues will need to do more than merely replace the polar paradox. They will have to develop a unified theory for different dispersed lipid systems, such as micelles, oil-in-water emulsions, complex food emulsions, liposomes, lipoprotein particles, and living cells. To do so, we need to find the smallest common structural trait that characterizes these different systems. This will be difficult because such a theory will have to consider the phospholipid membrane of a liposome, a lipoprotein particle, or a living cell as virtually equivalent to the lipid core of an oil-in-water emulsion or a micelle. The question that remains is whether the occurrence of lipid–water interfaces can be the common factor that gives all of these systems common properties. To conclude, location- and diffusion-based approaches should be fused into a unified approach that addresses the cut-off effect as originating from intermolecular interactions between antioxidants and lipid–water interfaces.

References

Abbot, J.; Waite, R. The effect of antioxidants on the keeping quality of whole milk powder. I. Flavones, gallates, butylhydroxyanisole and nordihydroquaiaretic acid. J. Dairy Res. 1962, 29, 55–61.

Ambrosi M.; Fratini, E.; Alfredsson, V.; Ninham, B. W.; Giorgi, R.; Lo Nostro, P.; Baglioni, P. Nanotubes from a vitamin C-based bolaamphiphile. J. Am. Chem. Soc. 2006, 128, 7209–7214.

An, C. B.; Li, D.; Liang, R.; Bu, Y. Z.; Wang, S.; Zhang, E. H.; Wang, P.; Ai, X. C.; Zhang, J. P.; Skibsted, L. H. Chain length effects in isoflavonoid daidzein alkoxy derivatives as antioxidants: A quantum mechanical approach. J. Agric. Food Chem. 2011, 59, 12652–12657.

Annuziata, O.; Costantino, L.; D'Errico, G.; Paduano, L.; Vitagliano, V. Transport properties for aqueous sodium sulfonate surfactants: 2. Intradiffusion measurements: Influence of the obstruction effect on the monomer and micelle mobilities. J. Colloid Interface Sci. 1999, 216, 16–24.

Busch, W.; Bastian, S.; Trahorsch, U.; Iwe, M.; Kühnel, D.; Meißner, T.; Springer, A.; Gelinsky, M.; Richter, V.; Ikonomidou, C.; et al. Internalisation of engineered nanoparticles into mammalian cells in vitro: Influence of cell type and particle properties. J. Nanopart. Res. 2011, 13, 293–310.

Capuzzi, G.; Lo Nostro, P.; Kulkarni, K.; Fernandez, J. E. Mixtures of stearoyl-6-*O*-ascorbic acid and α-tocopherol: A monolayer study at the gas/water interface. Langmuir 1996, 12, 3957–3963.

Castle, L.; Perkins, M. J. Inhibition kinetics of chain-breaking phenolic antioxidants in SDS micelles. Evidence that intermicellar diffusion rates may be rate-limiting for hydrophobic inhibitors such as α-tocopherol. J. Am. Chem. Soc. 1986, 108, 6381–6382.

Chaiyasit, C.; Elias, R. J.; McClements, D. J.; Decker, E. A. Role of physical structures in bulk oils on lipid oxidation. Crit. Rev. Food Sci. Nutr. 2007, 47, 299–317.

Chipault, J. R.; Mizuno, G. R.; Lundberg, W. O. The antioxidant properties of spices. Food Technol. 1956, 10, 209–212.

Cort, W. M. Antioxidant Properties of Ascorbic Acid in Foods. Ascorbic acids: Chemistry, Metabolism, and Uses. Seib, P. A., and Tolbert, B. M., Eds.; Advances in Chemistry Series 200; Washington, D.C.: American Chemical Society, 1982; 533–550.

Cort, W. M. Hemoglobin peroxidation test screens antioxidants. Food Technol. 1974, 28, 60–66.

Dragsten, P. R.; Blumenthal, R.; Handler, J. S. Membrane asymmetry in epithelia: Is the tight junction a barrier to diffusion in the plasma membrane? Nature 1981, 294, 718–722.

Frankel, E. N.; Huang, S.-W.; Kanner, J.; German, J. B. Interfacial phenomena in the evaluation of antioxidants: Bulk oils vs emulsion. J. Agric. Food Chem. 1994, 42, 1054–1059.

Gopinath, D.; Ravi, D.; Rao, B. R.; Apte, S. S.; Renuka, D.; Rambhau, D. Ascorbyl palmitate vesicles (Aspasomes): Formation, characterization and applications. Int. J. Pharm. 2004, 271, 95–113.

Gotoh, N.; Noguchi, N.; Tsuchiya, L.; Morita, K.; Sakai, H.; Shimasaki, H.; Niki, E. Inhibition of oxidation of low density lipoprotein by vitamin E and related compounds. Free Radic. Res. 1996, 24, 123–134.

Griffiths, P.; Stilbs, C. P.; Paulsen, K.; Howe, A. M.; Pitt, A. R. FT-PGSE NMR study of mixed micellization of an anionic and a sugar-based nonionic surfactant. J. Phys. Chem. B 1997, 101, 915–918.

Grosse, S.; Aron, Y.; Thevenot, G.; Francois, D.; Monsigny, M.; Fajac, I. Potocytosis and cellular exit of complexes as cellular pathways for gene delivery by polycations. J. Gene Med. 2005, 7, 1275–1286.

Gulati, A. S.; Subba Rao, B. C. Drug analogues from the phenolic constituents of cashewnut shell liquid. Indian J. Chem. 1964, 2, 337–338.

Heins, A.; McPhail, D. B.; Sokolowski, T.; Stöckmann, H.; Schwarz, K. The location of phenolic antioxidants and radicals at interfaces determines their activity. Lipids 2007, 42, 573–582.

Jacobsen, C.; Hartvigsen, K.; Lund, P.; Meyer, A. S.; Adler-Nissen, J.; Holstborg, J.; Hølmer, G. Oxidation in fish-oil-enriched mayonnaise 1. Assessment of propyl gallate as an antioxidant by discriminant partial least squares regression analysis. Eur. Food Res. Technol. 1999, 210, 13–30.

Jacobsen, C.; Hartvigsen, K.; Thomsen, M. K.; Hansen, L. F.; Lund, P.; Skibsted, L. H.; Hølmer, G.; Adler-Nissen, J.; Meyer, A. S. Lipid oxidation in fish oil enriched mayonnaise: Calcium disodium ethylenediaminetetraacetate, but not gallic acid, strongly inhibited oxidative deterioration. J. Agric. Food Chem. 2001, 49, 1009–1019.

Jacobsen, C.; Let, M. B.; Nielsen, N. S.; Meyer, A. S. Antioxidant strategies for preventing oxidative flavor deterioration of foods enriched with *n*-3 polyunsaturated lipids: A comparative evaluation. Trends Food Sci. Technol. 2008, 19, 76–93.

Kagan, V. E.; Quinn, P. J. The interaction of α-tocopherol and homologues with shorter hydrocarbon chains with phospholipids bilayer dispersions. Eur. J. Biochem. 1988, 171, 661–667.

Kagan, V. E.; Serbinova, E. A.; Bakalova, R. A.; Stoytchev, T. S.; Erin, A. N.; Prilipko, L. L.; Evstigneeva, R. P. Mechanisms of stabilization of biomembranes by alpha-tocopherol: The role of the hydrocarbon chain in the inhibition of lipid peroxidation. Biochemical Pharm. 1990a, 40, 2403–2413.

Kagan, V. E.; Serbinova, E. A.; Koynova, G. M.; Kitanova, S. A.; Tyurin, V. A.; Stoytchev, T. S.; Quinn, P. J.; Packer, L. Antioxidant action of ubiquinol homologues with different isoprenoid chain length in biomembranes. Free Radic. Biol. Med. 1990b, 9, 117–126.

Kagan, V. E.; Serbinova, E. A.; Packer, L. Recycling and antioxidant activity of tocopherol homologs of differing hydrocarbon chain lengths in liver microsomes. Arch. Biochem. Biophys. 1990c, 282, 221–225.

Kato, T.; Seki, K.; Kaneko, R. Insoluble monolayers of irisresorcinol at the air/water interface. Colloid Polym. Sci. 1990, 268, 934–937.

Kikuzaki, H.; Hisamoto, M.; Hirose, K.; Akiyama, K.; Taniguchi, H. Antioxidant properties of ferulic acid and its related compounds. J. Agric. Food Chem. 2002, 50, 2161–2168.
Kimura, Y.; Kanatani, H.; Shima, M.; Adachi, S.; Matsuno, R. Antioxidant activity of acyl ascorbates in intestinal epithelial cells. Biotechnol. Lett. 2003, 25, 1723–1727.
Kleszczynska, H.; Sarapuk, J.; Oswiecimska, M.; Witek, S. Antioxidative activity of pirolidinium salts in the erythrocyte membrane. Polish J. Environ. Stud. 2000, 9, 475–478.
Kozubek A. Detergent-like effect of phenolic lipids on biological membranes. Acta Univ. Wratisl. 1989, 868, 27–32.
Kozubek A. Interaction of alkylresorcinols with proteins. Acta Biochim. Pol. 1995, 42, 241–246.
Laguerre, M.; Bayrasy, C.; Panya, A.; Weiss, J.; McClements, D. J.; Lecomte, J.; Decker, E. A.; Villeneuve, P. What makes good antioxidants in lipid-based systems? The next theories beyond the polar paradox. Crit. Rev. Food Sci. Nutr. In press, 2013.
Laguerre, M.; Lecomte, J.; Villeneuve, P. Evaluation of the ability of antioxidants to counteract lipid oxidation: Existing methods, new trends and challenges. Prog. Lipid Res. 2007, 46, 244–282.
Laguerre, M.; López Giraldo, L. J.; Lecomte, J.; Figueroa-Espinoza, M.-C.; Baréa, B.; Weiss, J.; Decker, E. A.; Villeneuve, P. Chain length affects antioxidant properties of chlorogenate esters in emulsion: The cutoff theory behind the polar paradox. J. Agric. Food Chem. 2009, 57, 11335–11342.
Laguerre, M.; López Giraldo, L. J.; Lecomte, J.; Figueroa-Espinoza, M.-C.; Baréa, B.; Weiss, J.; Decker, E. A.; Villeneuve, P. Relationship between hydrophobicity and antioxidant ability of phenolipids in emulsion: A parabolic effect of the chain length of rosmarinate esters. J. Agric. Food Chem. 2010, 58, 2869–2876.
Laguerre, M.; Wrutniak-Cabello, C.; Chabi, B.; López Giraldo, L. J.; Lecomte, J.; Villeneuve, P.; Cabello, G. Does hydrophobicity always enhance antioxidant drugs? A cut-off effect of the chain length of functionalized chlorogenate esters on ROS-overexpressing fibroblasts. J. Pharm. Pharmacol. 2011, 63, 531–540.
Lee, J. H.; Panya, A.; Laguerre, M.; Bayrasy, C.; Lecomte, J.; Villeneuve, P.; Decker, E. A. Comparison of antioxidant capacities of rosmarinate alkyl esters in riboflavin photosensitized oil-in-water emulsions. J. Am. Oil Chem. Soc. 2013, 90, 225-232..
Lindman, B.; Puyal, M.-C.; Kamenka, N.; Brun, B.; Gunnarsson, G. Micelle formation of ionic surfactants. Tracer self-diffusion studies and theoretical calculations for sodium p-octylbenzenesulfonate. J. Phys. Chem. 1982, 86, 1702–1711.
Lo Nostro, P.; Capuzzi, G.; Romani, A.; Mulnacci, N. Self-assembly and antioxidant properties of octanoyl-6-O-ascorbic acid. Langmuir 2000, 16, 1744–1750.
Lucas, R.; Comelles, F.; Alcantara, D.; Maldonado, O. S.; Curcuroze, M.; Parra, J. L.; Morales, J. C. Surface-active properties of lipophilic antioxidants tyrosol and hydroxytyrosol fatty acid esters: A potential explanation for the nonlinear hypothesis of the antioxidant activity in oil-in-water emulsions. J. Agric. Food Chem. 2010, 58, 8021–8026.
Lue, B. M. Enzymatic lipophilization of bioactive compounds in ionic liquids. PhD thesis, Aarhus University, Denmark, 2009.
Maldonado, O. S.; Lucas, R.; Comelles, F.; González, M. J.; Parra, J. L.; Medina, I.; Morales, J. C. Synthesis and characterization of phenolic antioxidants with surfactant properties: Glucosyl- and glucuronosyl alkyl gallates. Tetrahedron 2011, 67, 7268–7279.
McClements, D. J. Lipid-Based Emulsions and Emulsifiers. Food Lipids: Chemistry, Nutrition, and Biotechnology, 3; Akoh, C. K. and Min, D. B., Eds.; CRC Press: Boca Raton, FL, 2008; 63–97.

Medina, I.; Lois, S.; Alcántara, D.; Lucas, R.; Morales, J. C. Effect of lipophilization of hydroxytyrosol on its antioxidant activity in fish oils and fish oil-in-water emulsions. J. Agric. Food Chem. 2009, 57, 9773–9779.

Miyashita, K.; Nara, E.; Ota, T. Oxidative stability of free fatty acids mixtures from soybean, linseed, and sardine oils in aqueous solution. Biosci. Biotechnol. Biochem. 1993, 60, 315–318.

Miyashita, K.; Tateda, T.; Ota, T. Oxidative stability of polyunsaturated fatty acids in aqueous solution. Biosci. Biotechnol. Biochem. 1994, 57, 1638–1640.

Niki, E. Mechanisms and dynamics of antioxidant action of ubiquinol. Molec. Aspects Med. 1997, 18, 63–70.

Niki, E.; Kawakami, A.; Saito, M.; Yamamoto, Y.; Tsuchiya, J.; Kamiya, Y. Effect of phytyl side chain of vitamin E on its antioxidant activity. J. Biol. Chem. 1985, 260, 2191–2196.

Niki, E.; Komuro, E.; Takahashi, M.; Urano, S.; Ito, E.; Terao, K. Oxidative hemolysis of erythrocytes and its inhibition by free radical scavengers. J. Biol. Chem. 1988, 263, 19809–19814.

Palma, S.; Manzo, R. H.; Allemandi, D.; Fratoni, L.; Lo Nostro, P. Coagels from ascorbic acid derivatives. Langmuir 2002a, 18, 9219–9224.

Palma, S.; Manzo, R. H.; Allemandi, D.; Fratoni, L.; Lo Nostro, P. Solubilization of hydrophobic drugs in octanoyl-6-*O*-ascorbic acid micellar dispersions. J. Pharm. Sci. 2002b, 91, 1810–1816.

Panya, A.; Laguerre, M.; Bayrasy, C.; Lecomte, J.; Villeneuve, P.; McClements, D. J.; Decker, E. A. An investigation of the versatile antioxidant mechanism of action of rosmarinate alkyl esters in oil-in-water emulsions. J. Agric. Food Chem. 2012, 60, 2692–2700.

Panya, A.; Laguerre, M.; Lecomte, J.; Weiss, J.; McClements, D. J.; Villeneuve, P.; Decker, E. A. Effects of chitosan and rosmarinate esters on the physical and oxidative stability of liposomes. J. Agric. Food Chem. 2010, 58, 5679–5684.

Porter, W. L. Recent Trends in Food Applications of Antioxidants. Autoxidation in Food and Biological Systems; Simic, M. G. and Karel, M. Eds.; Plenum Press: New York, 1980; 295–365.

Porter, W. L.; Black, E. D.; Drolet, A. M. Use of polyamide oxidative fluorescence test on lipid emulsions: Contrast in relative effectiveness of antioxidants in bulk versus dispersed systems. J. Agric. Food Chem. 1989, 37, 615–624.

Rejman, J.; Oberle, V.; Zuhorn, I. S.; Hoekstra, D. Size-dependent internalization of particles via the pathways of clathrin- and caveolae-mediated endocytosis. Biochem. J. 2004, 377, 159–169.

Richards, M. P.; Chaiyasit, W.; McClements, D. J.; Decker, E. A. Ability of surfactant micelles to alter the partitioning of phenolic antioxidants in oil-in-water emulsions. J. Agric. Food Chem. 2002, 50, 1254–1259.

Sánchez-Paz, V.; Pastoriza-Gallego, M. J.; Losada-Barreiro, S.; Bravo-Díaz, C.; Gunaseelan, K.; Romsted, L. S. Quantitative determination of α-tocopherol distribution in a tributyrin/Brij 30/water model food emulsion. J. Colloid Interface Sci. 2008, 320, 1–8.

Sasaki, K.; Alamed, J.; Weiss, J.; Villeneuve, P.; Lopez Giraldo L. J.; Lecomte, J.; Figueroa-Espinoza, M.-C.; Decker, E. A. Relationship between the physical properties of chlorogenic acid esters and their ability to inhibit lipid oxidation in oil-in-water emulsions. Food Chem. 2010, 118, 830–835.

Scott, J. W.; Cort, W. M.; Harley, H.; Parrish, D. R.; Saucy, G. 6-Hydroxychroman-2-carboxylic acids: Novel antioxidants. J. Am. Oil Chem. Soc. 1974, 51, 200–203.

Shahidi, F.; Zhong, Y. Revisiting the polar paradox theory: A critical overview. J. Agric. Food Chem. 2011, 59, 3499–3504.
Siderius, A.; Kehl, S. K.; Liaist, D. G. Surfactant diffusion near critical micelle concentrations. J. Solution Chem. 2002, 31, 607–625.
Simpson, T. H.; Uri, N. Hydroxyflavones as inhibitors of the aerobic oxidation of unsaturated fatty acids. Chem. Ind. (London) 1956, 956–957.
Söderman, O.; Stilbs, P.; Price, W. S. NMR studies of surfactants. Concepts Magn. Reson. 2004, 23, 121–135.
Sørensen, A.-D. M.; de Diego, S.; Petersen, L. K.; Nielsen, N. S.; Lue, B.-M.; Yang, Z.; Xu, X.; Jacobsen, C. The antioxidative effect of lipophilized rutin and dihydrocaffeic acid in fish-oil-enriched milk. Eur. J. Lipid Sci. Technol. 2012a, 114, 434–445.
Sørensen, A.-D. M.; Nielsen, N. S.; Yang, Z.; Xu, X.; Jacobsen, C. Lipophilization of dihydrocaffeic acid affects its antioxidative properties in fish-oil-enriched emulsion. Eur. J. Lipid Sci. Technol. 2012b, 114, 134–145.
Stasiuk, M.; Kozubek, A. Biological activity of phenolic lipids. Cell. Mol. Life Sci. 2010, 67, 841–860.
Stöckmann, H.; Schwarz, K.; Huynh-Ba, T. The influence of various emulsifiers on the partitioning and antioxidant activity of hydroxybenzoic acids and their derivatives in oil-in-water emulsions. J. Am. Oil Chem. Soc. 2000, 77, 535–542.
Takahashi, M.; Komuro, E.; Niki, E.; Tanaka, K. Action of fatty acid esters of L-ascorbic acid as antioxidants in phosphatidylcholine liposomal membranes. Bull. Chem. Soc. Jpn. 1992, 65, 679–684.
Tofani, D.; Balducci, V.; Gasperi, T.; Incerpi, S.; Gambacorta, A. Fatty acid hydroxytyrosyl esters: Structure/antioxidant activity relationship by ABTS and in cell-culture DCF assays. J. Agric. Food Chem. 2010, 58, 5292–5299.
Tournier, J. F.; Lopez, A.; Gas, N.; Tocanne, J. F. The lateral motion of lipid molecules in the apical plasma membrane of endothelial cells is reversibly affected by the presence of cell junctions. Exp. Cell Res. 1989, 181, 375–384.
Tyurin, V. A.; Kagan, V. E.; Serbinova, E. A.; Gorbunov, N. V.; Erin, A. N.; Prilipko, L. L.; Stoichev, T. S. Interaction of α-tocopherol with phospholipids liposomes: Absence of transbilayer mobility. Biomedical and Life Sciences. Bull. Exp. Biol. Med. 1987, 102, 1677–1680.
Uri, N. Mechanisms of antioxidation. In Autoxidation and Antioxidants, Lundberg, W. O., Ed. Interscience: New York, 1961, 1, 133–169.
Uri, N. The mechanism of the oxidation of linoleic acid with particular reference to metal catalysis. Proceedings of the Fourth International Conference on the Biochemical Problems of Lipids. Butterworths: London, 1958; 30.
von Gersdorff, K.; Sanders, N. N.; Vandenbroucke, R.; De Smedt, S. C.; Wagner, E.; Ogris, M. The internalization route resulting in successful gene expression depends on both cell line and polyethylenimine polyplex type. Mol. Ther. 2006, 14, 754–753.
Wang, S.-H.; Lee, C.-W.; Chiou, A.; Wei, P.-K. Size-dependent endocytosis of gold nanoparticles studied by three-dimensional mapping of plasmonic scattering images. Nanobiotechnology 2010, 8, 1–13.
Waraho, T.; McClements, D. J.; Decker, E. A. Mechanisms of lipid oxidation in food dispersions. Trends Food Sci. Tech. 2011, 22, 3–13.

Wen, X. L.; Zhang, J.; Liu, Z. L.; Han, Z. X.; Ricker, A. Micellar effects on the oxidative electrochemistry of lipophilic vitamin C derivatives. J. Chem. Soc. Perkin Trans. 1998, 2, 905–910.

Yuji, H., Weiss, J., Villeneuve, P., López Giraldo, L. J., Figueroa-Espinoza, M. C., Decker, E. A. Ability of surface-active antioxidants to inhibit lipid oxidation in oil-in-water emulsion. J. Agric. Food Chem. 2007, 55, 11052–11056.

Zana, R. Dynamic Properties of Interfaces and Association Structures; Pillai, V., Shah, D. O., Eds.; AOCS Press: Champaign, IL, 1996; 142.

Zana, R. Dynamics of Surfactant Self-Assemblies: Micelles, Microemulsions, Vesicles, and Lyotropic Phases; Zana, R. Ed.; Taylor and Francis: Boca Raton, FL, 2005; 75.

Zuev, Y. F.; Gnezdilov, O. I. ; Zueva, O. S.; Us'yarov O. G. Effective self-diffusion coefficients of ions in sodium dodecyl sulfate micellar solutions. Colloid J. 2011, 73, 59–64.

Zuev, Y. F.; Kurbanov, R. K; Idiyatullin, B. Z.; Us'yarov, O. G. Sodium dodecyl sulfate self-diffusion in premicellar and low-concentrated micellar solutions in the presence of a background electrolyte. Colloid J. 2007, 69, 444–449.

CHAPTER

9

Understanding Antioxidant and Prooxidant Mechanisms of Phenolics in Food Lipids

Lisa Zhou[1] and Ryan J. Elias[1]

Introduction

Phenolic compounds are secondary plant metabolites commonly encountered in fruits and vegetables. These compounds include at least one phenol group, a structure containing an aromatic benzene ring with at least one hydroxyl substituent (Fig. 9.1). Phenolics have long been added to lipid foods to inhibit oxidation reactions. However, in recent years, phenolics have become popular bioactive ingredients in formulated foods as studies have demonstrated that these compounds are beneficial to human health. Unfortunately, many phenolics are labile to oxidation, which complicates their addition to food products. Phenolics are often degraded as a result of nonenzymatic oxidation reactions over the life of food, and these reactions may compromise the oxidative stability of the surrounding food matrix under some conditions.

Applications of Phenolics as Antioxidants in Food Lipids

Phenolic compounds, especially di- and trihydroxyl substituted phenolics, are known to exhibit strong antioxidant activity in lipid-based foods (Table 9.A). With heightened consumer awareness of the potentially harmful effects of synthetic antioxidants (such as butylated hydroxytoluene and butylated hydroxyanisole) in the diet (Esterbauer, 1993), plant-derived phenolic extracts have become attractive nonsynthetic antioxidants in formulated lipid foods. Phenolic extracts from

[1]Department of Food Science, The Pennsylvania State University, University Park, PA, 16802, USA.

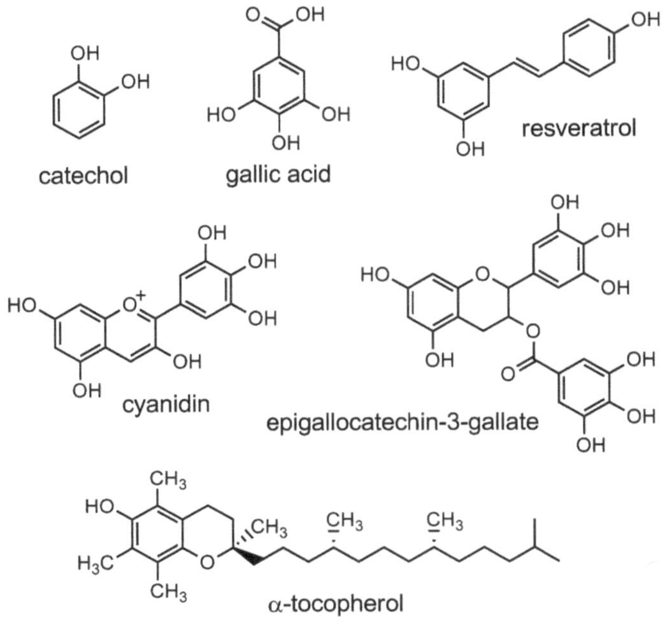

Fig. 9.1. Phenolic structures.

a host of plant-based foods and waste products (for example, skins, stems, and seeds) have been shown to exhibit equivalent or even higher antioxidant activity than conventional antioxidants, such as ascorbic acid and tocopherols, under many conditions (Kähkönen et al., 1999). Pure phenolic compounds have been shown to inhibit oxidation and discoloration in meat products, bulk oils, and lipid dispersions. The same is true for plant-derived phenolic extracts, which have also been proven to be effective antioxidants in foods.

Applications of Phenolics as Bioactive Ingredients in Food Lipids

Adding phenolics to formulated foods is a growing trend as consumers gain awareness of the purported health benefits of dietary phenolics. For example, the consumption of (-)-epigallocatechin-3-gallate (EGCG), the dominant polyphenol in tea, has been associated with the prevention of metabolic syndrome (Bose et al., 2008; Thielecke & Boschmann, 2009) and cancer (Yang et al., 2009). Cocoa polyphenols have been reported to reduce the risk of cardiovascular disease (Shrime et al., 2011). Many of these phenolics are available as dietary supple-

Table 9.A. Antioxidant Activity from Phenolic Compound and Extracts Observed in Lipid Systems

Food System	Phenolics	Reference
Raw, minced red meat; poultry muscle; and fish muscle	Green tea extract and α-tocopherol	Tang et al., 2001a
Minced beef patties	Green tea extract	Tang et al., 2006
Pork sausage	Rosemary extract	Sebranek et al., 2005
Turkey	Grape seed extract	Mielnik et al., 2006
Fish oil–enriched salad dressing	α-tocopherol	Let et al., 2007
Menhaden O/W emulsion mackerel mince	Catechin caffeic acid, ferulic acid, and tannic acid	Maqsood & Benjakul, 2010
Sunflower oil linoleate O/W emulsion phospho-lipid liposomes	Quercetin, α-tocopherol, and rutin	Becker et al., 2007
Rapeseed O/W emulsions	Gallic acid, cyanidin-3-glucoside, (+)-epicatechin, chlorogenic acid, genistein rutin, and α-tocopherol	Estevez et al., 2008
Fish-oil emulsion Bulk fish oil	Quercetin, quercetin-3-O-glycoside, and α-tocopherol	Huber et al., 2009

ments; however, high doses may lead to toxicity. Liver toxicity resulting from high EGCG intake has been reported (Lambert et al., 2007), and such high concentrations are easily achieved using dietary supplements. The delivery of biologically significant levels of phenolics in food appears to be a safer approach, with respect to mitigating the toxicity of these compounds.

Antioxidant Mechanisms of Phenolics in Foods

Phenolic compounds are known to have high antioxidant capacity, as seen in hydrogen abstraction assays, such as the oxygen radical absorbance capacity (ORAC) assay (Alarcón et al., 2008; Dudonné et al., 2009; Hurst et al., 2009; Ma et al., 2011) and in single electron transfer assays, such as the DPPH (1,1-diphenyl-2-picrylhydrazyl) assay (Nanjo et al., 1999; Roy et al., 2010) and the ABTS (2,2'-azino-bis(3-ethylbenzthiazoline-6-sulfonic acid) assay (Cai et al., 2006). Many phenolic compounds are also capable of serving as transition metal chelators (Khokhar & Owusu, 2003), which may augment their antioxidant activity in metal catalyzed oxidation reactions (Guo et al., 2007). The mechanism by which phenolic compounds scavenge lipid-derived radicals and chelate transition metals is shown in Fig. 9.2. Formulated foods commonly contain more than one antioxidant; therefore, it is important to understand the synergistic or antagonistic interactions between phenolics and nonphenolic antioxidants (Becker et al., 2007; Dai et al., 2008). Combinations of the phenolics—such as catechin, cyanidin, caffeic acid, quercetin, and ellagic acid—showed additive effects to the inhibition of copper-catalyzed human LDL oxidation, except in combinations with ellagic acid and catechin where antagonistic effects were observed (Meyer et al., 1998). Even among phenolic combinations found naturally in plants, antagonistic, additive, and synergistic interactions between phenolics may occur, as was observed in the ORAC assay with combinations of phenolics at their typical concentrations in navel oranges (examples include chlorogenic acid, hesperidin, luteolin, myricetin, naringenin, p-coumaric acid, and quercetin) (Freeman et al., 2010). Interactions between antioxidants may also result in a reversal in antioxidant activity. Using a Fe^{2+}/histidine-Fe^{3+} system, ascorbate alone (50–200 μM) was demonstrated to have prooxidative activity in liposomes, while ascorbate and tocopherols together demonstrated a synergistic antioxidant effect (Thomas et al., 1992). The lipid system also affects interactions. For example, α-tocopherol had a synergistic effect on quercetin in the metmyoglobin-initiated oxidation of methyl linoleate emulsions, but an antagonistic effect in bulk oil (high-oleic sunflower oil) (Becker et al., 2007).

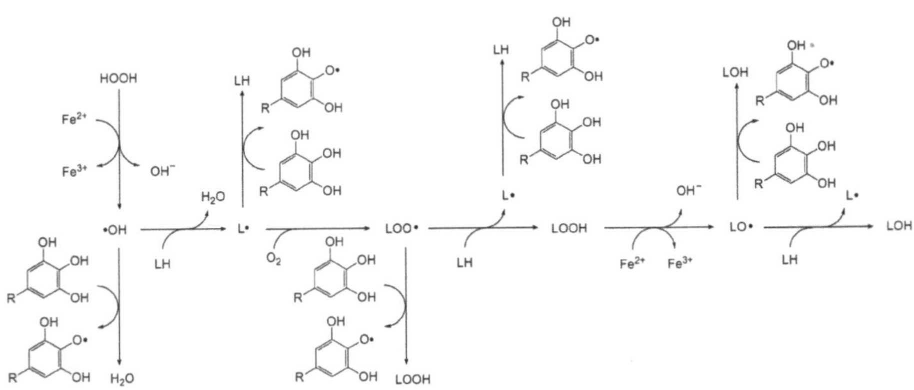

Fig. 9.2. Proposed prooxidant reaction mechanisms in lipid dispersions depicted with a galloyl group. Figure modified from Zhou and Elias, 2012. Abbreviations are as follows: hydroperoxyl radical (HOO•), hydroxyl radical (•OH), reduced lipid (LH), lipid alkyl radical (L•), lipid hydroperoxyl radical (LOO•), lipid hydroperoxide (LOOH), lipid alkoxyl radical (LO•).

Radical Scavenging Activity

Relationship between Phenolic Structure and Radical Scavenging Activity

The antioxidant activity of phenolic compounds is due primarily to resonance stabilization of the phenoxyl radical after oxidation. Polyphenols consisting of large, conjugated pi systems are typically more stable radicals, and the introduction of various substituents, such as electronegative carboxyl or hydroxyl groups, may also increase radical stability. The presence of a carboxyl group in monohydroxy benzoic acids causes electrons to withdraw, which significantly increases the antioxidant activity of the *meta*-substituted monohydroxy acid, compared to the *ortho* and *para* positions (Rice-Evans et al., 1996). Di- and trihydroxyl substitutions on phenol groups have greater antioxidant activity, especially with adjacent substitutions, because of resonance stabilization. This phenomenon has been seen with the substituted phenol group, presented in Fig. 9.3 (1,2,3-trihydroxybenzene (pyrogallol) > 1,2-dihydroxybenzene (catechol) > 1,3-dihydroxybenzene = 1,4-dihydroxybenzene > phenol) (Kumamoto et al., 2001). Blocking studies have also supported the importance of the adjacent di- and trihydroxyl group substitution on antioxidant activity. *O*-methylation of the catechol group significantly reduced the antioxidant activity of catechols (Dueñas et al., 2010). Similarly, a loss of resonance occurs in galloyl groups if the middle hydroxyl group is altered, which then causes a significant loss of antioxidant activity (Nanjo et al., 1996).

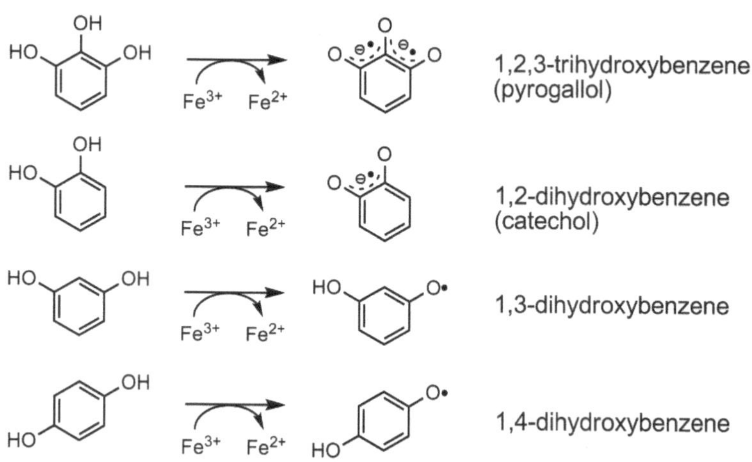

Fig. 9.3. Resonance stabilization of radical resulting from single electron transfer in di- and trihydroxl substituted phenol rings.

EGCG consists of two galloyl moieties and loses significant antioxidant activity only when both galloyl groups are glycosylated at the middle hydroxyl position (4' and 4'' hydroxyl positions). Though numerous studies have focused on the non-glycosylated species (that is, the aglycone), it is important to note that phenolic extracts may also contain phenolic glycosides. Phenolic glycosides may result in different antioxidant activities, as seen between quercetin and multiple quercetin glycosides in lipid systems (Huber et al., 2009).

Steric hindrance is another important factor that affects radical scavenging capacity as epimers with greater steric hindrance (EGCG, EGC, and EC versus GCG, GC, and C) tend to show less activity (Zhao et al., 2001). The effect of epimers was more pronounced at low concentrations and for larger free radicals, such as those generated from 2,2'-azobis(2-amidino-propane) dihydrochloride (AAPH) compared to superoxide (Guo et al., 1999). These may be important factors for scavenging lipid-derived radicals. Because phenolics represent a diverse class of compounds based on their structure, other features, such as additional double bonds, also impact antioxidant activity. The influence of these other structures will not be further discussed in this chapter; however, extensive studies have examined the effects of various structural motifs on phenolic antioxidant activity (Rice-Evans et al., 1996; Huvaere et al., 2009).

Relationship between Matrix pH and Radical Scavenging Activity

Matrix pH has been shown to greatly influence the antioxidant activity of phenolics as changes in reduction potentials appear to correlate with the deprotonation of phenolics. Relatively less energy is required to abstract an electron from the dissociated phenolic species. As such, some have argued that the radical scavenging activity of phenolics increases with increasing pH (Amorati et al., 2006).

Reduction or oxidation potentials are often seen as good predictors of antioxidant activity because higher oxidation potentials (that is, lower reduction potentials) correlate to an increased capability to lose an electron. A pH-dependent decrease in oxidation potential was observed in catechin (Janeiro & Oliveira Brett, 2004) as well as in tea catechins, epicatechin (EC), epigallocatechin (EGC), epicatechin-3-gallate (ECG), and EGCG (Kumamoto et al., 2001). Interestingly, oxidation potentials were pH-independent outside the pH range of 3–9 for catechin (Janeiro et al., 2004) and pH 5–8 for EC and EGCG (Kumamoto et al., 2001), which suggests that radical scavenging activity may not increase above those threshold pH values. The tea catechins showed strong antioxidant activity from pH 5–12; however, under strong alkaline conditions (pH 13), antioxidant activity decreased sharply, which was likely because increased phenolic oxidation leaves little of the unoxidized species intact to act as radical scavengers (Kumamoto et al., 2001).

Relationship between Phenolic Partitioning and Radical Scavenging Activity

Food systems are much more complex than the environments employed in the majority of radical scavenging assays. Radical scavenging activity data from assays, such as ORAC, do not always predict the efficacy of antioxidants like phenolics in complex, multiphase food systems (Alamed et al., 2009). The hydrophobicity of phenolic compounds has been suggested to be a major factor influencing a compound's activity, with respect to inhibiting its lipid oxidation (Aliaga et al., 2009; Iglesias et al., 2010). Polymerization of polyphenolic compounds has been found to increase antioxidant activity with mixed results; radical scavenging activity may increase, but antioxidant activity in emulsions may not change due to preferential partitioning into the lipid core instead of the interface as more units are polymerized (Iglesias et al., 2010; Pazos et al., 2010).

Surface-active phenolic derivatives (such as fatty acid–esterified phenolics) have also been the topic of much research in recent years, with the objective of increasing the antioxidant activity of phenolics by directing their partitioning to the sites of lipid oxidation reactions (that is, oil–water interfaces). The conjugation of alkyl groups to phenolics may lead to a decrease in radical scavenging ability. Peroxyl radical scavenging activity was observed as follows (in order of highest to lowest): free p-hydroxyphenylacetic acid (HPA) > dodecyl HPA > butyl HPA (Yuji et al., 2007). Similarly, fatty acid–esterified chlorogenic acid (CGA) exhibited a lower radical scavenging activity than free CGA; CGA showed the highest ORAC value, while dodecyl CGA showed the lowest (Sasaki et al., 2011). To account for the differences in radical scavenging activity, Sasaki et al. (2011) added varying concentrations of the conjugates to emulsions to achieve the same radical scavenging activity (that is, the dodecyl chlorogenic acid concentration was more than double the concentration of free chlorogenic acid). However, lipid oxidation in fish-oil (menhaden) emulsions showed antioxidant activity on the order of octyl CGA > butyl CGA > CGA > control = dodecyl CGA, suggesting that increasing chain-length and hydrophobicity do not necessarily predict greater antioxidant in emulsions, as would be expected. Also, dodecyl CGA greatly decreased the interfacial tension of the emulsion droplet; because it is located in the interface and not buried in the lipid core where its antioxidant activity may have been lowered, phenolic partitioning is not the sole explanation. Languere et al. (2009, 2010) have also produced multiple studies involving the antioxidant activity of phenolics conjugated to various alkyl chains (C1–C20) in CGA and rosmarinic acid. The authors also witnessed a linear increase in the antioxidant activity of emulsions (that is, medium-chained C8), up until a certain point. The decrease in observed antioxidant activity may be due to many factors, such as partitioning into the lipid droplet core, restricted mobility, steric hindrance, or increased solubilization by excess emulsifier for longer alkyl chains. From 0–17 μM Brij 35, C8–C20 show minimal aqueous phase phenolic concentration. Correspondingly, increasing the emulsifier content increased the partitioning of phenolics into the aqueous phase; C12–C20 conjugates also showed greater partitioning into the aqueous phase as a function of increasing emulsifier concentration, compared to C1–C8 conjugates.

Chelation of Transition Metals

The observed antioxidant activity of some polyphenolic compounds in lipid foods may also be related to their ability to chelate, or form complexes with, transition metal catalysts. Many phenolics have a strong capacity for binding ferric (Fe^{3+}) ions due to the presence of iron-binding motifs, such as *o*-dihydroxyl substituents (Khokhar & Owusu Apenten, 2003). Ligands with catechol-containing moieties can bind Fe^{3+} ions in a ligand:metal ratio from 1:1, 2:1, and 3:1 (Hider et al., 2001) depending on available ligand and metal concentrations. Flavonoids containing 3-hydroxy and 4-oxo, or 5-hydroxy and 4-oxo groups, are capable of binding ferrous (Fe^{2+}) ions (Leopoldini et al., 2006). Quercetin contains the 4-oxo, in addition to the 3- and 5-hydroxy groups, thus displaying a strong affinity for ferrous ions. As such, quercetin is capable of competing with the strong ferrous chelator Ferrozine [3-(2-pyridyl)-5, 6-bis(phenyl sulfonic acid)-1,2,4-triazine] for Fe^{2+} ions. In addition, it completely suppresses •OH radical generation (via the Fenton reaction), as measured by 2-deoxyribose degradation (Guo et al., 2007).

Prooxidant Mechanisms of Phenolics on Foods

Prooxidant activity due to metal-catalyzed phenolic oxidation has also been observed in lipid dispersions in a number of studies (Table 9.B). Green tea extracts (Frankel et al., 1997a) and tea catechins (Huang & Frankel, 1997a) were reported to produce prooxidative behavior in oil-in-water emulsions. Gallic acid exhibited higher peroxide values and yielded more oxidation products (such as oxidized triglyceride monomers, triglyceride dimers, and triglyceride oligomers) in sunflower oil-in-water emulsions at pH 7.5–7.8 (Velasco et al., 2004). The effect was also magnified in the presence of transition metals because ECG, EGCG, and myricetin exhibited prooxidant effects in the presence of added metals in sunflower oil-in-water emulsions at pH 5.5 (Roedig-Penman & Gordon, 1997). The factors that influence the prooxidative effects remain unclear but appear to be related to phenolic concentration and pH (Mei et al., 1999), as well as the influence of other matrix constituents. In the presence of phenolic compounds, Mei et al. demonstrated increased lipid oxidation markers in emulsions at pH 3 but increased antioxidant activity only at pH 7 (Mei et al., 1999). The authors reported varying effects de-

Table 9.B. Prooxidant Activity from Phenolic Compounds and Extracts Observed in Lipid Systems

Lipid Dispersion	Phenolic	Reference
Corn O/W emulsion No buffer, pH 4.8–5.0 +/– Cu^{2+} (10 μM)	Green tea extract	Frankel et al., 1997b
Corn O/W emulsion No buffer, pH 3.0–3.5	Catechin, epigallocatechin (EGC), epicatechin gallate (ECG), epigallocatechin gallate (EGCG), propyl gallate, and gallic acid	Huang & Frankel, 1997b
Sunflower O/W emulsion Acetate buffer (0.1 M), pH 5.5 + Fe^{3+} (50 μM)	Tea extract and rosemary extract	Roedig-Penman et al., 1997
Salmon O/W emulsion Acetate buffer (10 mM), pH 3.0	Gallamide, methyl gallate, and gallic acid	Mei et al., 1999
Flax O/W emulsion Phosphate buffer (10 mM), pH 3.0 + Fe^{3+} (25 μM)	EGCG	Zhou et al., 2012
Sunflower O/W emulsion Phosphate buffer (0.1 M), pH 7.5–7.8	Gallic acid	Velasco et al., 2004
Bulk fish oil	Grape pomace extract	Iglesias et al., 2010

pending on phenolic concentration: When the phenolics were present at high concentrations (500 μM), antioxidant activity was observed, while low phenolic concentrations (5 μM) resulted in prooxidant activity. The different effects may be due to competition between antioxidant and prooxidant reactions, as proposed in Figs. 9.2 and 9.4, respectively (Zhou & Elias, 2012). Prooxidant activity may result from the generation of reactive oxygen species (ROS) or the regeneration of transition metals to their catalytically active reduced states, which occurs because of the metal-catalyzed oxidation of phenolic compounds.

Metal-Catalyzed Oxidation of Phenolic Compounds

Many dietary phenolic compounds, particularly those bearing catechol or galloyl groups, are readily oxidized in foods, especially at neutral and alkaline pH values. These "autoxidative" reactions, which actually appear to be catalyzed by transition metals, are coupled to the reduction of oxygen to ROS (Li et al., 1994; Lapidot et

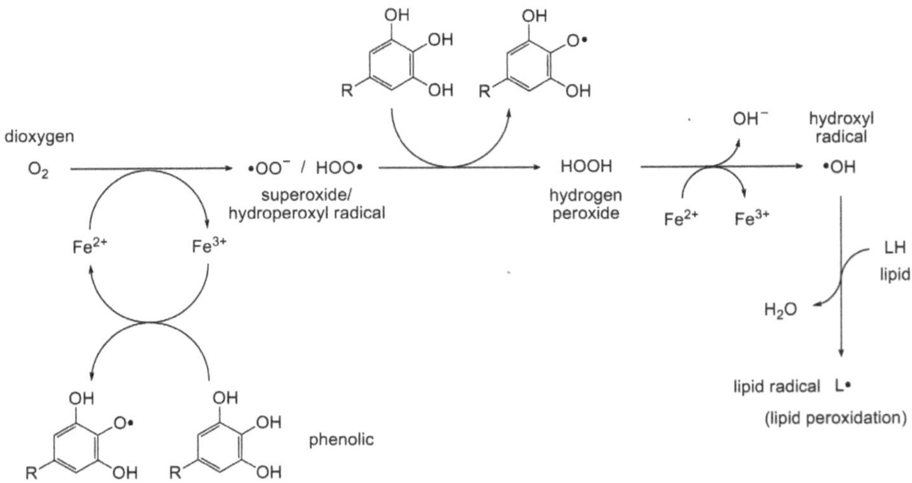

Fig. 9.4. Proposed antioxidant reaction mechanisms in lipid dispersions depicted with a galloyl group.

Figure modified from Zhou and Elias (2012).

al., 2002)—namely, superoxide ($O_2\bullet^-$), or its protonated form, the hydroperoxyl radical (•OOH) , which eventually leads to the formation of H_2O_2 (Wildenradt & Singleton, 1974; Danilewicz, 2003; Waterhouse & Laurie, 2006; Elias et al., 2009a) (Fig. 9.4). Though focus is placed on di- and trihydroxyl–substituted phenolics due to their greater ease of oxidation, the same oxidation pathway may exist in other species capable of reducing transition metals, just as ascorbic acid (Nappi & Vass, 1997) and tocopherols (Fukuzawa et al., 1988) have been shown to undergo the same metal-catalyzed reactions.

The oxidation of phenolic compounds is also affected by their concentration: Higher concentrations have greater stability against phenolic oxidation (Sang et al., 2005). In addition, oxidation products, such as quinones resulting from phenolic oxidation, may further propagate oxidation reactions by interacting with unoxidized phenolics to form dimers (Sang et al., 2007). In the presence of multiple catechol moieties, both rings are capable of undergoing oxidation. Both the B- and D-rings of EGCG are labile to oxidation: The B-ring was shown to be the primary site of metal-catalyzed oxidation for EGCG (Valcic et al., 2000; Severino et al., 2009), while the D-ring was the primary reaction site for $O_2\bullet^-$ (Severino et al., 2009).

Transition metals appear to be essential catalysts in the oxidation of phenolic compounds as their complete removal was demonstrated to prevent catechol oxidation (Nakagawa et al., 2002). In these reactions, transition metals assist in the transfer of electrons as a direct reaction between ground-state oxygen and catechol or gallate groups (or any organic compound, for that matter) is unfavorable. Oxygen in the ground state is referred to as triplet oxygen, a biradical with two unpaired valence electrons in separate orbitals. Phenolic compounds cannot directly donate an electron to oxygen as the overlapping of these orbitals would violate Pauli's exclusion principle, which states that no orbital can contain electrons with the same spin. Therefore, phenolic oxidation requires the presence of transition metals to initiate electron transfer between phenolics and triplet oxygen.

It has been suggested that the transfer of electrons is facilitated by the formation of a ternary phenolic-metal-oxygen complex in which electrons are transferred to oxygen by a concerted two-electron transfer (Bandy et al., 1990). Jarabak et al. provided evidence of ternary complex formation in a study of the redox cycling of polycyclic aromatic hydrocarbon (PAH) *o*-quinones in the presence of reductases (Jarabak et al., 1998). The investigators removed trace metals from buffers and reaction vessels using a metal chelating resin. Consequently, redox cycling of PAH-*o*-quinone (9,10-phenanthrenequinone) was inhibited by 90% in the presence of superoxide dismutase (SOD) (0.18 μM). In the presence of SOD, the initial rate of NADPH oxidation used for quinone reduction was identical to the control as PAH-*o*-quinone was added to initiate the reaction. However, within 10 m, the rate of NADPH loss was observed to decrease dramatically, indicating that in the presence of SOD, the limiting factor is the formation of substrates for reduction by hydroquinone oxidation. Upon addition of Cu^{2+} (1 μM), redox cycling was fully restored to control rates, while Fe^{3+} (10 μM) partially restored redox cycling (>50%). The addition of metals alone (no SOD) at the concentrations stated did not change the rate of redox cycling. This supports the formation of a ternary complex as SOD no longer significantly inhibited oxidation because no $O_2\bullet^-$ was released in the presence of free transition metals. The formation of a ternary complex is further supported by the introduction of metal chelators that occupy essential coordination sites. Bandy and Davidson (1987) demonstrated that in the presence of SOD, EDTA + SOD completely inhibited quinone formation from the oxidation of 6-hydroxydopamine. By itself, EDTA (1 mM) accelerated 6-hydroxydopamine oxidation, and SOD (20 U/mL) inhibited oxidation by 40%; however, EDTA + SOD together inhibited oxidation by 86%. Because EDTA + SOD showed greater inhibition of phenolic oxidation $O_2\bullet^-$ is expected to play a major role in phenolic oxidation in the presence of EDTA. When EDTA binds transition metals, it is likely to prevent the formation of a phenolic-metal-oxygen complex, leading to single electron transfers and the release of $O_2\bullet^-$.

Many studies have focused on the role of iron in phenolic oxidation reactions; however, other transition metal catalysts (e.g., copper, manganese) are also present in foods, albeit at relatively lower concentrations. Catechins are known chelators of Cu^{2+} (Kumamoto et al., 2001) and Fe^{3+} (Elhabiri et al., 2007) ions, but the reduction capacity for each of these metals differs, with most catechins exhibiting a higher capacity for reducing Cu^{2+} (Mira et al., 2002). This difference in reduction capacity may result in faster ROS generation and greater prooxidant activity, with the addition of Cu^{2+} compared to Fe^{3+} ions (Zhang et al., 1996; Kumamoto et al., 2001). The introduction of both Cu^{2+} and Fe^{3+} has also been shown to exert a synergistic effect with respect to phenolic oxidation rates in wine (Danilewicz, 2007; Danilewicz et al., 2008).

Effect of Matrix pH on Metal-Catalyzed Phenolic Oxidation

Matrix pH is an important factor influencing polyphenol stability. It has long been thought that the dissociated form of the phenolic is the species that readily undergoes oxidation; therefore, a matrix pH above the pKa of the phenolic compound should produce higher oxidation rates (Mochizuki et al., 2002). Kumamoto et al. (2001) have shown that other components, such as the presence of Fe^{3+} and Cu^{2+}, also lower the acid dissociation constant, potentially promoting higher oxidation rates at a lower pH. Higher pH values may stabilize intermediates from phenolic oxidation, such as the semiquinone radical (Mochizuki et al., 2002) and the $O_2\bullet^-$ radical (Knowles et al., 1969), thus promoting oxidation.

Matrix pH also leads to changes in transition metal properties and, consequently, reactivity. Metal speciation and solubility depend on pH, with lower pH stabilizing Fe^{2+} ions that would otherwise readily precipitate at neutral or alkaline pH. Phenolics may also influence iron speciation as the ratio of Fe^{2+} to total Fe is thought to depend on the phenolic compound in solution (epicatechin > catechol > pyrocatechinic acid > gallic acid) (Hider et al., 2001). With increasing pH, the Fe^{2+} to total Fe ratio decreased such that above pH 6 all iron was in the Fe^{3+} state, regardless of the different phenolics in solution. The effects of added metals are also less pronounced as pH increases due to oxidation rate increases from other factors, such as dissociation and decreased reduction potential, which we discussed above (Bandy et al., 2001).

Generation of Reactive Oxygen Species

The metal-catalyzed oxidation of phenolics yields H_2O_2, which alone is not a potent oxidant ($E^{0'}$= 320 mV at pH 7.0 for the H_2O_2, H+ / H_2O, •OH couple). The H_2O_2 generated from EGCG oxidation at pH 7 was shown to be stable in hexadecane emulsions for days (Zhou & Elias, 2011). Though H_2O_2 is relatively stable in the absence of scavengers, its reduction by transition metals via the Fenton reaction results in the production of highly oxidizing •OH radicals ($E^{0'}$ = 2310 mV at pH 7.0 for the •OH, H+ / H_2O couple). •OH radicals are capable of reacting with most organic matter (such as lipids, proteins, and DNA) at diffusion-limited rates

(Makrigiorgos et al., 1995; Elias et al., 2008a, 2009b; Elias & Waterhouse, 2010). Due to the high reactivity of •OH radicals, an increase in constituents in the lipid system may impede •OH radicals from reaching lipids to initiate oxidation. For instance, •OH radicals are capable of reacting with surfactant micelles; however, Bansal et al. (1971) observed different reaction rate constants, depending on the aqueous and critical micelle concentrations for the particular surfactant.

Phenolic Oxidation Products and Quality Loss

Quinones are electrophilic phenolic oxidation products that have been shown to exert biological damage by reacting with DNA (Lewis et al., 1996) and proteins, especially thiol groups (that is, cysteine residues). *o*-Quinones are also capable of isomerizing to *p*-quinone methides (Boersma et al., 2000), compounds that are relatively less stable and more electrophilic. The reaction between quinone methides and free nucleophilic amino acids, in order of increasing reactivity, is reported as follows: thiol group of cysteine > α-carbon of lysine and α-carbon of histidine > ε-carbon of lysine and π- and τ-carbon of histidine (Bolton et al., 1997). The formation of covalent bonds between quinones and sulfhydryl groups of membrane proteins may lead to the formation of protein aggregates with intermolecular cross-linking in EGCG and ECG (Chen et al., 2011). This, in turn, may have harmful effects on protein solubility or the physical stability of protein-stabilized emulsions.

A common sensory defect arising from phenolic oxidation is color changes caused by polymerization of phenolics. Polymerization resulting from phenolic oxidation yields high molecular weight tannins that may imbue a yellow, yellow-brown or yellow-red pigmentation. As tannins increase in size, visual faults such as precipitation and haze may result (Poncet-Legrand et al., 2003). Changes in color may also arise from reactions with anthocyanins. Anthocyanins are glucosides of anthocyanidins, a class of compounds that impart bright colors in plants like blueberries and grapes. However, undesired color loss may result from nucleophilic addition of flavanols to anthocyanins to form colorless adducts (Remy-Tanneau et al., 2003; Dueñas et al., 2006). These colorless adducts further oxidize to yield xanthylium cations (Timberlake & Bridle, 1976; Liao et al, 1992) as observed in anthocyanins in Roselle that changed from red to colorless and then to yellow after heating (Tsai & Huang, 2004).

Potential Solutions to Minimize Prooxidant Activity

We can employ various strategies to mitigate the prooxidant activity and quality loss in lipid foods that result from the metal-catalyzed oxidation of polyphenols

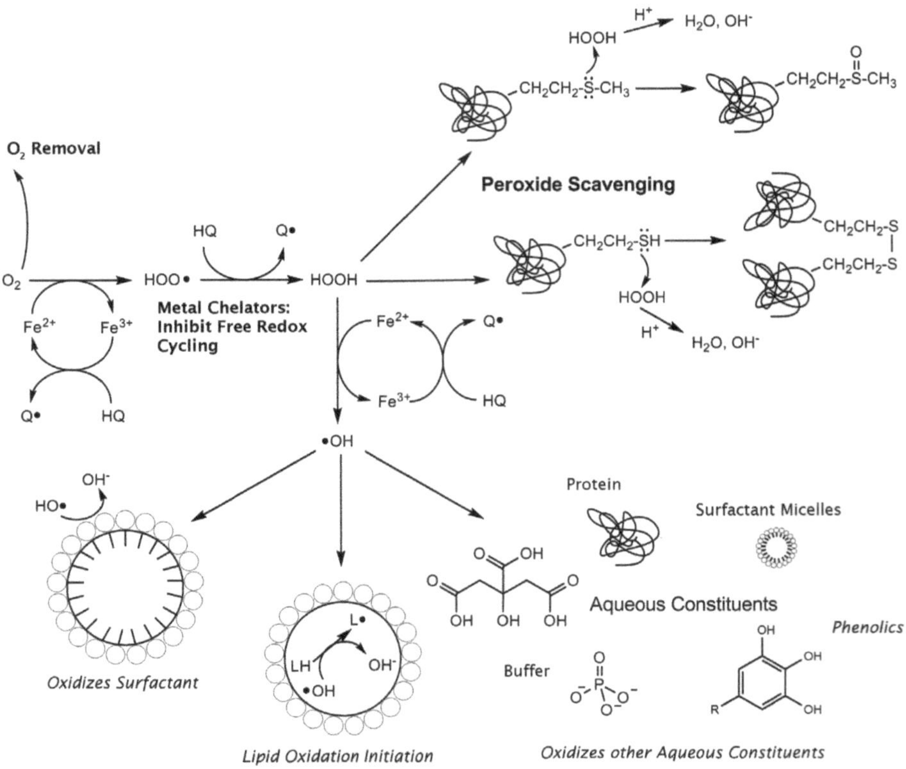

Fig. 9.5. Mechanistic model for strategies in preventing prooxidant activity due to metal-catalyzed oxidation of phenolics.

(as shown in Fig. 9.5). First, it is possible to exclude oxygen from—or at least minimize its presence in—the system, which would require us to limit headspace oxygen as well as dissolved oxygen in the system. Another strategy is to scavenge peroxides prior to •OH radical formation. Finally, the removal of metal catalysts, or the inactivation of redox active metals by suitable chelators, may also prove useful in inhibiting oxidation. While steps may be taken to reduce the levels of exogenous transition metals during food processing, it is impractical to completely remove metals (such as iron and copper) given that they are minor, endogenous components of foods. Therefore, changes to the redox activity of transition metals in polyphenol rich foods have the potential to greatly reduce prooxidant activity, especially if the chelator is capable of interfering with the redox cycling of metal catalysts.

Peroxide-Quenching Mechanisms

Due to the fact that •OH radicals are a highly reactive species, a better approach to limiting their prooxidant activity may lie in scavenging H_2O_2, the relatively stable precursor to •OH radicals. Proteins, in particular, are common food constituents that may scavenge peroxides via methionine and cysteine residues. Methionine residues in proteins are known to reduce H_2O_2 and lipid hydroperoxides to non-reactive hydroxides by two electron processes (Garner et al., 1998a, 1998b). Sulfhydryl groups may also reduce H_2O_2 by a similar mechanism (Denu & Tanner, 1998) as H_2O_2, generated from polyphenol oxidation, was effectively reduced in the presence of L-cysteine or glutathione (Aoshima & Ayabe, 2007). This may account for some of the antioxidant activity of proteins observed in lipid dispersions (Achyuthan et al., 2002; Elias et al., 2008b) and may be a viable route for scavenging H_2O_2 in systems that contain polyphenol. It is conceivable that the radical scavenging properties of polyphenols (that is, their ability to disrupt lipid peroxidation by quenching alkoxyl radicals) are revealed only when H_2O_2 is removed. For example, at pH 3 in a 5 wt% flax emulsion with added Fe^{3+} (25 μM), the introduction of β-lactoglobulin (1% w/v) and EGCG together (100 μM) effectively removed the prooxidant effect witnessed from EGCG treatment alone (Fig. 9.6; unpublished data). The combination of BLG + EGCG also showed greater antioxidant activity than the β-lactoglobulin control, suggesting that EGCG is a more effective antioxidant in the presence of food components like protein, which are capable of peroxide scavenging. Thus, the observation that proteins scavenge H_2O_2 may explain why tea catechins show antioxidant activity in protein-rich multiphase foods, such as meat systems (He & Shahidi, 1997; Tang et al., 2001b, 2002, 2006b; Yilmaz, 2006).

Sodium caseinate and whey protein isolate (WPI) were observed to effectively scavenge H_2O_2 in a hexadecane emulsion (Zhou et al., 2011). However, the peroxide-scavenging activity differed significantly between the two proteins, perhaps because of the content of labile amino acids and their relative accessibility to the solvent phase. Proteins may also influence phenolic oxidation as the introduction of sodium caseinate leads to a more rapid loss of EGCG in hexadecane emulsions (Zhou & Elias, 2011). Conversely, Ishii et al. (2011) demonstrated increased EGCG stability in the presence of human serum albumin. The effects of proteins on phenolic stability may also vary by phenolic type; the stabilization observed by Ishii et al. (2011) was significantly less for EGC compared to EGCG at pH 7.4.

The interaction between phenolic compounds and proteins may also be of interest as varying the protein in the continuous phase leads to different antioxidant activities (Estevez et al., 2008). Berry phenolics and proteins showed a synergistic antioxidant effect in emulsions (Salminen et al., 2010) and adducts resulting

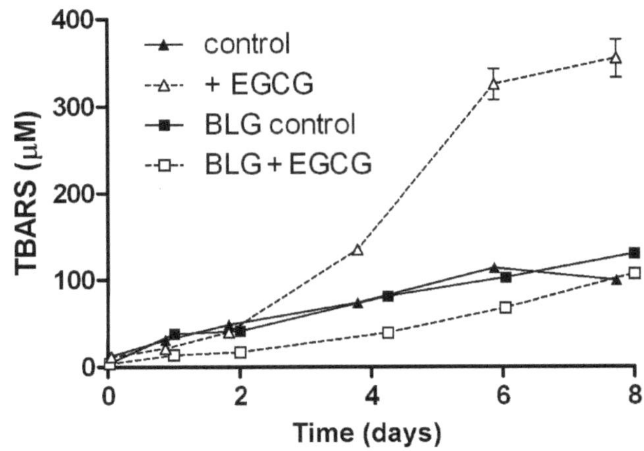

Fig. 9.6. TBARS concentration in 5 wt% flax emulsions stabilized with 1 wt% Tween 80 after the addition of EGCG (100 μM) and Fe^{3+} (25 μM) in 10 mM phosphate buffer pH 3 at 37 °C. (Unpublished results)

from polyphenolic compounds and proteins have also been shown to exert higher antioxidant activity (Almajano et al., 2007). The addition of phenolics that have already formed protein adducts may be an effective strategy, as the phenolic has already undergone oxidation, and ROS would have presumably been generated beforehand.

Transition Metal Chelator Strategies

Disrupting transition metal redox cycling is a potentially effective method for stabilizing phenolics in foods, thereby inhibiting lipid oxidation reactions. The reduction of metals by many phenolics (for example, catechols) occurs relatively quickly. For example, a 1:1 ratio of hydroquinone (HQ) to Cu^{2+} (100 μM) was able to reduce Cu^{2+} to Cu^{+} within 5 s, while conversion to the benzoquinone reached its highest level in 4 min (Li & Trush, 1993). However, it was also demonstrated that O_2 consumption and benzoquinone formation in the presence of Cu^{2+} could be inhibited by the copper chelator and bathocuproinedisulfonic acid, with the maximum effect observed at a metal:ligand ratio of 1:4.

As iron is the most ubiquitous transition metal in foods, this section will focus on iron chelator strategies. The effects of chelators on polyphenol oxidation have been mixed and depend on chelator type. Ligands with oxygen-based donor groups tend to bind Fe^{3+} ions, while nitrogen-based donor ligands often preferentially bind Fe^{2+} ions. Some ligands are capable of completely inactivating redox cycling. This may occur if the chelator occupies or blocks all of the metal's coordination sites, in which case oxidants like oxygen will be unable to bind. One such chelator is desferroxamine (DFO), a hexadentate ferric chelator with zero free coordination sites, which has been shown to prevent ascorbate oxidation and •OH radical formation (Mahoney & Graf, 1986). A strong binding affinity to a particular oxidation state, as we see in quercetin and Fe^{2+}, also leads to limited redox cycling. Other ligands, such as 2,2-bipyridine and EDTA, stabilize a specific redox state, thus influencing cycling between Fe^{3+} and Fe^{2+} states by promoting a preferred oxidation state.

However, it is important to note the effects of metal chelators on the overall oxidation mechanics, if the metal complex is still redox active. For instance, chelation by EDTA, a ferric chelator, results in one free coordination site that is capable of binding oxidants such as water, oxygen, or even H_2O_2, allowing the complex to readily generate •OH radicals by promoting Fe^{2+} oxidation. EDTA also stimulates the oxidation of phenolic compounds in the presence of trace metals (Bandy et al., 1987, 2001) by promoting transition metal oxidation. However, EDTA tends to have antioxidative effects on lipid oxidation (Cho et al., 2003; Hu et al., 2004), even in the presence of phenolic compounds (Zhou et al., 2012). Alternatively, 2,2-bipyridine is a ferrous chelator that stabilizes Fe^{2+} ions and has been shown to decrease metal-catalyzed phenolic oxidation. Additionally, it caused an increase in lipid oxidation due to an increase in catalytically active Fe^{2+} ions needed to generate •OH radicals or reduce lipid hydroperoxides. Many of these effects are also pH dependent. In the presence of EGCG at neutral pH, EDTA readily generated •OH radicals but was shown to inhibit the metal-catalyzed oxidation and subsequent generation of •OH radicals at acidic pH 3 (Zhou et al., 2012). Thus, the incorporation of metal chelators that do not fully inactivate transition metal cycling should be carefully studied to ensure that both metal-catalyzed phenolic oxidation and lipid oxidation are inhibited.

Oxygen Exclusion

An obvious yet extremely effective strategy for controlling the oxidation of phenolics is the removal or exclusion of oxygen. This may be accomplished by modified atmosphere packaging as well as vacuum packaging; both are known to greatly enhance lipid stability. The chemical stability of phenolics increases in low oxygen environments; ~95% EGCG (20 μM) was retained after 6 h in buffer flushed with gaseous N2 for 24 h, while nearly all EGCG was lost within 2 h in the same buffer system under normal atmospheric conditions (Sang et al., 2005). As would be expected, temperature can also influence metal-catalyzed phenolics oxidation. EGCG (500 μM) showed temperature-dependent increases in measured H_2O_2 concentrations, with an increase from 16 μM at 37 °C to more than 150 μM at 77 °C after only an 8 min incubation (Nakayama et al., 2002). Phenolic loss due to high temperature processing may be mitigated by lowering the matrix pH; green tea catechins in the buffer were stable (> 85% remaining) at pH 3 but were relative unstable at pH 6 (~20% remaining) after autoclaving for 20 min at 120 °C (Chen et al., 2001).

Conclusion

In order to deliver the benefits of phenolic compounds in lipid-based foods, we must take care to maintain their stability as well as to prevent potential prooxidative effects within the food matrix resulting from ROS (for example, H_2O_2 and •OH) generation. A better understanding of the conditions that promote phenolic instability and the resulting ROS production, and prooxidant activity, will enable manufacturers to formulate stable lipid foods containing biologically significant levels of phenolics.

References

Achyuthan, A. M.; Budi, E. H.; Elias, R.; Cardenas, M.; Cao, W.; McGinnis, J. F. Regulation of the rat recovering gene *in vivo* and in primary cell cultures. Invest. Ophthalmol Vis. Sci. 2002, 43, U324–U324.

Alamed, J.; Chaiyasit, W.; McClements, D. J.; Decker, E. A. Relationships between free radical scavenging and antioxidant activity in foods. J. Agric. Food Chem. 2009, 57, 2969–2976.

Alarcón, E.; Campos, A. M.; Edwards, A. M.; Lissi, E.; López-Alarcón, C. Antioxidant capacity of herbal infusions and tea extracts: A comparison of ORAC-fluorescein and ORAC-pyrogallol red methodologies. Food Chem. 2008, 107, 1114–1119.

Aliaga, C.; Rezende, M. C.; Arenas, A. How meaningful is the assessment of antioxidant activities in microheterogeneous media? Food Chem. 2009, 113, 1083–1087.

Almajano, M. P.; Delgado, M. E.; Gordon, M. H. Changes in the antioxidant properties of protein solutions in the presence of epigallocatechin gallate. Food Chem. 2007, 101, 126–130.

Amorati, R.; Pedulli, G. F.; Cabrini, L.; Zambonin, L.; Landi, L. Solvent and pH effects on the antioxidant activity of caffeic and other phenolic acids. J. Agric. Food Chem. 2006, 54, 2932–2937.

Aoshima, H.; Ayabe, S. Prevention of the deterioration of polyphenol-rich beverages. Food Chem. 2007, 100, 350–355.

Bandy, B.; Davison, A. J. Interactions between metals, ligands, and oxygen in the autoxidation of 6-hydroxydopamine: Mechanisms by which metal chelation enhances inhibition by superoxide dismutase. Arch. Biochem. Biophys. 1987, 259, 305–315.

Bandy, B.; Moon, J.; Davison, A. J. Multiple actions of superoxide dismutase: Why can it both inhibit and stimulate reduction of oxygen by hydroquinones? Free Radic. Biol. Med. 1990, 9, 143–148.

Bandy, B.; Walter, P. B.; Moon, J.; Davison, A. J. Reaction of oxygen with 6-hydroxydopamine catalyzed by Cu, Fe, Mn, and V complexes: Identification of a thermodynamic window for effective metal catalysis. Arch. Biochem. Biophys. 2001, 389, 22–30.

Bansal, K. M.; Patterson, L. K.; Fendler, E. J.; Fendler, J. H. Reactions of hydrated electrons, hydrogen atoms and hydroxyl radicals in micellar systems. Int. J. Radiat. Phys. and Chem. 1971, 3, 321–331.

Becker, E. M.; Ntouma, G.; Skibsted, L. H. Synergism and antagonism between quercetin and other chain-breaking antioxidants in lipid systems of increasing structural organisation. Food Chem. 2007, 103, 1288–1296.

Boersma, M. G.; Vervoort, J.; Szymusiak, H.; Lemanska, K.; Tyrakowska, B.; Cenas, N.; Segura-Aguilar, J.; Rietjens, I. M. C. M. Regioselectivity and reversibility of the glutathione conjugation of quercetin quinone methide. Chem. Res. Toxicol. 2000, 13, 185–191.

Bolton, J. L.; Turnipseed, S. B.; Thompson, J. A. Influence of quinone methide reactivity on the alkylation of thiol and amino groups in proteins: Studies utilizing amino acid and peptide models. Chem. Biol. Interact. 1997, 107, 185–200.

Bose, M.; Lambert, J. D.; Ju, J.; Reuhl, K. R.; Shapses, S. A.; Yang, C. S. The major green tea polyphenol, (-)-epigallocatechin-3-gallate, inhibits obesity, metabolic syndrome, and fatty liver disease in high-fat-fed mice. J. Nutr. 2008, 138, 1677–1683.

Cai, Y.-Z.; Mei, S.; Jie, X.; Luo, Q.; Corke, H. Structure-radical scavenging activity relationships of phenolic compounds from traditional Chinese medicinal plants. Life Sciences 2006, 78, 2872–2888.

Cartagena, L. G.; Perez-Zuniga, F. J.; Abad, F. B. Interactions of some environmental and chemical parameters affecting the color attributes of wine. Am. J. Enol. Vitic. 1994, 45, 43–48.

Chen, R.; Wang, J.-B.; Zhang, X.-Q.; Ren, J.; Zeng, C.-M. Green tea polyphenol epigallocatechin-3-gallate (EGCG) induced intermolecular cross-linking of membrane proteins. Arch. Biochem. Biophys. 2011, 507, 343–349.

Chen, Z.-Y.; Zhu, Q. Y.; Tsang, D.; Huang, Y. Degradation of green tea catechins in tea drinks. J. Agric. Food Chem. 2000, 49, 477–482.

Cho, Y.-J.; Alamed, J.; McClements, D. J.; Decker, E. A. Ability of chelators to alter the physical location and prooxidant activity of iron in oil-in-water emulsions. J. Food Sci. 2003, 68, 1952–1957.

Clark, A. C.; Scollary, G. R. Copper(II)-mediated oxidation of (+)-catechin in a model white wine system. Aust. J. Grape Wine Res. 2002, 8, 186–195.

Dai, F.; Chen, W.-F.; Zhou, B. Antioxidant synergism of green tea polyphenols with α-tocopherol and l-ascorbic acid in SDS micelles. Biochimie 2008, 90, 1499–1505.

Danilewicz, J. C. Interaction of sulfur dioxide, polyphenols, and oxygen in a wine-model system: Central role of iron and copper. Am. J. Enol. Vitic. 2007, 58, 53–60.

Danilewicz, J. C. Review of reaction mechanisms of oxygen and proposed intermediate reduction products in wine: Central role of iron and copper. Am. J. Enol. Vitic. 2003, 54, 73–85.

Danilewicz, J. C.; Seccombe, J. T.; Whelan, J. Mechanism of interaction of polyphenols, oxygen, and sulfur dioxide in model wine and wine. Am. J. Enol. Vitic. 2008, 59, 128–136.

Denu, J. M.; Tanner, K. G. Specific and reversible inactivation of protein tyrosine phosphatases by hydrogen peroxide: Evidence for a sulfenic acid intermediate and implications for redox regulation. Biochemistry 1998, 37, 5633–5642.

Dudonné, S. P.; Vitrac, X.; Coutiére, P.; Woillez, M.; Mérillon, J.-M. Comparative study of antioxidant properties and total phenolic content of 30 plant extracts of industrial interest using DPPH, ABTS, FRAP, SOD, and ORAC assays. J. Agric. Food Chem. 2009, 57, 1768–1774.

Dueñas, M.; Fulcrand, H.; Cheynier, V. Formation of anthocyanin–flavanol adducts in model solutions. Analytica Chimica Acta 2006, 563, 15–25.

Dueñas, M.; González-Manzano, S.; González-Paramás, A.; Santos-Buelga, C. Antioxidant evaluation of O-methylated metabolites of catechin, epicatechin and quercetin. J. Pharm. Biomed. Anal. 2010, 51, 443–449.

Elhabiri, M.; Carrër, C.; Marmolle, F.; Traboulsi, H. Complexation of iron(III) by catecholate-type polyphenols. Inorganica Chimica Acta 2007, 360, 353–359.

Elias, R. J.; Andersen, M. L.; Skibsted, L. H.; Waterhouse, A. L. Identification of free radical intermediates in oxidized wine using electron paramagnetic resonance spin trapping. J. Agric. Food Chem. 2009a, 57, 4359–4365.

Elias, R. J.; Andersen, M. L.; Skibsted, L. H.; Waterhouse, A. L. Key factors affecting radical formation in wine studied by spin trapping and EPR spectroscopy. Am. J. Enol. Vitic. 2009b, 60, 471–476.

Elias, R. J.; Kellerby, S. S.; Decker, E. A. Antioxidant activity of proteins and peptides. Crit. Rev. Food Sci. Nutr. 2008b, 48, 430–441.

Elias, R. J.; Laurie, V. F.; Ebeler, S. E.; Wong, J. W.; Waterhouse, A. L. Analysis of selected carbonyl oxidation products in wine by liquid chromatography with diode array detection. Analytica Chimica Acta 2008b, 626, 104–110.

Elias, R. J.; Waterhouse, A. L. Controlling the fenton reaction in wine. J. Agric. Food Chem. 2010, 58, 1699–1707.

Es-Safi, N.-E.; Cheynier, V. R.; Moutounet, M. Role of aldehydic derivatives in the condensation of phenolic compounds with emphasis on the sensorial properties of fruit-derived foods. J. Agric. Food Chem. 2002, 50, 5571–5585.

Esterbauer, H. Cytotoxicity and genotoxicity of lipid-oxidation products. Am. J. Clin. Nutr. 1993, 57, 779S–785S.

Estevez, M.; Kylli, P.; Puolanne, E.; Kivikari, R.; Heinonen, M. Oxidation of skeletal muscle myofibrillar proteins in oil-in-water emulsions: Interaction with lipids and effect of selected phenolic compounds. J. Agric. Food Chem. 2008, 56, 10933–10940.

Frankel, E.; Huang, S. W.; Aeschbach, R. Antioxidant activity of green teas in different lipid systems. J. Am. Oil Chem. Soc. 1997, 74, 1309–1315.

Freeman, B. L.; Eggett, D. L.; Parker, T. L. Synergistic and antagonistic interactions of phenolic compounds found in navel oranges. J. Food Sci. 2010, 75, C570–C576.

Fukuzawa, K.; Kishikawa, K.; Tadokoro, T.; Tokumura, A.; Tsukatani, H.; Gebicki, J. M. The effects of alpha-tocopherol on site-specific lipid peroxidation induced by iron in charged micelles. Arch. Biochem. Biophys. 1988, 260, 153–160.

Garner, B.; Waldeck, A. R.; Witting, P. K.; Rye, K. A.; Stocker, R. Oxidation of high density lipoproteins. II. Evidence for direct reduction of lipid hydroperoxides by methionine residues of apolipoproteins AI and AII. J. Biol. Chem. 1998a, 273, 6088–6095.

Garner, B.; Witting, P. K.; Waldeck, A. R.; Christison, J. K.; Raftery, M.; Stocker, R. Oxidation of high density lipoproteins. I. Formation of methionine sulfoxide in apolipoproteins AI and

AII is an early event that accompanies lipid peroxidation and can be enhanced by alpha-tocopherol. J. Biol. Chem. 1998b, 273, 6080–6087.

Guo, M.; Perez, C.; Wei, Y.; Rapoza, E.; Su, G.; Bou-Abdallah, F.; Chasteen, N. D. Iron-binding properties of plant phenolics and cranberry's bio-effects. Dalton Trans. 2007, 43, 4951–4961.

Guo, Q.; Zhao, B.; Shen, S.; Hou, J.; Hu, J.; Xin, W. ESR study on the structure-antioxidant activity relationship of tea catechins and their epimers. BBA—Gen. Subj. 1999, 1427, 13–23.

He, Y. H.; Shahidi, F. Antioxidant activity of green tea and its catechins in a fish meat model system. J. Agric. Food Chem. 1997, 45, 4262–4266.

Hider, R. C.; Liu, Z. D.; Khodr, H. H.; Lester, P. Metal Chelation of Polyphenols. Methods in Enzymology. Academic Press: San Diego, CA, 2001; 355, 190–203.

Hu, M.; Julian McClements, D.; Decker, E. A. Impact of chelators on the oxidative stability of whey protein isolate-stabilized oil-in-water emulsions containing Ω-3 fatty acids. Food Chem. 2004, 88, 57–62.

Huang, S. W.; Frankel, E. N. Antioxidant activity of tea catechins in different lipid systems. J. Agric. Food Chem. 1997, 45, 3033–3038.

Huber, G. M.; Vasantha Rupasinghe, H. P.; Shahidi, F. Inhibition of oxidation of omega-3 polyunsaturated fatty acids and fish oil by quercetin glycosides. Food Chem. 2009, 117, 290–295.

Hurst, W. J.; Payne, M. J.; Miller, K. B.; Stuart, D. A. Stability of cocoa antioxidants and flavan-3-ols over time. J. Agric. Food Chem. 2009, 57, 9547-9550.

Huvaere, K.; Olsen, K.; Skibsted, L. H. Quenching of triplet-excited flavins by flavonoids. Structural assessment of antioxidative activity. J. Org. Chem. 2009, 74, 7283–7293.

Iglesias, J.; Pazos, M.; Lois, S.; Medina, I. Contribution of galloylation and polymerization to the antioxidant activity of polyphenols in fish lipid systems. J. Agric. Food Chem. 2010, 58, 7423–7431.

Ishii, T.; Ichikawa, T.; Minoda, K.; Kusaka, K.; Ito, S.; Suzuki, Y.; Akagawa, M.; Mochizuki, K.; Goda, T.; Nakayama, T. Human serum albumin as an antioxidant in the oxidation of (-)-epigallocatechin gallate: Participation of reversible covalent binding for interaction and stabilization. Biosci. Biotechnol. Biochem. 2011, 75, 100–106.

Janeiro, P.; Oliveira Brett, A. M. Catechin electrochemical oxidation mechanisms. Anal. Chim. Acta 2004, 518, 109–115.

Jarabak, R.; Harvey, R. G.; Jarabak, J. Redox cycling of polycyclic aromatic hydrocarbon o-quinones: Metal ion-catalyzed oxidation of catechols bypasses inhibition by superoxide dismutase. Chem-Biol. Interact. 1998, 115, 201–213.

Kähkönen, M. P.; Hopia, A. I.; Vuorela, H. J.; Rauha, J.-P.; Pihlaja, K.; Kujala, T. S.; Heinonen, M. Antioxidant activity of plant extracts containing phenolic compounds. J. Agric. Food Chem. 1999, 47, 3954–3962.

Khokhar, S.; Owusu Apenten, R. K. Iron binding characteristics of phenolic compounds: Some tentative structure-activity relations. Food Chem. 2003, 81, 133–140.

Knowles, P.; Gibson, J.; Pick, F.; Bray, R. Electron-spin-resonance evidence for enzymic reduction of oxygen to a free radical, the superoxide ion. Biochemical Journal 1969, 111, 53–58.

Kumamoto, M.; Sonda, T.; Nagayama, K.; Tabata, M. Effects of pH and metal ions on antioxidative activities of catechins. Biosci. Biotechnol. Biochem. 2001, 65, 126–132.

Laguerre, M. L.; López Giraldo, L. J.; Lecomte, J.; Figueroa-Espinoza, M.-C.; Baréa, B.; Weiss, J.; Decker, E. A.; Villeneuve, P. Chain length affects antioxidant properties of chlorogenate esters in emulsion: The cutoff theory behind the polar paradox. J. Agric. Food Chem. 2009, 57, 11335–11342.

Laguerre, M. L.; López Giraldo, L. J.; Lecomte, J.; Figueroa-Espinoza, M.-C.; Baréa, B.; Weiss, J.; Decker, E. A.; Villeneuve, P. Relationship between hydrophobicity and antioxidant ability of "phenolipids" in emulsion: A parabolic effect of the chain length of rosmarinate esters. J. Agric. Food Chem. 2010, 58, 2869–2876.
Lambert, J. D.; Sang, S.; Yang, C. S. Possible controversy over dietary polyphenols: Benefits vs risks. Chem. Res. Toxicol. 2007, 20, 583–585.
Lapidot, T.; Walker, M. D.; Kanner, J. Can apple antioxidants inhibit tumor cell proliferation? Generation of H_2O_2 during interaction of phenolic compounds with cell culture media. J. Agric. Food Chem. 2002, 50, 3156–3160.
Leopoldini, M.; Russo, N.; Chiodo, S.; Toscano, M. Iron chelation by the powerful antioxidant flavonoid quercetin. J. Agric. Food Chem. 2006, 54, 6343–6351.
Let, M. B.; Jacobsen, C.; Meyer, A. S. Ascorbyl palmitate, γ-tocopherol, and EDTA affect lipid oxidation in fish oil enriched salad dressing differently. J. Agric. Food Chem. 2007, 55, 2369–2375.
Lewis, M. A.; Graff Yoerg, D.; Bolton, J. L.; Thompson, J. A. Alkylation of 2'-deoxynucleosides and DNA by quinone methides derived from 2,6-di-tert-butyl-4-methylphenol. Chem. Res. Toxicol. 1996, 9, 1368–1374.
Li, Y.; Trush, M. A.; Yager, J. D. DNA damage caused by reactive oxygen species originating from a copper-dependent oxidation of the 2-hydroxy catechol of estradiol. Carcinogenesis 1994, 15, 1421–1427.
Li, Y. B.; Trush, M. A. Oxidation of hydroquinone by copper: Chemical mechanism and biological effects. Arch. Biochem. Biophys. 1993, 300, 346–355.
Liao, H.; Cai, Y.; Haslam, E. Polyphenol interactions. Anthocyanins: Co-pigmentation and colour changes in red wines. J. Sci. Food Agric. 1992, 59, 299-305.
Ma, X.; Wu, H.; Liu, L.; Yao, Q.; Wang, S.; Zhan, R.; Xing, S.; Zhou, Y. Polyphenolic compounds and antioxidant properties in mango fruits. Scientia Horticulturae 2011, 129, 102–107.
Mahoney, J. R.; Graf, E. Role of alpha-tocopherol, ascorbic acid, citric acid and EDTA as oxidants in model systems. J. Food Sci. 1986, 51, 1293–1296.
Makrigiorgos, G. M.; Bump, E.; Huang, C.; Baranowska-Kortylewicz, J.; Kassis, A. I. A fluorimetric method for the detection of copper-mediated hydroxyl free radicals in the immediate proximity of DNA. Free Radic. Biol. Med. 1995, 18, 669–678.
Maqsood, S.; Benjakul, S. Comparative studies of four different phenolic compounds on in vitro antioxidative activity and the preventive effect on lipid oxidation of fish oil emulsion and fish mince. Food Chem. 2010, 119, 123–132.
Mei, L.; McClements, D. J.; Decker, E. A. Lipid oxidation in emulsions as affected by charge status of antioxidants and emulsion droplets. J. Agric. Food Chem. 1999, 47, 2267–2273.
Meyer, A. S.; Heinonen, M.; Frankel, E. N. Antioxidant interactions of catechin, cyanidin, caffeic acid, quercetin, and ellagic acid on human LDL oxidation. Food Chem. 1998, 61, 71–75.
Mielnik, M. B.; Olsen, E.; Vogt, G.; Adeline, D.; Skrede, G. Grape seed extract as antioxidant in cooked, cold stored turkey meat. LWT—Food Science and Technology 2006, 39, 191–198.
Mira, L.; Fernandez, M. T.; Santos, M.; Rocha, R.; Florêncio, M. H.; Jennings, K. R. Interactions of flavonoids with iron and copper ions: A mechanism for their antioxidant activity. Free Radical Res. 2002, 36, 1199–1208.
Mochizuki, M.; Yamazaki, S.-I.; Kano, K.; Ikeda, T. Kinetic analysis and mechanistic aspects of autoxidation of catechins. Biochimica et Biophysica Acta (BBA)—General Subjects 2002, 1569, 35–44.

Nakagawa, H.; Wachi, M.; Woo, J.-T.; Kato, M.; Kasai, S.; Takahashi, F.; Lee, I.-S.; Nagai, K. Fenton reaction is primarily involved in a mechanism of (-)-epigallocatechin-3-gallate to induce osteoclastic cell death. Biochem. Biophys. Res. Commun. 2002, 292, 94–101.

Nakayama, T.; Ichiba, M.; Kuwabara, M.; Kajiya, K.; Kumazawa, S., Mechanisms and structural specificity of hydrogen peroxide formation during oxidation of catechins. Food Sci. Technol. Res. 2002, 8, 261-267.

Nanjo, F.; Goto, K.; Seto, R.; Suzuki, M.; Sakai, M.; Hara, Y. Scavenging effects of tea catechins and their derivatives on 1,1-diphenyl-2-picrylhydrazyl radical. Free Radic. Biol. Med. 1996, 21, 895–902.

Nanjo, F.; Mori, M.; Goto, K.; Hara, Y. Radical scavenging activity of tea catechins and their related compounds. Biosci., Biotechnol., Biochem. 1999, 63, 1621–1623.

Nappi, A. J.; Vass, E. Comparative studies of enhanced iron-mediated production of hydroxyl radical by glutathione, cysteine, ascorbic acid, and selected catechols. Biochimica et Biophysica Acta (BBA)—General Subjects 1997, 1336, 295–302.

Oszmianski, J.; Cheynier, V. R.; Moutounet, M. Iron-catalyzed oxidation of (+)-catechin in model systems. J. Agric. Food Chem. 1996, 44, 1712–1715.

Pazos, M.; Iglesias, J.; Maestre, R.; Medina, I. Structure–activity relationships of polyphenols to prevent lipid oxidation in pelagic fish muscle. J. Agric. Food Chem. 2010, 58, 11067–11074.

Poncet-Legrand, C. L.; Cartalade, D.; Putaux, J.-L.; Cheynier, V.; Vernhet, A. Flavan-3-ol aggregation in model ethanolic solutions: Incidence of polyphenol structure, concentration, ethanol content, and ionic strength. Langmuir 2003, 19, 10563-10572.

Remy-Tanneau, S.; Le Guernevé, C.; Meudec, E.; Cheynier, V. Characterization of a colorless anthocyanin-flavan-3-ol dimer containing both carbon-carbon and ether interflavanoid linkages by NMR and mass spectrometry. J. Agric. Food Chem. 2003, 51, 3592-3597.

Rice-Evans, C. A.; Miller, N. J.; Paganga, G. Structure-antioxidant activity relationships of flavonoids and phenolic acids. Free Radic. Biol. Med. 1996, 20, 933–956.

Roedig-Penman, A.; Gordon, M. H. Antioxidant properties of catechins and green tea extracts in model food emulsions. J. Agric. Food Chem. 1997, 45, 4267–4270.

Roy, M. K.; Koide, M.; Rao, T. P.; Okubo, T.; Ogasawara, Y.; Juneja, L. R. ORAC and DPPH assay comparison to assess antioxidant capacity of tea infusions: Relationship between total polyphenol and individual catechin content. Int. J. Food Sci. Nutr. 2010, 61, 109–124.

Salminen, H.; Heinonen, M.; Decker, E. A. Antioxidant effects of berry phenolics incorporated in oil-in-water emulsions with continuous phase β-lactoglobulin. J. Am. Oil Chem. Soc. 2010, 87, 419–428.

Sang, S.; Lee, M.-J.; Hou, Z.; Ho, C.-T.; Yang, C. S. Stability of tea polyphenol (-)-epigallocatechin-3-gallate and formation of dimers and epimers under common experimental conditions. J. Agric. Food Chem. 2005, 53, 9478–9484.

Sang, S.; Yang, I.; Buckley, B.; Ho, C.-T.; Yang, C. S. Autoxidative quinone formation in vitro and metabolite formation *in vivo* from tea polyphenol (-)-epigallocatechin-3-gallate: Studied by real-time mass spectrometry combined with tandem mass ion mapping. Free Radic. Biol. Med. 2007, 43, 362–371.

Sasaki, K.; Alamed, J.; Weiss, J.; Villeneuve, P.; López Giraldo, L. J.; Lecomte, J.; Figueroa-Espinoza, M.-C.; Decker, E. A. Relationship between the physical properties of chlorogenic acid esters and their ability to inhibit lipid oxidation in oil-in-water emulsions. Food Chem. 2011, 118, 830–835.

Sebranek, J. G.; Sewalt, V. J. H.; Robbins, K. L.; Houser, T. Comparison of a natural rosemary extract and BHA/BHT for relative antioxidant effectiveness in pork sausage. Meat Sci. 2005, 69, 289–296.

Severino, J. F.; Goodman, B. A.; Kay, C. W. M.; Stolze, K.; Tunega, D.; Reichenauer, T. G.; Pirker, K. F. Free radicals generated during oxidation of green tea polyphenols: Electron paramagnetic resonance spectroscopy combined with density functional theory calculations. Free Radic. Biol. Med. 2009, 46, 1076-1088.

Shrime, M. G.; Bauer, S. R.; McDonald, A. C.; Chowdhury, N. H.; Coltart, C. E. M.; Ding, E. L. Flavonoid-rich cocoa consumption affects multiple cardiovascular risk factors in a meta-analysis of short-term studies. J. Nutr. 2011, 141, 1982–1988.

Tang, S.; Sheehan, D.; Buckley, D. J.; Morrissey, P. A.; Kerry, J. P. Anti-oxidant activity of added tea catechins on lipid oxidation of raw minced red meat, poultry and fish muscle. Int. J. Food Sci. Technol. 2001, 36, 685–692.

Tang, S. Z.; Kerry, J. P.; Sheehan, D.; Buckley, D. J. Antioxidative mechanisms of tea catechins in chicken meat systems. Food Chem. 2002, 76, 45–51.

Tang, S. Z.; Ou, S. Y.; Huang, X. S.; Li, W.; Kerry, J. P.; Buckley, D. J. Effects of added tea catechins on colour stability and lipid oxidation in minced beef patties held under aerobic and modified atmospheric packaging conditions. J. Food Eng. 2006, 77, 248–253.

Thielecke, F.; Boschmann, M. The potential role of green tea catechins in the prevention of the metabolic syndrome—A review. Phytochemistry 2009, 70, 11–24.

Thomas, C.; McLean, L.; Parker, R.; Ohlweiler, D. Ascorbate and phenolic antioxidant interactions in prevention of liposomal oxidation. Lipids 1992, 27, 543–550.

Timberlake, C. F.; Bridle, P. Interactions between anthocyanins, phenolic compounds, and acetaldehyde and their significance in red wines. Am. J. Enol. Vitic. 1976, 27, 97–105.

Tsai, P.-J.; Huang, H.-P. Effect of polymerization on the antioxidant capacity of anthocyanins in roselle. Food Res. Int. 2004, 37, 313–318.

Valcic, S.; Burr, J. A.; Timmermann, B. N.; Liebler, D. C. Antioxidant chemistry of green tea catechins. New oxidation products of (-)-epigallocatechin gallate and (-)-epigallocatechin from their reactions with peroxyl radicals. Chem. Res. Toxicol. 2000, 13, 801–810.

Velasco, J.; Dobarganes, M. C.; Márquez-Ruiz, G. Antioxidant activity of phenolic compounds in sunflower oil-in-water emulsions containing sodium caseinate and lactose. Eur. J. Lipid Sci. Technol. 2004, 106, 325–333.

Waterhouse, A. L.; Laurie, V. F. Oxidation of wine phenolics: A critical evaluation and hypotheses. Am. J. Enol. Vitic. 2006, 57, 306–313.

Wildenradt, H. L.; Singleton, V. L. The production of aldehydes as a result of oxidation of polyphenolic compounds and its relation to wine aging. Am. J. Enol. Vitic. 1974, 25, 119–126.

Yang, C. S.; Wang, X.; Lu, G.; Picinich, S. C. Cancer prevention by tea: Animal studies, molecular mechanisms and human relevance. Nat. Rev. Cancer 2009, 9, 429–439.

Yilmaz, Y. Novel uses of catechins in foods. Trends Food Sci. Tech. 2006, 17, 64-71.

Yuji, H.; Weiss, J.; Villeneuve, P.; López Giraldo, L. J.; Figueroa-Espinoza, M.-C.; Decker, E. A. Ability of surface-active antioxidants to inhibit lipid oxidation in oil-in-water emulsion. J. Agric. Food Chem. 2007, 55, 11052-11056.

Zhang, L.; Bandy, B.; Davison, A. J. Effects of metals, ligands and antioxidants on the reaction of oxygen with 1,2,4-benzenetriol. Free Radic. Biol. Med. 1996, 20, 495–505.

Zhao, B.; Guo, Q.; Xin, W.; Lester, P. Free Radical Scavenging by Green Tea polyphenols. Methods in Enzymology; Academic Press: San Diego, CA, 2001; Vol. 335, 217–231.
Zhou, L.; Elias, R. J. Investigating the hydrogen peroxide quenching capacity of proteins in polyphenol-rich foods. J. Agric. Food Chem. 2011, 59, 8915–8922.
Zhou, L.; Elias, R. J. Factors influencing the antioxidant and prooxidant activity of polyphenols in oil-in-water emulsions. J. Agric. Food Chem. 2012, 60, 2906–2915.

CHAPTER

10

Antioxidant Evaluation and Antioxidant Activity Mechanisms

Dejian Huang[1] and Ik Chian Wong[1]

Introduction

Mounting evidences suggest that oxidative stress causes chronic diseases, including cardiovascular diseases (Stocker & Keaney, 2004), cancer (Valko et al., 2006), diabetes (Rains & Jain, 2011), neurodegenerative diseases (Jomova et al., 2010), and aging related ailments (Sohal & Orr, 2012). Preventing these chronic diseases by using naturally occurring antioxidants (or radical scavengers) has become a very attractive idea, as evidenced by a large number of published scientific articles. These papers broadly cover the efficacy and reaction mechanisms of dietary antioxidants for reducing oxidative stress in human clinical trials (Tinkel et al., 2012), animal models (Styskal et al., 2012), cell lines (Speciale et al., 2011), and chemical systems (Liu, 2010). A large number of assays have been developed to evaluate the activity of dietary antioxidants *in vitro,* which experts in the field have reviewed (Huang et al., 2005; Takayuki, 2009; Etsuo, 2010). From these reviews, we see a lack of correlation between the radical- (or oxidant-) scavenging activity measured by these "one-dimensional" chemical assays and their performance *in vivo* or in food systems. As a result, other researchers have debated the validity of these assays (Yoo et al., 2007). In this chapter, we provide an overview of lipid peroxidation chemistry. Then, we transition into a brief summary of existing methods for evaluating antioxidants, along with their mechanisms, scopes, and limitations as they correlate with biological and food lipid systems. Finally, we give a detailed

[1]Food Science and Technology Program, Department of Chemistry, National University of Singapore, 3 Science Drive 3, Singapore 117543, Singapore.

account of the fluorescent methods recently used to evaluate antioxidant capacity for protection against lipid oxidation.

Cause and Prevention of Lipid Autoxidation

Organic biomolecules are sensitive to oxidation, which causes food products to lose their biological function and quality, in terms of their nutritional value and safety. Among proteins, DNA/RNA, and lipids, protecting the latter from oxidative damage is perhaps the number-one challenge for food technologists dealing with oxidative stability of foods. When an oxidant initiates a radical chain reaction, it can damage lipid molecules, particularly polyunsaturated fatty acids. In addition, the reactive secondary oxidation products from lipids (such as α, β-unsaturated aldehydes) can have toxic effects as they chemically modify proteins. Yet, multiple factors can initiate lipid autoxidation, as Fig. 10.1 shows.

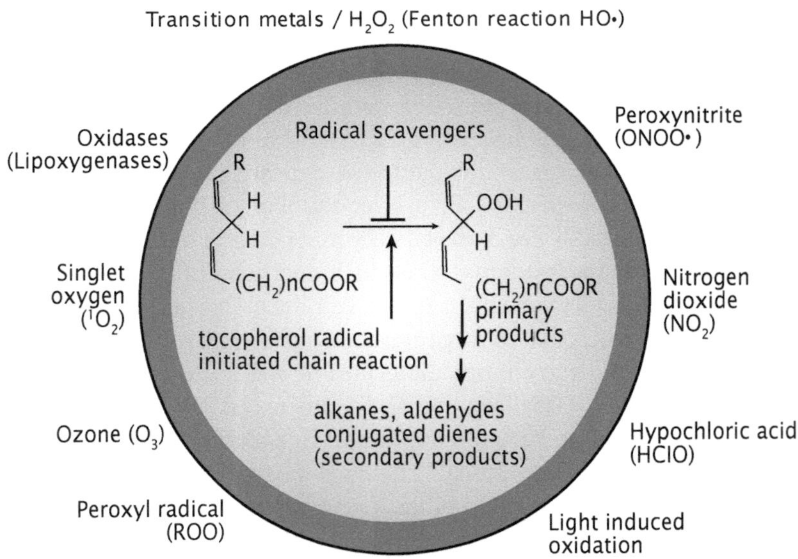

Fig. 10.1. Multiple pathways leading to lipid autooxidation make antioxidant evaluation a highly complex task. Oxidants include enzymes, light, a wide range of reactive oxygen species, and the combination of transition metal ions and peroxides. The outer circle represents antioxidants' actions in preventing these oxidants from initiating lipid autoxidation. Radical chain breakers work within the oil phase to stop the reaction.

In biological systems, oxidases (such as 13-lipoxygenase in soybeans) catalyze lipid peroxidation in response to oxidative stress or inflammation. In combination with hydrogen peroxide or oxygen, redox-active transition metal ions [such as Fe(II)] (Carlsen et al., 2005) can catalyze lipid oxidation in the presence of reductants (or prooxidants, in this situation). Research has shown that the semiquinone radicals generated through hydrogen-atom abstraction from hydroquinones lead to the formation of hydroperoxyl radicals (Valgimigli et al., 2008), and tocopheryl radicals formed after hydrogen atom abstraction from tocopherol initiate low-density lipoprotein (LDL) autoxidation (Alessi et al., 2002). In addition, autoxidation can be initiated by reactive oxygen species, including peroxynitrite (Rubbo et al., 2009), nitrogen dioxide (Pryor et al., 2006), singlet oxygen (Uemi et al., 2011), peroxyl radicals, hypochlorous acid (Ford, 2010), ozone (Wisthaler & Weschler, 2010), and UV-Vis light (Dalsgaard et al., 2011). Therefore, it is a complicated task to comprehensively assess antioxidant capacity for lipid protection. In fact, the limitations of current methods are apparent when considering the complexity of lipid peroxidation. Radical chain breakers are most effective at preventing oxidative damage because one chain reaction can lead to oxidation of many lipid molecules. It is no surprise that antioxidant research has largely been focused on the capacity of antioxidants to act as radical scavengers, and less attention has been paid to preventing radical-chain initiations.

Common Antioxidant Evaluation Methods Based on Chemical Reactions

There are many chemical methods used for measuring antioxidant capacity. Based on chemical reaction mechanisms, these methods fall into two basic groups: electron transfer (ET) reaction-based redox titration methods and hydrogen atom transfer (HAT) reaction-based methods, which use synthetic molecular probes to measure antioxidants (Huang et al., 2005). Proponents of ET-based titration methods assume that antioxidation is equivalent to reduction (Benzie et al., 1999). Obviously, they have oversimplified the situation, at least in the case of lipid peroxidation. ET-based assays come in different forms that depend on the oxidants used. Popular examples include total phenolic assay using a Folin–Ciocalteu reagent (Subramanian et al., 1965), ferric [Fe(III)] ion–reducing antioxidant power (FRAP) assay (Benzie & Strain, 1996), 2,2-diphenyl-1-picrylhydrazyl (DPPH) assay (Boguth & Repges, 1969), 2,2'-azino-bis(3-ethylbenzothiazoline-6-sulphonic acid) (ABTS) assay (Miller & Rice Evans, 1996), cupric-reducing antioxidant capacity (CUPRAC) (Apak et al., 2004), and cerium(IV)-based reducing antioxidant capacity (CERAC) assay (Ozyurt et al., 2010). The reactions of these single-electron oxidants with reductants present in the sample cause color changes. A spectrophotometer can conveniently determine the endpoint of the reaction. The degree of color change is correlated to the concentrations of "antioxidants" in the sample. Due to their simplicity, these assays have been broadly applied for measuring antioxidant activity in food and biological samples.

The HAT-based assays are represented by the oxygen radical absorbance capacity (ORAC) assay (Cao et al., 1993), which has been gradually upgraded through automation (Cao et al., 1995), replacement of fluorescent proteins with small molecular probes (such as fluorescein) (Ou et al., 2001), and the realization of high-throughput screening using a 96-well microplate reader (Huang et al., 2002). The U.S. Department of Agriculture adopted the high-throughput ORAC assay as its method of choice for generating data for its databases (Wu et al., 2004). The ORAC assay applies competitive reaction kinetics (Eqs. 10.1 to 10.4):

$$R-N=N-R \rightarrow N_2 + R\bullet \qquad \text{(Eq. 10.1)}$$

$$R\bullet + O_2 \rightarrow ROO\bullet \qquad \text{(Eq. 10.2)}$$

$$ROO\bullet + \text{fluorescein} \rightarrow ROOH + \text{fluorescein dimer} \qquad \text{(Eq. 10.3)}$$

$$ROO\bullet + \text{antioxidant} \rightarrow ROOH + \text{oxidized antioxidant} \qquad \text{(Eq. 10.4)}$$

Peroxyl radicals are generated by respective diazo precursors through thermal decomposition, producing an alkyl radical (Eq. 10.1), which reacts rapidly with molecular oxygen dissolved in the reaction mixture to form a peroxyl radical (ROO•) (Eq. 10.2). The radical reacts with a fluorescent probe (such as fluorescein; Eq. 10.3) and an antioxidant (Eq. 10.4) in the same reaction solution. The kinetics of fluorescence decay are monitored over time until the fluorescence signal decreases to the baseline. The sample's ability to protect the fluorescein from damage by radicals is quantified in reference to Trolox, a water-soluble vitamin E analogue, and expressed as $ORAC_{FL}$ values.

Scope and Limitation of ET- and HAT-Based Antioxidant Activity Assays

These assays (particularly the DPPH, $ORAC_{FL}$, and total phenolic assays) have been broadly applied for screening the antioxidant activity of food and biological samples, but disputes have arisen surrounding the lack of correlation between dietary antioxidant activity measured using these assays and the performance of these antioxidants in biological and food systems (Alamed et al., 2009). These systems are more complicated than in those of the chemical assays used. We must account for many more factors in order to fully evaluate antioxidant performance even in a chemical system (Huang et al., 2002; Yoo et al., 2007). As a chemical kinetic property of antioxidants, antioxidant activity depends on many factors, such as media, temperature, the nature of oxidants, and microenvironments in the biological and food systems. The results obtained from these assays can only serve as parameters for the comparison of different samples in the context of individual assay conditions. It is important that we do not overinterpret data obtained from these assays.

It sounds reasonable, but not necessarily scientific, to expect that chemical assay results will correlate well with antioxidant performance in actual food or biological systems. In chemical assays, the antioxidants are measured by challenging antioxidants with excessive synthetic oxidants [such as DPPH, Fe(III)/Cu(II) salts, and diazo radical precursors] using accelerated oxidation. The reactions normally finish within an hour, so the assay covers a reasonable time frame. Air oxidation of the antioxidants is much slower and does not pose a problem during the assay. However, oxidative stress occurs in a real food or biological system without the forced addition of reactive oxygen species or oxidants. Antioxidants will be consumed slowly (for days or months) by molecular oxygen dissolved in the system or by oxidants randomly formed by the light or chemicals present. In biological systems, antioxidants could be oxidized by various oxidases or reactive oxygen species generated in the oxygen metabolism. Potent antioxidants in food systems are often oxygen sensitive and lead to slow oxidation (covering hours or days) with molecular oxygen and the formation of oxygen radicals and hydrogen peroxide. For example, tea and coffee degrade over time to form hydrogen peroxide, which can accumulate up to 100 μM (Long et al., 1999). This is a problem particularly under a physiological pH of 7.40, when some acidic phenolic compounds are deprotonated and the phenolates become more reactive towards molecular oxygen via the electron transfer reaction (Long et al., 2000; Quek & Huang, 2011). Therefore, antioxidants may act as prooxidants in food and biological systems, although they may give high antioxidant values in chemical assays.

Because of their high reactivity, many reactive oxygen species have a short lifetime and do not diffuse far from their sites of generation. Therefore, it is critical to place antioxidants close to the radical generation site to control the ROS and protect biomolecules. For example, carotenoids are plants' first line of defense against singlet oxygen toxicity in a light-harvesting complex photosystem

II (LHC-II) (Liu et al., 2004). The LHC-II is the principal solar energy collector during the photosynthesis of green leaves. The X-ray structure of LHC-II reveals that each monomer of the tetramer membrane proteins contains four carotenoids, which act as antioxidants that quench singlet oxygen through both physical and chemical pathways (Ramel et al., 2012). In food systems, the phase distribution of antioxidants and interactions with other food components (such as the binding of polyphenolic compounds with proteins, starch, and fiber) may alter the availability of antioxidants to neutralize radicals. However, this is not an issue in typical antioxidant capacity assays carried out in homogenous solutions. Based on the above rationale, it is not surprising that there is poor correlation between the "antioxidant capacity" of a sample measured using chemical reactions and "antioxidant performance" in food and biological systems.

Antioxidant Evaluation Methods Based on Lipid Peroxidation Model

Controlling lipid autoxidation has been a theme of research among food scientists for more than a century due to the need to prolong the shelf life of food products and to develop products containing desirable nutrients and functional ingredients, such as n-3 fatty acids. Furthermore, it would be ideal to replace synthetic antioxidants, such as butylated hydroxytoluene (BHT) and butylated hydroxyanisole (BHA), with naturally occurring radical scavengers that are high-performance, cost-effective, and beneficial to human health. Several analytical methods exist to quantify an antioxidant's capacity to delay lipid peroxidation. In 2007, Villeneuve and coworkers reviewed this topic in great detail (Laguerre et al., 2007); therefore, this section will review related methods published after 2007.

Lipid peroxidation (Scheme 10.1) leads to a drop in oxygen concentration and the accumulation of lipid peroxide. To monitor the reaction progress, one must quantify the concentrations of conjugated diene hydroperoxide (A_{max} = 234 nm) or the oxygen consumption rate (for example, Rancimat® or oxygen electrode). The primary products of linoleate peroxidation (L-H = linoleate) also include a kinetic product, 11-hydroperoxy-linoleate (Brash, 2000). At high concentrations, α-tocopherol (0.1–1 M) facilitates the formation of this isomer over the conjugated diene peroxide isomer (Tallman et al., 2001). At low tocopherol concentration, which most antioxidant assays use (< 100 μM), conjugated diene peroxides are the dominant primary oxidation products (Pratt et al., 2011).

At 234 nm, the absorbance wavelength of conjugated dienes is short and easily interfered by other components, particularly phenolic compounds and natural

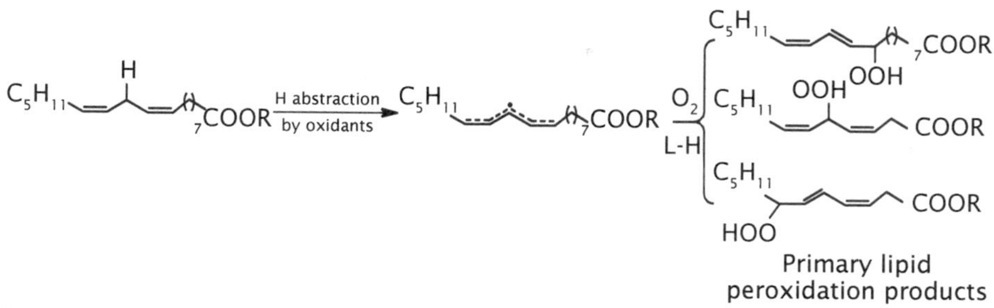

Scheme 10.1.

pigments, such as carotenoids. An oxygen electrode can only be applied in an aqueous environment, which means that it has a limited scope. To solve these problems, researchers have developed improved methods for monitoring lipid peroxidations in the past five years, which we review next.

Conjugated Autoxidizable Triene (CAT) Assay

Villeneuve and coworkers (2007) developed this assay by taking advantage of the unique triacylglycerols (TAGs) found in *Aleurites fordii* seed oil (tung oil) as an oxidizable substrate (Laguerre et al., 2008). The TAGs in tung oil contain ~86% el-eostearic acids (mainly, Δ9c, Δ11t, Δ13t isomer). As a conjugated triene, it has an absorption peak at 273 nm. The tung oil was carried out in an oil-in-water emulsion at 37°C with a 2,2'-azobis(2-amidinopropane) dihydrochloride (AAPH) precursor for peroxy radicals, which destroyed the conjugated trienes; the absorbance of the peak decreased as the reaction progressed. In the CAT assay, the antioxidant is approximately 100 times less concentrated than the substrate (Tung oil); thus, the interference due to UV-Vis absorption at 273 nm can be minimized. In addition, the molar ratio between the antioxidant and the substrate fits the definition of an antioxidant, which demands a small amount of antioxidant compounds for the protection of a large amount of substrate (such as vitamin E in plant oil). The quantification of the CAT assay is similar to that of the ORACFL assay where the area under the curve (AUC) correlated with the antioxidant's concentration (Fig. 10.2). Additionally, they found good linearity between antioxidant concentration and the AUC. The assay has since been used to measure the antioxidant activity of olive-leaf phenolics (Laguerre et al., 2009). It remains to be seen if the antioxidant activity measured by CAT assay correlates with antioxidant performance in real food and biological systems.

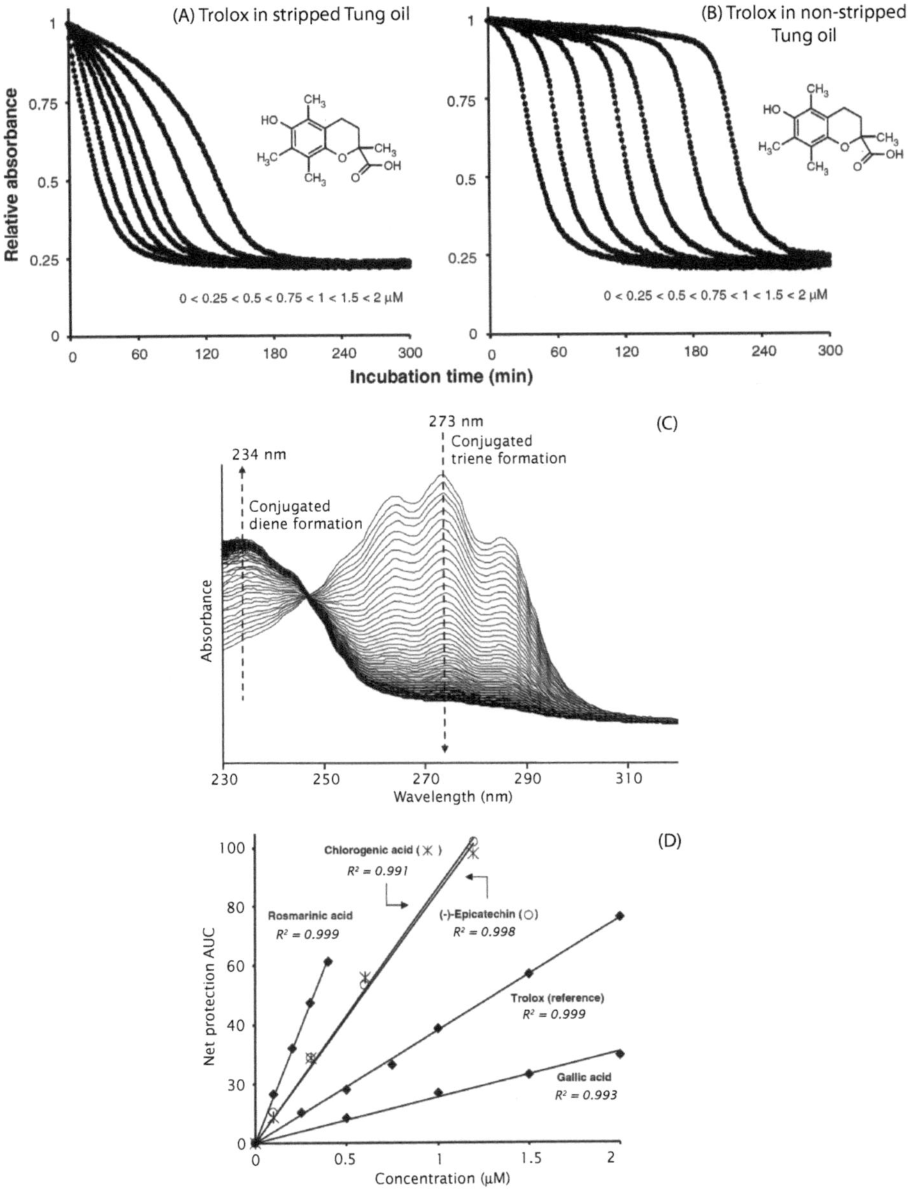

Fig. 10.2. Typical kinetic curves of CAT assay: (A) The unhibited and inhibited peroxidation of stripped Tung oil by Trolox; (B) The oxidation kinetics of nonstripped Tung oil in the presence of Trolox; (C) The UV spectra changes upon oxidation of Tung oil; (D) Dose response plot of area under the curve (AUC) versus selected antioxidants.

Adapted from Laguerre, M., Lopez-Giraldo, L. J. et al., Anal. Biochem. 380 (2): 282–290, with permission.

Oxygen Radical Absorbance Capacity in Bulk Oil—ORAC$_{oil}$ Assay

Besides conjugated lipid peroxides, the oxygen consumption rate could also be monitored in a closed system to determine the progress of autoxidation. To this end, luminescent ruthenium and iridium probes have great potential because the intensity of their phosphorescence is sensitive to molecular oxygen dissolved in the solution (Fig. 10.3) (Ruggi et al., 2011). Due to the long duration of their triplet excited states, these complexes can interact strongly with dioxygen and quench the phosphorescence. This phenomenon that can be readily exploited for oxygen sensing during lipid peroxidation events, as well as in other biomedical systems where oxygen consumption is involved. For high-throughput screening purposes, the Ru(II)(BIPY)$_3$ derivatives were embedded at the bottom of 96-well microplates, which are commercially available and used in studies of cytotoxicity and cell proliferation (Wodnicka et al., 2000) to monitor cell growth dynamics under various conditions.

We used these microplates to develop an ORAC$_{oil}$ assay to monitor methyl linoleate autoxidation initiated by 2,2'-azobis(2,4-dimethylvaleronitrile) (AMVN). Antioxidant activity was quantified and ranked for tocopherols and cold-pressed, organically grown vegetable seed oils (Hay et al., 2006). We monitored reaction kinetics until the oxygen in the oil was completely consumed. The phosphorescence signal indensity values can transform oxygen concentrations to give oxygen consumption kinetic during the oxidation reaction. The net area under the kinetic curve is integrated and plotted against the concentration of antioxidant (Fig. 10.4F). The results of ORAC$_{oil}$ assay is expressed as Trolox equivalents (TE), as is the case for the ORAC$_{FL}$ assay. The difference between ORAC$_{oil}$ and ORAC$_{FL}$ is that the former measures the inhibition of lipid autoxidation in an oil-phase oxidizable substrate—methyl linoleate—in great excess as compared to radical scavengers (such as Trolox). In contrast, the oxidizable substrate (fluorescein) has much lower concentration than that of antioxidants in the ORAC$_{FL}$ assay, and the

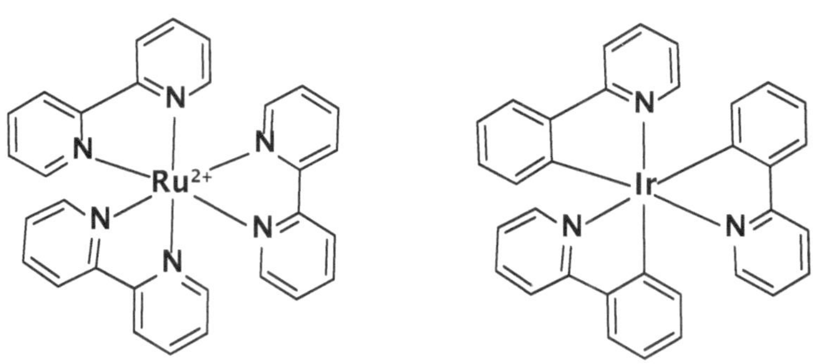

Fig. 10.3. Chemical structure of phosphorescence probes, Ru(BIPY)$_3$ and Ir(BPY)$_3$.

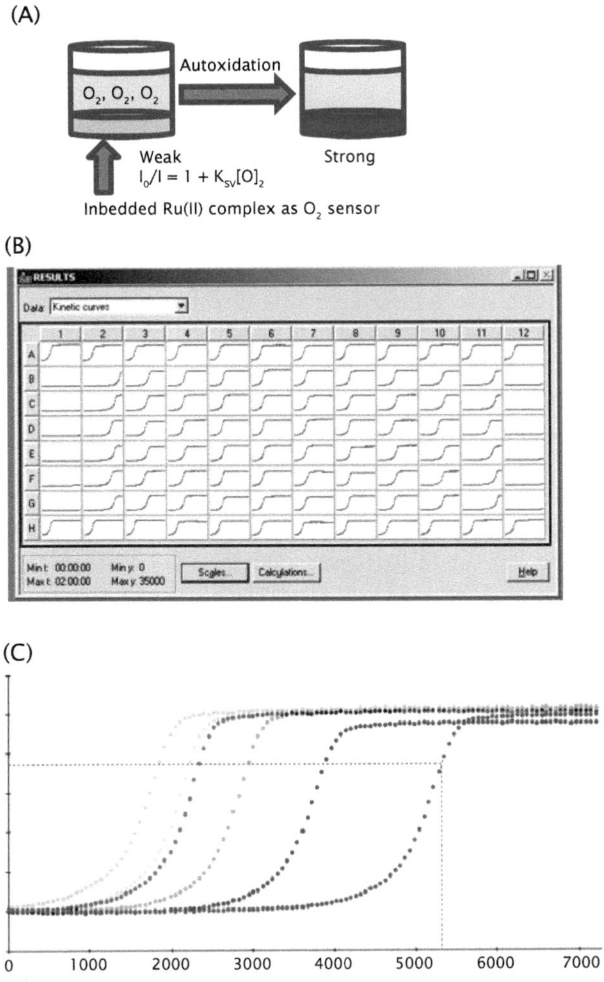

Fig. 10.4. $ORAC_{oil}$ assay principle and representative results. (A) 96-microplate well (cartoon) is coated with $Ru(BIPY)_3$ probe dispersed in air permeable silicon polymer. When the solution is saturated with air, the phosphorescence of signal is weak because O_2 is a good quencher. If the oxygen is removed, the phosphorescence signal increase by about five fold. (B) Typical kinetic curves of a $ORAC_{oil}$ assay on a 96-well microplate. (C) Dose response of kinetic curves and the concentration of antioxidant (Trolox). (D) The kinetic curves of a blank well and that of Trolox based on the phosphorescence intensity. (E) Converted curve of D to oxygen concentration changes over time. (F) Standard curve obtained from Trolox versus the net area under the curve.

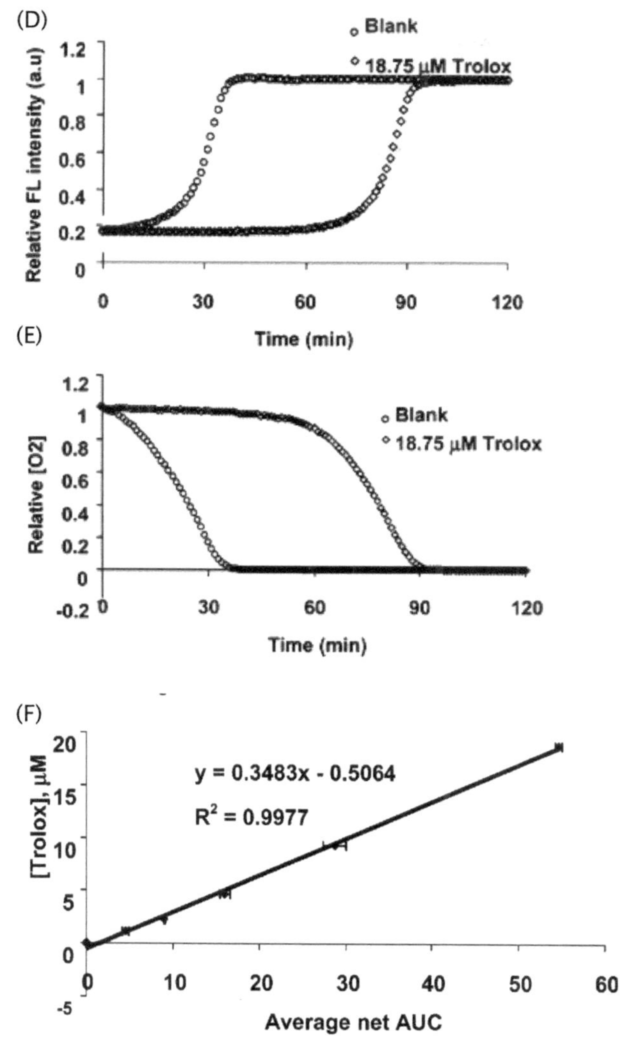

Fig. 10.4. *Continued*

radical-scavenging activity of peroxyl is measured in a buffered aqueous system. Since no direct contact takes place between the reaction solution and the luminescent probe, this method can also be applied to monitor oxygen consumption rates in turbid samples and give more flexibility, compared to assays that require a clear solution.

Oxygen Radical Absorbance Capacity in Microemulsion—$ORAC_E$ Assay

We further extended the microplate's application to measure antioxidant capacities in a microemulsion system (Sim et al., 2009). Emulsions are commonly applied to disperse food lipids. Because oil droplets have more surface area than bulk edible oil, lipid autoxidation in an emulsion system is more likely to happen by oxidants generated in the aqueous phase. Thus, antioxidants that perform well in bulk oil may not be as potent in an emulsion system. Instead, the phase distribution of antioxidants is a critical factor for their effectiveness. However, regular emulsions are not ideal because they are thermodynamically unstable and their consistency as a model system for the evaluation of antioxidants is difficult to maintain. On the other hand, a microemulsion is thermodynamically stable and presents an ideal model system for the analytical method. We applied an oil-in-water (o/w) microemulsion consisting of methyl linoleate (12%), Tween-20 (29%), n-butanol (15%), and a potassium phosphate buffer (10 mM, pH 7.04, 44%). The stable microemulsion yielded mean oil droplet sizes of 19.3 nm, with polydispersity at 0.103. Using AAPH as a radical initiator, we monitored the oxidation of methyl linoleate in the absence and presence of polyphenolic antioxidants using 96-well oxygen biosensor systems. From oxygen consumption's kinetic curves, we calculated the antioxidant capacity using Trolox as the standard and measured the synergistic effect of hydrophilic antioxidants and α-tocopherol for representative flavonoids. Chlorogenic acid had the best synergistic effect, at 44.8%. We call this assay $ORAC_E$, which stands for "oxygen radical absorbance capacity in microemulsion." $ORAC_E$ provides a model system convenient for quantifying antioxidant capacity and protecting against the oxidation of polyunsaturated fatty acids in microemulsions.

A comparison between $ORAC_E$ and $ORAC_{FL}$ values showed that compounds with high-antioxidant activity do not necessarily have high $ORAC_{FL}$ values. This result is not surprising because the phase distributions of antioxidants play a critical role in preventing lipid oxidation in emulsion systems. We expect that the type of surfactants (that is, anionic, nonionic, or cationic) and the radical initiator used [such as R_2N_2, $Fe(II)/H_2O_2$] will also influence the antioxidant's performance in the microemulsion system. Future work is necessary to address these issues. Therefore, it is imperative that antioxidant capacity is measured in a system that closely approximates a real food system. The advantage of oxygen sensor microplates is that they allow the alteration of emulsion compositions and radical initiators. However, microplates' high cost may limit their broad application.

Fluorescent Molecular Probes as Trackers for Antioxidant and Lipid Oxidation

Fluorescence probes are becoming more popular in antioxidant capacity assays because of their sensitivity and because they allow fluorescence imaging to track the lipids' oxidative stress. In this section, we selectively review the working principles of several probes that can be used for the activity quantitation of lipid peroxides and antioxidants.

BODIPY-Based Probes for Antioxidant and Lipid Peroxide Sensing

In comparison to other fluorophores (such as fluorescein and rhodamine), 4,4-difluoro-4-bora-3a,4a-diaza-s-indacene (also known as BODIPY) has very attractive features, such as high-quantum yield. Additionally, it is readily modified for different functional groups with a wide range of applications, including bioimaging and fluorescent sensors (Boens et al., 2012). The latest developments on BODIPY-based probes were summarized regarding their use for tracking reactive oxygen species and lipid oxidation.

C_{11}-BODIPY 581/591

A lipid soluble probe, C_{11}-BODIPY581/591, shifts fluorescence from red to green when its olefinic double bond is oxidized by a peroxyl radical. It was first used to measure antioxidant activity in a dioleoylphosphatidyl choline (DOPC) liposomal suspension, with 2,2'-azobis-2,4-dimethyl valeronitrile (AMVN) as a peroxyl radical generator (Naguib, 1998). It was later applied in a number of systems, including the imaging of oxidative events in cell membranes (Drummen et al., 2002). The selectivity of the probe was not high because other reactive oxygen species, such as peroxynitrite and NO_2, could react with BODIPY581/591. The reaction leads to oxidation of the conjugated diene interconnection between the BODIPY core and the terminal phenyl moiety, which explains the blue shift in fluorescence emission (Nicolescu et al., 2006). LC/MS analysis confirmed that the peroxyl radical–induced irreversible fluorescence decay of the BODIPY11 fluorophore occurs because of the oxidative cleavage of the activated phenyldiene side chain (Fig. 10.5) (Drummen et al., 2004). Another BODIPY probe (λ_{ex} = 510 nm, λ_{em} =

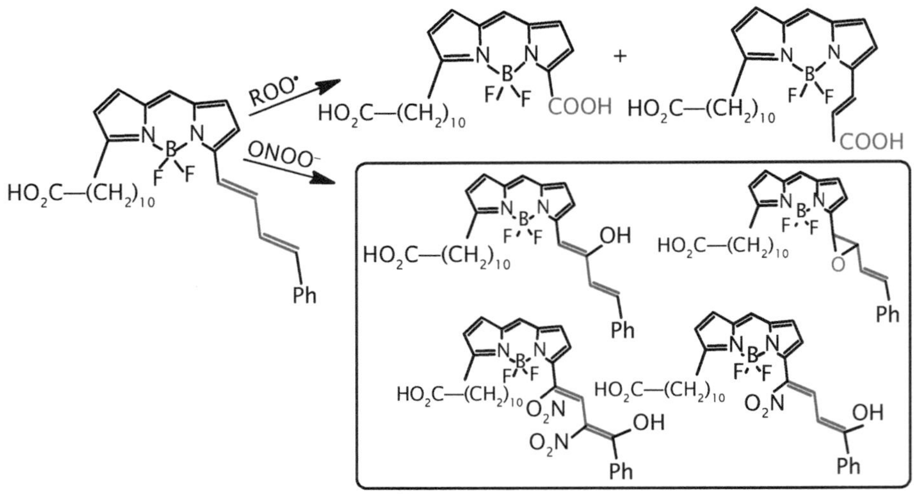

Fig. 10.5. BODIPY probe, C_{11}-BODIPY 581/591 for tracking lipid peroxidation. Upon oxidative modification by peroxyl radical, the conjugation system is shortened and the flurorescence signal blue-shifts. In the case of peroxynitrite, the reaction products are quite complex.

520 nm) was applied to study lipid peroxidation, but it showed that the suppression of fluorescence emission by antioxidants did not always correlate quantitatively with the suppression of lipid peroxidation (Itoh et al., 2007). α-Tocopherol suppressed BODIPY fluorescence but enhanced the peroxidation of plasma lipids in the absence of ascorbic acid. 2,2,5,7,8-Pentamethyl-6-chromanol, a vitamin E analogue without a phytyl tail, almost completely suppressed both fluorescence emission and lipid peroxidation in the plasma. These results show that BODIPY may be used as a convenient probe for radical detection, but using it to evaluate antioxidant capacity requires more research (Itoh et al., 2007).

Antioxidant-Functionalized BODIPY Probe

Krumova and coworkers conducted an interesting investigation that linked Trolox to BODIPY fluorophore (B-TOH), as shown in Fig. 10.6 (Krumova et al., 2009). B-TOH has weak fluorescence, but the fluorescence is turned on after peroxyl radicals oxidize the phenolic moiety. On the other hand, the 4-hydroxybenzoic acid ester of BODIPY (B-BHB) retains fluorescence but reacts slowly with a peroxyl radical. Intramolecular photo-induced electron transfer (PeT) causes the fluorescence quenching of B-TOH because the highest-occupied molecular orbital (HOMO) of Trolox has more energy than that of BODIPY. A PeT is not possible for

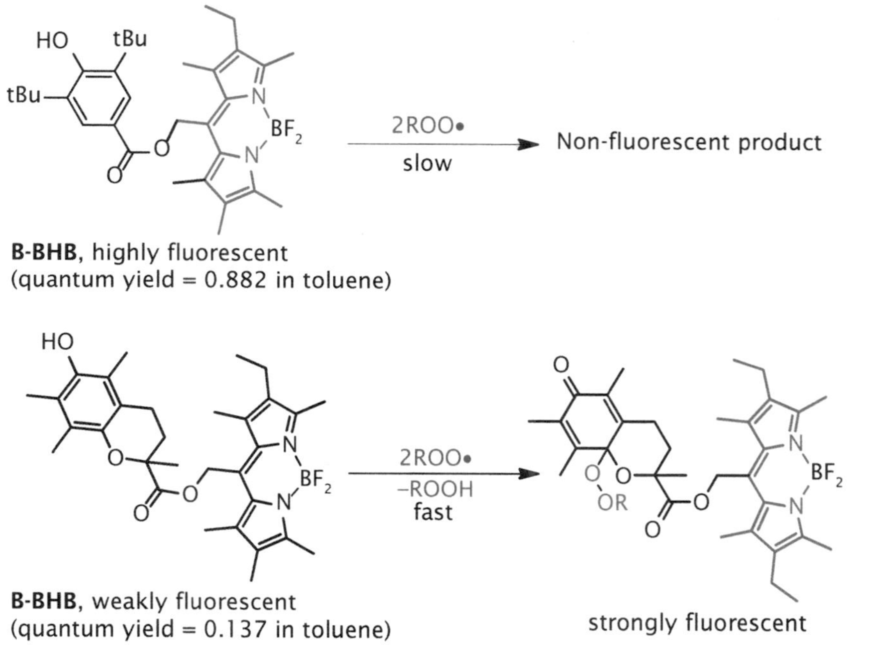

Fig. 10.6. Radical scavenger-linked BODIPY probes.

B-BHB because the HOMO of BHB is lower than that of BODIPY. B-TOH was successfully applied as a fluorescent probe in tracking styrene polymerization, which was triggered by peroxyl radicals generated by 2,2'-azobis-(2-methylpropionitrile) (AIBN). A good linear response occurred between the lag phase and the amount of α-tocopherol added to the solution. B-TOH was applied in imaging of cellular lipid peroxidation events triggered by oxidative stress factors (Khatchadourian et al., 2009).

Lipid Peroxide Probes with Organophosphine as Quencher

Using the photo-induced electron transfer (PeT) mechanism, we found the Ph_2P moiety to be a good fluorescence quencher due to its lone electron pair on phosphine. The fluorescence turns on with the oxidation of Ph_2P moiety by lipid peroxide. Thus far, four Ph_2P groups containing probes have been reported, with fluorophores ranging from UV (380 nm) in diphenyl-1-pyrenylphosphine (DPPP) to long yellow (535 nm) in SPY-Ox (Fig. 10.7). DPPP was applied in the quantification of lipid peroxides in triglycerides (Akasaka et al., 1992; Ohshima

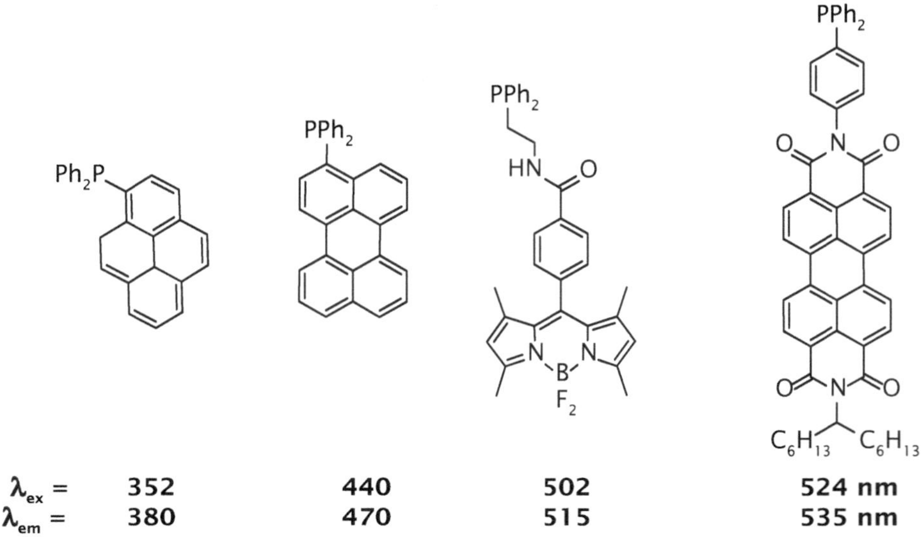

Fig. 10.7. Structures of diphenylphosphine containing fluorescent probes for lipid peroxide detection. Upon oxidation of the Phosphine atom to Phosphine oxide, the fluorescent is turned on by lipid peroxide.

et al., 1996), LDL (Okimoto et al., 2003), and the cell membrane (Takahashi et al., 2001).

The longer wavelength analog 3-perylenyldiphenylphosphine (3-PeDPP) was also applied to quantify lipid hydroperoxides in homogeneous solutions (with chloroform and methanol as solvents) and liposome model systems induced by AAPH or light (Chotimarkorn et al., 2005). They observed a good linear correlation between fluorescence intensity and the concentration of hydroperoxides. Refined tuna-oil lipid peroxides were determined by PeDPP, and the results correlated well with a ferric-xylenol orange (FOX) assay (R^2 = 0.984) (Chotimarkorn et al., 2005). Ph_2P-linked BODIPY emits green fluorescence at 515 nm upon oxidation by methyl linoleate hydroperoxide (Inoue et al., 2009). Perylene 3,4,9,10-tetracarboxyl bisimide—a yellow, fluorescent, Ph_2P-linked perylene derivative—emits light at 535 nm and has been applied to detect hydroperoxides (Soh et al., 2006) and pinpoint peroxidation mechanisms in a photosystem II depleted with manganese (Khorobrykh et al., 2011). The probe distinguished lipid-soluble peroxides from hydrophilic ones, which react much more slowly with the probe.

Although the Ph_2P-linked fluorescence probes have interesting applications in the detection of lipid peroxides, only limited information is available on their potential interference. It is foreseeable that lipid permeable oxidants like hypochlorous acid, NO_2, and HO radicals would react with Ph_2P to turn on the fluorescence. Because they are a soft Lewis base, phosphines are good ligands for soft Lewis acids like late transition metal ions, such as Zn(II), Cu(II), Ni(II), Co(II), and Fe(II). Nonetheless, we do not know whether coordination with these Lewis acids will also enchance to fluorescence. It would also be interesting to know the longest emission wavelength of a fluorophore that Ph_2P can turn off because utilization of Ph_2P as a redox-sensitive quencher of NIR probes would allow their usage in the bioimaging of the oxidative stress status of experimental animal models.

Conclusion

CAT, $ORAC_{oil}$, and $ORAC_E$ assays utilize polyunsaturated fatty acids as the substrate, and the antioxidant concentrations used in the assays were smaller than those of the substrates. Hence, these assays more closely mimic the chemistry of lipid autoxidation than other assays, such as $ORAC_{FL}$, DPPH, and FRAP. It remains to be seen whether the antioxidant capacity measured using these methods can be correlated to their performance in food lipid systems. These assays apply radical-induced oxidation reactions; thus, prooxidants will likely act as antioxidants under these assays. Lipid peroxide–sensitive fluorescent probes provide convenient tools for monitoring lipid peroxidation events, but we have not seen the extent of their application in assessing antioxidant capacity. It would be interesting to explore the potential advantages these probes may offer in terms of convenience, selectivity, and sensitivity, as compared to the existing array of antioxidant capacity methods. Because a wide range of oxidation mechanisms can trigger lipid peroxidation, as shown in Fig. 10.1, it is imperative to comprehensively evaluate antioxidant performance by accounting for the multifaceted nature of antioxidant action mechanisms, which include metal-chelating activity, scavenging activity against different reactive oxygen species that are biologically relevant, and inhibitory activity of oxidative enzymes.

References

Akasaka, K.; Ijichi, S.; Watanabe, K.; Ohrui, H.; Meguro, H. High-performance liquid-chromatography and postcolumn derivatization with diphenyl-1-pyrenylphosphine for fluorometric-determination of triacylglycerol hydroperoxides. J. Chromatogr. 1992, 596 (2), 197–202.

Alamed, J.; Chaiyasit, W.; Alamed, J.; Chaiyasit, W.; McClements, D. J.; Decker, E. A. Relationships between free radical scavenging and antioxidant activity in foods. J. Agric. and Food Chem. 2009, 57 (7), 2969–2976.

Alessi, M.; Paul, T.; Scaiano, J. C.; Ingold, K. U. The contrasting kinetics of peroxidation of vitamin E-containing phospholipid unilamellar vesicles and human low-density lipoprotein. J. Am. Chem. Soc. 2002, 124 (24), 6957–6965.

Apak, R.; Güçlü, K.; Ozyürek, M.; Karademir, S. E. Novel total antioxidant capacity index for dietary polyphenols and vitamins C and E, using their cupric ion reducing capability in the presence of neocuproine: CUPRAC method. J. Agric. Food Chem. 2004, 52 (26), 7970–7981.

Benzie, I. F. F.; Chung, W. Y.; Strain, J. J. "Antioxidant" (reducing) efficiency of ascorbate in plasma is not affected by concentration. J. Nutri. Biochem. 1999, 10 (3), 146–150.

Benzie, I. F. F.; Strain, J. J. The ferric reducing ability of plasma (FRAP) as a measure of "antioxidant power": The FRAP assay. Anal. Biochem. 1996, 239 (1), 70–76.

Boens, N.; Leen, V.; Dehean, W. Fluorescent indicators based on BODIPY. Chem. Soc. Rev. 2012, 41 (3), 1130–1172.

Boguth, W.; Repges, R. Reaction of alpha-tocopherol with 1,1-diphenyl-2-picrylhydrazyl. 3. Photometric determination of alpha-tocopherol with DPPH. Int. Z. Vitaminforsch 1969, 39 (3), 289–295.

Brash, A. R. Autoxidation of methyl linoleate: Identification of the bis-allylic 11-hydroperoxide. Lipids 2000, 35 (9), 947–952.

Cao, G.; Verdon, C. P; Wu, A. H.; Wang, H.; Prior, R. L. Automated-assay of oxygen radical absorbency capacity with the cobas fara-II. Clin. Chem. 1995, 41 (12), 1738–1744.

Cao, G. H.; Alessio, H. M.; Cutler R. G. Oxygen-radical absorbency capacity assay for antioxidants. Free Radical Biol. Med. 1993, 14 (3), 303–311.

Carlsen, C. U.; Møller, J. K. S.; Skibsted, L. H. Heme-iron in lipid oxidation. Coord. Chem. Rev. 2005, 249 (3–4), 485–498.

Chotimarkorn, C.; Nagasaka, R.; Ushio, H.; Ohshima, T.; Matsunaga S. Development of novel fluorescent probe 3-perylene diphenylphosphine for determination of lipid hydroperoxide with fluorescent image analysis. Biochem. Biophys. Res. Commun. 2005, 338 (2), 1222–1228.

Chotimarkorn, C.; Ohshima, T.; Ushio, H. Fluorometric and fluorescent image analysis methods for determination of lipid hydroperoxides in oil models with 3-perylene diphenylphosphine (3-PeDPP). J. Agric. Food Chem. 2005, 53 (19), 7361–7366.

Dalsgaard, T. K.; Sørensen, J.; Bakman, M.; Nebel, C.; Albrechtsen, R.; Vognsen L.; Nielsen, J. H. Light-induced protein and lipid oxidation in low-fat cheeses: Whey proteins as antioxidants. Dairy Science & Tech. 2011, 91 (2), 171–183.

Drummen, G. P.; Gadella, B. M.; Post, J. A.; Brouwers, J. F. Mass spectrometric characterization of the oxidation of the fluorescent lipid peroxidation reporter molecule C11-BODIPY(581/591). Free Radical Biol. Med. 2004, 36 (12), 1635–1644.

Drummen, G. P.; van Liebergen, L. C.; Op den Kamp, J. A.; Post, J. A. C11-BODIPY(581/591), an oxidation-sensitive fluorescent lipid peroxidation probe: (Micro)spectroscopic characterization and validation of methodology. Free Radical Biol. Med. 2002, 33 (4), 473–490.

Etsuo, N. Assessment of antioxidant capacity of natural products. Curr. Pharm. Biotech. 2010, 11 (8), 801–809.

Ford, D. A. Lipid oxidation by hypochlorous acid: Chlorinated lipids in atherosclerosis and myocardial ischemia. Clin. Lipidol. 2010, 5 (6), 835–852.

Hay, K. X.; Waisundara, V. Y.; Timmins, M., Ou, B.; Pappalardo, K.; McHale, N.; Huang, D. High-throughput quantitation of peroxyl radical scavenging capacity in bulk oils. J. Agric. Food Chem. 2006, 54 (15), 5299–5305.

Huang, D.; Ou, B.; Hampsch-Woodill, M.; Flanagan, J. A.; Prior, R. L. High-throughput assay of oxygen radical absorbance capacity (ORAC) using a multichannel liquid handling system coupled with a microplate fluorescence reader in 96-well format. J. Agric. Food Chem. 2002, 50 (16), 4437–4444.

Huang, D.; Ou, B.; Prior, R. L. The chemistry behind antioxidant capacity assays. J. Agric. Food Chem. 2005, 53 (6), 1841–1856.

Inoue N.; Suzuki, Y.; Yokoyama, K.; Karube, I. Novel fluorescent probe for analysis of hydroperoxides based on boron dipyrromethane fluorophore. Biosci. Biotechnol. Biochem. 2009, 73 (5), 1215–1217.

Itoh, N.; Cao, J.; Chen, Z. H.; Yoshida, Y., Niki, E. Advantages and limitation of BODIPY as a probe for the evaluation of lipid peroxidation and its inhibition by antioxidants in plasma. Bioorg. Med. Chem. Lett. 2007, 17 (7), 2059–2063.

Jomova, K.; Vondrakova, D.; Lawson, M.; Valko, M. Metals, oxidative stress and neurodegenerative disorders. Mol. Cell. Biochem. 2010, 345 (1–2), 91–104.

Khatchadourian, A.; Krumova, K.; Boridy, S.; Ngo, A. T.; Maysinger, D.; Cosa, G. Molecular imaging of lipid peroxyl radicals in living cells with a BODIPY-alpha-tocopherol adduct. Biochemistry 2009, 48 (24), 5658–5668.

Khorobrykh, S. A.; Khorobrykh, A. A.; Yanykin, D. V.; Ivanov, B. N.; Klimov, V. V.; Mano, J. Photoproduction of catalase-insensitive peroxides on the donor side of manganese-depleted photosystem II: Evidence with a specific fluorescent probe. Biochemistry 2011, 50 (49), 10658–10665.

Krumova, K.; Oleynik, P.; Karam, P.; Cosa, G. Phenol-based lipophilic fluorescent antioxidant indicators: A rational approach. J. Org. Chem. 2009, 74 (10), 3641–3651.

Laguerre, M.; Giraldo, L. J.; Lecomte, J.; Figueroa-Espinoza, M. C.; Baréa, B.; Weiss, J., Decker, E. A.; Villeneuve, P. Characterization of olive-leaf phenolics by ESI-MS and evaluation of their antioxidant capacities by the CAT assay. J. Am. Oil Chem. Soc. 2009, 86 (12), 1215–1225.

Laguerre, M.; Lecomte, J.; Villeneuve, P. Evaluation of the ability of antioxidants to counteract lipid oxidation: Existing methods, new trends and challenges. Prog. Lipid Res. 2007, 46 (5), 244–282.

Laguerre, M.; López-Giraldo, L. J.; Lecomte, J.; Baréa, B.; Cambon, E.; Tchobo, P. F.; Barouh, N.; Villeneuve, P. Conjugated autoxidizable triene (CAT) assay: A novel spectrophotometric method for determination of antioxidant capacity using triacylglycerol as ultraviolet probe. Anal. Biochem. 2008, 380 (2), 282–290.

Liu, Z.; Yan, H.; Wang, K.; Kuang, T.; Zhang, J.; Gui, L.; An, X.; Chang, W. Crystal structure of spinach major light-harvesting complex at 2.72 angstrom resolution. Nature 2004, 428 (6980), 287–292.

Liu, Z. Q. Chemical methods to evaluate antioxidant ability. Chem. Rev. 2010, 110 (10), 5675–5691.

Long, L. H; Clement, M. V.; Halliwell, B. Artifacts in cell culture: Rapid generation of hydrogen peroxide on addition of (-)-epigallocatechin, (-)-epigallocatechin gallate, (+)-catechin, and quercetin to commonly used cell culture media. Biochem. Biophys. Res. Commun. 2000, 273 (1), 50–53.

Long, L. H.; Lan, A. N.; Hsuan, F. T.; Halliwell, B. Generation of hydrogen peroxide by "antioxidant" beverages and the effect of milk addition. Is cocoa the best beverage? Free Radical Res. 1999, 31 (1), 67–71.

Miller, N. J.; Rice-Evans, C. A. Spectrophotometric determination of antioxidant activity. Redox Report 1996, 2 (3), 161–171.

Naguib, Y. M. A fluorometric method for measurement of peroxyl radical scavenging activities of lipophilic antioxidants. Anal. Biochem. 1998, 265 (2), 290–298.

Nicolescu, A. C.; Li, Q.; Brown, L.; Thatcher, G. R. Nitroxidation, nitration, and oxidation of a BODIPY fluorophore by RNOS and ROS. Nitric Oxide 2006, 15 (2), 163–176.

Ohshima, T.; Hopia, A.; German, J. B.; Frankel, E. N. Determination of hydroperoxides and structures by high-performance liquid chromatography with post-column detection with diphenyl-1-pyrenylphosphine. Lipids 1996, 31 (10), 1091–1096.

Okimoto, Y.; Warabi, E.; Wada, Y.; Niki, E.; Kodama, T.; Noguchi, N. A novel method of following oxidation of low-density lipoprotein using a sensitive fluorescent probe, diphenyl-1-pyrenylphosphine. Free Radical Biol. Med. 2003, 35 (6), 576–585.

Ou, B.; Hampsch-Woodill, M.; Prior, R. L. Development and validation of an improved oxygen radical absorbance capacity assay using fluorescein as the fluorescent probe. J. Agric. Food Chem. 2001, 49 (10), 4619–4626.

Ozyurt, D.; Demirata, B.; Apak, R. Modified cerium(IV)-based antioxidant capacity (CERAC) assay with selectivity over citric acid and simple sugars. J. Food Compos. Anal. 2010, 23 (3), 282–288.

Pratt, D. A.; Tallman, K. A.; Porter, N. A. Free radical oxidation of polyunsaturated lipids: New mechanistic insights and the development of peroxyl radical clocks. Acc. Chem. Res. 2011, 44 (6), 458–467.

Pryor, W. A.; Houk, K. N.; Foote, C. S.; Fukuto, J. M.; Ignarro, L. J.; Squadrito, G. L.; Davies, K. J. Free radical biology and medicine: It's a gas, man! Am. J. Physiol. Regul. Integr. Comp. Physiol. 2006, 291 (3), R491–511.

Quek, Y. L.; Huang, D. J. Hydroethidine as a probe for measuring superoxide formation rates during air oxidation of myricetin and quercetin. Tetrahedron Lett. 2011, 52 (41), 5384–5387.

Rains, J. L.; Jain, S. K. Oxidative stress, insulin signaling, and diabetes. Free Radical Biol. Med. 2011, 50 (5), 567–575.

Ramel, F.; Birtic, S.; Cuiné, S.; Triantaphylidès, C.; Ravanat, J. L.; Havaux, M. Chemical quenching of singlet oxygen by carotenoids in plants. Plant Physiol. 2012, 158 (3), 1267–1278.

Rubbo, H.; Trostchansky, A.; O'Donnell, V. B. Peroxynitrite-mediated lipid oxidation and nitration: Mechanisms and consequences. Arch. Biochem. Biophys. 2009, 484 (2), 167–172.

Ruggia, A.; van Leeuwenb, F. W. B.; Veldersa, A. H. Interaction of dioxygen with the electronic excited state of Ir(III) and Ru(II) complexes: Principles and biomedical applications. Coord. Chem. Rev. 2011, 255 (21–22), 2542–2554.

Sim, W. L.; Han, M. Y.; Huang, D. Quantification of antioxidant capacity in a microemulsion system: Synergistic effects of chlorogenic acid with alpha-tocopherol. J. Agric. Food Chem. 2009, 57 (9), 3409–3414.

Soh, N.; Ariyoshi, T.; Fukaminato, T.; Nakano, K.; Irie, M.; Imato, T. Novel fluorescent probe for detecting hydroperoxides with strong emission in the visible range. Bioorg. Med. Chem. Lett. 2006, 16 (11), 2943–2946.

Sohal, R. S.; Orr, W. C. The redox stress hypothesis of aging. Free Radical Biol. Med. 2012, 52 (3), 539–555.

Speciale, A.; Chirafisi, J.; Saija, A.; Cimino, F. Nutritional antioxidants and adaptive cell responses: An update. Curr. Mol. Med. 2011, 11 (9), 770–789.

Stocker, R.; Keaney, J. F., Jr. Role of oxidative modifications in atherosclerosis. Physiol. Rev. 2004, 84 (4), 1381–1478.

Styskal, J.; Van Remmen, H.; Richardson, A.; Salmon, A. B. Oxidative stress and diabetes: What can we learn about insulin resistance from antioxidant mutant mouse models? Free Radical Biol. Med. 2012, 52 (1), 46–58.

Subramanian, K. N.; Padmanaban, G.; Sarma, P. S. Folin-Ciocalteu reagent for the estimation of siderochromes. Anal. Biochem. 1965, 12 (1), 106–112.

Takahashi, M.; Shibata, M.; Niki, E. Estimation of lipid peroxidation of live cells using a fluorescent probe, diphenyl-1-pyrenylphosphine. Free Radical Biol. Med. 2001, 31 (2), 164–174.

Takayuki, M. J.-K. S. Assays for plant and food components. J. Agric. Food Chem. 2009, 57 (5), 1655–1666.

Tallman, K. A.; Pratt, D. A.; Porter, N. A. Kinetic products of linoleate peroxidation: Rapid beta-fragmentation of nonconjugated peroxyls. J. Am. Chem. Soc. 2001, 123 (47), 11827–11828.

Tinkel, J.; Hassanain, H.; Khouri, S. J. Cardiovascular antioxidant therapy: A review of supplements, pharmacotherapies, and mechanisms. Cardiol. Rev. 2012, 20 (2), 77–83.

Uemi, M.; Ronsein, G. E.; Prado, F. M.; Motta, F. D.; Miyamoto, S.; Medeiros, M. H.; Di Mascio, P. Cholesterol hydroperoxides generate singlet molecular oxygen [O(2) ((1)Delta(g))]: Near-IR emission, (18)O-labeled hydroperoxides, and mass spectrometry. Chem. Res. Toxicol. 2011, 24 (6), 887–895.

Valgimigli, L.; Amorati, R.; Fumo, M. G.; DiLabio, G. A.; Pedulli, G. F.; Ingold, K. U.; Pratt D. A. The unusual reaction of semiquinone radicals with molecular oxygen. J. Org. Chem. 2008, 73 (5), 1830–1841.

Valko, M.; Rhodes, C. J.; Moncol, J.; Izakovic, M.; Mazur, M. Free radicals, metals and antioxidants in oxidative stress-induced cancer. Chem. Biol. Interact. 2006, 160 (1), 1–40.

Wisthaler, A.; Weschler, C. J. Reactions of ozone with human skin lipids: Sources of carbonyls, dicarbonyls, and hydroxycarbonyls in indoor air. Proc. Natl. Acad. Sci. USA 2010, 107 (15), 6568–6575.

Wodnicka, M.; Guarino, R. D.; Hemperly, J. J.; Timmins, M. R.; Stitt, D.; Pitner, J. B. Novel fluorescent technology platform for high throughput cytotoxicity and proliferation assays. J. Biomol. Screen. 2000, 5 (3), 141–152.

Wu, X.; Beecher, G. R.; Holden, J. M.; Haytowitz, D. B.; Gebhardt, S. E.; Prior, R. L. Lipophilic and hydrophilic antioxidant capacities of common foods in the United States. J. Agric. Food Chem. 2004, 52 (12), 4026–4037.

Yoo, K. M.; Kim, D.-O.; Lee, C. Y. Evaluation of different methods of antioxidant measurement. Food Science and Biotechnology 2007, 16 (2), 177–182.

CHAPTER

11

Strategies to Minimize Oxidative Deterioration in Aquatic Food Products

Application of Natural Antioxidants from Edible Mushrooms

Huynh Nguyen Duy Bao[1] and Toshiaki Ohshima[2]

Introduction

Lipids of aquatic food products are highly unsaturated in nature. Omega-3 fatty acids have been thought to positively affect various cardiovascular diseases, such as blood clotting disorders and high blood pressure (Uauy & Valenzuela, 2000). Eicosapentaenoic acid (EPA) and docosahexaenoic acid (DHA), both being present in relatively higher concentration in fish lipids, are typical examples of such polyunsaturated fatty acids (PUFA). However, the susceptibility of PUFAs to attack by oxygen increases with an increasing degree of unsaturation. Thus, these PUFAs readily undergo oxidation during postmortem storage. A feature that increases this risk for oxidation is the close proximity between lipids and an array of prooxidative systems in the fish tissue. Various naturally occurring antioxidative compounds can balance the effect of these prooxidants *in vivo*. However, this balance is disturbed immediately after the catch, and various postmortem changes initiate lipid oxidation in fish muscle. Initiation of lipid oxidation is followed by propagation, chain branching, and termination, all steps driven by free-radical chain reactions (Porter et al., 1995). Under most circumstances, the production of off-flavor compounds constitutes the primary decrease in quality observed during lipid oxidation, although lowered nutritional value, changed texture, and develop-

[1]Faculty of Food Technology, Nha Trang University, Vietnam; [2]Department of Food Science and Technology, Graduate School, Tokyo University of Marine Science and Technology, Japan.

ment of color may also result from lipid oxidation (Love & Pearson, 1971). The time lag before these quality changes occur depends on the type of post-harvest handling and processing of the fish. These factors also decide whether lipid oxidation becomes a process that limits shelf life. During storage at ambient or chilled temperatures, bacterial growth is usually faster than rancid development in fresh fish. In frozen, salted or dried fish, where microbial growth is suppressed, rancidity usually limits shelf life (Cho et al., 1989; Losada et al., 2007).

This chapter deals with the factors affecting the development of oxidative deterioration in aquatic food products and the efficacies of natural antioxidants in mushroom extracts applied to the products.

Lipid Oxidation and Quality Deterioration in Aquatic Food Products

In live animals, the ingestion and regeneration of antioxidative compounds prevent excessive oxidative deterioration of important biological components. In postmortem tissue, protective systems are depleted and unable to regenerate. Thus, many biochemical components, including the lipids in tissues of fish as well as other aquatic species, are liable to react with oxygen when they are exposed to the air. Because of the presence of highly unsaturated lipids, aquatic food products contain both powerful catalytic and antioxidative systems in close proximity. During postmortem storage, several biochemical processes take place, leading to the initiation of lipid oxidation.

Basic Reactions in Lipid Oxidative Processes

The lipid oxidation process is a chain reaction that takes place through several intermediate steps, which are usually categorized as three reaction steps—initiation, propagation and termination—according to the following scheme:

Initiation:
$$LH + X\bullet \rightarrow L\bullet + HX \quad \text{(Eq. 11.1)}$$

Propagation:
$$L\bullet + O_2 \rightarrow LOO\bullet \quad \text{(Eq. 11.2)}$$

$$LOO\bullet + LH \rightarrow L\bullet + LOOH \quad \text{(Eq. 11.3)}$$

$$LOOH \rightarrow LO\bullet + \bullet OH \quad \text{(Eq. 11.4)}$$

$$LOOH + LO\bullet \rightarrow LOO\bullet + LOH \quad \text{(Eq. 11.5)}$$

$$LO\bullet + LH \rightarrow L\bullet + LOH \quad \text{(Eq. 11.6)}$$

$$\bullet OH + LH \rightarrow L\bullet + H_2O \quad \text{(Eq. 11.7)}$$

Termination:

$$L\bullet + L\bullet \rightarrow L\text{-}L \quad \text{(Eq. 11.8)}$$

$$L\bullet + LOO\bullet \rightarrow LOOL \quad \text{(Eq. 11.9)}$$

$$LOO\bullet + LOO\bullet \rightarrow LOOL + O_2 \quad \text{(Eq. 11.10)}$$

The alkyl radical (L•), generated during an initiation step, is converted quickly to a peroxyl radical (LOO•) by combining with coexisting oxygen. Subsequently, the LOO• abstracts a hydrogen atom from another fatty acid to form a lipid hydroperoxide molecule (LOOH) and a newly generated alkyl radical. Decomposition of the LOOH, in turn, is responsible for further propagation of the free-radical process. When the free-radical content becomes large enough, two free radicals may combine to terminate the free-radical process. Lipid oxidation is a very complex process, not only in food but also in biological systems. In particular, the reactions involved in the initiation, propagation, and termination steps are not separated from each other. The chemical and enzymatic reactions existing in living tissue occur during post-harvest storage; therefore, initiation reactions exist throughout the entire storage life of the products.

Factors Affecting Lipid Oxidation

Whether it occurs through enzymatic or nonenzymatic pathways, lipid oxidation in aquatic food products is affected by numerous factors. The main factors that determine the rate and extent of oxidative deterioration in aquatic food products include the lipid substrate, presence of oxygen, prooxidants, and inhibitors, as well as storage temperature.

Lipids

The total lipid content varies widely among and within species of fish, as well as different types of muscle. Lean fish, such as cod and haddock, usually contain less than 2% of lipids, whereas in fatty species, such as mackerel and herring, lipid contents range between 0.5 and 30%. The largest fat deposits are found in the liver, belly flap, dark muscle, connective tissue, and under the skin. Lipid oxidation will proceed much more rapidly in fatty species because the higher lipid content might cause lipid oxidation in fatty dark muscle of fish to proceed more rapidly than that in lean ordinary muscle (Sohn et al., 2007). Lipids present in aquatic food products form roughly two groups: neutral and polar lipids. Neutral lipids consist of triglycerides (TGs), which are present primarily in adipose tissue and generally serve as an energy source. Therefore, the total amount of neutral lipid correlates positively to the total fat content of fish muscle and varies widely according to species, season, diet, and geographical origin. On the other hand, polar lipids consist mainly of phospholipids (PLs) and glycolipids and serve a structural role in membranes. The total polar lipid content is relatively constant between different fish and independent of environmental factors (Ackman & Ratnayake, 1992). The PLs of muscle membranes are considered most susceptible to lipid oxidation. This hypothesis explains that PLs have more PUFA compared to TGs and that the large surface area of membranes effectively increases the exposure of PLs to aqueous prooxidants (Huang et al., 1993). Furthermore, the involvement of membranes in electron transport processes has a high possibility of electron leakage. The reaction of these electrons with molecular oxygen produces certain reactive oxygen species, such as superoxide and hydrogen peroxide (Hultin, 1995).

During the storage of postmortem fish, free fatty acids (FFA) are successively hydrolyzed from both TG and PL by lipases (EC.3.1.1.3) and phospholipase A2 (EC.3.1.1.4), respectively (Izquierdo & Henderson, 1998; Nayak et al., 2003). This phenomenon is more profound in ungutted than in gutted fish, probably because of the involvement of digestive enzymes (Jorgensen, 1995). Although it is generally accepted that FFAs are oxidized more rapidly than esterified fatty acids (Labuza, 1971; Miyashita & Takagi, 1986), the correlation between lipolysis and lipid oxidation in fish meat has not been firmly established.

Oxygen

Oxygen is an essential component for lipid oxidation. According to the principle of spin conservation, two unpaired electrons in the outer orbital of ground-state oxygen (triplet oxygen, $^3O_2$) prevent its direct reaction with an intact unsaturated fatty-acid molecule. However, a number of compounds present in the cell can activate $^3O_2$ to produce activated oxygen, which can cause lipid oxidation. One such mechanism for activation is the production of reduced forms of oxygen, such as the superoxide anion radical, $O_2\bullet^-$. The production of this active oxygen species can occur in three ways: (i) *in vivo* during energy metabolism when ground-state oxygen is converted to water through a series of reductions, (ii) during the autoxidation of oxymyoglobin and oxyhemoglobin, or (iii) from the reaction between ferrous iron and $^3O_2$. The $O_2\bullet^-$ itself is not reactive enough to abstract hydrogen from a fatty-acid molecule and also cannot cross membranes due to its charged nature (Harris & Tall, 1989). This is in contrast to its protonated form, the hydroperoxyl radical ($HO_2\bullet$). However, $HO_2\bullet$ can react further to produce hydrogen peroxide (H_2O_2), which, in turn, can react with ferrous iron (Fe^{2+}) or with $O_2\bullet^-$ to produce hydroxyl radicals ($\bullet OH$) (Kubow, 1992). Due to the high reactivity, membrane permeability, and membrane production in the cell's aqueous phase, both $HO_2\bullet$ and $\bullet OH$ are oxygen species that have the potential to cause oxidative deterioration in postmortem fish (Hultin, 1994). Another type of $^3O_2$ activation is the production of singlet oxygen ($^1O_2$). The reaction rate of the $^1O_2$ oxidation of linoleic acid is at least 1500 times faster than that of $^3O_2$ (Korycha-Dahl & Richardson, 1978), based on the oxygen uptake measurement. The production of $^1O_2$

may occur after exposing a photosensitizer to light in the presence of $^3O_2$, as schematically illustrated in Fig. 11.1. In contrast to the inability of intact fatty acids to react with $^3O_2$, fatty-acid radicals readily undergo such a reaction, leading to the formation of hydroperoxides. Thus, the low importance of $^3O_2$ is only true for the initiation step. Once propagation and chain-branching have started, the level of $^3O_2$ is remarkably significant for the rate of lipid oxidation.

Three principle sources of oxygen required for the progress of lipid oxidation in postmortem fish meat are: oxygen present *in situ* at the reaction sites, oxygen that has penetrated throughout the whole muscle, and oxygen incorporated through processing methods, such as skinning, filleting, and mincing (Flink & Goodhard, 1978). This is a key point contributing to the development of strategies for minimizing lipid oxidation in post-harvest aquatic food products.

Prooxidants

A number of components that are naturally present in the tissue of fish and other aquatic species can serve as prooxidants interfering with the reactants at various stages of the oxidation chain. These prooxidants are both nonenzymatic and enzymatic in nature, such as redox-active transition metals, reducing agents, hemeproteins, lipoxygenases, cyclooxygenase, and peroxidases.

Nonenzymatic mechanism

Redox-active transition metals are major factors catalyzing lipid oxidation in biological systems. At concentrations found in the low molecular weight fraction of fish

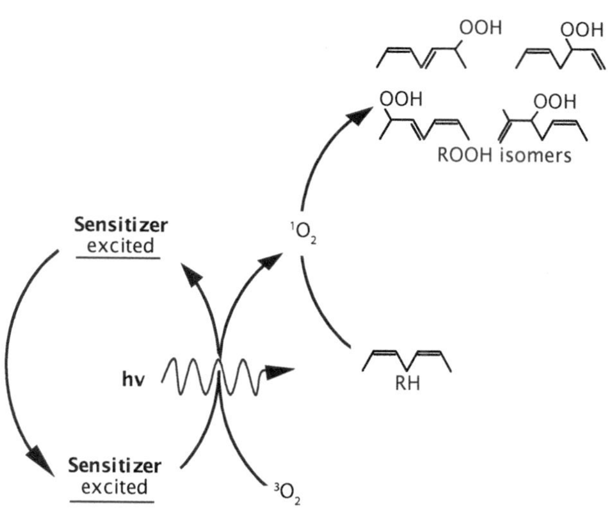

Fig. 11.1. The involvement of singlet oxygen in lipid oxidation.

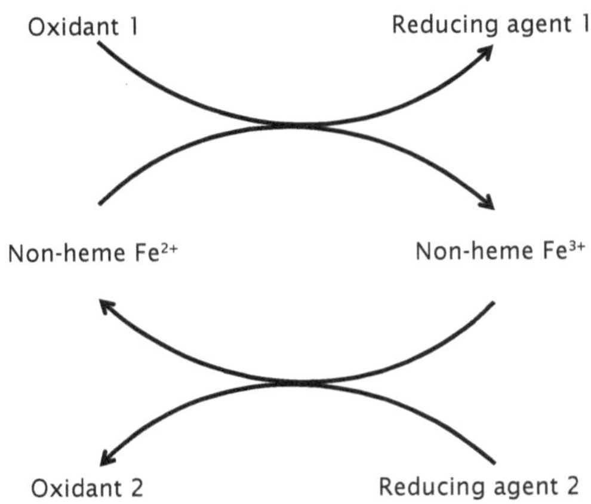

Fig. 11.2. Non-heme iron form participates in the production of reactive oxygen species.

muscle, both iron (2.2 μM) and copper (1.4 μM) have been found to catalyze lipid oxidation in the sarcoplasmic reticulum of fish muscle (Decker et al., 1989). Particularly, heme and non-heme irons are well-known catalysts of lipid oxidation with their contribution to the decomposition of lipid hydroperoxides. Non-heme iron participates in the production of reactive oxygen species, schematically illustrated in Fig. 11.2. The classical Fenton reaction (Eq. 11.11) is an example in which the production of the hydroxyl radical (•OH) is catalyzed by some non-heme iron species.

$$Fe^{2+} + H_2O_2 \rightarrow Fe^{3+} + \bullet OH + OH^- \qquad \text{(Eq. 11.11)}$$

The Fe^{3+} that was formed through reaction (Eq. 11.11) can be recycled to Fe^{2+} by various reducing agents, such as superoxide or ascorbate.

Numerous studies demonstrate the ability of heme proteins to promote lipid oxidation in aquatic food products (Lee et al., 2003; Sohn et al., 2005; Grunwald & Richards, 2006; Bao et al., 2008). These studies suggest different mechanisms for the catalysis of lipid oxidation by heme iron, based on studies with mainly myoglobin (Mb), hemoglobin (Hb), and also with non-protein-bound heme iron. Non-protein-bound heme iron may be released from Mb or Hb under certain storage conditions. The basic mechanism of oxidative reactions between lipids and Mb, which was proposed by Reeder and Wilson (1998), is an obvious example for illustration.

$$\text{deoxy}(Fe^{+2})\text{Mb} + \text{LOOH} \rightarrow \text{met}(Fe^{+3})\text{Mb} + \text{LO}\bullet \qquad \text{(Eq. 11.12)}$$

$$\text{met}(Fe^{+3})\text{Mb} + \text{LOOH} \rightarrow \text{ferryl}(Fe^{+4})\text{Mb} + \text{LO}\bullet \qquad \text{(Eq. 11.13)}$$

$$LO\bullet \rightarrow epoxyL\bullet \quad \text{(Eq. 11.14)}$$

$$epoxyL\bullet + O2 \rightarrow LOO\bullet \quad \text{(Eq. 11.15)}$$

$$epoxyL\bullet + ferryl(Fe^{+4})Mb \rightarrow met(Fe^{+3})Mb + epoxyLOH \quad \text{(Eq. 11.16)}$$

Usually, Mb exists in three forms, which include deoxymyoglobin [deoxy(Fe^{+2}) Mb], oxymyoglobin [oxy(Fe^{+2})Mb], and metmyoglobin [met(Fe^{+3})Mb] in fresh meat. The metMb is an undesirable form not only because of its brown color but also because of its catalytic effect on the oxidation of unsaturated lipids (Grunwald & Richards, 2006). The reaction between metMb and lipid hydroperoxides, as described by Eq. 11.13, generates ferrylMb, which has the ability to abstract a bis-allylic hydrogen atom from lipid (LH) to form L•. In the presence of oxygen, L• then forms LOO•, which can, in turn, abstract a bis-allylic hydrogen atom from another lipid molecule, resulting in the production of lipid hydroperoxides. When lipid hydroperoxides form in the meat, they are continuously involved in the redox reactions of Mb to generate lipid alkoxy radicals (LO•), as described in Eqs. 11.12 and 11.13). Once under way, LO• formed in Eq. 11.12 and Eq. 11.13 rearranges to form an epoxyalkyl radical (epoxyL•) (Eq. 11.14), which reacts with molecular oxygen to form LOO• (Eq. 11.15). EpoxyL• can also react with ferrylMb to form metMb, which plays a continuous role as an intermediate to oxidize lipid in meat (Baron & Andersen, 2002; Grunwald & Richards, 2006).

Enzymatic mechanism

Direct enzymatic initiation of lipid oxidation in aquatic animal tissue can be catalyzed by lipoxygenases (EC.1.13.11), which have been identified in various tissues, such as the skin (German et al., 1985) and gills (German et al., 1986) of trout, skin of sardines (Mohri et al., 1990, 1992), muscle of lake herring (Wang et al., 1991) and mackerel (Harris & Tall, 1994), as well as the haemolymph of shrimp (Kuo & Pan, 1992). Lipoxygenases are iron-containing enzymes that catalyze the oxidation not only of free and esterified fatty acids but also that of lipoprotein (Pan & Kuo, 2000). Lipoxygenase present in fish skin tissue generates hydroperoxy arachidonic acid and 14-hydroperoxy DHA. The alpha-cleavage at either side of the carbon atoms bearing the hydroperoxy groups can result in the volatile compounds responsible for the typical oxidative fishy odor, such as 2-nonenal from hydroperoxy arachidonic acid and 3,6-nonadienal from 14-hydroperoxy DHA (German et al., 1985; Hsieh & Kinsella, 1989).

Peroxidase enzymes are another group critical for the onset of lipid oxidation in fish. It has been reported that myeloperoxidase (EC.1.11.1.7) isolated from trout leukocytes can initiate lipid peroxidation in the presence of hydrogen peroxide and halides, such as bromide and iodide. This initiation might be critical during the processing of fish when the contact between air and blood increases (Kanner & Kinsella, 1983).

Inhibitors

Aquatic animal tissues contain numerous inhibitors that directly or indirectly inhibit the initiation and propagation steps of lipid oxidation. Those interfering with the propagation step by converting free radicals to stable compounds are generally referred to as primary antioxidants, and those interfering with the initiation step are considered secondary antioxidants or preventive inhibitors.

The predominant function of a primary antioxidant is donating hydrogen to a lipid free radical and becoming a free antioxidant radical (A•) (Eq. 11.17). A• can subsequently combine either with another LOO• or with another A• to terminate a series of reactions (Eqs. 11.19 and 11.20).

$$LOO\bullet + AH \rightarrow LOOH + A\bullet \quad \text{(Eq. 11.17)}$$

$$L\bullet + AH \rightarrow LH + A\bullet \quad \text{(Eq. 11.18)}$$

$$LOO\bullet + A\bullet \rightarrow LOOA \quad \text{(Eq. 11. 19)}$$

$$A\bullet + A\bullet \rightarrow A\text{-}A \quad \text{(Eq. 11.20)}$$

Secondary antioxidants include both oxygen scavengers and chelators. The endogenous antioxidants found in aquatic food products include tocopherols (Hung et al., 1981; Syvaoja & Salminen, 1985), ubiquinol (Ramasarma, 1982; Petillo et al., 1998), carotenoids (Miki, 1991), ascorbate, organic acids, glutathione peroxidase (Nakano et al., 1992; Jia et al., 1996; Watanabe et al., 1996), superoxide dismutase (Aksnes & Njaa, 1981), catalase, peroxidases (Kanner & Kinsella, 1983), ferroxidases (Kanner et al., 1988), nucleotides, peptides, amino acids (Boldyrev et al., 1987), and phospholipids (Lee et al., 1984; Ohshima et al., 1993).

Storage Temperature

It has generally been accepted that lipid oxidation in aquatic food products accelerates much more rapidly at higher temperatures, when the products are stored above 0 °C. However, the rate of lipid oxidation increases when the products are stored between −10 and 0 °C. The concentrations of both prooxidants and lipids in the products correlate with the number of ice crystals, which accelerate the lipid oxidation rate. Furthermore, the cell membranes are damaged due by the larger ice crystals formed under this temperature zone, leading to an increase in the lipid oxidation rate. The structure of frozen fish meat depends on the freezing rate, while the amount of ice formed depends on the temperature reached during freezing. The size and distribution of ice crystals in the intra- or extracellular space vary not only with the freezing rate but also with the thawing rate, also influencing the quality of thawed fish meat. High freezing rates induce the formation of many small intracellular ice crystals; at intermediate freezing rates, crystal size increases and the crystal number decreases, and at slow freezing rates, the ice crystals only form extracellularly. The latter induces dehydration and distortion of the fibres (Bevilacqua & Zaritzky, 1982). Similar results have been reported when freezing cod muscle (Love, 1964). Since the cell membranes are damaged, the lipid is more exposed to initiators, which accelerates the lipid oxidation rate.

Typical Oxidative Deterioration in Aquatic Food Products

The onset of lipid oxidation reduces the quality of aquatic food products. Loss of quality is usually evident in the later stages of lipid oxidation and associated with the attributes of flavor, color, and nutritional value. The impact of lipid oxidation on the quality of aquatic food products is schematically illustrated in Fig. 11.3.

Rancid Off-Flavor Development

It is well known that the development of rancid off-flavors in post-harvest aquatic food products can be attributed to lipid oxidation because both enzymatic and autoxidative mechanisms result in the formation of short-chain or intermediate-chain aldehydes and ketones (Cadwallader, 2000). The volatile compounds that are responsible for the characteristic rancid off-flavor of oxidized foods are formed when lipid hydroperoxides decompose into secondary oxidative products, including aldehydes, ketones, alcohols, hydrocabons, furans, and acids. Their impact on food flavor depends both on the concentration and threshold value of the compound in the food (Brewer & Vega, 1995). Therefore, the development of rancid off-flavor has a significant correlation with the lipid hydroperoxide content in fish flesh during postmortem storage (Sohn et al., 2005). A positive correlation was found between the lipid content and the increase in solubility of many rancid volatiles, and one would expect a decrease in sensory perception for a given concentration of volatiles (Roozen et al., 1994a, 1994b); however, no relationship was found between lipid contents and the rate of lipid oxidation in the flesh of different fish species during iced storage (Sohn et al., 2005), as seen in Fig. 11.4 on page 355.

Discoloration of Fish Meat

Discoloration is another manifestation of lipid oxidation in post-harvest aquatic food products. Discoloration is a major problem because consumers have their first impression of a food's acceptability based on its color. In tuna meat and red meat of other fish species, the principal pigments that contribute their red color are heme-pigments, especially Mb. During low-temperature storage, this fish meat gradually discolors, going from red to dark brown, due to oxidation of deoxyMb and oxyMb

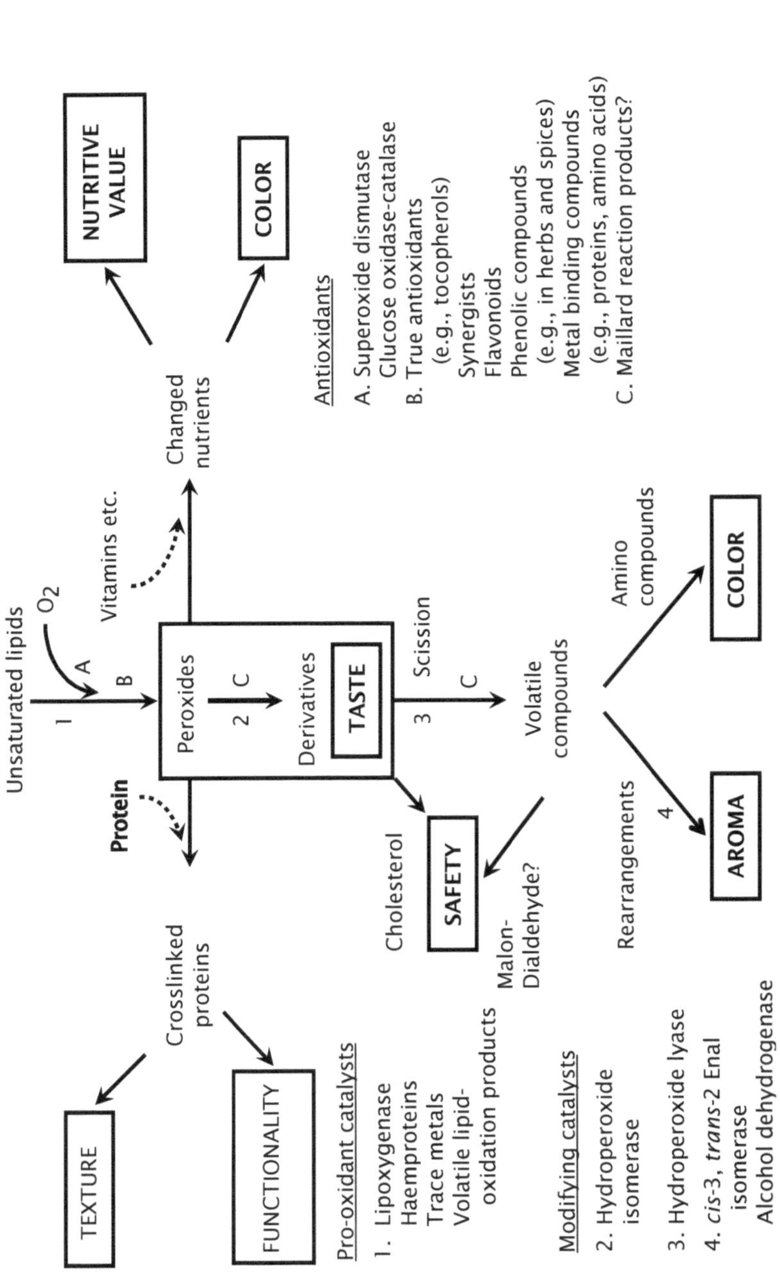

Fig. 11.3. The impact of lipid oxidation on the quality of aquatic food products.

Adapted from Eriksson (1982).

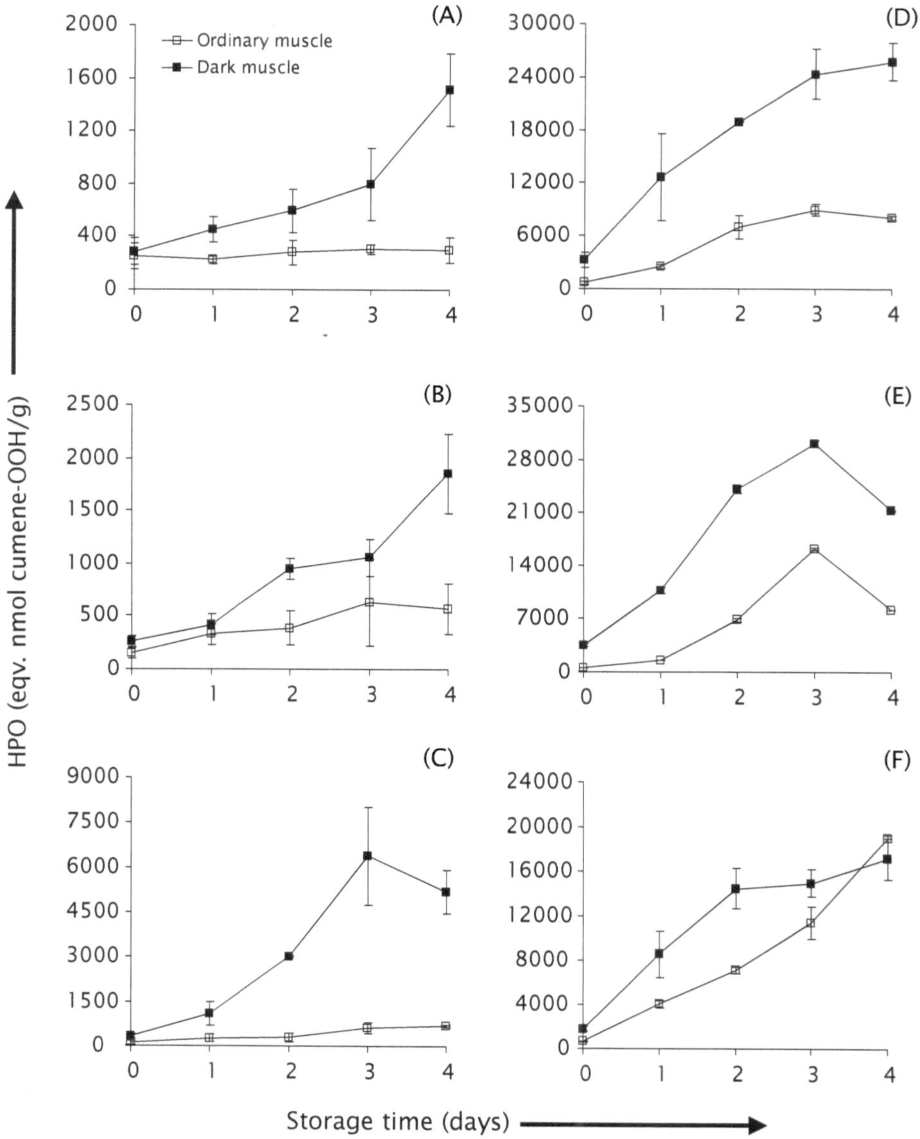

Fig. 11.4. Changes in the content of total lipid hydroperoxides in ordinary and dark muscles in ice storage. (A) Yellowtail (*Seriola quinqueradiata*); (B) amberjack (*Seriola purpurascens*); (C) Japanese butterfish (*Hyperoglyphe japonicus*); (E) Pacific saury (*Cololabis saira*); (E) Japanese Spanish mackerel (*Scomberromorus niphonius*); (F) chub mackerel (*Scomber japonicus*). Error bars present as mean ± S.D. ($n = 3$)

Adapted from Sohn et al. (2005).

to metMb. Several reports have suggested the cooxidation of fish meat Mb and lipids (Lee et al., 2003; Grunwald & Richards, 2006). The metMb formation in fish meat is related to the accumulation of lipid hydroperoxides when fresh fish meat was stored on ice (Sohn et al., 2005; Bao et al., 2009), as seen in Fig. 11.5. Lipid oxidation, along with Mb oxidation, has a high correlation with the discoloration of tuna meat during ice storage (Bao et al., 2009). It has been hypothesized that this relationship explains that free radicals generated through lipid oxidation may initiate Mb oxidation (Reeder & Wilson, 1998) and that metMb formation also promotes lipid oxidation (Baron & Andersen, 2002). Thus, the discoloration occurs when the amounts of metMb accumulated in fish meat reach certain levels (Bao et al., 2009).

Changes in Texture and Functionality of Protein

The interaction of lipids with proteins and other muscle constituents has numerous implications with regard to the nutrient content and characteristics of muscle food. Free fatty-acid levels in fish, which increase during storage, also cause protein denaturation and adversely affect texture and water-holding capacity, also known as "drip loss" (Love & Elerian, 1964; Anderson & Ravesi, 1969). The increase in free fatty-acid levels has been attributed to phospholipase A2 (EC.3.1.1.4) and a lysosomal lipase (Audley et al., 1978; Geromel & Montgomery, 1980).

Lipid hydroperoxides (LOOH) are intermediates of peroxidation that accumulate in peroxidized lipids, and they react with transition metals to produce LO•

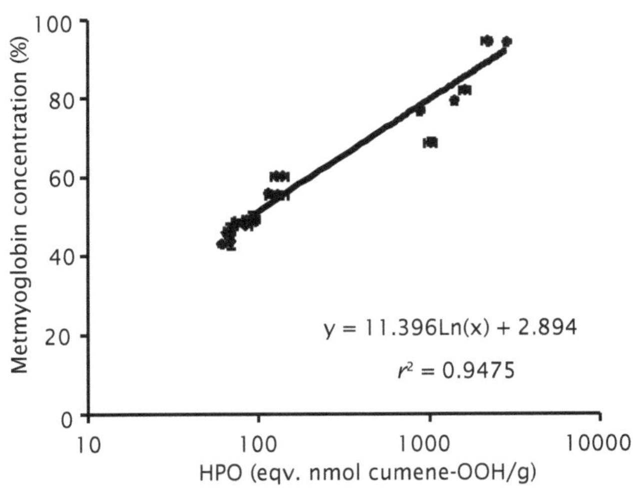

Fig. 11.5. Correlation between the total lipid hydroperoxide contents and metmyoglobin concentration in the total myoglobin of minced bigeye tuna meat in ice storage. Error bars present as mean ± S.D. (n = 3)
Adapted from Bao et al. (2009).

and LOO•. The latter radical can be responsible for structural changes in proteins, such as cross-linking and fragmentation. The resulting denatured proteins may alter cellular function, or radicals may produce tissue degeneration.

Loss of Nutrients

As the major substrate for lipid oxidation, the decline of total PUFAs in fish meat significantly correlates with the rate of lipid oxidation under ice storage conditions. Significant changes were found only in certain PUFAs, including 16:4n-3, 20:4n-6, 20:4n-3, 20:5n-3, 22:5n-6, 22:5n-3, and 22:6n-3, in the tuna and/or mackerel meat after seven days of ice storage (Bao et al., 2008), as shown in Table 11.A.

Both types of reactions involving free radicals and secondary products produced from the lipid oxidation process may possibly lead to protein damage. The reaction of proteins with lipid radicals and/or lipid hydroperoxides generates protein radicals, which eventually cause protein damage, such as the formation of cross-linkages between protein molecules or protein and lipids, the scission of proteins, and the damage of amino acids. Most of the aldehydes formed from lipid oxidation are probably involved in the protein damage as they form fluorescent or cross-linkages with proteins (Gardner, 1979; Schaich, 1980).

The loss of water- and lipid-soluble antioxidants due to lipid oxidation in fish meat was remarkable during ice storage. High correlations between the decline of antioxidants—including tocopherol, ubiquinone-10, ascorbate, and glutathione—and the accumulation of TBARS in light and dark mackerel muscles were observed when the muscles were stored on ice for 11 days (Petillo et al., 1998). This result suggests that free radicals and peroxides are produced during lipid oxidation, which oxidizes or destroys these antioxidants.

Application of Natural Antioxidants from Edible Mushrooms for Controlling Oxidative Deterioration in Aquatic Food Products

Many researchers have believed oxidative deterioration to be one of the factors limiting the shelf life of foods (Jacobsen, 1999), particularly that of aquatic food products. It is a complex process whereby lipids—particularly, PUFAs—are degraded via free radical formation, causing the deterioration of flavor, texture, color, aroma, taste, consistency, nutritional benefits, and to some extent, the safety of foodstuffs for human consumption.

To improve the quality of aquatic food products, oxidation should be mitigated or delayed. A number of potential strategies, which have been extensively studied, may help to achieve this goal. In recent years, critical problems of food safety have begun to change the food industry worldwide. Most consumers tend to

Table 11.A. Fatty Acid Contents of Minced Bigeye Tuna (*Thunnus obesus*) and Pacific Mackerel (*Scomber japonicus*) Meats Before and After Ice Storage

(mg/g oil)

	Bigeye tuna		Pacific mackerel	
Fatty acids	**Day 0**	**Day 7**	**Day 0**	**Day 7**
Saturated fatty acids				
C14:0	19.13	14.98	16.84	16.88
C15:0	7.03	5.51	2.65	2.74
Iso 16:0	7.00	4.56	2.07	1.20
C16:0	252.77	201.53	147.56	148.05
Iso 17:0	6.06	4.57	3.17	2.70
C17:0	10.83	8.28	4.64	4.84
Iso 18:0	1.18	0.71	0.50	0.44
C18:0	68.55	55.61	65.00	66.18
C19:0	4.26	2.90	1.71	3.17
C20:0	2.49	1.84	3.30	3.45
C22:0	1.55	1.30	1.27	1.29
C24:0	1.19	1.14	1.35	1.39
Total	382.05 a	302.92 a	250.07 a	252.32 a
Monounsaturated fatty acids				
C14:1n-7	2.23	1.92	1.20	1.00
C14:1n-5	0.42	0.36	0.94	1.02
C15:1n-6	0.55	0.27	0.58	0.55
C16:1n-9	0.80	1.07	2.23	2.18
C16:1n-7	44.64	33.12	17.63	17.99
C16:1n-5	1.41	1.15	1.12	1.04
C18:1n-9	209.03	154.02	180.26	187.11
C18:1n-7	10.81	13.40	1.36	1.54
C18:1n-5	1.28	1.00	1.38	1.18
C20:1n-9	17.89	12.81	40.18	42.25
C20:1n-7	1.36	0.98	2.14	2.25
C22:1n-11	1.35	0.95	22.06	22.58
C22:1n-9	2.21	1.74	4.30	4.50
Total	293.98 a	222.79 a	275.37 a	285.19 a
Polyunsaturated fatty acids				
C16:2n-6	1.01 a	0.75 a	1.36 a	1.26 a
C16:3n-6	8.57 a	6.37 a	3.76 a	2.43 b
C16:3n-4	7.75 a	5.65 a	2.76 a	2.82 a
C16:3n-1	2.16 a	1.59 a	1.88 a	1.94 a
C16:4n-3	0.85 a	0.35 a	1.53 a	1.06 a
C18:2n-6	10.57 a	11.24 a	10.60 a	6.88 a

Table 11.A. *Continued*

(mg/g oil)

	Bigeye tuna		Pacific mackerel	
Fatty acids	**Day 0**	**Day 7**	**Day 0**	**Day 7**
C18:2n-4	3.11 a	2.01 a	1.05 a	0.85 a
C18:3n-6	3.47 a	2.31 a	2.14 a	2.09 a
C18:3n-4	1.40 a	0.94 a	1.02 a	1.05 a
C18:3n-3	3.51 a	4.74 a	4.85 a	1.75 a
C18:4n-3	1.73 a	1.11 a	2.07 a	1.94 a
C20:2n-6	2.30 a	1.59 a	1.21 a	1.35 a
C20:3n-6	1.31 a	0.51 a	1.34 a	1.38 a
C20:4n-6	28.11 a	20.43 a	16.55 a	13.84 b
C20:3n-3	1.84 a	1.34 a	0.47 a	0.52 a
C20:4n-3	3.70 a	2.50 a	2.46 a	2.20 b
C20:5n-3	49.12 a	35.49 a	35.04 a	27.61 b
C22:4n-6	4.29 a	2.75 a	2.38 a	2.11 a
C22:5n-6	15.93 a	11.94 b	6.96 a	5.62 a
C22:5n-3	15.59 a	11.45 b	13.99 a	12.04 a
C22:6n-3	256.27 a	191.18 b	173.03 a	127.38 a
Total	422.6 a	316.22 b	286.44 a	218.06 a

Data are presented as mean values of three replicates for each sample. Different letters within the rows indicate significant differences between samples ($p < 0.05$).

Adapted from Bao et al. (2008).

prefer foods containing natural additives, such as extracts from fruits, vegetables, and other edible materials, over those with artificial food additives. The application of natural antioxidants for controlling oxidative deterioration in aquatic food products is, therefore, necessary to meet the demands of these consumers.

Natural Antioxidants from Edible Mushrooms

Mushrooms have been known to be a potential source of antioxidants. The antioxidative properties of mushroom extracts have been reported in several studies (Fu et al., 2002; Mau et al., 2002; Jang et al., 2004; Cheung & Cheung, 2005; Puttaraju et al., 2006; Elmastas et al., 2007). Among the active compounds, ergothioneine (ESH) and phenolic compounds have been known to be potent antioxidants, which exist in the fruiting body of certain mushroom species, as shown in Table 11.B. The distribution of ESH and phenolics in different parts of the mushroom and the spent mushroom culture solid medium (Fig. 11.6) are shown

Table 11.B. Total Phenolics, Ergothioneine Contents, and DPPH Radical Scavenging Activity of Hydrophilic Extracts Prepared from the Fruiting Body of Edible Mushrooms and Waste of *Flammulina velutipes*.

Extracts	Ergothioneine contents (mg mL^{-1})	Total phenolics (eqv. mg mL^{-1} of gallic acid)	DPPH radical scavenging EC_{50}* (mg of wet weight basis)
Fruiting body			
Flammulina populicola	0.48 ± 0.01	2.30 ± 0.07	0.98 ± 0.04
Flammulina velutipes	2.98 ± 0.02	2.91 ± 0.04	0.81 ± 0.03
Grifola frondosa	0.30 ± 0.01	4.26 ± 0.20	1.41 ± 0.01
Grifola frondosa (white)	0.08 ± 0.01	3.61 ± 0.06	1.45 ± 0.01
Hypsizygus tessellates	0.06 ± 0.01	6.06 ± 0.28	1.65 ± 0.02
Hypsizygus tessulatus	Tr†	4.14 ± 0.15	3.33 ± 0.06
Lentinula edodes	2.84 ± 0.04	6.72 ± 0.25	0.56 ± 0.04
Pholiota nameko	0.02 ± 0.01	1.21 ± 0.06	1.48 ± 0.03
Pleurotus cornucopiae	20.82 ± 0.12	6.72 ± 0.19	0.29 ± 0.01
Pleurotus eryngii	3.17 ± 0.11	2.95 ± 0.09	0.81 ± 0.02
Waste			
Flammulina velutipes	5.08 ± 0.08	6.16 ± 0.07	0.71 ± 0.01

*Amount of extract necessary to scavenge 50% of DPPH radical.
†Trace.

Adapted from Bao et al. (2010a).

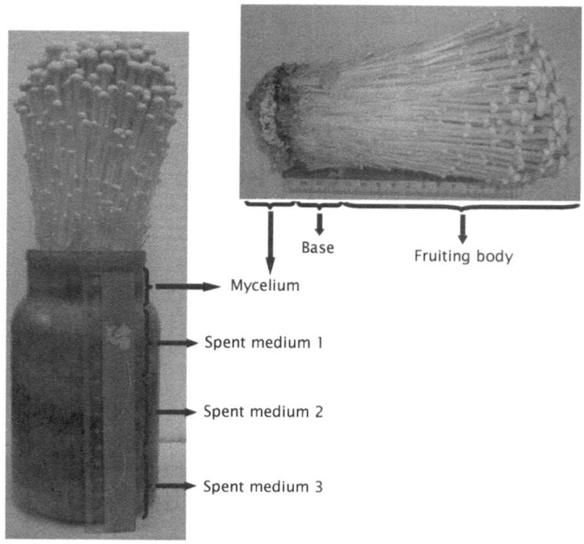

Fig. 11.6. Fruiting body, base, mycelium, and spent culture medium of the mushroom (*Flammulina velutipes*).

Adapted from Bao et al. (2010)b.

Table 11.C. Total Phenolic and Ergothioneine Contents of Hydrophilic Extracts Prepared from the Fruiting Body, Base, Mycelium, Spent, and Fresh Culture Media of the Mushroom (*Flammulina velutipes*)

Extracts	Ergothioneine contents (mg/mL)	Total phenolic contents (eqv. mg gallic acid/mL)
Fruiting body	0.38 ± 0.01	3.75 ± 0.13
Base	0.56 ± 0.07	6.31 ± 0.09
Mycelium	0.06 ± 0.02	11.48 ± 0.19
Spent medium 1	TR[a]	15.74 ± 0.41
Spent medium 2	ND[b]	9.87 ± 0.08
Spent medium 3	ND[b]	8.51 ± 0.12

[a]Trace.
[b]No detection.

Adapted from Bao et al. (2010b).

in Table 11.C. It was reported that the highest amount of ESH was distributed at the bottom of the *F. velutipes,* and the antioxidative activities of hydrophilic extracts from the mushroom's fruiting body and base were mainly caused by ESH in the extracts. The extracts from the spent culture solid medium of *F. velutipes* have strong ability against lipid oxidation, as well as oxyMb autoxidation (Fig. 11.7) and phenolic compounds, due mainly to the extracts' antioxidative activities (Bao et al., 2010).

Preventing Fish Meat Discoloration and Lipid Oxidation by Natural Antioxidants from Edible Mushrooms

Color is one of the principal criteria used to evaluate the freshness of fish meat. Because of its higher Mb content, tuna meat or the dark muscle of other fish species has a typically red color. During low-temperature storage, discoloration of fish meat from red to dark brown proceeds gradually due to oxidation of deoxyMb and oxyMb to metMb (Bito, 1965). It has been shown that Mb oxidation in fish meat is related to lipid oxidation when fresh fish meat is chilled (Lee et al., 2003). This relationship has been proposed to explain that free radicals generated through lipid oxidation may initiate Mb oxidation (Render & Wilson, 1998) and also that metMb formation promotes lipid oxidation (Baron & Andersen, 2002). During storage, the amount of lipid oxidation and Mb oxidation products increase in meat; therefore, the addition of antioxidants to meat is effective in reducing the deteriorative processes of lipids and Mb by preventing the discoloration of meat (Hamre et al., 2003; Lee et al., 2003; Georgantelis et al., 2007).

There are several categories of antioxidants that exist for application. However, the selection of antioxidants for food applications is a serious concern because the use of antioxidants in foods is strictly controlled by legislation. Generally, the antioxidants involved in food additives must be effective at low doses, have no impact on flavor, and be nontoxic. For these reasons, the natural antioxidants are generally preferred for food applications.

Jang et al. (2004) reported that the hydrophilic extract prepared from *F. velutipes* successfully prevented oxidation of PUFAs in cod liver oil-in-water emulsion systems, as shown in Fig. 11.8 (p. 364). Later, the antioxidative activity and antidiscoloration efficacy of ESH in the extract were found by adding the extract to beef and fish meat (Bao et al., 2008). The addition of 15 mg crude ESH in 5 mL of the extract, prepared from *F. velutipes,* to 100 g of minced bigeye tuna meat was more effective at oxidizing lipids in tuna meat than adding sodium ascorbate (500 ppm) or α-tocopherol (500 ppm), as shown in Fig. 11.9 (p. 365). Lipid oxidation significantly correlated with metMb formation, as well as color change, in the tuna meat. Incorporating the crude mushroom ESH prolonged the fresh-color

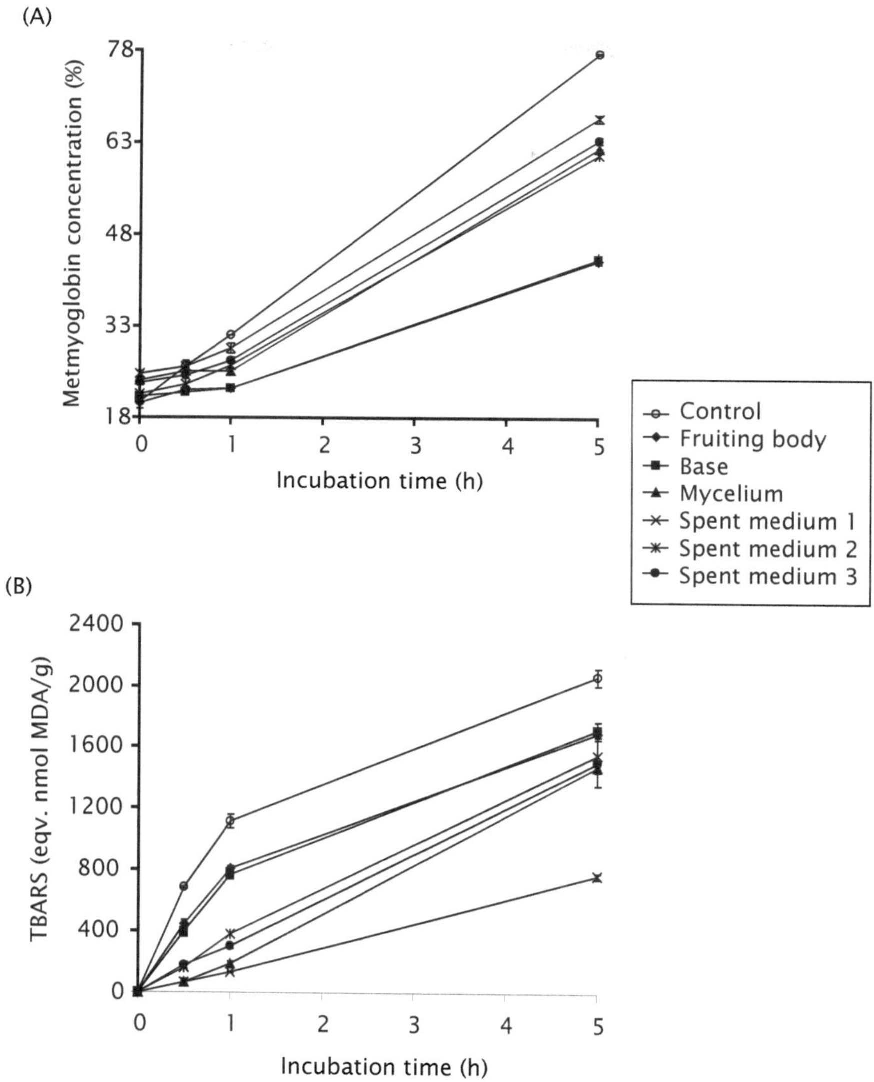

Fig. 11.7. Effect of hydrophilic extracts prepared from the fruiting body, base, mycelium, and spent solid culture medium of the mushroom (*Flammulina velutipes*) on the auto-oxidation of oxymyoglobin purified from yellowtail dark muscle incubation at 37 °C (A) and on the formation of thiobabituric acid reactive substances (TBARS) in yellowtail dark muscle homogenates incubated at 37 °C (B). Data are presented as mean ± S.D. ($n = 3$).

Adapted from Bao et al. (2010b).

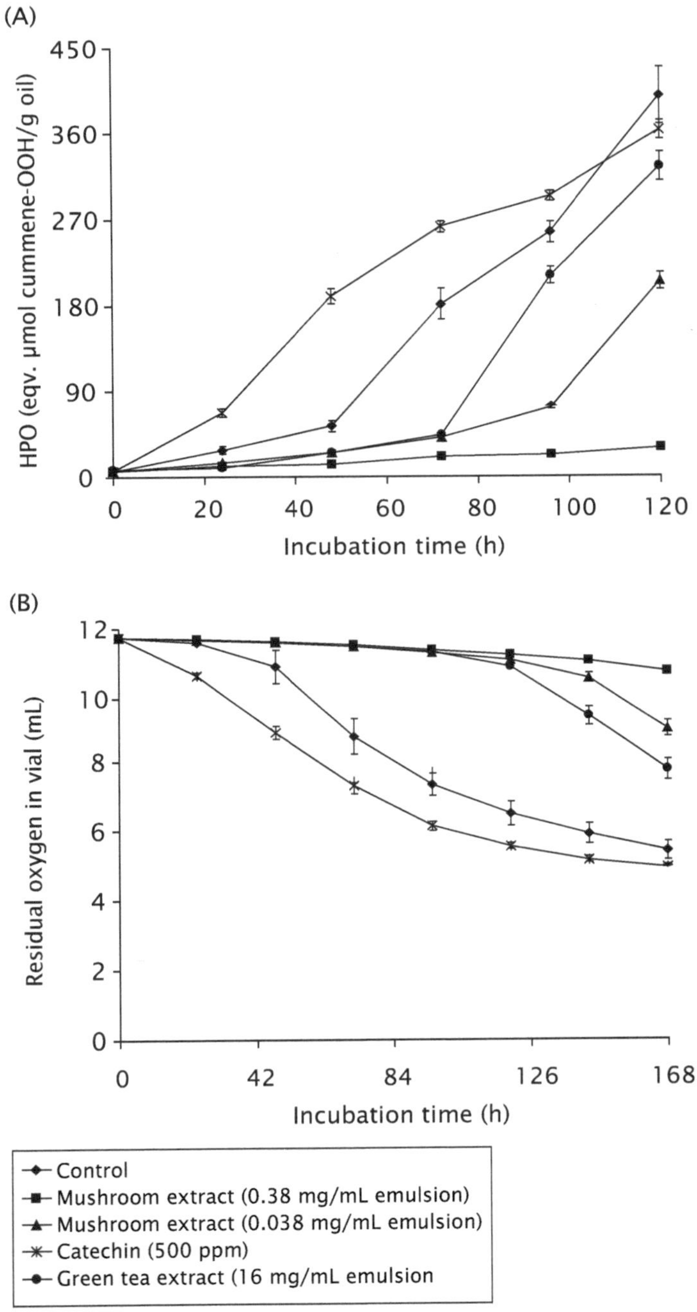

Fig. 11.8. Effect of the mushroom extract on the total lipid hydroperoxide formation (A) and the residual amount of oxygen absorbed (B) of cod liver oil in water emulsion. Error bars present as mean ± S.D. (n = 5).

Adapted from Jang et al. (2004).

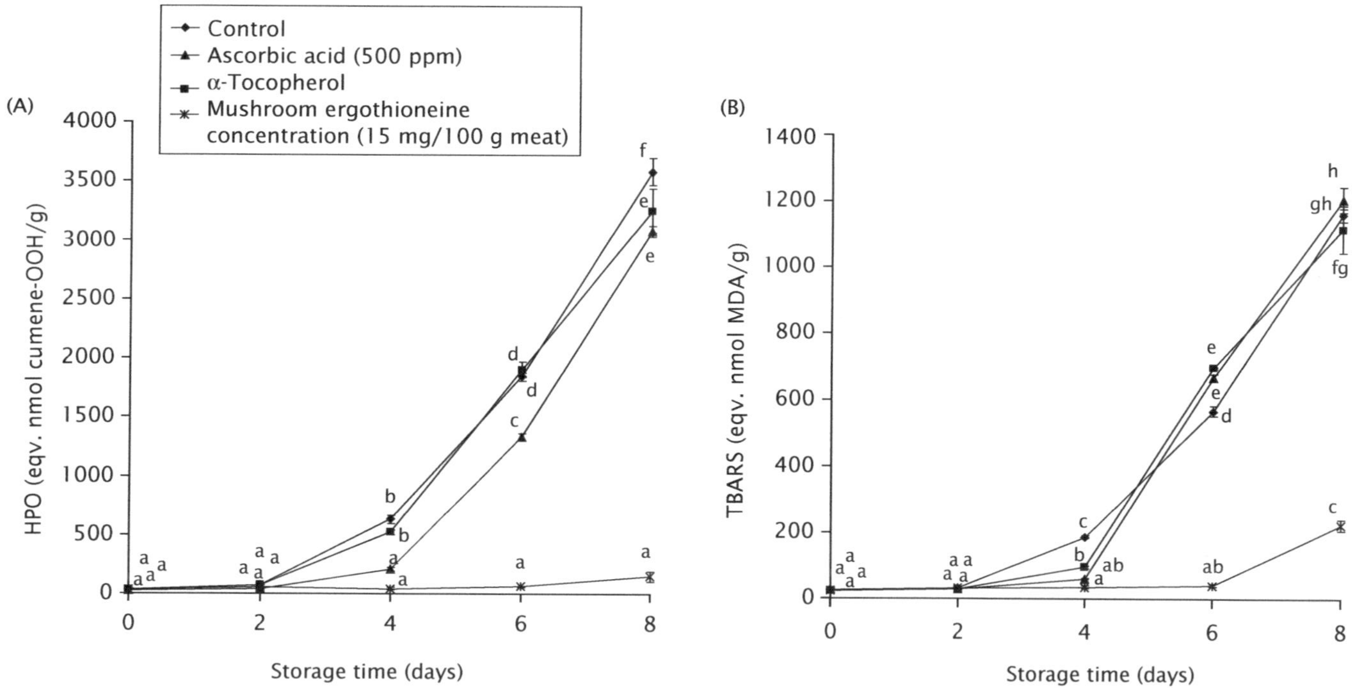

Fig. 11.9. Changes in the total lipid hydroperoxides (A) and thiobarbituric acid reactive substances (B) of minced bigeye tuna meat with different antioxidants added: (⁎), 15 mg of mushroom ergothioneine added to 100 g meat; (▪), 500 ppm of α-tocopherol added to 100 g of meat; (▲), 500 ppm of ascorbic acid sodium salt added to 100 g meat; (♦), control with 5 mL of distilled water added to 100 g of meat. Error bars present as mean ± S.D. ($n = 3$). Values with different superscript letters represent significant difference ($p < 0.05$).

Adapted from Bao et al. (2009a).

shelf life of minced bigeye tuna meat during ice storage, when compared with the control tuna meat without the mushroom extract, as shown in Fig.11.10 (Bao et al., 2009a).

In live fish and other aquatic animals, the catalytic and antioxidative systems in their tissues are balanced. However, this balance does not continue during postmortem storage, and lipid oxidation begins when several biochemical processes take place (Hultin, 1992). Once lipid oxidation has already occurred, none of the above methods may be able to improve the quality of meat. Therefore, mushroom extract should be used to control fish meat's discoloration and lipid oxidation for original or fresh products.

The dietary supplementation of antioxidants has generally been believed to be an effective method for stabilizing lipids and meat color (Corino et al., 1999; Philips et al., 2001; Yang et al., 2002). This area has been more extensively studied in the production of meat from farmed animals and, to a lesser extent, in aquaculture. This research may provide useful learning applicable for the farming of fatty fish, such as yellowtail, salmonoids, and the Pangasius species. Dietary composition influences the composition and stability of animal tissue after slaughter, including its antioxidant and fatty-acid contents (Joo et al., 2002; Turchini et al., 2003). Many studies have shown that elevating the antioxidant content of muscle tissue via increased dietary supplementation improves the oxidative stability of stored meat and fish (Mansour et al., 2006; Chen et al., 2008). Dietary tocopherol is one of the most studied supplemented antioxidants, and it has been reported to be one of the most potent stabilizers for lipids in meat (Faustman et al., 1998; Corino et al., 1999; Castellini et al., 2001; Phillips et al., 2001; Yang et al., 2002). Dietary α-tocopherol provides greater protection of meat lipids and color, compared to α-tocopherol applied postmortem, because dietary antioxidants are incorporated into the membranes. In comparison, postmortem-applied antioxidants are exogenous. Hamre et al. (2000) investigated the turnover of α-, γ- and δ-tocopherols in Atlantic salmon and the effects of enriching Atlantic salmon fillets with α-, γ- and δ-tocopherols via diet on their oxidative stability during chilled and frozen storage. They found that the salmon muscle incorporated γ-tocopherol at a similar rate as α-tocopherol, while the incorporation rate of δ-tocopherol was approximately 50%. Non α-tocopherol was prooxidative during storage on ice but effective at reducing oxidation in a dose-dependent manner during frozen storage at –30°C over 48 weeks storage. They concluded that α-tocopherol may be better suited than tocopherol mixes for optimizing the oxidative stability of salmon fillets. In some studies, the positive effects of dietary tocopherol supplementation were only observed when the antioxidants were at high levels. A study by Raskin et al. (1998) showed that tocopherol incorporation was 10 to 20 times the normal dietary requirement, which significantly increased feeding costs. They reported

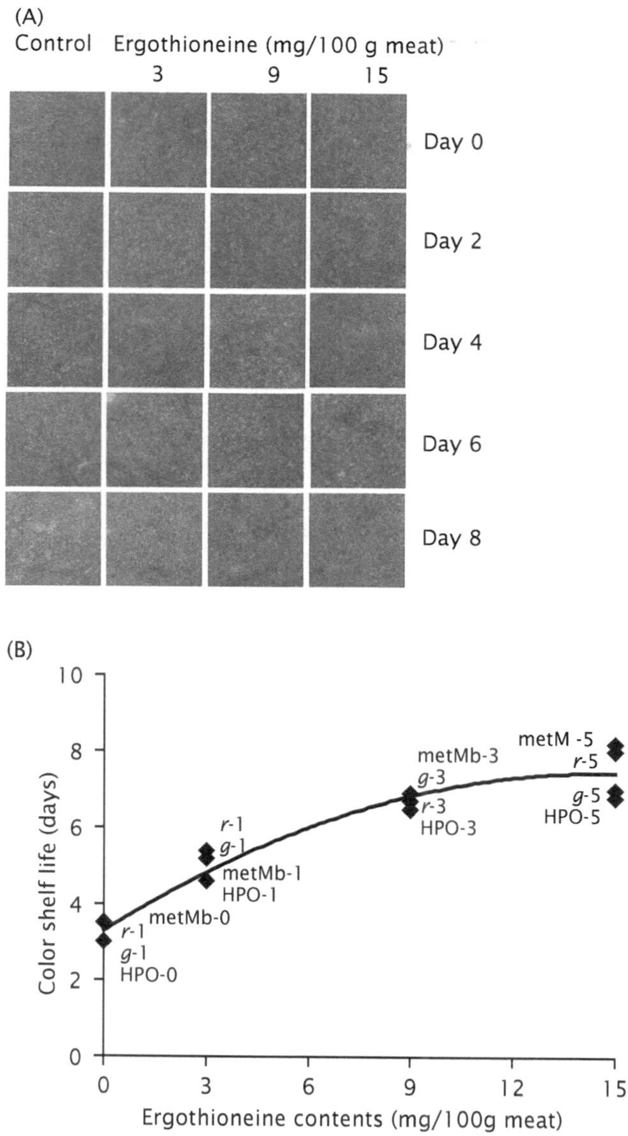

Fig. 11.10. Changes in visual color (A) and the correlation of regression models (B) loadings based on total lipid hydroperoxide content (HPO), metmyoglobin concentration in total myoglobin (metMb), and *r* and *g* values of minced bigeye tuna meat with different amounts of mushroom ergothioneine added.

Adapted from Bao et al. (2009a).

that dietary supplementation of α-tocopherol did not protect the white muscle lipids of rainbow trout from rancidity, although a slight protective effect was reported for dark muscle lipids. The use of carotenoids, such as astaxanthin, in aquaculture has been widely studied, mainly in salmonoids. Lipid oxidation was significantly suppressed after rainbow trout fed a high astaxanthin (100 ppm) diet was stored for 18 months compared to 40 ppm diets (Jensen et al., 1998). The opposite was true with the same type but a greater quantity of processed fish, where lipid oxidation was more advanced. Raskin et al. (1998) concluded that the carotenoid was important for controlling the early stages of lipid oxidation.

Dietary supplementation of antioxidants extracted from plants, mushrooms, and microbial biomass is another potential strategy to minimize the oxidative deterioration of aquatic food products. Yellowtail (*Seriola quinqueradiata*) supplementarily fed extract prepared from a spent solid culture medium of *F. velutipes* demonstrated a significantly expanded shelf life of their dark muscle in terms of lipid and color stability under chilled storage conditions (Bao et al., 2009b). Changes in the total lipid hydroperoxide (HPO) of yellowtail dark muscle during chilled storage are shown in Fig. 11.11. HPO accumulation in the dark muscle of nonsupplemented (control) yellowtail increased as storage time was prolonged. After feeding yellowtail the extract supplement for one week, Bao et al. (2009) found that the amounts of HPO accumulated in the dark muscle were significantly suppressed when compared to those of the control group under chilled storage conditions. After two and four weeks of continuous feeding, HPO accumulation in the dark muscle of yellowtail was remarkably controlled during four days of chilled storage. Fig. 11.12 (p. 370) shows changes in the percentage of metMb in yellowtail dark muscle during chilled storage for four days. The formation of metMb in the dark muscle of yellowtail fed the extract supplement was significantly delayed during chilled storage. After two weeks of feeding the extract supplement, the percentage of metMb in the dark muscle of yellowtail remained unchanged in both the 1% and 10% groups for four days of chilled storage. In contrast, the percentage of metMb in the dark muscle of yellowtail in the control group significantly increased under similar storage conditions. After two weeks of the extract supplement, the dark muscle color of yellowtail in both the 1% and 10% groups stabilized remarkably during four days of chilled storage. It had a similar effect, preventing lipid oxidation in the ordinary muscle of jack mackerel (*Trachurus japonicus*) fed this extract supplementation (Encarnacion et al., 2008), as seen in Fig. 11.13 (p. 371). Because the extract from a spent culture medium of *F. velutipes* contains ESH at a level of 340 mg/L, researchers have seen positive effects for the extract's dietary supplementation due to contribution of ESH (Bao et al., 2009b). Thus far, biosynthesis of ESH has only been noted in certain fungi and Actinomycetales bacteria (Chaudiere & Ferrari-Iliou, 1999). Although ESH is present in

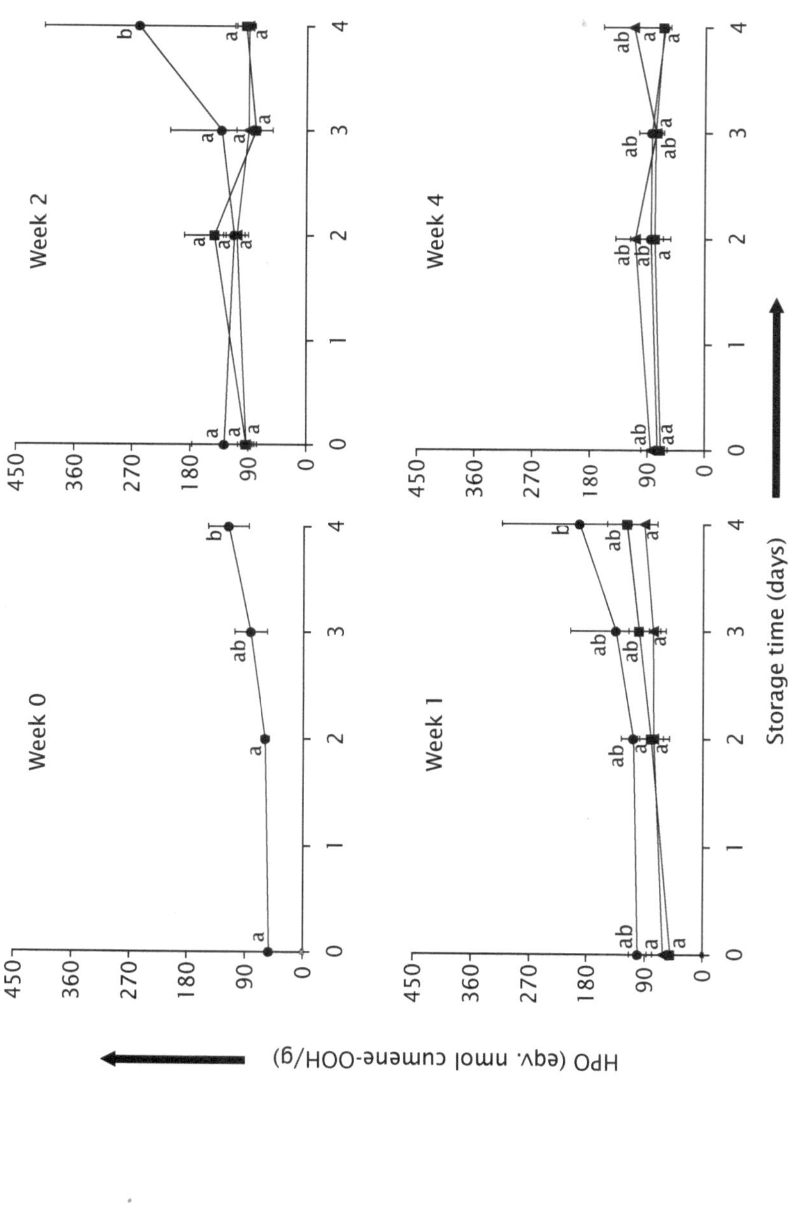

Fig. 11.11. Changes in the total lipid hydroperoxides of yellowtail (*Seriola quinqueradiata*) dark muscle during chilled storage: (●) fed without concentrated extract supplementation; (■) fed with 1% concentrated extract supplementation; (▲) fed with 10% concentrated extract supplementation. Data are presented as mean ± S.D. ($n = 5$). Values with different superscript letters are significantly different ($P < 0.05$).

Adapted from Bao et al. (2009b).

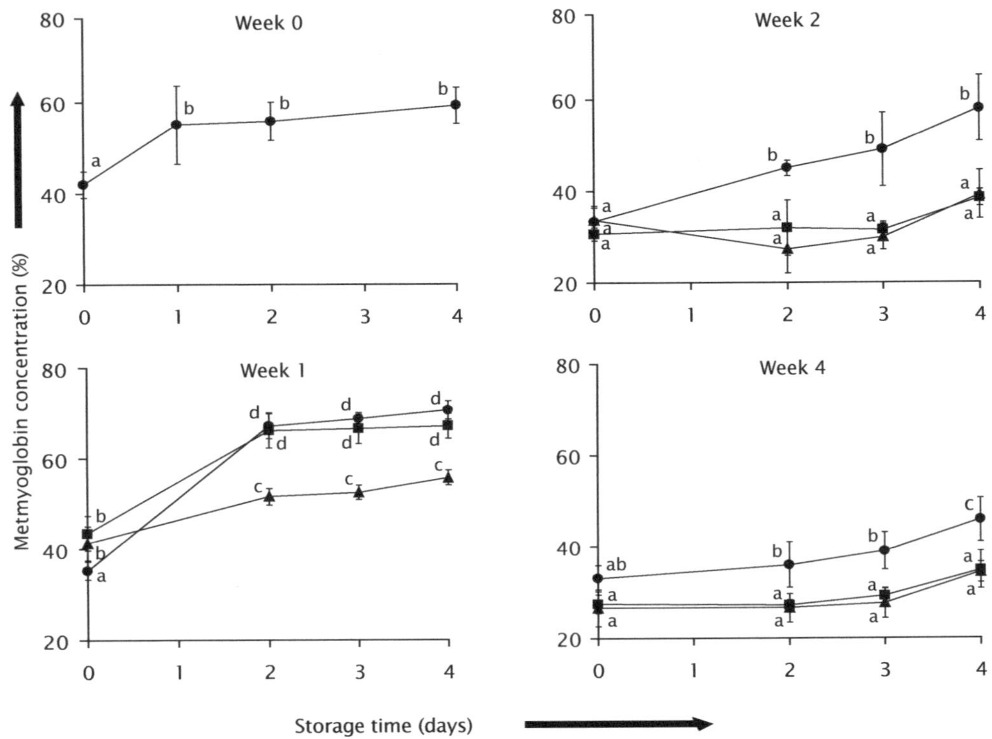

Fig. 11.12. Changes in the metmyoglobin concentrations of yellowtail (*Seriola quinqueradiata*) dark muscle during chilled storage: (●) fed without concentrated extract supplementation; (■) fed with 1% concentrated extract supplementation; (▲) fed with 10% concentrated extract supplementation. Data are presented as mean ± S.D. ($n = 5$). Values with different superscript letters are significantly different ($P < 0.05$).

Adapted from Bao et al. (2009b).

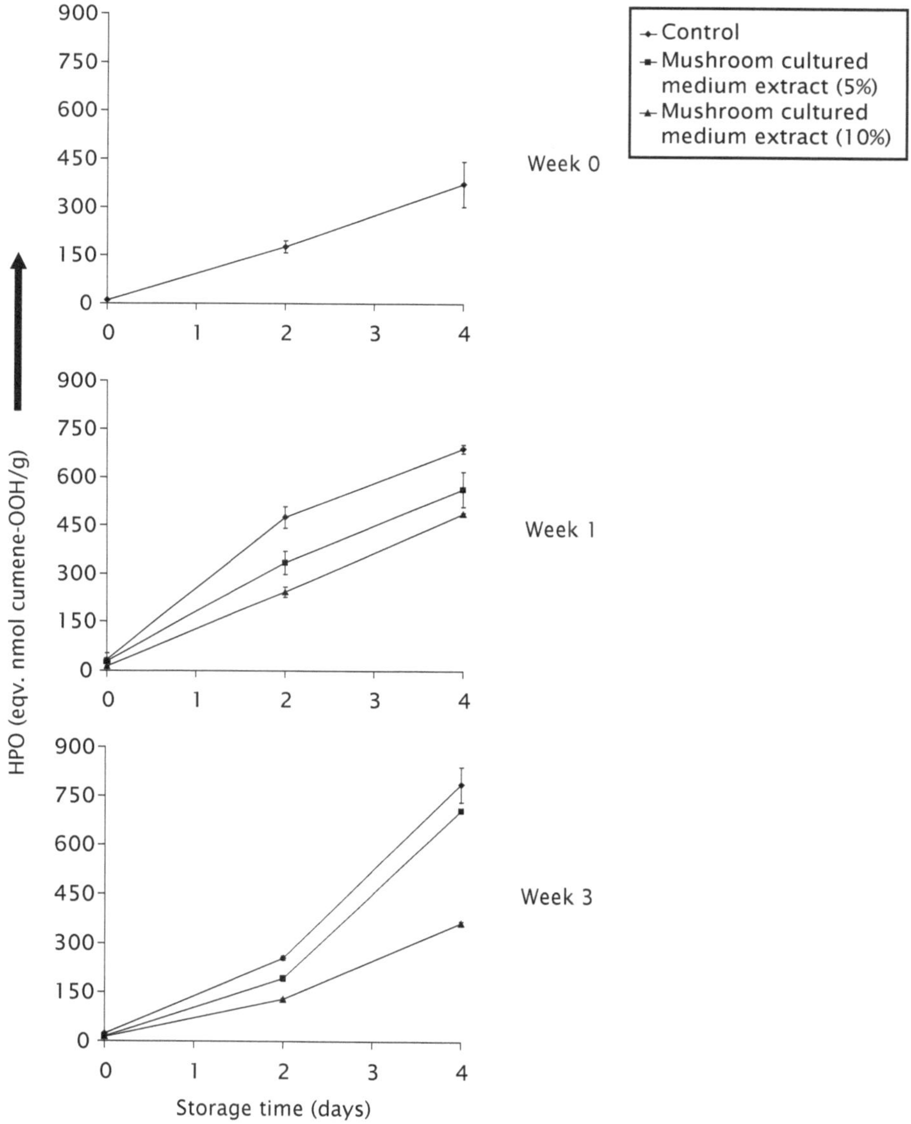

Fig. 11.13. Changes the in total lipid hydroperoxides of jack mackerel (*Trachurus japonicus*) ordinary muscle during chilled storage: (◆) fed without concentrated extract supplementation; (■) fed with 5% mushroom-cultured, medium-extract supplementation; (▲) fed with 10% mushroom-cultured, medium-extract supplementation. Error bars present as mean ± S.D. ($n = 5$).

Adapted from Encarnacion et al. (2008).

certain tissues of higher plants and animals, the existence of a biosynthetic pathway in these species has not been demonstrated (Misiti et al., 2001). Hence, it has been suggested that the ESH existing in plants may result from the absorption of ESH produced by microorganisms in the soil through a plant's roots. Consequently, the ESH content of animal cells would come from ESH-containing plant and animal foods (Ey et al., 2007). Baldridge (1954) reported that the level of ESH in the blood of rats fed with oats correlated to the percentage of oats in their diet. This occurred because oats contain appreciable amounts of ESH (Melville & Eich, 1956). ESH is usually found when tissues that include the liver, kidneys, heart, lung, central nervous system, ocular tissues, seminal fluid, skeletal muscle, and red blood cells are exposed to oxidative stress (Briggs, 1972; Kawano et al., 1982; Shires et al., 1997; Ey et al., 2007). However, levels of ESH vary among species and type of tissues. Kawano et al. (1983) demonstrated that ESH has an inhibitory effect on lipid peroxide formation in mouse liver. Franzoni et al. (2006) believe that this effect is due to the radical scavenging behavior of ESH. Furthermore, ESH functions not only to protect an organism's organs against lipid peroxidation but also to conserve endogenous glutathione and α-tocopherol (Deiana et al., 2004). Dietary ESH supplementation may be a potential solution for stabilization of fish meat color because ESH has no ability to reduce metMb to deoxyMb but converts ferrylMb to metMb when it is exposured to hydrogen peroxide (Arduini et al., 1990). Therefore, the delay in metMb formation in the dark muscle of yellowtail that was fed on the extract prepared from spent culture medium of *F. velutipes*, might result from both suppression of lipid oxidation and conservation of endogenous antioxidants.

Mechanism Underlying Antioxidation and Antidiscoloration Efficacies of the ESH and Phenolics in Mushroom Extract Applied to Fish Meat

Antioxidants can perform multiple actions, which include sequestering catalytic metal ions, decreasing oxygen concentration, quenching singlet oxygen and superoxide anions, decomposing primary oxidation products to nonvolatile compounds, preventing first-chain initiation by scavenging initially generated radicals, and chain-breaking. Among these, studies have discussed the chain-breaking mechanism for several antioxidants (Roginsky & Lissi, 2005). The antioxidant can donate a hydrogen atom to a lipid peroxyl radical, forming an antioxidant radical (Eq. 11.17), and the antioxidant radical can subsequently combine either with another lipid peroxyl radical or another antioxidant radical to terminate the reaction (Eq. 11.19, Eq. 11.20).

Both ESH and phenolic compounds in mushroom extracts have been documented as potent antioxidants not only *in vitro* but also *in vivo*. Arduini et al. (1990) demonstrated that ferrylMb, which formed when deoxyMb and metMb were exposed to hydrogen peroxide, was reduced to metMb in the presence of ESH. Additionally, ESH has been known as a powerful scavenger of hydroxyl radical (•OH) as well as an inhibitor of •OH generation from hydrogenperoxide, which is catalyzed by iron or copper ions (Akanmu et al., 1991). Compared with traditional antioxidants—such as glutathione, uric acid, and 6-hydroxy-2,5,7,8-tetramethylchroman-2-carboxylic acid (Trolox®)—ESH is the most powerful scavenger of free radicals (Franzoni et al., 2006). We propose that the mushroom extract delays the oxidation of lipids as well as Mb in fish meat by ESH, as seen in Fig. 11.14. LO• and •OH radicals generated from the reaction between hydrogen peroxide, lipid hydroperoxides, and metMb in meat could be scavenged by ESH existing in the mushroom extract; ferryl Mb and perferryl Mb formed from these reactions received electrons from the thiol group of ESH and reduced to metMb (Bao et al., 2009a).

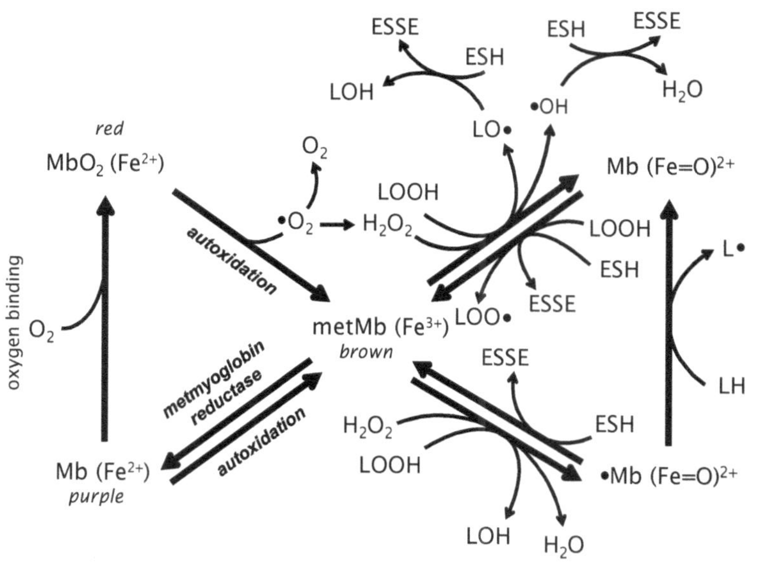

Fig. 11.14. Proposed mechanism for delaying the oxidation of lipids and myoglobins in minced bigeye tuna meat by adding ergothioneine in mushroom extract. Mb (Fe^{2+}), deoxymyoglobin; MbO_2 (Fe^{2+}), oxymyoglobin; Mb (Fe^{3+}), metmyoglobin; Mb $(Fe{=}O)^{2+}$, ferrylmyoglobin; •Mb $(Fe{=}O)^{2+}$, perferrylmyoglobin; ESH, ergothioneine; ESSE, ergothioneine disulfide.

Adapted from Bao et al. (2009a).

A class of phenolic or polyphenolic compounds is a group of antioxidants identified in extracts from various mushrooms (Cheung et al., 2003; Puttaraju et al., 2006; Kim et al., 2008). Phenolic compounds are highly effective antioxidants because of their multifunctional modes of action, including reactions with •OH, superoxide anions, singlet oxygen, and chelating transition metal. A relationship exists between the chemical formula and antioxidant activity of phenolic compounds. Many studies have reported the antioxidant action of phenolic compounds because of their high tendency to chelate metals. Phenolics possess hydroxyl and carboxyl groups, which bind iron and copper, particularly (Michalak, 2006). Thus, iron ions released from hemoglobin (Hb) or Mb in meat are chelated, leading to the suppression of the superoxide-driven Fenton reaction. Another mechanism underlies their antioxidant ability. Metal ions decompose lipid hydroperoxide (LOOH) by the hemolytic cleavage of the O-O bond and produce lipid alkoxyl radicals, which initiates free-radical chain oxidation. Phenolic antioxidants suppress lipid peroxidation by trapping the lipid alkoxyl radical. This activity depends on the molecules' structure, along with the number and position of the hydroxyl group in the molecules (Millic et al., 1998). The chemical activity of phenolic compounds, in terms of their reducing properties as hydrogen- or electron-donating agents, predicts their potential for action as free-radical scavengers. Phenolic compounds possess the ideal structural chemistry for free-radical scavenging activities, and they have been shown to be more effective antioxidants *in vitro* than vitamins E and C on a molar basis (Rice-Evan et al., 1996, 1997).

Conclusion

Lipid oxidation is one of the major causes of quality loss in aquatic food products. Quality losses are usually evident in the later stages of lipid oxidation and are associated with the attributes of flavor, color, and nutritional value. As the industry has developed, a number of strategies have been established to prevent oxidative deterioration in aquatic food products. Among them, the use of antioxidants has been traditionally believed to be almost the only effective way. Although the most effective function of antioxidants is preventing oxidative deterioration in aquatic food products, the application of antioxidants to food has to be carefully evaluated in order to achieve the most effective and safest applications. Oxidative spoilage occurs quite rapidly in aquatic food products; therefore, post-harvest application of antioxidants is usually too late to control the action of oxidative compounds in these products. To enhance this method, the combination of antioxidants with other techniques, such as packaging or lower-temperature storage, has been tailored and continually developed. Investigations of antioxidative activity and the antidiscoloration efficacy of mushroom extracts will contribute to strategies for stabilizing fresh meat color during lower-temperature storage. Extracts prepared from fruiting bodies and the spent culture medium of *F. velutipes* are potent antioxidants that can be applied to control the oxidation not only of lipids but also of myoglobin in meat. Both approaches, which include the direct addition of extracts to meat and feeding the extracts to fish as dietary supplements, had clear positive effects on extending the color shelf-life of fish meat in terms of delaying lipid oxidation and metMb formation. The strongly antioxidative activities of the extracts prepared from the fruiting body, base part, mycelium, and spent culture medium of *F. velutipes* clearly showed that *F. velutipes* and its spent culture medium are promising bioresources for natural antioxidants that will stabilize fish meat color. Particularly, the value-added utilization of the spent-culture medium from *F. velutipes* cultivation, which has been largely discarded or merely used as a fertilizer of low-cost performance, in aquaculture would be more valuable not only from an economic perspective but also as an environmentally friendly resolution to industrial waste treatment.

References

Ackman R. G.; Ratnayake, W. M. H. Non-Enzymatic Oxidation of Seafood Lipids. Flick, G. J. and Martin, R. E.; Eds.; Technomic Publishing Co. Inc.: Lancaster, PA,1992; 245–267.

Akanmu, D.; Cecchini, R.; Aruoma, O. I.; Halliwell, B. The antioxidant action of ergothioneine. Arch. Biochem. Biophys. 1991, 288 (1), 10–16.

Aksnes, A.; Njaa, L. R. Catalase, glutathione peroxidase and superoxide dismutase in different fish species. Comp. Biochem. Physiol. B Biochem. Mol. Biol. 1981, 69 (4), 893–896.

Anderson, M. L.; Ravesi, E. M. Reaction of free fatty acids with protein in cod muscle frozen and stored at –29°C after aging in ice. J. Fish. Res. Board Can. 1969, 26, 272–277.

Arduini, A.; Eddy, L.; Hochstein, P. The reduction of ferryl myoglobin by ergothioneine: A novel function for ergothioneine. Arch. Biochem. Biophys. 1990, 281 (1), 41–43.

Audley, M. A.; Shetty, K. J.; Kinsella, J. E. Isolation and properties of phospholipase A from pollock muscle. J. Food Sci. 1978, 43 (6), 1771–1775.

Baldridge, R. C. Blood ergothioneine and dietary oats. J. Nutr. 1954, 56 (1), 107–113.

Bao, H. N. D.; Ushio, H.; Ohshima, T. Antioxidative activities of mushroom (*Flammulina velutipes*) extract added to bigeye tuna meat: Dose-dependent efficacy and comparison with other biological antioxidants. J. Food Sci. 2009, 74 (2), C162–C169.

Bao, H. N. D.; Ushio, H.; Ohshima, T. Antioxidative activity and antidiscoloration efficacy of ergothioneine in mushroom (*Flammulina velutipes*) extract added to beef and fish meats. J. Agric. Food Chem. 2008, 56 (21), 10032–10040.

Baron, C. P.; Andersen, H. J. Myoglobin-induced lipid oxidation. A review. J. Agric. Food Chem. 2002, 50 (14), 3887–3897.

Bevilacqua, A. E.; Zaritzky, N. E. Ice recrystallisation in frozen beef. J. Food Sci. 1982, 47 (5), 1410–1414.

Bito, M. Studies on the retention of meat color of frozen tuna. II. Effect of storage temperature on preventing discoloration of tuna meat during freezing storage. Bull. Jpn. Soc. Sci. Fish. 1965, 31, 534–539.

Boldyrev, A. A.; Dupin, A. M.; Bumin, A. Y.; Babizhaev, M. A.; Severin, S. E. The antioxidative properties of carnosine, a natural histidine containing dipeptide. Biochem. Int. 1987, 15 (6), 1105–1113.

Brewer, S. M.; Vega, J. D. Detectable odor thresholds of selected lipid oxidation compounds in a meat model system. J. Food Sci. 1995, 60 (3), 592-595.

Briggs, I. Ergothioneine in the central nervous system. J. Neurochem. 1972, 19 (1), 27–35.

Cardwallader, K. R. Enzymes and flavor biogenesis in fish. Haard, N. F. and Simpson, B. K.; Eds.; Marcel Dekker: New York, 2000; 365–383.

Casttellini, C.; Bosco, A. D.; Bernardini, M. Improvement of lipid stability of rabbit meat by vitamin E and C administration. J. Sci. Food Agric. 2001, 81 (1), 46–53.

Chaudiere, J.; Ferrari-Iliou, R. Intracellular antioxidants: From chemical to biochemical mechanisms. Food Chem. Toxicol. 1999, 37 (9–10), 949-962.

Chen, Y. C.; Nguyen, J.; Semmens, K.; Beamer, S.; Jaczynski, J. Effects of dietary alpha-tocopheryl acetate on lipid oxidation and alpha-tocopherol content of novel omega-3-enhanced farmed rainbow trout (*Oncorhynchus mykiss*) fillets. LWT—Food Sci. Technol. 2008, 41 (2), 244–253.

Cheung, L. M.; Cheung, P. C. K. Mushroom extracts with antioxidant activity against lipid peroxidation. Food Chem. 2005, 89, 403–409.

Cheung, L. M.; Cheung, P. C. K.; Ooi, V. E. C. Antioxidant activity and total phenolics of edible mushroom extracts. Food Chem. 2003, 81, 249–255.

Cho, S. Y.; Endo, Y.; Fujimoto, K.; Kaneda, T. Oxidative deterioration of lipids in salted and dried sardine during storage at 5°C. Nippon Suisan Gakkaishi. 1989, 55 (3), 541–544.

Corino, C.; Pastorelli, G.; Pantaleo, L.; Oriani, G.; Salvatori, G. Improvement of color and lipid stability of rabbit meat by dietary supplementation with vitamin E. Meat Sci. 1999, 52 (3), 285–289.

Decker, E. A.; Huang, C. H.; Osinchak, J. E.; Hultin, H. O. Iron and copper: Role in enzymic lipid oxidation of fish sarcoplasmic reticulum at in situ concentrations. J. Food Biochem. 1989, 13 (3), 179–186.

Deiana, M.; Rosa, A.; Casu, V.; Piga, R.; Dessi, A. M.; Aruoma, O. I. L-Ergothioneine modulate oxidative damage in the kidney and liver of rats *in vivo*: Studies upon the profile of polyunsaturated fatty acids. Clin. Nutr. 2004, 23 (2), 183–193.

Elmastas, M.; Isildak, O.; Turkekul, I., Temur, N. Determination of antioxidant activity and antioxidant compounds in wild edible mushrooms. J. Food Compos. Anal. 2007, 20, 337–345.

Encarnacion, A. B.; Kamei, M.; Matsumoto, H.; Ohshima, T. *In vivo* application of "Enokitake" extract to control lipid oxidation in post-mortem fish. Presented at the 5th World Fisheries Congress, Yokohama, Japan, October 20–24, 2008.

Ey, J.; Schömig, E.; Taubert, D. Dietary sources and antioxidant effects of ergothioneine. J. Agric. Food Chem. 2007, 55 (16), 6466–6474.

Faustman, C.; Chan, W. K. M.; Schaefer, D. M.; Havens, A. Beef color update: The role for vitamin E. J. Anim. Sci. 1998, 76 (4), 1019-1026.

Flink, J. M.; Goodhart, M. Transport of oxygen through ice and frozen minced fish. J. Food Process. Preserv. 1978, 2 (3), 229–248.

Franzoni, F.; Colognato, R.; Galetta, F.; Laurenza, I.; Barsotti, M.; Stefano, R. D.; Bocchetti, R.; Regoli, F.; Carpi, A.; Balbarini, A.; et al. An *in vitro* study on the free radical scavenging capacity of ergothioneine comparison with reduced glutathione, uric acid and trolox. Biomed. Pharmaco. 2006, 60 (8), 45–47.

Fu, H. Y.; Shieh, D. E.; Ho, C. T. Antioxidant and free radical scavenging activities of edible mushrooms. J. Food Lipid 2002, 9 (1), 35–46.

Gardner, H. W. Lipid hydroperoxide reactivity with proteins and amino acids: A review. J. Agric. Food Chem. 1979, 27 (2), 220–229.

Georgantelis, D.; Blekas, G.; Katikuo, P.; Ambrosiadis, I.; Fletouris, D. J. Effect of rosemary extract, chitosan and α-tocopherol on lipid oxidation and colour stability during frozen storage of beef burgers. Meat Sci. 2007, 75 (2), 256–264.

German, J. B.; Chen, S. E.; Kinsella, J. E. Lipid oxidation in fish tissue. Enzymatic initiation via lipoxygenase. J. Agric. Food Chem. 1985, 33 (4), 680–683.

German, J. B.; Geza, G.; Kinsella, J. E. Lipoxygenase in trout gill tissue acting on arachidonic, eicosapentaenoic and docosahexaenoic acids. Biochim. Biophys. Acta. 1986, 875 (1), 12–20.

Geromel, E. J.; Montgomery, M. W. Lipase release from lysosomes of rainbow trout muscle subjected to low temperatures. J. Food Sci. 1980, 45 (3), 412–415.

Grunwald, E. W.; Richards, M. P. Mechanisms of heme protein-mediated lipid oxidation using hemoglobin and myoglobin variants in raw and heated washed muscle. J. Agric. Food Chem. 2006, 54 (21), 8271–8280.

Hamre, K.; Berge R. K.; Lie, Ø. Turnover of tocopherols in Atlantic salmon (*Salmo salar, L.*). BioFactor. 2000, 11 (1–2), 15–17.

Harris, P.; Tall, J. Rancidity in fish. Allen, J. C.; Hamilton, R. J., Eds.; Blackie A&P: London, 1989; 256–272.

Hsieh, R. J.; Kinsella, J. E. Lipoxygenase generation of specific volatile flavor carbonyl compounds in fish tissues. J. Agric. Food Chem. 1989, 37 (2), 280–286.

Huang, C. H.; Hultin, H. O.; Jafar, S. S. Some aspects of Fe^{2+}-catalyzed oxidation in fish sarcoplasmatic reticular lipid. J. Agric. Food Chem. 1993, 41 (11), 1886–1892.

Hultin, H. O. Biochemical deterioration in fish muscle. Huss H. H.; Jakobsen, M; Liston, J., Eds.; Elsevier Science Publ. B. V.: Amsterdam, 1992; 125–138.

Hultin, H. O. Oxidation of lipids in seafoods. Shahidi, F.; Botta, J. R., Eds.; Blackie A&P: London, 1994; 49–74.

Hultin, H. O. Role of membranes in fish quality. Nordic Council of Ministers: Copenhagen, TemaNord, 1995; 13–35.

Hung, S. S. O.; Cho, C. Y.; Slinger, S. J. Effect of oxidized fish oil, DL-alpha-tocopheryl acetate and ethoxyquin supplementation on the vitamin E nutrition of rainbow trout (*Salmo gairneri*) fed practical diets. J. Nutr. 1981, 111 (4), 648–657.

Izquierdo, M. S.; Henderson, R. J. The determination of lipase and phospholipase activities in gut contents of turbot (*Scophthalmus maximus*) by fluorescence-based assays. Fish Physiol. Biochem. 1998, 19 (2), 153–162.

Jacobsen, C. Sensory impact of lipid oxidation in complex food systems. Lipid 1999, 101 (5), 484–490.

Jang, M. S.; Eun, J. B.; Ushio, H.; Ohshima, T. Antioxidative properties of mushroom (*Flammulina Velutipes*) crude extract on the oxidation of cod liver oil in emulsion. Food Sci. Biotechnol. 2004, 13 (2), 215–218.

Jensen, C.; Birk, E.; Jokumsen, A.; Skibsted, L. H.; Bertelsen, G. Effect of dietary levels of fat, α-tocopherol and astaxanthin on colour and lipid oxidation during storage of frozen rainbow trout (*Oncorhynchus mykiss*) and during chill storage of smoked trout. Z. Lebensm. Unters. Forsch. A 1998, 207 (3), 189–196.

Jia, T. D.; Kelleher, S. D; Hultin, H. O.; Petillo, D.; Maney, R.; Krzynowek, J. Comparison of quality loss and changes in the glutathione antioxidant system in stored mackerel and bluefish muscle. J. Agric. Food Chem. 1996, 44 (5), 1195–1201.

Joo, S. T.; Lee, J. I.; Ha, Y. L.; Park, G. B. Effects of dietary conjugated linoleic acid on fatty acid composition, lipid oxidation, color, and water-holding capacity of pork loin. J. Anim. Sci. 2002, 80 (1), 108–112.

Jorgensen, B. Lipid oxidation and hydrolysis. Huss, H. H., Ed.; FAO, Rome, 1995; 64–67.

Kanner, J.; Kinsella, J. E. Initiation of lipid oxidation by a peroxidase/hydrohen peroxide/halide system. Lipid 1983, 18 (3), 204–210.

Kawano, H.; Murata, H.; Iriguchi, S.; Mayumi, T.; Hama, T. Studies on ergothioneine. XI. Inhibitory effect on lipid peroxide formation in mouse liver. Chem. Pharm. Bull. 1983, 31 (5), 1682–1687.

Kawano, H.; Otani, M.; Takeyama, K.; Kawai, Y.; Mayumi, T.; Hama, T. Studies on ergothioneine. VI. Distribution and fluctuations of ergothioneine in rats. Chem. Pharm. Bull. 1982, 30 (5), 1760–1765.

Kim, M. Y; Seguin, P.; Ahn, J. K.; Kim, J. J.; Chun, S. C.; Kim, E. H.; Seo, S. H.; Kang, E. Y.; Kim, S. L.; Park, Y. J.; Ro, H. M.; Chung, I. M. Phenolic compound concentration and antioxidant activities of edible and medicinal mushrooms from Korea. J. Agric. Food Chem. 2008, 56, 7265–7270.

Korycha-Dahl, M. B.; Richardson T. Activated oxygen species and oxidation of food constituents. CRC Crit. Rev. Food Sci. Nutr. 1978, 10, 209–241.

Kubow, S. Routes of formation and toxic consequences of lipid oxidation products in foods. Free Rad. Biol. Med. 1992, 12 (1), 63–81.

Kuo, J. M.; Pan, B. S. Occurrence and properties of 12-lipoxygenase in the hemolymph of shrimp (*Paenaesus japonicus Bate*). J. Chin. Biochem. Soc. 1992, 21, 9–16.

Labuza, T. P. Kinetics of lipid oxidation in foods. CRC Crit. Rev. Food Technol. 1971, 2, 355–405.

Lee, J. H.; Fujimoto, K.; Kaneda, T. Importance of oxidative degradation on peroxide-decomposition activities of Antarctic krill phospholipids. Nippon Suisan Gakkashi 1984, 50 (11), 1863–1868.

Lee, S.; Joo, S. T.; Alderton, A. L.; Hill, D. W.; Faustman, C. Oxymyoglobin and lipid oxidation in yellowfin tuna (*Thunnus albacares*) loins. J. Food Sci. 2003, 68 (5), 1664–1668.

Losada, V.; Barros-Velázquez, J.; Aubourg, S. P. Rancidity development in frozen pelagic fish: Influence of slurry ice as preliminary chilling treatment. LWT—Food Sci. Technol. 2007, 40 (6), 991–999.

Love, J. D.; Pearson, A. M. Lipid oxidation in meat and meat products—A review. J. Amer. Oil Chem. Soc. 1971, 48 (10), 547–549.

Love, R. M.; Elerian, M. K. Protein denaturation in frozen fish. VIII.—The temperature of maximum denaturation in cod. J. Sci. Food Agric. 1964, 15 (11), 805–809.

Mansour, N.; McNiven, M. A.; Richardson, G. F. The effect of dietary supplementation with blueberry, α-tocopherol or astaxanthin on oxidative stability of Arctic char (*Salvelinus alpinus*) semen. Theriogenol. 2006, 66 (2), 373–382.

Mau, J. L.; Lin, H. C.; Chen, C. C. Antioxidant properties of several medicinal mushrooms. J. Agric. Food Chem. 2002, 50, 6072–6077.

Melville, D. B.; Eich, S. The occurrence of ergothioneine in plant material. J. Biol. Chem. 1956, 218 (2), 647–651.

Michalak, A. Phenolic compounds and their antioxidant activity in plants growing under heavy metal stress. Polish J. Environ. Stud. 2006, 15 (4), 523–530.

Miki, W. Biological functions and activities of animal carotenoids. Pure Appl. Chem. 1991, 63 (1), 141–146.

Milic, B. L.; Djilas, S. M.; Canadanovic, B. J. M. Antioxidative activity of phenolic compounds on the metal-ion break down of lipid peroxidation system. Food Chem. 1998, 61, 443–447.

Misiti, F.; Castagnola, M.; Zuppi, C.; Giardina, B.; Messana, I. Role of ergothioneine on S-nitrosoglutathione catabolism. Biochem. J. 2001, 356, 799–804.

Miyashita, K.; Takagi, T. Study on the oxidative rate and prooxidant activity of free fatty acids. J. Am. Oil Chem. Soc. 1986, 63, 1380–1384.

Mohri, S.; Cho, S. Y.; Endo, Y.; Fujimoto, K. Linoleate 13(S)-lipoxygenase in sardine skin. J. Agric. Food Chem. 1992, 40 (4), 573–567.

Mohri, S.; Cho, S. Y.; Endo, Y.; Fujimoto, K. Lipoxygenase activity in sardine skin. Agric. Biol. Chem. 1990, 54 (7), 1889–1891.

Nakano, T.; Sato, M.; Takeuchi, M. Glutathione peroxidase of fish. J. Food Sci. 1992, 57 (5), 1116–1119.

Nayak, J.; Nair, P. G. V.; Ammu, K.; Mathew, S. Lipase activity in different tissues of four species of fish: Rohu (*Labeo rohita* Hamilton), oil sardine (*Sardinella longiceps* Linnaeus), mullet (*Liza subviridis* Valenciennes) and Indian mackerel (*Rastrelliger kanagurta* Cuvier). J. Sci. Food Agric. 2003, 83 (11), 1139–1142.

Ohshima, T.; Fujita, Y.; Koizumi, C. Oxidative stability of sardine and mackerel lipids with reference to synergism between lipids and α-tocopherol. J. Amer. Oil Chem. Soc. 1993, 70 (3), 269–276.

Pan, B. S.; Kuo, J. M. Lipoxigenases. Norman, F. H.; Benjamin, K.S., Eds.; Marcel Dekker: New York, 2000; 317–336.

Petillo, D.; Hultin, H. O.; Krzynowek, J.; Autio, W. R. Kinetics of antioxidant loss in mackerel light and dark muscle. J. Agric. Food Chem. 1998, 46 (10), 4128–4137.

Phillips, A. L.; Faustman, C.; Lynch, M. P.; Govoni, K. E.; Hoagland, T. A.; Zinn, S. A. Effects of dietary α-tocopherol supplementation on color and lipid stability in pork. Meat Sci. 2001, 58 (4), 389–393.

Porter, N. A.; Caldwell, S. E.; Mills, K. A. Mechanisms of free radical oxidation of unsaturated lipids. Lipid 1995, 30 (4), 277–290.

Puttaraju, N. G.; Venkateshaiah, S. U.; Dharmesh, S. M.; Urs, S. M.; Somasundaram, R. Antioxidant activity of indigenous edible mushrooms. J. Agric. Food Chem. 2006, 54, 9764–9772.

Ramasarma, T. Generation of H_2O_2 in biomembranes. Biochem. Biophys. Acta. 1982, 694 (1), 69–93.

Reeder, B. J.; Wilson M. T. Mechanism of reaction of myoglobin with the lipid hydroperoxide hydroperoxyoctadecadienoic acid. Biochem. J. 1998, 330, 1317–23.

Rice-Evans, C.; Miller, N.; Paganga, G. Antioxidant properties of phenolic compounds. Trend. Plant Sci. 1997, 2 (4), 152–159.

Rice-Evans, C. A.; Miller, N. J.; Paganga, G. Structure-antioxidant activity relationships of flavonoids and phenolic acids. Free Rad. Biol. Med. 1996, 20 (7), 933–956.

Roginsky, V.; Lissi, E. A. Review of methods to determine chain-breaking antioxidant activity in food. Food Chem. 2005, 92 (2), 235–254.

Roozen, J. P.; Frankel, E. N.; Kinsella J. E. Enzymic and autoxidation of lipids in low fat foods: Model of linoleic acid in emulsified hexadecane. Food Chem. 1994a, 50 (1), 33–38.

Roozen, J. P.; Frankel, E N.; Kinsella J. E. Enzymic and autoxidation of lipids in low fat foods: Model of linoleic acid in emulsified triolein and vegetable oils. Food Chem. 1994b, 50 (1), 39–43.

Schaich, K. M. Free radical initiation in proteins and amino acids by ionizing and ultraviolet radiations and lipid oxidation—Part III: Free radical transfer from oxidizing lipids. Crit. Rev. Food Sci. Nutr. 1980, 13 (3), 189–244.

Shires, T. K.; Brummel, M. C.; Pulido, J. S.; Stegink, L. D. Ergothioneine distribution in bovine and porcine ocular tissues. Compar. Biochem. Physiol. C. 1997, 177 (1), 117–120.

Sohn J. H.; Taki, Y.; Ushio, H.; Kohata, T.; Shioya, I.; Ohshima, T. Lipid oxidations in ordinary and dark muscles of fish: Influences on rancid off-odor development and color darkening of yellowtail flesh during ice storage. J. Food Sci. 2005, 70 (7), S490–S496.

Sohn, J. H.; Ushio, H.; Ishida, N.; Yamashita, M.; Terayama, M.; Ohshima, T. Effect of bleeding treatment and perfusion of yellowtail on lipid oxidation in post-mortem muscle. Food Chem. 2007, 104 (3), 962–970.

Syvaoja, E. L.; Salminen, K. Tocopherols and tocotrienols in finfish foods: Fish and fish products. J. Amer. Oil Chem. Soc. 1985, 62 (8), 1245–1248.

Turchini, G. M.; Mentasti, T.; Frøyland, L.; Orban, E.; Caprino, F.; Morettia, V. M.; Valfre, F. Effects of alternative dietary lipid sources on performance, tissue chemical composition, mitochondrial fatty acid oxidation capabilities and sensory characteristics in brown trout (*Salmo trutta L.*). Aquaculture 2003, 225 (1–4), 251–267.

Uauy, R.; Valenzuela, A. Marine oils: The dealth benefits of n-3 fatty acids. Nutr. 2000, 16 (7–8), 680–684.

Wang, Y. J.; Miller, L. A.; Addis, P. B. Effect of heat inactivation of lipoxygenase on lipid oxidation in lake herring (*Coregonus artrerii*). J. Amer. Oil Chem. Soc. 1991, 68 (10), 752–758.

Watanabe, F.; Goto, M.; Abe, K.; Nakano, Y. Glutathione peroxidase activity during storage of fish muscle. J. Food Sci. 1996, 61 (4), 734–735 & 782.

Yang, A.; Lanari, M. C.; Brewster, M.; Tume, R. K. Lipid stability and meat colour of beef from pasture- and grain-fed cattle with or without vitamin E supplement. Meat Sci. 2002, 60 (1), 41–50.

CHAPTER

12

The Natural Antioxidant Ergothioneine

Resources, Chemical Characterization, and Applications

The Han Nguyen[1], Reiko Nagasaka[1], and Toshiaki Ohshima[1]

Introduction

Lipid oxidation contributes to the deterioration of the quality and flavor of foods containing lipids. It is also responsible for changes in color, texture, and nutritional value, as well as the wholesomeness of quality muscle food (Shahidi, 1992). Therefore, preventing the oxidation of lipids during food processing and storage has been a challenge for the food industry, especially for food scientists. It is also well known that free radicals produced during the lipid oxidation process that occur in living organisms are responsible for health damage and accelerated aging (Finley & Otterburn, 1993; Aruoma, 1998). Thus, lipid peroxidation is a problem encountered not only in food processing but also in animal tissues. Antioxidants are generally classified into two groups—exogenous (that is, natural and synthetic) and endogenous compounds—both of which are responsible for removing free radicals, scavenging reactive oxygen species (ROS) or their precursors, inhibiting ROS formation, and binding metal ions needed for the catalysis of ROS generation and activation of endogenous oxidants (Gilgun-Sherki et al., 2002).

A wide range of antioxidants, both natural and synthetic, have been proposed to prevent the postharvest deterioration of foods. They can arrest lipid oxidation and free-radical formation in foods during storage and/or processing and reduce

[1]Department of Food Science and Technology, Tokyo University of Marine Science and Technology, Konan 4, Minato-ku, Tokyo 108-8477, Japan.

the concentration of free radicals in living organisms after food ingestion (Pokorný, 2007). However, the use of synthetic antioxidants is restricted by legislative rules because of their carcinogenicity; teratogenicity; high, acute toxicity; long degradation periods; environmental pollution effects; effects on food; and other side effects on humans (Lingk, 1991; Unnikrishnan & Nath, 2002). In addition, increasing regulatory attention and consumer interest in natural food additives has created a need for safe and effective alternatives for use in foods (McEvily et al., 1991). Therefore, interest in the use of antioxidants of natural origin has increased. The well-known natural antioxidants that have been used in the food industry and for human health are tocopherols, vitamin C, carotenoids, and phenolic compounds (Shahidi & Rubin, 1987; Ramanathan & Das, 1992; Granit et al., 2001; Dimakou & Oreopoulou, 2011). Although consumers readily accept the antioxidants naturally present in foods, some antioxidants, such as those obtained from spices and herbs (oregano, thyme, dittany, marjoram, lavender, and rosemary, for example), have limited applications because they transfer their flavor to the food (Moure et al., 2001). Subsequently, these antioxidants require deodorization steps for additional applications. De-flavored antioxidants such as rosemary extract exist in the market. Further, some evidence has indicated that foods rich in protein may react with added phenolic antioxidants, causing the loss of nutritional and sensory food value as well as the reduction of phenolic antioxidant capacity (Naczk et al., 1996; Rohn et al., 2004). In addition, heat-sensitive antioxidants, such as vitamin C, may be less effective at the high temperatures necessary for food processes like frying. Hence, many researchers have begun to investigate and identify new natural antioxidants.

The betaine of 2-mercapto-l-histidine (ergothioneine) (Fig. 12.1) was first isolated from ergot, a fungal infection of rye grain (Tanret, 1909). Ergothioneine is a naturally occurring amino-acid analog, which is synthesized only by some bacteria and fungi. A substantial body of evidence has demonstrated that animals cannot synthesize ergothioneine but can accumulate it through diet via an organic cation transporter, OCTN1 (Melville et al., 1955; Gründemann et al., 2005). It has been well documented that it is a stable and powerful antioxidant found in most plants and animals (Hartman, 1990; Aruoma et al., 1999). Evidence also exists that the

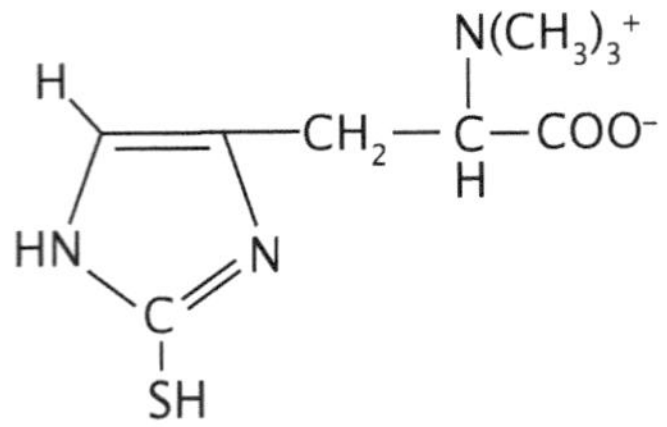

Fig. 12.1. Structure of ergothioneine.

cellular antioxidant status, which is reflected in the ergothioneine content (Meister, 1992), is correlated with resistance to lipid peroxidation in atherosclerosis, smoking, and diabetes (Diaz et al., 1997). Thus, ergothioneine is a promising, safe, and effective antioxidant for controlling lipid oxidation–induced changes in food quality. It can also be useful for preventing various oxidative diseases associated with stress. Given the importance of this compound, fast, sensitive, and accurate quantitative determination is essential to understand its existence and bioavailability in foods and biological materials. Since the discovery of ergothioneine, numerous methods have been devised for its detection and quantification in biological samples, which Cheah and Halliwell (2012) have reviewed. To evaluate the antioxidant activity of pure ergothioneine and ergothioneine-containing extracts, several methods have been used. Though the colorimetric method for evaluation of antioxidant activity is simple, it is not applicable to colored materials due to interference by pigments (Chandrasekar et al., 2006). Bao and colleagues (2008a) pointed out that impurities in the mushroom extract affected the activity of ergothioneine on the DPPH radical-scavenging activity and total reducing power assays. In addition, there is growing interest in investigating ergothioneine's presence and role as an antioxidant in novel resources. This interest has raised the demand for the antioxidant's rapid screening in complex mixtures. An online HPLC–DPPH method has been widely applied for rapid screening antioxidant compounds in a large number of extracts. The method is based on a post-column reaction of the HPLC separated antioxidants with the DPPH radical. Consequently, it provides a good analytical tool for screening ergothioneine in biological samples.

In this chapter, we provide an overview of the powerful antioxidant ergothioneine and its potential application to prevent lipid oxidation in foods and biological systems. Additionally, we discuss the developmental history of the methods used for its analysis in biological materials, its bioavailability in humans and animals, and its role in preventing oxidative stress. Finally, we review the online HPLC–DPPH method for the evaluation of radical-scavenging compounds in complex mixtures and its application for detecting ergothioneine in mushroom samples.

Sources of Ergothioneine

Ergothioneine is present throughout the plant and animal kingdoms. Its concentration in human and mammalian tissues has been estimated to be 1–2 mM (Melville, 1959; Hartman, 1990). Its subsequent discovery free form in blood has greatly stimulated interest in this compound (Kawano et al., 1982). Kumosani (2001) investigated the ergothioneine's concentration in the male's blood at different stages of life. The author found that the concentration is 1.5–2.0 mg/100 mL at the early stages of life (1–10 years). It increases gradually between 11 and 18 years of age and reaches the maximum value of 3.7 mg/100 mL at the age of 18 years. It has also been found in semen and various mammalian tissues, principally, the liver (Melville, 1959), kidney (Mayumi et al., 1978), and whole blood of pigs and rats (Heath et al., 1953; Kawano et al., 1982). In addition, research has found that certain mushrooms contain higher concentrations of ergothioneine than do any other foods (Ey et al., 2007) (Table 12.A). Dubost and colleagues (2007c) reported that ergothioneine makes up from 210 to 2,590 mg/kg of the dry weight of most commonly consumed mushrooms in the United States, including *Agaricus bisporus, Lentinula edodes, Pleurotus ostreatus,* and *Grifola frondosa.* Bao and colleagues (2010b) determined that the ergothioneine content of various mushrooms cultivated in Japan varied between trace amounts and 2,080 mg/kg wet weight. After harvesting the edible fruiting body of mushrooms, the solid culture waste (that is, the mycelia and spent culture medium) is usually discarded. Interestingly, it contains a considerable amount of ergothioneine (Bao et al., 2009, 2010a, 2010b). We would like to note that the amount of ergothioneine in an extract prepared from the waste culture medium of the mushroom *Flammulina velutipes* was 1.7 times that present in an extract prepared directly from the mushroom's fruiting body (Bao et al., 2010a) (Table 12.A). Lee and colleagues (2009) analyzed ergothioneine's production in both the fruiting bodies and mycelia of various mushroom species from different geographical locations in Korea. The concentrations ranged between 60 and 5,540 mg/kg dry weight of the mushroom samples. Higher levels were found in the mycelia than in the fruiting bodies of some mushroom species, such as *Ganoderma neojaponicum, Ganoderma applanatum,* and *Paecilomyces tenuipes.* Similarly, Chen and colleagues (2012) and Tepwong and colleagues (2012) reported that most mushroom fruiting bodies and mycelia collected from Japan, Korea, and Taiwan contained considerable amounts of the antioxidant (Table 12.A). Therefore, it is reasonable to conclude that mushrooms are the best-known source of ergothioneine. The limitation of sources for naturally derived ergothioneine has prompted research for enhancing its production in mushrooms. Several methods have been developed to synthetically or naturally produce ergothioneine. The amino acids histidine, cysteine, and methionine act as precursors for the synthesis of this compound (Askari & Melville, 1962). Dubost and colleagues (2007b) found that adding histidine to compost increased the ergothioneine level in the fruiting bodies of the mushroom *Agaricus bisporus.* Lee and colleagues (2009) and Tepwong and colleagues (2012) reported that addition of methionine to a mycelial culture medium using submerged fermentation improved the ergothioneine content in most mushroom species tested.

Aside from its presence in mushrooms, ergothioneine has also been observed in common foods of Western or Mediterranean diets, as shown in Table 12.A (Ey et al., 2007). Its concentrations varied greatly not only between food groups

Table 12.A. Ergothioneine Concentrations in Selected Foods and Beverages

Sources	Ergothioneine Content
Mushroom fruiting body[a] (Ey et al., 2007; Bao et al., 2010)	
Flammulina populicola	48
Flammulina velutipes	298
Grifola frondosa	30
Grifola frondosa (white)	8
Hypsizygus tessellates	6
Lentinula edodes	284
Pholiota nameko	2
Pleurotus cornucopiae	2082
Pleurotus eryngii	317
Boletus edulis	528.14
Pleurotus ostreatus	118.91
Agaricus bisporus (brown)	0.93
Agaricus bisporus (white)	0.46
Cantharellus cibarius	0.06
Mushroom waste[a] (Bao et al., 2010)	
Flammulina velutipes	508
Mushroom mycelia[b] (Chen et al., 2012)	
Agaricus blazei	79.6
Agrocybe cylindracea	279.4
Antrodia camphorata	281.6
Antrodia salmonea	7.6
Armillariella mellea	219.6
Coprinus comatus	399
Cordyceps cicadae	588
Cordyceps militaris (strain cm1)	215
Cordyceps militaris (strain cm5)	785.1
Cordyceps militaris (strain 419)	123.4
Cordyceps sinensis	142
Coriolus versicolor	13
Ganoderma lucidum	16.5
Grifola frondosa	296.2
Hericium erinaceus	376.2
Hypsizigus marmoreus (white)	221.4
Hypsizigus marmoreus (normal)	206.7

(continued)

Table 12.A. *Continued*

Sources	Ergothioneine Content
Mushroom mycelia[b] (continued)	
Inonotus obliquus	252.1
Phellinus linteus	181.8
Pleurotus eryngii	1514.6
Meats, sausages[a] (Ey et al., 2007)	
Chicken, Liver	10.78
Pork, Liver	8.71
Pork, Kidney	7.66
Pork, Loin Fillet	1.68
Beef, Loin Steak	1.33
Lamb, Loin Fillet	1.2
Chicken, Breast	1.15
Ham	1.12
Blood Sausage (from Pork)	1.08
Liver Sausage (from Pork)	1.03
Salami (from Pork)	0.51
Aspic (from Pork)	0.46
Fish and Seafoods (Ey et al., 2007)	
Trout[a]	0.07
Cyanobacterial preparations[b] (Pfeiffer et al., 2011)	ND
Aphanizomenon flos-aquae tablets (Naturscheune)	600
Spirulina granules (Dr. Niedermaier)	600
Eggs[a]	
Egg Yolk	0.68
Egg White	0.38
Grains and grain products, seeds, nuts[a] (Ey et al., 2007)	
Oat Bran	4.41
Pumpkin Seed	1.49
Wheat Bran	0.84
Wheat Germ	0.63
Spelt	0.61
Whole Grain Wheat Bread	0.53
Whole Grain Rye Bread	0.47
Brown Rice	0.04
Vegetables, Fruits	

Table 12.A. *Continued*

Sources	Ergothioneine Content
Grains and grain products, seeds, nuts (continued)	
Black Turtle Bean	13.49
Red Kidney Bean	4.52
Garlic	3.11
Broccoli	0.24
Onion	0.23
Spinach	0.11
Celery	0.08
Beverages[c] (Ey et al., 2007)	
Beer	0.02

[a]Values are expressed as mg/kg wet weight
[b]Values are expressed as mg/kg dry weight
[c]Values are expressed as mg/L

but also within a given food group (Table 12.A). It has been well documented that ergothioneine is only biosynthesized from fungi (including edible mushrooms) and mycobacteria (Melville, 1959; Hartman, 1990). More recently, however, Pfeiffer and colleagues (2011) demonstrated that several species of cyanobacteria can produce large amounts (up to 800 mg/kg dry mass) of this compound. In samples of cyanobacterial preparations (examples include *Aphanizomenon flos-aquae* and *Spirulina platensis*) used as dietary supplements for humans, ergothioneine's concentration has been measured to be up to 600 mg/kg dry mass (Table 12.A).

Antioxidant Properties of Ergothioneine and Its Application to Foods

Ergothioneine has been recognized as an excellent antioxidant after demonstrating various antioxidant properties such as scavenging and quenching strong pro-oxidants like hydroxyl radicals, hypochlorite, and peroxynitrite (ONOO); chelating redox-active bivalent cations, such as Cu^{2+}, Hg^{2+}, Zn^{2+}, Cd^{2+}, Co^{2+}, Fe^{2+}, and Ni^{2+}; and playing a key role in the oxidation reactions mediated by various hemoproteins (Akanmu et al., 1991; Melville et al., 1955). It mainly exists in thione form in aqueous solutions. The standard redox potential of its thiol/disulfide system (–0.02 V) is much higher than that of other naturally occurring thiols (between -0.2 and -0.32 V) (Jocelyn, 1972). These attributes make ergothioneine a very stable antioxidant at physiological pH, which means that it does not promote the generation of hydroxyl radicals from hydrogen peroxide (H_2O_2) and Fe^{2+} ions (Grigat et al., 2007). In contrast, thiol-containing compounds like glutathione are readily oxidized to generate free radicals (Misra, 1974). A comparative study showed that ergothioneine is the most powerful scavenger of hydroxyl, peroxyl, and peroxynitrite free radicals when compared with traditional antioxidants, such as glutathione, Trolox®, and uric acid (Franzoni et al., 2006) (Fig. 12.2). More recently, a report by Liao and colleagues (2012) showed that it had much higher DPPH radical-scavenging activity than histidine. Ergothioneine is the betaine of histidine, with a sulfur atom attached at position 2 in the imidazole ring (Fig. 12.1), which suggests that its thiol group has a critical effect on its scavenging of DPPH radicals. In some cases, modifying the functional groups of ergothioneine may not affect its antioxidant activity. For example, the antioxidant activity of an ergothioneine analog—namely, β-hydroxyergothioneine, isolated from the mushroom *Lyophyllum connatum*—is almost the same as that of ergothioneine in terms of their radical-scavenging activity (Kimura et al., 2005). This suggests the important role that ergothioneine's thiol group has in its antioxidant effect.

The role of ergothioneine as a cellular protector against oxidative damage has also been well documented. Aruoma and colleagues (1999) investigated its effect on oxidative damage in cells. The compound showed potential in inhibiting DNA oxidation caused by peroxynitrite, as well as cell death caused by H_2O_2 in a human neuronal hybridoma cell line, N-18-RE-105. The authors suggested that ergothioneine may serve as a nontoxic, thiol-buffering antioxidant *in vivo*. Compelling evidence shows that oxidative and/or nitrosative stress mediates Abeta-induced cytotoxicity. In one study, exposure to ergothioneine effectively inhibited Abeta-induced oxidative cell death in rat PC12 cells by suppressing peroxynitrite formation and the subsequent nitration of protein tyrosine residues (Jang et al., 2004). Interleukin-8 (IL-8) is a proinflammatory chemokine produced by macrophages, epithelial cells, and other somatic cells. Its induction is associated with the activation of nuclear transcription factors, such as nuclear factor κB (NF-κB), activator protein 1 (AP-1), and the nuclear factor for IL-6 (NF-IL6). The inhibitory effects of ergothioneine on the H_2O_{2}- and tumor necrosis factor α (TNF-α)-mediated activation of NF-κB, as well as the release of IL-8 in human alveolar epithelial cells (A549), have been reported by Rahman and colleagues (2003), who revealed that pretreating A549 cells with ergothioneine for 2 h could inhibit both TNF-α- and H_2O_2-induced IL-8 release in the cells. Moreover, the release of IL-8 was suppressed in these cells, as opposed to the untreated control cells. Ergothioneine also has the ability to protect the skin from oxidative damage caused by exposure to ultraviolet (UV) radiation. Obayashi and colleagues (2005) demonstrated in chemical studies that it could scavenge the superoxide anion radical ($O_2\bullet^-$) and

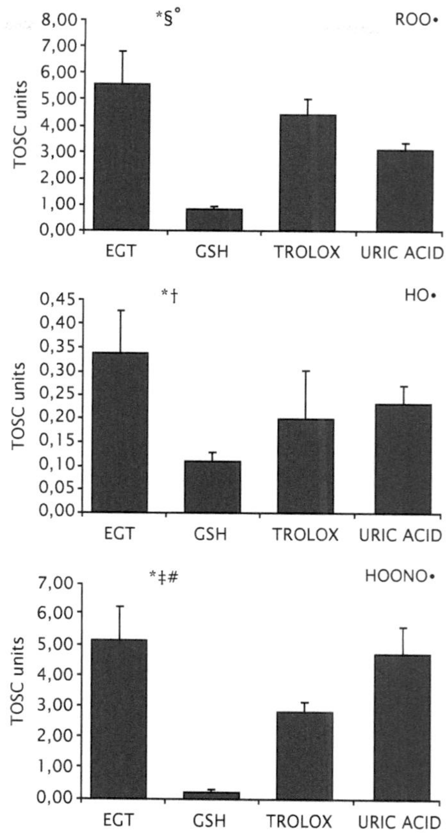

Fig. 12.2. Total antioxidant capacity of ergothioneine (EGT) toward the peroxyl radical (ROO•), hydroxyl radical (HO•), and peroxynitrate (HOONO•), compared with glutathione (GSH), Trolox®, and uric acid.* $P < 0.0001$ versus GSH antioxidant activity toward ROO•; $^{§}P < 0.001$ versus uric acid antioxidant activity toward ROO•; $^{\circ}P < 0.01$ versus Trolox® antioxidant activity toward ROO•; $^{\dagger}P < 0.001$ versus Trolox® and uric acid antioxidant activity toward HO•; $^{\ddagger}P < 0.001$ versus Trolox® antioxidant activity toward HOONO•; and $^{\#}P < 0.05$ versus uric acid antioxidant activity toward HOONO•.

singlet oxygen ($^1O_2$), as well as downregulate TNF-α upregulation by UVB irradiation in cultured fibroblasts. Hence, it may protect human skin from the harmful effects of UV radiation. Another study by Markova and colleagues (2009) provided evidence that skin cells and tissues can take up, accumulate, and utilize ergothioneine, which consequently reduces ROS levels and damage to DNA, protein, and lipids in keratinocytes subjected to solar-simulating UV oxidative stress.

Lipid oxidation is one of the major causes of postharvest deterioration in muscle foods because it results in the formation of undesirable off-flavors and unhealthy compounds, as well as the loss of nutritional value. Antioxidants have been widely applied to inhibit lipid oxidation in raw materials and processed foods (Erickson, 2002) as current consumer concerns about synthetic additives have fueled interest in the identification and use of naturally occurring antioxidants. The potent antioxidant properties of ergothioneine have prompted investigators to begin exploring its novel application to food and nonfood products. As previously mentioned, mushrooms are an excellent source of ergothioneine and could be a good candidate for controlling the oxidative deterioration of lipids. Humans have consumed wild or cultivated mushrooms for centuries worldwide, appreciating this food source not only for its texture and flavor but also for its functional characteristics (Manzi et al., 2001). Although mushroom extracts have strong antioxidant activity *in vitro* (Cheung et al., 2003), examples of food supplementation with mushroom extracts have been rare. Despite this fact, researchers have concluded preliminary studies on the application of mushroom extracts for controlling the lipid oxidation of cod-liver oil in oil-in-water emulsions (Jang et al., 2004) and in rat brain homogenates (Cheung & Cheung, 2005). The same researchers investigated the effect of *F. velutipes* extract on polyunsaturated oils in oil-in-water emulsion (Jang et al., 2009). They found that adding the mushroom extract had a significant effect in the oil-in-water emulsion system on the suppression of lipid oxidation and prevention of decreases in eicosapentaenoic acid (EPA) and docosahexaenoic acid (DHA). However, the active compound(s) remained unidentified in these studies. Recently, ergothioneine present in edible mushrooms has been found to act as a highly effective antioxidant (Dubost et al., 2007c; Lee et al., 2009). Mushroom extracts containing ergothioneine have potential uses in preventing postharvest lipid oxidation in farmed fish and crustaceans (Bao et al., 2008, 2009; Encarnacion et al., 2010). Both *in vitro* and *in vivo* experiments have demonstrated that ergothioneine-rich extracts prepared from the fruiting bodies and processing waste of the edible mushroom *F. velutipes* significantly suppress lipid oxidation in beef, bigeye tuna, and yellowtail meats, as well as shrimp muscles. A related study has deduced that lipid oxidation induces myoglobin oxidation, which is closely related to the rate of meat discoloration (Yin & Faustman, 1993). In particular, the oxidation of muscle tissue lipids leads to the formation of chemical species that can exacerbate the oxidation of myoglobin. Additionally, oxidized lipids can react with amines, amino acids, and proteins to form brown macromolecular products (Frankel, 1998). It has also been documented that the degree of fatty-acid unsaturation is an important factor affecting color formation in meats (Pan et al., 2004). On the other hand, Faustman and colleagues (1992) indicated that high concentrations of iron and myoglobin are associated with high rates of lipid oxidation. Other studies have presented similar findings (Baron &

Andersen, 2002; Carlsen et al., 2005). For example, Bao and colleagues (2008) found a significant correlation between metmyoglobin concentration and hydroperoxide content in tuna meat stored on ice. Therefore, controlling both lipid and myoglobin oxidation in meats should be simultaneously considered.

It has previously been reported that ergothioneine reduces ferrylmyoglobin, a potent oxidant formed by the reaction between metmyoglobin and H_2O_2 (Arduini et al., 1990). Ashida and colleagues (2005) demonstrated that extracts from certain edible mushrooms containing high levels of ergothioneine inhibit the oxidation of oxymyoglobin isolated from cattle meat *in vitro*. These results were later confirmed by Bao and colleagues (2008), who reported that both authentic ergothioneine and mushroom extract arrest the formation of metmyoglobin in beef and bigeye tuna meats. Consequently, these substances successfully controlled the oxidation of lipids and myoglobins and the discoloration of meats during low-temperature storage. The antioxidant potency of ergothioneine could be the basis for its application in food products. The novel use of this compound as a nutritional additive and preservative in foods and beverages has been patented (Beelman & Hausman, 2010). According to the invention, this powerful antioxidant is very effective and stable over time, without having deleterious effects on the taste of foods and beverages.

Chemical Analysis of Ergothioneine

Despite a considerable amount of research, ergothioneine's exact mode of action in various organisms and its origin in the animal body remain unclear. One factor that has hindered the elucidation of these factors is the lack of a convenient and rapid assay method for tracing ergothioneine in biological materials. Several methods—including spectrophotometry (Hunter, 1930; Lawson et al., 1950; Melville, 1958), thin-layer chromatography (Kaneko et al., 1980), HPLC (Mayumi et al., 1978), and HPLC-MS (Dubost et al., 2007a)—have been developed for analyzing ergothioneine. Each method has its own advantages and disadvantages, as we describe below.

Spectrophotometric Method

Bulmer and colleagues (1925) developed the first spectrophotometric method for estimating ergothioneine, based on its oxidation with arsenophosphotungstic acid. Subsequently, many researchers have attempted to improve the sensitivity and selectivity of this method. In general, all proposed spectrophotometric methods principally aim at removing interfering substances before applying the color reaction, without a loss of ergothioneine. For example, a method developed by Behre and Benedict (1929) used silver to precipitate uric acid and ergothioneine in blood filtrates. The ergothioneine in the residues was then determined colorimetrically with the uric acid reagent. One of the shortcomings of the method, as Melville (1959) indicated, is the nonspecific oxidizing nature of the reagent used. To overcome this drawback, Hunter (1930) established a method using the diazo procedure, in which he added diazotized sulfanilic acid and strong alkali to a deproteinized solution before measuring the intensity of the resulting bluish-red color. Melville and Lubschez (1953), however, pointed out that Hunter's method was inadequate for locating ergothioneine in blood and other tissues because many other compounds, such as tyrosine, histidine, L-cysteine, glutathione, and purines, either inhibit the development of the red color or mask it by combining with a reagent to produce differently colored complexes (phenols, histidine, and uric acid, for example). Many attempts have been made to overcome such drawbacks (Lawson et al., 1950; Touster, 1951; Jocelyn, 1958; Ackermann et al., 1959).

To eliminate the interfering compounds, subsequent investigations performed various pretreatments on the samples. The most satisfactory method, which uses spectrophotometric analysis, was developed by Carlsson and colleagues (1974). The assay is based on the resistance of ergothioneine to Cu^{2+}-catalyzed oxidation in alkaline media and its very rapid reaction with 2,2'-dipyridyl disulfide (2-Py-S-S-2-Py) at low pH to produce an essentially stoichiometric release of chromophoric 2-thiopyridone (Py-2-SH). Two research teams, Melville and Horner (1953) and Heath and Wildy (1956), devised a chromatographic method that involves the separation of ergothioneine from interfering compounds by prior chromatography on an alumina column. The antioxidant was then determined by a modification of the diazotization reaction, as Hunter (1930) previously described. Although the method allows good recovery of added ergothioneine, these assay techniques are too laborious and time consuming to be routine tests or for application in other labs (Jocelyn, 1958). In general, to achieve precise values using the spectrophotometric method, several sample pretreatments are necessary in order to eliminate interfering compounds. As a result, a loss of ergothioneine occurs during sample preparation because of various removal processes, particularly in the precipitation of proteins (Lawson et al., 1950), which usually leads to under- or overestimation of ergothioneine levels.

Chromatographic Methods

HPLC Method

HPLC has become the most prominent analytical technique applied in the last 30 years. Mayumi and colleagues (1978) first used HPLC with UV detection to analyze ergothioneine in rat blood and tissues, a method that required various pretreatment processes. Shires and colleagues (1997) used a similar method to quantify the amount of ergothioneine in bovine and porcine ocular tissues. Recently, HPLC, coupled with a UV-Vis detector, has been widely applied to the analysis of erogthioneine in mushrooms (Lee et al., 2007; Chen et al., 2012). HPLC with electrochemical detection (ECD) has been an important tool for analyzing redox-reactive compounds, such as thiols and disulfides, which include ergothioneine (Rabenstein & Saetre, 1977). Kuninori and Nishiyama (1991) developed a rapid and sensitive method for the measurement of picomole levels, not only for ergothioneine but also for other biological thiols, including glutathione, cysteine, penicillamine, and cysteamine. The method was based on a combination of HPLC and ECD, in which the analytes, separated isocratically on a reversed-phase C18 column by ion-pair chromatography, were combined with silver nitrate dissolved in a 30 M ammonium nitrate buffer at pH 10.5. The authors detected the number of silver ions consumed in the reaction with thiols using a platinum disk electrode at –0.1 V versus Ag/AgCl. They concluded that the method could be used for assaying thiols and disulfides in biological materials. Subsequently, Kleinman and Richie Jr. (1995) developed a simple method for the simultaneous measurement of thiols and disulfides, including ergothioneine, in biological samples by using HPLC–ECD. In some cases, however, the HPLC method is limited by its low sensitivity and/or selectivity, particularly when quantitatively determining traces of compounds in complex matrices. Thus, it has remained a challenging task to establish a suitable chromatographic separation technique without interference from co-eluting compounds found in biological materials.

HPLC Post-Column Derivation Method

Online post-column reaction detection has become more popular as its sensitivity and/or selectivity in HPLC has improved. Post-column derivatization involves modifications to the chromatographic system that allow the reaction to take place before entry of a reactive derivative into the detector. Beesley and colleagues (2001) accomplished this technique by inserting a post-column reactor between the column and the detector. Online post-column reaction detection has several main advantages. First, the reaction does not need to be completed and the reaction products need not be stable, with reproducibility as the only requirement. Second, the analytes are separated in their original form, which often permits researchers to adopt the reported separation procedures. Third, the formation of artifacts plays a minor role, and finally, because the procedure is carried out online, it is easily automated (Brinkman, 1987; Barceló & Hennion, 1995). Thus, the post-column reaction method is expected to eliminate chromatographic interference from the compounds co-eluting with ergothioneine in biological samples. Reed and colleagues (1980) described a method that used *S*-carboxymethylation of compounds followed by derivatization with Sanger's reagent for the determination of various thiol-containing compounds. However, this method could not detect some thiols, which lack amine groups like ergothioneine. Consequently, Nishiyama and Kuninori (1984) developed another method in which they combined reversed-phase HPLC with a post-column reaction of thiols performed with 6,6'-dithiodinicotinic acid. Although the method located glutathione, cysteine, cysteamine, homocysteine, and penicillamine at nanomole levels, it could not measure ergothioneine. Newton and colleagues (1981) reported a third method involving the derivatization of ergothioneine with monobromobimane and separation of the derivatives by HPLC using fluorescence detection. The method was effectively applied for antioxidant determination in red blood cells.

Recently, Nguyen and colleagues (2012a) reported a rapid HPLC post-column method for analyzing ergothioneine on the basis of a selective post-column reaction of chromatographically separated ergothioneine with 2-Py-S-S-2-Py under strongly acidic conditions. A corresponding reactive derivative, Py-2-SH, was detected at 343 nm for estimating ergothioneine concentration (Carlsson et al., 1974). The system is presented schematically in Fig. 12.3. The optimized conditions for acquiring intense peak signals were as follows: a 30 m–long reaction coil; 0.23 mM solution of 2-Py-S-S-2-Py, pH 1.0, used at a flow rate of 0.15 mL/min; reaction temperature of 25 °C; and mobile phase (10% MeOH) used at a flow rate of 0.2 mL/min. Recovery and reproducibility assays indicated that HPLC post-column reaction analysis is a satisfactory method for the quantitative determination of ergothioneine in certain biological materials. The optimized post-column reaction assay for ergothioneine was compared with traditional spectrophotometric and advanced MS methods. A highly linear correlation for the authentic standard was observed over a wide range of ergothioneine concentration (1–100 μg/mL) for both the spectrophotometric assay and the HPLC post-column reaction method; however, the former overestimated the ergothioneine levels in the bi-

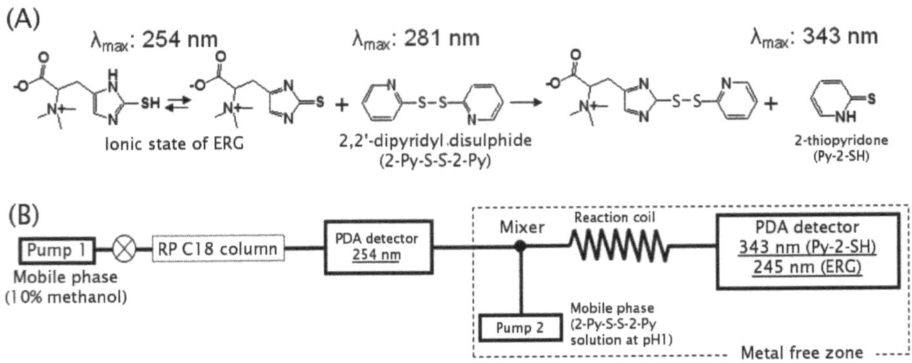

Fig. 12.3. Schematic representation of the reaction between ergothioneine (ERG) with 2,2'-dipyridyl disulfide (2-Py-S-S-2-Py) (A) and a schematic diagram of a new system for the quantitative determination of ERG by HPLC post-column reaction analysis (B).

ological samples. Several biologically abundant compounds (such as L-cysteine, glutathione, purines, tyrosine, and histidine) can undergo the spectrophotometric assay reaction and/or inhibit its development (Melville & Lubschez, 1953). On the other hand, MS using electrospray ionization (ESI) revealed a highly linear correlation for both the ergothioneine standard and the ergothioneine in the biological samples by monitoring the molecular ion $[M + H]^+$ of *m/z* 230 and sodiated cluster ions $[M+Na]^+$ of *m/z* 252 and $[2M+Na]^+$ of *m/z* 459 (Nguyen et al., 2012a). The results clearly indicate that the current post-column reaction assay for ergothioneine is as precise as advanced MS analysis. Under the optimized conditions of the post-column reaction assay, the authors evaluated authentic ergothioneine, biological ergothioneine from blood samples, and internal standards, as shown in Fig. 12.4. The results indicated that the post-column reaction can successfully eliminate the interference of unresolved co-eluting compounds monitored at 254 nm, as Fig. 12.4(C) clearly shows. In addition, spiking the biological samples with authentic ergothioneine also confirmed the elution time and reactivity towards 2-Py-S-S-2-Py. It is important to note that the stability of 2-Py-S-S-2-Py was highly vulnerable to ferrous ions, with Py-2-SH forming rapidly in the presence of metal ions at pH 1.0 (data not shown). To avoid this problem, we recommend using a metal-free/Teflon®-coated instrument system in the post-column reaction analysis [Fig. 12.3.(B)]. Under optimized conditions, the total ergothioneine content in several species of edible mushrooms and in the blood of several animals and muscles of cattle were evaluated. The ergothioneine content of the fruiting bodies of

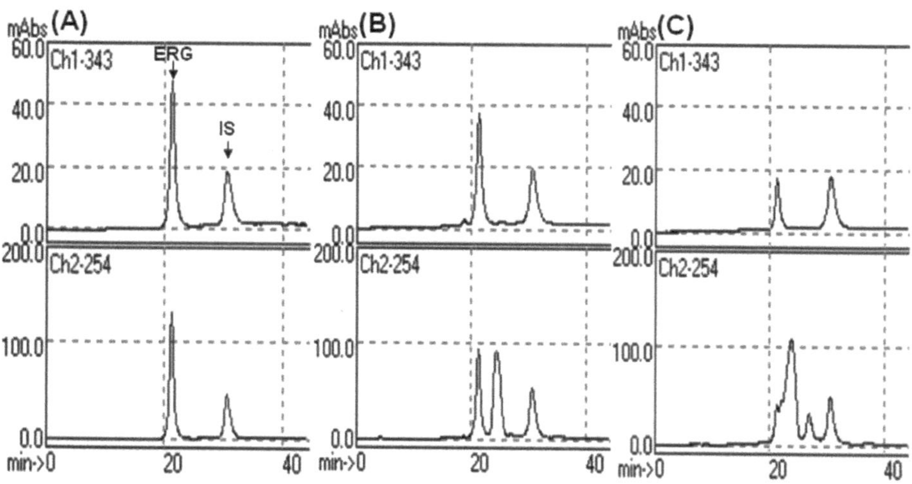

Fig. 12.4. Typical HPLC chromatograms of 60 μg/mL authentic ergothioneine (A), yellowtail blood sample (B), and porcine blood sample (C) after post-column derivatization. ERG, ergothioneine; IS, internal standard (methimazole).

common edible mushrooms cultivated in Japan ranged between 12.69 and 234.85 mg/kg wet weight. Dietary supplementation with extracts from mushroom processing waste indicated an improvement of ergothioneine bioavailability in the blood of yellowtail fish and the muscle tissue of cattle (Fig. 12.5). In summary, we found the developed and validated post-column reaction assay to be a reliable, simple, fast, precise, accurate, and sensitive method for the routine analysis of ergothioneine in biological materials.

Thin-Layer Chromatography (TLC) Method

Kaneko and colleagues (1980) developed a reliable TLC-densitometry method for the quantitative analysis of ergothioneine in biological samples. Ergothioneine was separated from TLC, and the researchers quantified the spot corresponding to the compound by scanning it at 256 nm with a densitometer. The specificity, sensitivity, and accuracy of the TLC-densitometry method are comparable to the same measurements for the HPLC method. However, the TLC-densitometry method is time consuming and laborious, and the instrument still needs improvement. For these reasons, the method has not been widely applied to routine analysis.

Liquid Chromatography–Mass Spectrometry (LC-MS) Method

Dubost and colleagues (2007a) applied liquid chromatography–mass spectroscopy (LC–MS) for the quantification of ergothioneine in mushrooms. Ey and col-

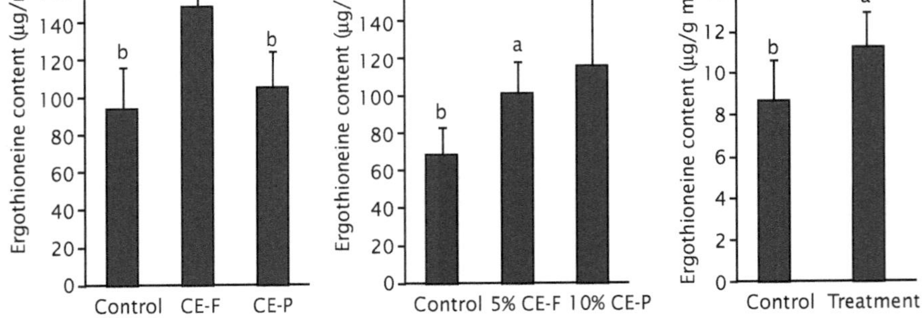

Fig. 12.5. Ergothioneine content in the whole blood of yellowtails and cattle muscle, quantitatively determined under optimized conditions. Yellowtail fed extracts from the processing waste of different mushroom species (CE-F, *F. velutipes;* CE-P, *Pleurotus cornucopiae*) (A); yellowtail fed diets supplemented with 5% or 10% CE-F (B); cattle fed 40 g/(cattle•day) of CE-F powder (C). Results are presented as mean ± standard deviation (SD). Values indicated with different superscript letters were significantly different ($P < 0.05$).

leagues (2007) used the same technique for analyzing the ergothioneine content in common foods. The ergothioneine assay that uses MS is a satisfactory quantitative method. However, the HPLC–MS system is costly to run and maintain. Moreover, sodiated and/or potassiated adducts of the ergothioneine molecule distributed in biological materials or generated during extraction and/or analysis often create difficulty for the quantification of total ergothioneine content by MS. Therefore, the HPLC–MS method is not suitable for the rapid and routine analysis of ergothioneine in biological materials.

Enhancement of Ergothioneine and Its Protective Effects on Oxidative Stress in Animals

Ergothioneine Accumulation in Animal Tissues

Antioxidant nutrients accumulated in animal tissues can enhance the nutritive value and functional properties of food materials. So far, there is no evidence that humans and other vertebrates can synthesize this antioxidant, even though it accumulates at high levels in some mammalian cells as a result of dietary intake (Gründemann, 2012). Since its discovery, different mechanisms have been proposed to explain its presence in animal tissues. Based on our knowledge that only bacteria and fungi can synthesize ergothioneine, Melville and Horner (1953) tried to determine whether intestinal microorganisms can synthesize it. They carried out a preliminary experiment with rats, which were fed a stock diet of 2% sulfasuxidine for six weeks. No significant difference was observed in the blood ergothioneine concentrations between the control and the germ-free animal group. A later study on chickens further confirmed that most of the ergothioneine present in chicken blood does not arise from synthesis by intestinal bacteria (Melville & Horner, 1953). Therefore, the authors concluded that ergothioneine in the blood must derive from diet and/or synthesis in the tissues. Cumulative data have demonstrated that ergothioneine is concentrated in cells and tissues that are frequently subjected to a high degree of oxidative stress, such as the blood, lens of the eye, liver, bone marrow, and seminal fluid. This finding suggests ergothioneine's functional role protecting against oxidative damage. Recently, Gründemann and colleagues (2005) discovered a highly specific transporter of ergothioneine in human tissues, OCTN1 (human gene, *SLC22A4*). This transporter mediates the absorption, distribution, and retention of ergothioneine in all vertebrates; cells and tissues that lack the transporter do not accumulate ergothioneine because the plasma membrane is virtually impermeable to this hydrophilic compound. In contrast, cells expressing OCTN1 accumulate high levels of ergothioneine. In humans, the expression of OCTN1 can vary, depending on the tissue, with the small intestine, kidney, erythrocyte progenitor cells in bone marrow, and monocytes showing high levels of OCTN1 expression (Bacher et al., 2009). Paul and Snyder (2010) reported that the depletion of endogenous OCTN1 in HeLa cells causes severe oxidative stress. Furthermore, an *in vivo* study on knockout mice confirmed that a deficiency of ergothioneine produces increased susceptibility to oxidative stress, as indicated by its high lethality in *octn1*$^{-/-}$ mice (Kato et al., 2010). These results strongly reinforce evidence of ergothioneine's role as a physiological antioxidant. The discovery of its transporter is expected to be a useful tool in studies of the metabolism and role of ergothioneine in humans and animals.

In an early study by Heath and colleagues (1952), authentic ergothioneine accumulated in the tissues when they administered it to rats orally or subcutaneously. The initial concentration in the whole blood of all the rats tested was less than 1 mg/100 mL. The concentration in the treated rats, which were fed a diet containing 0.1% ergothioneine, slowly increased, reaching 2 mg/100 mL after 4 days and 10 mg/100 mL after 14 days of administration. The rate of ergothioneine accumulation differed among tissues; it was absorbed rapidly in the liver but more slowly in the erythrocytes. In addition, it could not be detected in the adrenal or thyroid glands. These results were confirmed by another study in which one week of administration (1.6 mg/100 g body weight/day) increased ergothioneine content in the blood and liver of rats by 12 and 7 times, respectively (Mayumi et al., 1978). Further, the metabolism of ergothioneine administered to the rats was very slow; no decline in concentration was observed, either in the blood or liver after fasting for one week. Similarly, Kawano and colleagues (1982) demonstrated that after ingestion, ergothioneine in rat tissues is rapidly cleared from the circulation and then avidly retained in the body with minimal metabolism. In the rats fed an ergothioneine-supplemented diet, the compound accumulated with time; blood and liver levels reached up to 85 μg/mL and 270 μg/mg, respectively, in 10 weeks.

It has been demonstrated that diet affects the level of ergothioneine in animal tissues. Baldridge and Lewis (1953) observed that its concentration increased from four- to sixfold in the blood of rabbits when the animals were fed a diet of oats and cabbage for several weeks. Consequently, Baldridge (1955) studied the effect of different amounts of dietary oats and cabbage on the ergothioneine content of the blood of rabbits and rats. The blood level of rabbits fed either a cabbage-plus-casein diet or an oats-plus-casein diet increased from 4.6 mg% to 9.4 mg% during 10 weeks of feeding. However, after 10 weeks, a small change was observed in the blood ergothioneine level of rabbits fed cabbage plus a diet of oats and casein. With respect to rats, the authors observed that the average blood ergothioneine concentration was related to the amount of oats in their diet. In particular, its level in the blood of rats fed a diet containing 50% oats increased from 0.2 mg% to 1.6 mg% after the four-day feeding period, whereas that in the blood of rats fed a 100% oat diet increased from 0.2 mg% to 8.2 mg% after a similar period. Because mushrooms contain high levels of ergothioneine, their consumption may supply dietary ergothioneine to humans and animals. Moreover, understanding the absorption, metabolism, and function of ergothioneine from mushroom-based extracts and/or foods can be of special interest. Recently, Nguyen and colleagues (2012a) showed that dietary supplementation with extracts prepared from mushroom processing waste significantly improved ergothioneine bioavailability in the blood of yellowtail fish and muscle tissues of cattle. The mean content in yellowtail fed diets containing 5% and 10% extracts was 1.7 and 1.5 times higher, respectively, than that in yellowtails fed the control diet without extract supplementation [Fig. 12.5(B)]. Further, dietary supplementation with mushroom waste extract powder (1 g/kg diet) significantly increased the ergothioneine content in cattle muscles [Fig. 12.5(C)]. A similar result was obtained in another investigation, in which the ergothioneine content in the muscles of kuruma shrimp (*Marsupenaeus japonicus*) fed a supplemented diet containing mushroom extract for one week was 4.5 times higher than that of the control group (Encarnacion et al., 2010). Typically, the amount of ergothioneine in shrimp muscles has been reported as negligible, below the limit of detection (Ey et al., 2007). Thus, the study's results strongly suggest that ergothioneine from the mushroom extract added to the diet accumulated in shrimp muscles during feeding. The bioavailability and antioxidant effects of mushroom-derived ergothioneine were first investigated in humans by Weigand-Heller and colleagues (2012). Ergothioneine was administered through a mushroom test meal containing 8 g and 16 g of mushroom powder, and its concentration in postprandial red blood cells was determined. The concentration increased after both the 8-g and 16-g treatments. Notably, ergothioneine was rapidly absorbed; after 2 h of consuming the 16-g dose, the treated group had a significantly higher concentration of ergothioneine than the control group. The authors concluded that the bioavailability of ergothioneine in red blood cells warrants further study to evaluate the antioxidant's health-promoting properties and to explain its mechanisms of action.

From a legal point of view, toxicological studies are crucial for determining the safety profile of an antioxidant and acceptable daily intake (ADI) levels (Miková, 2001). Even though extensive animal toxicity studies have not been reported, a number of investigators have indicated that ergothioneine ingestion by humans or animals has no deleterious effects (OXIS, 1997). Early pharmacological studies have not shown that the antioxidant has any side effects at physiologically relevant concentrations (Tainter, 1926; Trabucchi, 1936). Recent experiments indicated that yellowtail fish and shrimp, which were fed diets containing significant levels of ergothioneine, did not exhibit a significant mortality rate (Bao et al., 2009; Encarnacion et al., 2010). In addition, the hemocyte counts in shrimps fed a diet supplemented with mushroom extract did not differ significantly from those in the controls (Encarnacion et al., 2010). These findings suggest that the intake of extracts containing ergothioneine has no adverse effects on the immune systems of shrimp and yellowtail fish. In a recent study on humans, Weigand-Heller and colleagues (2012) reported that men who were administered a 16-g mushroom dose (8.8 mg ergothioneine) showed no adverse effects.

Protective Effect of Ergothioneine against Oxidative Stress *in vivo*

In animal models, oxidative damage is associated with free-radical formation and causes deterioration in both health and food quality. Therefore, dietary supplementation with antioxidants not only protects animals from oxidative stress but also improves food quality by preventing the oxidative deterioration of lipids in postharvest storage and processing of foods. Because ergothioneine has been shown to be a stable, safe, and powerful antioxidant, it obviously can be expected to provide these benefits. Deiana and colleagues (2004) examined the effect that orally administering ergothioneine had on oxidative damage *in vivo,* which was caused by ferric nitrilotriacetate via the Fenton reaction. They found that ergothioneine supplementation not only protected the liver and kidney against lipid peroxidation, but also prevented the depletion of endogenous glutathione and α-tocopherol. Intriguingly, the concentration of ergothioneine in the kidneys was directly related to their oxidative status; this suggested increased metabolism of ergothioneine, presumably reflecting increased utilization to protect the kidneys against oxidative damage. Another *in vivo* study confirmed that ergothioneine administered orally at 2.0 and 8.0 mg/kg/day prevented brain lipid peroxidation and maintained the glutathione/glutathione disulfide ratio in the brain tissues of mice treated with cisplatin (Song et al., 2010).

Recently, Mishra and Singh (2010) reported that rats fed dried mushrooms and mushroom extracts have an improved antioxidant status during aging and fewer age-associated disorders due to the involvement of free radicals. A similar result was reported by Giannenas and colleagues (2010) for broiler chickens. They revealed that dietary inclusion of *Agaricus bisporus* mushroom reduced malondialdehyde production in liver, breast, and thigh tissues; elevated glutathione peroxidase; and reduced glutathione, glutathione reductase, and glutathione S-transferase levels. This suggested that dietary supplementation of *A. bisporus* delayed lipid oxidation of poultry meat. These results were presumably obtained because of the fact that a considerable amount of ergothioneine from the mushroom (Ey et al., 2007) accumulated in the animal tissues and exhibited antioxidant properties. Bao and colleagues (2009) demonstrated that dietary supplementation with ergothioneine-containing extracts prepared from the processing waste of *F. velutipes* could prevent lipid oxidation and the discoloration of fish meat on an industrial scale. The results from these studies emphasize that supplementing ergothioneine with mushroom extract can increase its content in the blood and muscles of animals. Therefore, this may be a novel strategy for preventing the oxidation of certain animal muscles in postharvest storage.

Detection of Antioxidant Compounds Using Online HPLC–DPPH Method: Application to Ergothioneine in Mushrooms

Colorimetric DPPH Method

With increasing interest in antioxidants, many analytical methods have been developed to assess the antioxidant capacity of pure compounds and mixtures *in vitro*. These assays are satisfactory for determining the antioxidant activity of individual pure antioxidants or the total antioxidant activity of mixtures of antioxidants or antioxidants in complex matrices (Niederländer et al., 2008). The two stable free radicals that are commonly used to assess antioxidant activity *in vitro* are 2,2'-azino-bis(3-ethylbenzothiazoline-6-sulfonic acid) ($ABTS^{\bullet +}$) and DPPH•. The DPPH• method was first developed by Blois (1958), who revealed that DPPH was reduced by the thiol-containing amino acid cysteine and other active compounds. The method is based on the reduction of an alcoholic DPPH solution at 517 nm in the presence of an antioxidant donating hydrogen or an electron (Fig. 12.6). It is a simple and economical method to evaluate the scavenging activity of antioxidants in fruit and vegetable juices or extracts because radical generation is not required, as it is in other scavenging assays (Singh & Singh, 2008). The method, however, has some limitations because the interference of absorbance

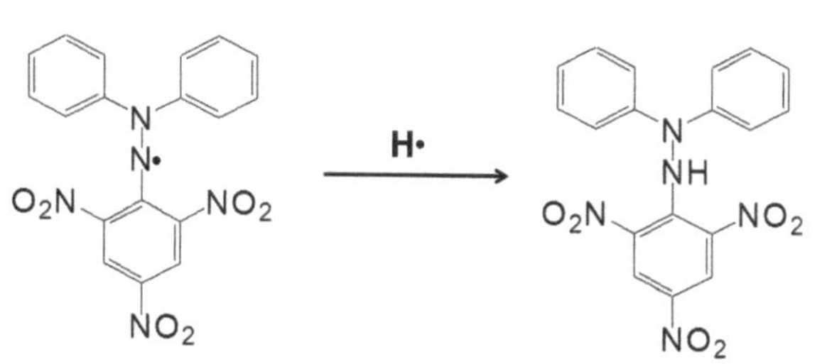

Fig. 12.6. Principle of measurement of antioxidant activity according to the DPPH method.

from sample compounds could be problematic for quantitative determinations (Arnao, 2000). For example, colored compounds, such as anthocyanins and carotenoids, present in test samples also have a strong absorbance at 515 nm, which overlap with that of DPPH and thereby interfere with optical density measurements (Prior et al., 2005; Pérez-Jiménez et al., 2008). In addition, it is difficult to assess the contribution of individual antioxidants in a mixture by the DPPH spectrophotometric assay. Consequently, an antioxidant activity–guided fractionation, which includes extraction, compound isolation, structure determination, and bioassay, is commonly used. But, the bioactivity-guided fractionation of plant extracts is a time-consuming, labor-intensive, and expensive process. Moreover, the activity of natural antioxidants often decreases during their isolation and purification because of decomposition (Hostettmann et al., 2001; Bandonienė et al., 2002). Therefore, a rapid and cost-effective method for screening and evaluating the activity of samples is required. A technique that combines separation and activity evaluation would be useful for such investigations.

Online HPLC–DPPH Method

Koleva and colleagues (2000) first developed a rapid online method for the screening of complex mixtures for radical-scavenging components using the stable free radical DPPH•. In this method, analytes are separated by HPLC and then subjected to post-column reaction with the free radical DPPH•. Next, an absorbance detector measures the bleaching induced as a negative peak at 517 nm. Several parameters, including the concentration of the DPPH• solution, antioxidant-DPPH• reaction time, and composition and pH of the HPLC mobile phase, are optimized to achieve a high signal-to-noise ratio and thus maximum sensitivity. The authors found that a DPPH stock solution concentration of 10^{-5} M in MeOH and a reaction time of 30 s yielded the best results. They concluded that the online HPLC–DPPH method has a broad range of applications in which (i) isocratic and gradient runs with mobile phases of different compositions and pH can be used; (ii) compounds with a broad range of polarity and pK_a values can be evaluated; and (iii) compounds with different behaviors toward the radical can be determined. Therefore, the method can be used for the screening as well as the quantitative analysis of antioxidant compounds in complex samples, such as plant extracts, foods, and drugs. Owing to its many advantages, the HPLC–DPPH method has been applied for screening radical-scavenging compounds in various complex extracts and foodstuffs. Bandonienė and colleagues (2002) applied the online HPLC–DPPH technique to detect radical-scavenging constituents in the crude extracts of *Salvia* species, namely, *Salvia officinalis, S. glutinosa, S. sclarea,* and *S. aethiopis.* The UV (positive) and DPPH radical-quenching (negative) chromatograms indicated the presence of several radical-scavenging compounds in the extracts. The authors concluded that the finding would be useful for further investigation on the identification of active constituents in the four *Salvia* species that they studied. Kosar and colleagues (2003) used the online HPLC–DPPH method for the qualitative and quantitative analyses of *Lamiaceae* plant extracts. Several compounds with radical-scavenging activities were detected in the UV and DPPH-quenching chromatograms. The authors identified Rosmarinic acid and carnosic acid as the dominant radical scavengers in the extracts. Additionally, they found a strong correlation between the scavenging (negative) peak area and the concentration of the radical-scavenging reference substances used. Later, the same group applied the online HPLC–DPPH method to simultaneously determine the qualitative–quantitative composition and identify polar and nonpolar radical-scavenging compounds in water extracts of selected *Mentha* samples (Kosar et al., 2004). Caffeic acid, eriocitrin (eriodictyol-7-O-rutinoside), luteolin-7-*O*-glucoside, and rosmarinic acid were identified as the dominant radical scavengers in the extracts. Quantitative analysis of active phenols indicated a significant linear correlation between the concentration and negative peak areas. Notably, a high correlation was observed between the half-maximal inhibitory concentration (IC_{50}) values measured by spectrophotometric analysis and the total negative peak areas obtained in the HPLC–DPPH assay. Thus, the authors concluded that correlated total negative peak areas and IC_{50} values can be used to calculate the IC_{50} values of different *Mentha* extracts. Giri and colleagues (2011) developed an online HPLC–DPPH method by using an 85% ethanolic solution of the DPPH free radical for screening radical-scavenging compounds in a newly developed fermented paste, squid miso. They used the HPLC–DPPH method to monitor the development of peptides with radical-scavenging activities and their molecular mass distributions during fermentation (Fig. 12.7). The DPPH radical-scavenging activity of the paste extracts increased because peptides developed as the fermentation period was prolonged. Certain peptides with low molecular mass significantly contributed to the radical-scavenging activity. Protein standards of different molecular masses—namely, cytochrome c (12,500 Da), aprotinin (6500 Da), bacitracin (1450 Da), tetrapeptide GGYR (451 Da), and tripeptide GGG (189 Da)—were used for the preparation of calibration standards (Giri et al., 2012). A high positive correlation was observed between the radical-scavenging activity peak area and the peptide peak area, and Giri and colleagues (2012) used the method for exploring the development of antioxidant phenolics during *koji* fermentation for a period of 12 months. Through online HPLC–DPPH analysis, the development of antioxidant compounds and their relative contribution to the radical-scavenging activity were determined. Several phenolic compounds, including gallic acid, kojic acid, protocatechuic acid, ferulic acid, and vanillic acid, were found during *koji*

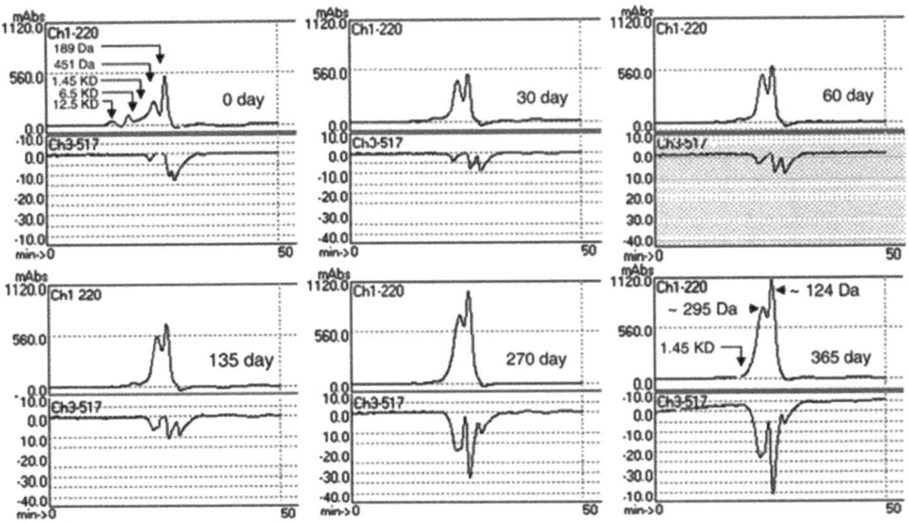

Fig. 12.7. Changes in molecular mass distributions of peptides and radical-scavenging capacity of extracts prepared from fermented squid meat under different fermentation periods, measured with an online HPLC–DPPH system.

fermentation by *Aspergillus oryzae* and identified as the extract's dominant radical scavengers. Thus, the proposed screening method, which uses online HPLC–DPPH assay, can also monitor antioxidant activity during the fermentation period of fish products. Pukalskas et al. (2005) developed a method to rapidly identify antioxidant compounds in complex extracts. The method was based on the combination of an online HPLC–DPPH assay with a diode array detection (DAD)-solid phase extraction (SPE)-nuclear magnetic resonance (NMR) system (HPLC-DPPH-UV-NMR system). The advantage of this technique was that the SPE unit allowed temporary peak parking without peak broadening; thus, NMR only measured the active compounds. Using the HPLC–DPPH-UV-NMR system, Pukalskas and colleagues (2005) used authentic standards to identify five main antioxidants from a rosemary extract without prior isolation: carnosol, carnosic acid, carnosaldehyde, 12-methoxycarnosic acid, and epiisorosmanol. Later, Nuengchamnong and Ingkaninan (2010) developed a technique that coupled reversed-phase HPLC online with a radical-scavenging detection system and MS/MS to combine separation, activity determination, and structural identification of antioxidants in complex mixtures into one run. The analytes were separated on an HPLC column, and the eluate split into two flows. The major portion was subjected to an ESI–MS/MS system, while the minor portion was mixed with a methanolic DPPH solution. The technique permitted the rapid determination of antioxidant activity and structural identification of the antioxidant compounds involved. It was applied for the identification of antioxidants in a Thai medicinal herb, *Butea superba,* and in *Orthosiphon grandiflorus* tea.

Online HPLC–DPPH Method for Evaluating the Antioxidant Ergothioneine in Mushrooms

Ergothioneine has been well documented *in vitro* as a potent radical scavenger (Akanmu et al., 1991). Antioxidant activity–guided fractionation of hot water extracts prepared from 11 species of mushrooms indicated that ergothioneine is the most active antioxidant compounds (Ito et al., 2011). However, some investigators showed that the DPPH radical-scavenging activity of various mushroom extracts, as measured by a spectrophotometric method, had a weak correlation with the ergothioneine content (Dubost et al., 2007c; Bao et al., 2010b). This suggested that some impurities in mushroom extracts may affect the activity of ergothioneine and/or ergothioneine-containing extracts and, subsequently, the measured DPPH radical-scavenging activity (Bao et al., 2010b). As mentioned earlier, certain colored compounds present in extracts may interfere with the absorption of DPPH in the colorimetric method, leading to a failure in the detection of small changes in DPPH absorbance. Chandrasekar and colleagues (2006) demonstrated that the colorimetric method fails to estimate the free radical-scavenging activity of commercial polyherbal formulations. In particular, the IC_{50} values estimated by colorimetric assay were much higher than those obtained from the online HPLC–DPPH method. Bhandari and colleagues (2010) reported a similar result for the herb *Picrorhiza kurroa.* These findings indicate that there is some interference in the colorimetric-UV method that diminished the antioxidant activity of the plant extract. In these particular cases, the online HPLC–DPPH method can be a useful tool for the assessment of DPPH radical-scavenging compounds in mushroom extracts.

Tepwong and colleagues (2012) used this method to qualitatively monitor radical-scavenging compounds in mushroom mycelia. They first optimized flow injection analysis conditions, such as suitable pre-column and post-column solvent concentrations, flow rates, reaction temperatures, and reaction time. On the basis of the optimized conditions, a standard calibration curve was obtained for a series of concentrations (5–50 μg/mL) of authentic ergothioneine. The curve showed linearity ($r^2 = 0.9948$) between the peak area of ergothioneine and the corresponding negative peak area at 517 nm (Fig. 12.8). The limit of quantification, defined as the lowest concentration that could be determined with the linear calibration curve, was 2.12 μg/mL. To evaluate the recovery, these researchers applied the standard addition technique and achieved a recovery of 97.9%. The method's precision for both intraday and interday reproducibility was also evaluated by measuring the relative standard deviation (RSD, %) for several concentrations with

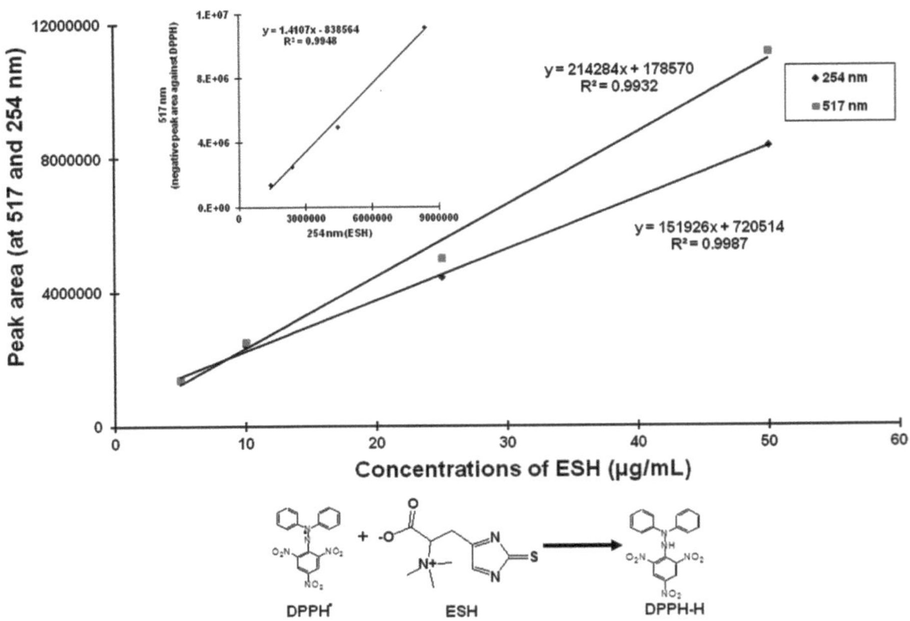

Fig. 12.8. Linearity between 2,2-diphenyl-1-picrylhydrazyl (DPPH) radical-scavenging activity and ergothioneine (ESH) content, estimated by an online flow injection analysis system. The reaction is also explained in brief.

multiple injections. The intraday and interday precisions were 3.9% and 4.3%, suggesting high accuracy, reliability, stability, and reproducibility for the quantitative determination of ergothioneine. The optimized HPLC–DPPH post-column reaction system applied to the mycelial extract detected several compounds at 254 m [Fig. 12.9(A)]. Monitoring the negative peak at 517 nm clearly demonstrated that ergothioneine was the predominant compound with radical-scavenging activity in the extract [Fig. 12.9(B)]. Selected ion monitoring in ESI-MS, under chromatographic conditions different from those of the HPLC–DPPH post-column reaction system [Fig. 12.9(C)], supported that the compound was ergothioneine. Thus, the online HPLC–DPPH method provides reliable evidence that ergothioneine is the most abundant antioxidant, which contributes most to the antioxidant activity of mushroom fruiting bodies, as well as mycelia. Nguyen and colleagues (2012b) applied this technique for screening antioxidants in an extract prepared from the edible mushroom *F. velutipes*. Several compounds in the extract were detected at 254 nm, and the prominent negative peak at 517 nm was ergothioneine, which demonstrates that ergothioneine is a major contributor to the antioxidant potential of the mushroom fruiting body.

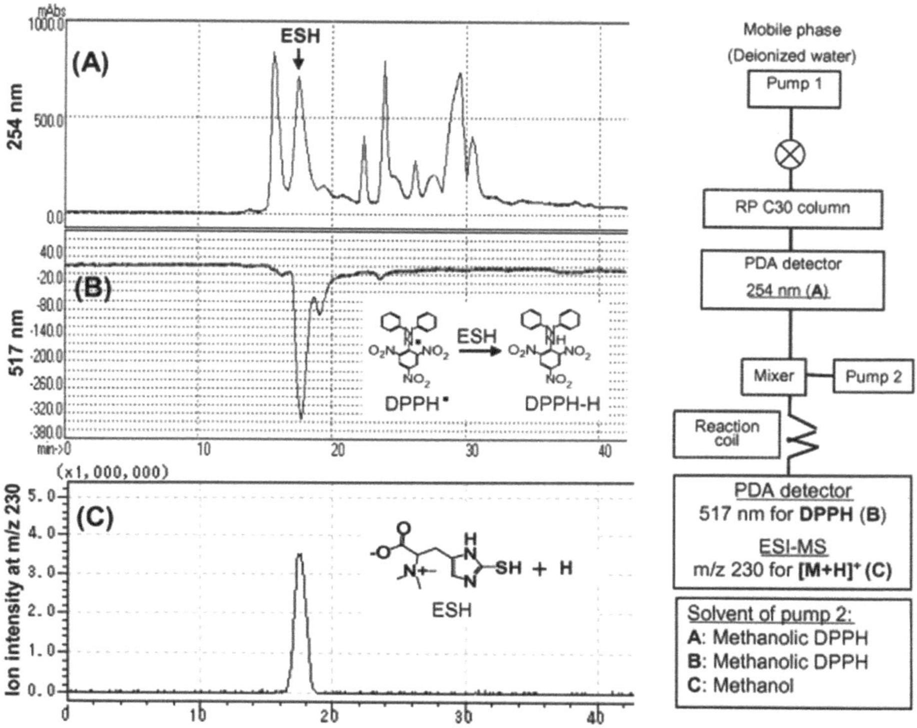

Fig. 12.9. Characterization and identification of ergothioneine (ESH) with an online flow injection analysis system. Chromatogram of the shiitake mushroom (*Lentinus edodes*) mycelial extract monitored at 254 nm (A); DPPH-scavenging activity monitored at 517 nm (B); and selected ion monitoring at *m/z* 230 (C). The right side shows the schematic diagram of the flow injection analysis system in brief. RP, reversed phase; PDA, photodiode array detector; ESI-MS, electrospray ionization-mass spectrometry.

Conclusion

The use of natural antioxidants in food has received special attention because of the worldwide inclination to avoid or minimize the use of synthetic food additives. This chapter reviewed the potential sources, antioxidant properties, methods of analysis, health benefits, and potential applications of the natural antioxidant ergothioneine. In addition to mushrooms, certain cyanobacteria have also been considered as novel and promising sources of ergothioneine for humans. The antioxidant efficacies of this compound depend on the purity of mushroom extract; thus, it is necessary to purify crude extract to use mushroom ergothioneine as a food antioxidant. This approach, however, is restricted because we lack a convenient method for the purification of ergothioneine. Future study should focus on developing an efficient purification method for its high recovery from known resources. In addition, previous studies have reported that the application of ergothioneine-containing extracts has mainly been obtained with raw materials, and further investigation is necessary for processed foods, such as cooked meats and deep-fried food, regarding the thermal degradation and antioxidant stability of ergothioneine. Future research should also examine the interaction of ergothioneine and/or ergothioneine-containing extracts with other food components and the sensory evaluation of such products before practical use of this product begins. This review has indicated that because of their valuable antioxidant functions, ergothioneine-rich materials such as mushrooms may also have the potential to be used as active ingredients in the preparation of various functional foods and pharmaceutical products. We can expect a range of commercial products containing ergothioneine-rich extracts/materials to enter the market. Various reports reviewed in this chapter have provided evidence that ergothioneine can be absorbed, distributed, and retained in different animal and human tissues. Ergothioneine accumulated in animal organs not only protects them from oxidative damage but also enriches their nutritional and pharmaceutical value. Further studies need to investigate these promising aspects. Finally, although preliminary studies reported no side effects in both humans and animals after the ingestion of ergothioneine-containing extracts/materials, comprehensive toxicity studies of these extracts/materials should be conducted prior to their application.

References

Ackermann, D.; List, P.; Meussen, H. G. Über das Vorkommen von Herzynin neben Ergothionein in der Samenflussigkeit des Ebers sowie in Rindererythrocyten und die biologische Beziehung der beiden Basen zueinander. Z. Physiol. Chem. 1959, 314, 33–37.

Akanmu, D.; Cecchini, R.; Aruoma, O. I.; Halliwell, B. The antioxidant action of ergothioneine. Arch. Biochem. Biophys. 1991, 288, 10–16.

Arduini, A.; Eddy, L.; Hochstein, P. The reduction of ferryl myoglobin by ergothioneine: A novel function for ergothioneine. Arch. Biochem. Biophys. 1990, 281, 41–43.

Arnao, M. B. Some methodological problems in the determination of antioxidant activity using chromogen radicals: A practical case. Trends Food Sci. Technol. 2000, 11, 419–421.

Aruoma, O. I. Free radicals, oxidative stress, and antioxidants in human health and disease. Am. Oil Chem. Soc. 1998, 75, 199–212.

Aruoma, O. I.; Spencer, J. P.; Mahmood, N. Protection against oxidative damage and cell death by the natural antioxidant ergothioneine. Food Chem. Toxicol. 1999, 37, 1043–1053.

Ashida, S.; Ryoichi, S.; Minoru, T. Screening of edible plants for reducing activity by monitoring their effects on the oxidation of oxymyoglobin. Food Sci. Technol. Res. 2005, 11, 349–354.

Askari, A.; Melville, D. B. The reaction sequence in ergothioneine biosynthesis: Hercynine as an intermediate. J. Biol. Chem. 1962, 237, 1615–1618.

Bacher, P.; Giersiefer, S.; Bach, M.; Fork, C.; Schömig, E.; Gründemann, D. Substrate discrimination by ergothioneine transporter SLC22A4 and carnitine transporter SLC22A5: Gain-of-function by interchange of selected amino acids. Biochim. Biophys. Acta. 2009, 1788, 2594–2602.

Baldridge, R. C. Blood ergothioneine and dietary oats. J. Nutr. 1955, 56, 107–113.

Baldridge, R. C.; Lewis, H. B. Diet and the ergothioneine content of blood. J. Biol. Chem. 1953, 202, 169–176.

Bandonienė, D.; Murkovic, M.; Pfannhauser, W.; Venskutonis, P.; Grudzienė, D. Detection and activity evaluation of radical scavenging compounds by using DPPH free radical and on-line HPLC–DPPH methods. Eur. Food Res. Technol. 2002, 214, 143–147.

Bao, H. N. D.; Ochiai, Y.; Ohshima, T. Antioxidative activities of hydrophilic extracts prepared from the fruiting body and spent culture medium of *Flammulina velutipes*. Biores. Technol. 2010a, 101, 6248–6255.

Bao, H. N. D.; Osaka, K.; Ohshima, T. Value-added use of mushroom ergothioneine as a colour stabilizer in processed fish meats. J. Sci. Food Agric. 2010b, 90, 1634–1641.

Bao, H. N. D.; Shinomiya, Y.; Ikeda, H.; Ohshima, T. Preventing discoloration and lipid oxidation in dark muscle of yellowtail by feeding an extract prepared from mushroom (*Flammulina velutipes*) cultured medium. Aquaculture 2009, 295, 243–249.

Bao, H. N. D.; Ushio, H.; Ohshima, T. Antioxidative activity and antidiscoloration efficacy of ergothioneine in mushroom (*Flammulina velutipes*) extract added to beef and fish meats. J. Agric. Food Chem. 2008, 56, 10032–10040.

Barceló, D.; Hennion, M. C. On-line sample handling strategies for the trace-level determination of pesticides and their degradation products in environmental waters. Analytica. Chimica. Acta. 1995, 318, 1–41.

Baron, C. P.; Andersen, H. J. Myoglobin-induced lipid oxidation. A review. J. Agric. Food. Chem. 2002, 50, 3887–3897.

Beelman, R. B.; Hausman, M. S. Use of ergothioneine as a preservative in foods. U.S. Patent 0,076,093 A1, 3/25/2010.

Beesley, T. E.; Buglio, B.; Scott, R. P. W. Quantitative Chromatographic Analysis; Marcel Dekker, Inc.: New York, 2001.

Behre, J. A.; Benedict, S. R. The occurrence and determination of thioneine (ergothioneine) in human blood. J. Biol. Chem. 1929, 82, 11–15.

Bhandari, P.; Kumar, N.; Singh, B.; Ahuja, P. S. Online HPLC–DPPH method for antioxidant activity of *Picrorhiza kurroa* Royle ex Benth. and characterization of kutkoside by ultra-performance LC-electrospray ionization quadrupole time-of-flight mass spectrometry. Indian J. Exp. Biol. 2010, 48, 323–328.

Blois, M. S. Antioxidant determinations by the use of a stable free radical. Nature 1958, 181, 1199–1200.

Brinkman, U. A. Th. A review of reaction detection in HPLC. Chromatographia 1987, 24, 190–200.

Bulmer, F. M. R.; Eagles, B. A.; Hunter, G. Uric acid determinations in blood. J. Biol. Chem. 1925, 63, 17–35.

Carlsen, C. U.; Moller, J. K. S.; Skibsted, L. H. Heme-iron in lipid oxidation. Coord. Chem. Rev. 2005, 249, 485–498.

Carlsson, J.; Kierstan, M. P. J.; Brocklehurst, K. A convenient spectrophotometric assay for the determination of L-ergothioneine in blood. Biochem. J. 1974, 139, 237–242.

Chandrasekar, D.; Madhusudhana, K.; Ramakrishna, S.; Diwan, P. V. Determination of DPPH free radical scavenging activity by reversed-phase HPLC: A sensitive screening method for polyherbal formulations. J. Pharm. Biomed. Anal. 2006, 40, 460–464.

Chaudière, J.; Ferrari-Iliou, L. Intracellular antioxidants: From chemical to biochemical mechanisms. Food Chem. Toxicol. 1999, 37, 949–962.

Cheah, I. K.; Halliwell, B. Ergothioneine: Antioxidant potential, physiological function and role in disease. Biochim. Biophys. Acta. 2012, 1822, 784–793.

Chen, S. Y.; Ho, K. J.; Hsieh, Y. J.; Wang, L.T.; Mau, J. L. Contents of lovastatin, γ-aminobutyric acid and ergothioneine in mushroom fruiting bodies and mycelia. LWT-Food Sci. Technol. 2012, 47, 274–278.

Cheung, L. M.; Cheung, P. C. K. Mushroom extracts with antioxidant activity against lipid peroxidation. Food Chem. 2005, 89, 403–409.

Cheung, L. M.; Cheung, P. C. K.; Ooi, V. E. C. Antioxidant activity and total phenolics of edible mushroom extracts. Food Chem. 2003, 81, 249–255.

Çoyan, K.; Bas¸pınar, N.; Bucak, M. N.; Akalın, P. P. Effects of cysteine and ergothioneine on post-thawed Merino ram sperm and biochemical parameters. Cryobiology 2011, 63, 1–6.

Deiana, M.; Rosa, A.; Casu, V.; Piga, R.; Dessi, M.; Aruoma, O. I. L-Ergothioneine modulates oxidative damage in the kidney and liver of rats *in vivo:* Studies upon the profile of polyunsaturated fatty acids. Clin. Nutr. 2004, 23, 183–193.

Diaz, M. N.; Frei, B.; Vita, J. A.; Keaney, J. F. Mechanisms of disease antioxidants and atherosclerotic heart disease. N. Engl. J. Med. 1997, 337, 408–416.

Dimakou, C.; Oreopoulou, V. Antioxidant activity of carotenoids against the oxidative destabilization of sunflower oil-in-water emulsions. LWT-Food Sci. Technol. 2011, 46, 393–400.

Dubost, N. J.; Beelman, R.; Peterson, D.; Royse, D. Identification and quantification of ergothioneine in cultivated mushrooms using liquid chromatography-mass spectroscopy. Int. J. Med. Mushrooms 2007a, 8, 215–222.

Dubost, N. J.; Beelman, R. B.; Royse, D. J. Influence of selected cultural factors and postharvest storage on ergothioneine content of common button mushroom *Agaricus bisporus* (J. Lge) imbach (*Agaricomycetideae*). Int. J. Med. Mushrooms 2007b, 9, 163–176.

Dubost, N. J.; Ou, B.; Beelman, R. B. Quantification of polyphenols and ergothioneine in cultivated mushrooms and correlation to total antioxidant capacity. Food Chem. 2007c, 105, 727–735.

Encarnacion, A. B.; Fagutao, F.; Hirono, I.; Ushio, H.; Ohshima, T. Effects of ergothioneine from mushrooms (*Flammulina velutipes*) on melanosis and lipid oxidation of kuruma shrimp (*Marsupenaeus japonicus*). J. Agric. Food Chem. 2010, 58, 2577–2585.

Erickson, M. C. Lipid Oxidation of Muscle Foods. Food Lipids: Chemistry, Nutrition and Biotechnology; Akoh, C. C., Min, D. B., Eds.; Marcel Dekker: New York, 2002; 365–411.

Ey, J.; Schömig, E.; Taubert, D. Dietary sources and antioxidant effects of ergothioneine. J. Agric. Food Chem. 2007, 55, 6466–6474.

Faustman, C.; Yin, M.; Nadeau, D. Color stability, lipid stability, and nutrient composition of red and white veal. J. Food Sci. 1992, 57, 302–304.

Finley, J. W.; Otterburn, M. S. Consequences of Free Radicals in Foods. Antioxidants: Chemical, Physiological, Nutritional and Toxicological Aspects; Williams, G. M., Ed.; Princeton Scientific Publishing Co., Inc.: Princeton, NJ, 1993; 77–91.

Frankel, E. N. Antioxidants. Lipids Oxidation; Frankel, E. N., Ed.; The Oily Press: Dundee, Scotland, 1998; 209–258.

Franzoni, F.; Colognato, R.; Galetta, F.; Laurenza, I.; Barsotti, M.; Stefano, R. D.; Bocchetti, R.; Regoli, F.; Carpi, A.; Balbarini, A.; et al. An in vitro study on the free radical scavenging capacity of ergothioneine comparison with reduced glutathione, uric acid and trolox. Biomed. Pharmacother. 2006, 60, 45–47.

Giannenas, I.; Pappas, I. S.; Mavridis, S.; Kontopidis, G.; Skoufos, J.; Kyriazakis I. Performance and antioxidant status of broiler chickens supplemented with dried mushrooms (*Agaricus bisporus*) in their diet. Poult. Sci. 2010, 89, 303–311.

Gilgun-Sherki, Y.; Rosenbaum, Z.; Melamed, E.; Offen, D. Antioxidant therapy in acute central nervous system injury: Current state. Pharmacol. Rev. 2002, 54, 271–284.

Giri, A.; Osako, K.; Okamoto, A.; Okazaki, E.; Ohshima, T. Antioxidative properties of aqueous and aroma extracts of squid miso prepared with Aspergillus oryzae-inoculated koji. Food Res. Int. 2011, 44, 317–325.

Giri, A.; Osako, K.; Okamoto, A.; Okazaki, E.; Ohshima, T. Effects of Koji fermented phenolic compounds on the oxidative stability of fish miso. J. Food Sci. 2012, 77, C228–C235.

Gole, M. D.; Souza, J. M.; Choi I.; Hertkorn, C.; Malcolm, S.; Foust III, R. F.; Finkel, B.; Lanken, P. N.; Ischiropoulos, H. Plasma proteins modified by tyrosine nitration in acute respiratory distress syndrome. Am. J. Physiol. Lung Cell Mol. Physiol. 2000, 278, L961–L967.

Granit, R.; Angel, S.; Akiri, B.; Holzer, Z.; Aharoni, Y.; Orlov, A.; Kanner, J. Effects of vitamin E supplementation on lipid peroxidation and color retention of salted calf muscle from a diet rich in polyunsaturated fatty acids. J. Agric. Food Chem. 2001, 49, 5951–5956.

Grigat, S.; Harlfinger, S.; Pal, S.; Striebinger, R.; Golz, S.; Geerts, A.; Lazar, A.; Schömig, E.; Gründemann, D. Probing the substrate specificity of the ergothioneine transporter with methimazole, hercynine, and organic cations. Biochem. Pharmacol. 2007, 74, 309–316.

Gründemann, D. The ergothioneine transporter controls and indicates ergothioneine activity—A review. Prev. Med. 2012, 54, S71–S74.

Gründemann, D.; Harlfinger, S.; Golz, S.; Geerts, A.; Lazar, A.; Berkels, R.; Jung, N.; Rubbert, A.; Schömig, E. Discovery of the ergothioneine transporter. Proc. Natl. Acad. Sci. USA. 2005, 102, 5256–5261.

Hartman, P. E. Ergothioneine as antioxidant. Meth. Enzymol. 1990, 186, 310–318.

Heath, H.; Rimington, C.; Glover, T.; Mann, T.; Leone, E. Studies using radioactive sulphur on ergothioneine formation in the pig. Biochem. J. 1953, 54, 606–611.

Heath, H.; Rimington, C.; Searle, C. E. Some effects of administering ergothioneine to rats. Biochem. J. 1952, 50, 530–533.

Heath, H.; Wildy, J. The biosynthesis of ergothioneine and histidine by *Claviceps purpurea*. 1. The incorporation of [2-^{14}C]acetate. Biochem. J. 1956, 64, 612–620.

Hostettmann, K.; Wolfender, J. L.; Terreaux, C. Modern screening techniques for plant extracts. Pharm. Biol. 2001, 39, 18–32.

Hunter, G. Colour reactions of thioglyoxalines (thioimidazoles) with sodium diazobenzene-p-sulphonate. J. Chem. Soc. 1930, 2343–2346.

Ito, T.; Kato, M.; Tsuchida, H.; Harada, E.; Niwa, T.; Osawa, T. Ergothioneine as an anti-oxidative/anti-inflammatory component in several edible mushrooms. Food. Sci. Technol. Res. 2011, 17, 103–110.

Jang, J. H.; Aruoma, O. I.; Jen, L. S.; Chung, H. Y.; Surh, Y. J. Ergothioneine rescues PC12 cells from beta-amyloid-induced apoptotic death. Free Radic. Biol. Med. 2004, 36, 288–299.

Jang, M. S.; Eun, J.; Ushio, H.; Ohshima, T. Antioxidative properties of mushroom (*Flammulina velutipes*) crude extract on the oxidation of cod liver oil in emulsion. Food Sci. Biotechnol. 2004, 13, 215–218.

Jang, M. S.; Park, H. Y.; Ushio, H.; Ohshima, T. Antioxidative effects of mushroom *Flammulina velutipes* extract on polyunsaturated oils in oil-in-water emulsion. Food Sci. Biotechnol. 2009, 18, 604–609.

Jansen, H.; Brinkman, U. A. Th.; Frei, R. W. Miniaturization of solid-phase reactors for on-line post-column derivatization in narrow-bore liquid chromatography. Chromatographia 1985, 20, 485–460.

Jocelyn, P. C. Biochemistry of the SH Group: The Occurrence, Chemical Properties, Metabolism and Biological Function of Thiols and Disulphides; Academic Press: London, 1972.

Jocelyn, P. C. The distribution of ergothioneine in blood as determined by a new method of estimation. Biochem. J. 1958, 70, 656–666.

Kaneko, I.; Takeuchi, Y.; Yamaoka, Y.; Tanaka, Y.; Fukuda, T.; Fukumori, Y.; Mayumi, T.; Hama, T. Quantitative determination of ergothioneine in plasma and tissues by TLC-densitometry. Chem. Pharmaceut. Bull. 1980, 10, 3093–3097.

Kato, Y.; Kubo, Y.; Iwata, D.; Kato, S.; Sudo, T.; Sugiura, T.; Kagaya, T.; Wakayama, T.; Hirayama, A.; Sugimoto, M. Gene knockout and metabolome analysis of carnitine/organic cation transporter OCTN1. Pharm. Res. 2010, 27, 832–840.

Kawano, H.; Otani, M.; Takeyama, K.; Kawai, Y.; Mayumi, T.; Hama, T. Studies on ergothioneine. VI. Distribution and fluctuations of ergothioneine in rats. Chem. Pharm. Bull. 1982, 30, 1760–1765.

Kimura, C.; Nukina, M.; Igarashi, K.; Sugawara, Y. Beta-hydroxyergothioneine, a new ergothioneine derivative from the mushroom *Lyophyllum connatum,* and its protective activity against carbon tetrachloride-induced injury in primary culture hepatocytes. Biosci. Biotechnol. Biochem. 2005, 69, 357–363.

Kleinman, W. A.; Richie, J. P. Determination of thiols and disulfides using high-performance liquid chromatography with electrochemical detection. J. Chromatogr. B. 1995, 672, 73–80.

Koleva, I. I.; Niederländer, H. A.; van Been, T. A. An on-line HPLC method for detection of radical scavenging compounds in complex mixtures. Anal. Chem. 2000, 72, 2323–2328.

Kollef, M. H.; Schuster, D. P. The acute respiratory distress syndrome. N. Engl. J. Med. 1997, 332, 27–37.

Kosar, M.; Dorman, H. J. D.; Bachmayer, O.; Baser, K. H. C.; Hiltunen, R. An improved on-line HPLC–DPPH method for the screening of free radical scavenging compounds in water extracts of Lamiaceae plants. Chem. Nat. Comp. 2003, 39, 161–166.

Kosar, M.; Dorman, H. J. D.; Başer, K. H. C.; Hiltunen, R. Screening of free radical scavenging compounds in water extracts of mentha samples using a postcolumn derivatization method. J. Agric. Food Chem. 2004, 52, 5004–5010.

Kumosani, T. A. l-Ergothioneine level in red blood cells of healthy human males in the Western province of Saudi Arabia. Exp. Mol. Med. 2001, 33, 20–22.

Kuninori, T.; Nishiyama, J. Measurement of biological thiols and disulfides by high-performance liquid chromatography and electrochemical detection of silver mercaptide formation. Anal. Biochem. 1991, 197, 19–24.

Lawson, A.; Morley, H. V. The determination of ergothioneine. The non-occurrence of ergothioneine in urine. Biochem. J. 1950, 47, 513–518.

Lee, W. Y.; Park, E. J.; Ahn, J. K.; Ka, K. H. Ergothioneine contents in fruiting bodies and their enhancement in mycelia cultures by the addition of methionine. Mycobiology 2009, 37, 43–47.

Liao, W. C.; Wu, W. H.; Tsai, P. C.; Wang, H. F.; Liu, Y. H.; Chan, C. F. Kinetics of ergothioneine inhibition of mushroom tyrosinase. Appl. Biochem. Biotechnol. 2012, 166, 259–267.

Lingk, W. Health risk evaluation of pesticide contamination in drinking water. Gesunde Pflangen 1991, 43, 21–25.
Louie, S.; Halliwell, B.; Cross, C. E. Adult respiratory distress syndrome: A radical perspective. Adv. Pharmacol. 1997, 38, 457–490.
Manzi, P.; Aguzzi, A.; Pizzoferrato, L. Nutritional value of mushrooms widely consumed in Italy. Food Chem. 2001, 73, 321–325.
Markova, N. G.; Karaman-Jurukovska, N.; Dong, K. K.; Damaghi, N.; Smiles, K. A.; Yarosh, D. B. Skin cells and tissue are capable of using L-ergothioneine as an integral component of their antioxidant defense system. Free Radic. Biol. Med. 2009, 46, 1168–1176.
Mayumi, T.; Kawano, H.; Sakamoto, Y.; Suehisa, E.; Kawai, Y.; Hama, T. Studies on ergothionine. V. Determination by high performance liquid chromatography and application to metabolic research. Chem. Pharm. Bull. 1978, 26, 3772–3778.
McEvily, A. J.; Iyengar, R.; Otwell, S. Sulfite alternative prevents shrimp melanosis. Food Technol. 1991, 80, 82–86.
Meister, A. On the antioxidant effects of ascorbic acid and glutathione. Biochem. Pharmacol. 1992, 44, 1905–1915.
Melville, D. B. Ergothioneine. Vitamins and Hormones: Advances in Research and Applications; Harris, R. S., Marrian, G. F., and Thimann, K. V., Eds.; Academic Press: New York, 1959; 155–204.
Melville, D. B.; Horner, W. H. Blood ergothioneine in the germ-free chicken. J. Biol. Chem. 1953, 202, 187–191.
Melville, D. B.; Horner, W. H.; Otken, C. C.; Ludwig, M. L. Studies on the origin of ergothioneine in animals. J. Biol. Chem. 1955, 213, 61–68.
Melville, D. B.; Lubschez, R. A method for the determination of ergothioneine in blood. J. Biol. Chem. 1953, 200, 275–285.
Miková, K. Practical Applications. Antioxidants in Food: Practical Applications; Pokorny, J., Ed.; CRC Press: Cambridge, England, 2001; 267–283.
Mishra, S.; Singh, R. B. Effect of mushroom on the lipid profile, lipid peroxidation and liver functions of aging Swiss albino rats. Open Nutraceuticals J. 2010, 3, 248–253.
Misra, H. P. Generation of superoxide free radical during the autoxidation of thiols. J. Biol. Chem. 1974, 249, 2151–2155.
Mitsuyama, H.; May, J. M. Uptake and antioxidant effects of ergothioneine in human erythrocytes. Clin. Sci. 1999, 97, 407–411.
Moure, A.; Cruz, J. M.; Franco, D.; Dominguez, J. M.; Sineiro, J.; Dominguez, H.; Nuñez, M. J.; Parajó, J. C. Natural antioxidants from residual sources. Food Chem. 2001, 72, 145–171.
Naczk, M.; Oickle, D.; Pink, D.; Shahidi, F. Protein precipitating capacity of crude canola tannins, effect of pH, tannin, and protein concentrations. J. Agric. Food Chem. 1996, 44, 2144–2148.
Newton, G. L.; Dorian, R.; Fahey, R. C. Analysis of biological thiols: Derivatization with monobromobimane and separation by reverse-phase high-performance liquid chromatography. Anal. Biochem. 1981, 114, 383–387.
Niederländer, H. A.; van Beek, T. A.; Bartasiute, A.; Koleva, I. I. Antioxidant activity assays online with liquid chromatography. J. Chromatogr. A. 2008, 1210, 121–134.
Nishiyama, J.; Kuninori, T. Assay of biological thiols by a combination of high-performance liquid chromatography and postcolumn reaction with 6,6'-dithiodinicotinic acid. Anal. Biochem. 1984, 138, 95–98.
Nguyen, T. H.; Giri, A.; Ohshima, T. A rapid HPLC post-column reaction analysis for the quantification of ergothioneine in edible mushrooms and in animals fed a diet supplemented

with extracts from the processing waste of cultivated mushrooms. Food Chem. 2012a, 133, 585–591.

Nguyen, T. H.; Nagasaka, R.; Ohshima, T. Effects of extraction solvents, cooking procedures and storage conditions on the contents of ergothioneine and phenolic compounds and antioxidative capacity of the cultivated mushroom *Flammulina velutipes*. Int. J. Food Sci. Tech. 2012b, 47, 1193–1205.

Nuengchamnong, N.; Ingkaninan, K. On-line HPLC–MS–DPPH assay for the analysis of phenolic antioxidant compounds in fruit wine: *Antidesma thwaitesianum* Muell. Food Chem. 2010, 118, 147–152.

Obayashi, K.; Kurihara, K.; Okano, Y.; Masaki, H.; Yarosh, D. B. L-Ergothioneine scavenges superoxide and singlet oxygen and suppresses TNF-alpha and MMP-1 expression in UV-irradiated human dermal fibroblasts. J. Cosmet. Sci. 2005, 56, 17–27.

OXIS. Image L-Ergothioneine: Compound Monograph. Technical Document for OXIS Health Products, Portland, OR, USA, 1997.

Pan, X.; Hideki, U.; Ohshima, T. Photo-oxidation of lipids impregnated on the surface of dried seaweed (*Porphyra yezoenosis* Ueda): Hydroperoxide distribution. J. Am. Oil Chem. Soc. 2004, 81, 765-771.

Paul, B. D.; Snyder, S. H. The unusual amino acid L-ergothioneine is a physiologic cytoprotectant. Cell Death Differ. 2010, 17, 1134–1140.

Pérez-Jiménez, J.; Aranz, S.; Tabernero, M.; Díaz-Rubio, M. E.; Serrano, J.; Goñi, I.; Saura-Calixto, F. Updated methodology to determine antioxidant capacity in plant foods, oils and beverages: Extraction, measurement and expression of results. Food Res. Int. 2008, 41, 274–285.

Pfeiffer, C.; Bauer, T.; Surek, B.; Schömig, E.; Gründemann, D. Cyanobacteria produce high levels of ergothioneine. Food Chem. 2011, 129, 1766–1769.

Pokorný, J. Are natural antioxidants better—and safer—than synthetic antioxidants? Eur. J. Lipid. Sci. Tech. 2007, 109, 629-642.

Prior, R. L.; Wu, X.; Schaich, K. Standardized methods for the determination of antioxidant capacity and phenolics in food and dietary supplements. J. Agric. Food Chem. 2005, 53, 4290–4302.

Pukalskas, A.; van Beek, T. A.; de Waard, P. Development of a triple hyphenated HPLC-radical scavenging detection-DAD-SPE-NMR system for the rapid identification of antioxidants in complex plant extracts. J. Chromatogr. A. 2005, 1074, 81–88.

Rabenstein, D. L.; Saetre, R. Mercury-based electrochemical detector of liquid chromatography for the detection of glutathione and other sulfur-containing compounds. Anal. Chem. 1977, 49, 1036–1039.

Rahman, I.; Gilmour, P. S.; Jimenez, L. A.; Biswas, S. K.; Antonicelli, F.; Aruoma, O. Ergothioneine inhibits oxidative stress- and TNF-alpha-induced NF-kappa B activation and interleukin-8 release in alveolar epithelial cells. Biochem. Biophys. Res. Commun. 2003, 302, 860–864.

Ramanathan L.; Das, N. P. Studies on the control of lipid oxidation in ground fish by some polyphenolic natural products. J. Agric. Food Chem. 1992, 40, 17–21.

Reed, D. J.; Babson, J. R.; Beatty, P. W.; Brodie, A. E.; Ellis, W. W.; Potter, D. W. High-performance liquid chromatography analysis of nanomole levels of glutathione, glutathione disulfide and related thiols and disulfides. Anal. Biochem. 1980, 106, 55–62.

Rohn, S.; Rawel, H. M.; Kroll, J. Antioxidant activity of protein-bound quercetin. J. Agric. Food Chem. 2004, 54, 4725–4729.

Shahidi, F. Prevention of Lipid Oxidation in Muscle Foods by Nitrite and Nitrite-Free Composition. Lipid Oxidation in Food; St. Angelo, A. J., Ed.; American Chemical Society: Washington, DC, 1992; 161–182.

Shahidi, F.; Rubin, L. J. Control of lipid oxidation in cooked meats by combinations of antioxidants and chelators. Food Chem. 1987, 23, 151–157.

Shires, T. K.; Brummel, M. C.; Pulido, J. S.; Stegink, L. D. Ergothioneine distribution in bovine and porcine ocular tissues. Comp. Biochem. Physiol. 1997, 117, 117–120.

Singh, S.; Singh, R. P. In vitro methods of assay of antioxidants: An overview. Food Rev. Int. 2008, 24, 392–415.

Song, T. Y.; Chen, C. L.; Liao, J. W.; Ou, H. C.; Tsai, M. S. Ergothioneine protects against neuronal injury induced by cisplatin both *in vitro* and *in vivo*. Food Chem. Toxicol. 2010, 48, 3492–3499.

Tainter, M. L. Notes on the pharmacology of L-Ergothioneine. Proc. Soc. Exp. Biol. Med. 1926, 24, 621.

Tanret, C. Sur une base nouvelle retiree du seigle ergote, L-Ergothioneine. C. R. Acad. Bulg. Sci. 1909, 149, 222–224.

Tepwong, P.; Giri, A.; Sasaki, F.; Fukui, R.; Ohshima, T. Mycobial enhancement of ergothioneine by submerged cultivation of edible mushroom mycelia and its application as an antioxidative compound. Food Chem. 2012, 131, 247–258.

Touster, O. Estimation of blood ergothioneine by determination of brominelabile sulfur. J. Biol. Chem. 1951, 188, 371-377.

Trabucchi, E. Richerche farmacologiche sulla ergothioneine. Boll.—Soc. Ital. Biol. Sper. 1936, 11, 117.

Unnikrishnan, V.; Nath, B. S. Hazardous chemicals in foods. Indian J. Dairy Biosci. 2002, 11, 155–158.

Uttara, B.; Singh, A. V.; Zamboni, P.; Mahajan, R. T. Oxidative stress and neurodegenerative diseases: A review of upstream and downstream antioxidant therapeutic options. Curr. Neuropharmacol. 2009, 7, 65–71.

Ware, L. B.; Matthay, M. A. The acute respiratory distress syndrome. N. Engl. J. Med. 2000, 342, 1334–1349.

Weigand-Heller, A. J.; Kris-Etherton, P. M.; Beelman, R. B. The bioavailability of ergothioneine from mushrooms (*Agaricus bisporus*) and the acute effects on antioxidant capacity and biomarkers of inflammation. Prev. Med. 2012, 54, S75–S78.

Yin, M. C.; Faustman, C. The influence of temperature, pH, and phospholipid composition upon the stability of myoglobin and phospholipida liposome model. J. Agric. Food Chem. 1993, 41, 853–857.

CHAPTER

13

Rosemary and Green Tea Extracts as Natural Antioxidants

Chemistry, Technology, and Applications

S. P. J. Namal Senanayake[1]

Introduction

The food and feed industries have long recognized the usefulness of antioxidants for improving the shelf life of various products. Lipid oxidation is one of the most significant quality-deterioration processes in lipid-containing foods, and synthetic antioxidants are routinely used to retard the oxidation process. However, some consumers have concerns about the safety of these synthetic antioxidants. Consequently, there is growing interest in using plant-derived natural extracts that exhibit potential antioxidant activity to replace or minimize the use of chemical antioxidants. Consumers now demand clean labels with no artificial ingredients and retailers expect longer shelf lives and more stability, and food processors can meet the needs of both groups by using plant-derived natural extracts. Rosemary and green tea extracts are the most important commercially available natural extracts that exhibit antioxidant properties. This chapter discusses the antioxidant mechanisms, processing, and application of rosemary and green tea extracts in foods.

Rosemary Extract

Rosemary extract is derived from the leaves of the rosemary plant (*Rosmarinus officinalis* L.), an evergreen perennial shrub from the *Limiaceae* (Labiatae) family

[1]DuPont Nutrition & Health, Four New Century Parkway, New Century, KS 66031.

that grows in warm, arid climates worldwide. Its name derives from the Latin "*ros marinus*," or "dew of the sea," and there are hundreds of varieties, each with its own aesthetic value, growing characteristics, seasoning profile, and functional properties. Growing to a height of approximately 1.8 meters, the rosemary plant is characterized by needle-shaped leaves. It grows well at temperatures between 9 and 28 °C in areas with an annual precipitation of 6 to 54 inches and a soil pH of 4.5 to 8.7. The pale-blue or deep-blue flowers may develop throughout the growing season, although blooming occurs predominantly during late winter or early spring. The plant is typically harvested once or twice a year, depending on the geographical area and whether it is grown for plant material or essential oil. Rosemary leaves are used in several food applications as an aromatic agent, while rosemary oil and extract are used extensively in soaps, detergents, cosmetics, and perfumes. Rosemary oil is colorless to slightly yellow, with a very low viscosity, and can be extracted from flowering tops, stems, and leaves by steam distillation or the use of organic solvents. While recreational gardeners and farm-to-market suppliers grow rosemary for the restaurant and culinary industries, large-scale cultivation occurs in only a few regions of the world—most notably in Mediterranean countries such as Morocco, Spain, and Turkey (Aldrich, 2010). In the United States, agricultural production of rosemary is confined to a small region, primarily in the Southwest.

Manufacturing Process

Commercial rosemary farms harvest the leaves and stem tips in a way similar to hedge pruning and ship the trimmings to extraction facilities. Most industrial-scale extraction plants use organic solvents, such as acetone, hexane, or methanol, in a process like that of extracting oil from soybeans or canola. The resulting extract is a complex mixture of various compounds, rather than a specifically defined chemical entity. The dark black-green extract may contain antioxidants (such as carnosic acid, carnosol, rosmarinic acid, and their metabolites) and aromatic "essential oils" (such as cineol or eucalyptol, pinene, and camphor). This crude extract is further processed to a uniform consistency with standardized antioxidant activity. Hence, rosemary extract is derived from the rosemary plant with subsequent extraction using solvents and standardization of active components to ensure consistency in product quality. The phenolic compounds extracted from rosemary are responsible for the antioxidant activity of this plant. Its antioxidant properties are based on the content of phenolic diterpenes, which are fat soluble; examples of these compounds include carnosic acid and carnosol. Rosemary also liberates a water-soluble extract, rosmarinic acid, which is widely used by the food industry.

Method of Incorporation

Commercial extracts from rosemary are available in two forms: liquids or fine powders. Liquid products are soluble/dispersible by either oil or water. Oil-soluble rosemary extract provides a number of benefits, including easy handling and pumping, carrier-assisted incorporation, and homogeneous distribution of antioxidants (Fig. 13.1). Rosemary extract may also be dispersed in water before adding it to foods and has a low viscosity for easy spraying and homogeneous distribution. Moreover, very little energy is required to disperse it in water (Fig. 13.2). On the other hand, rosemary extract in powder form is often preferred for formulations containing dry ingredients as it facilitates blending. The powder carrier helps to incorporate and distribute the antioxidant homogeneously (Fig. 13.3). Where dry ingredients are part of the formulation, a powdered rosemary extract is often preferred for easier blending.

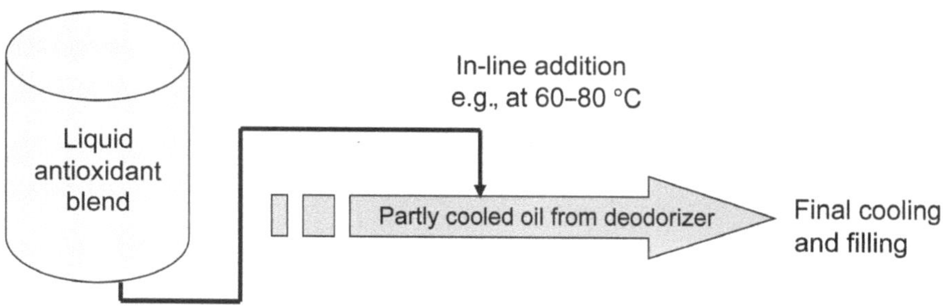

Fig. 13.1. Addition of oil-soluble/dispersible liquid antioxidants into freshly deodorized oil.

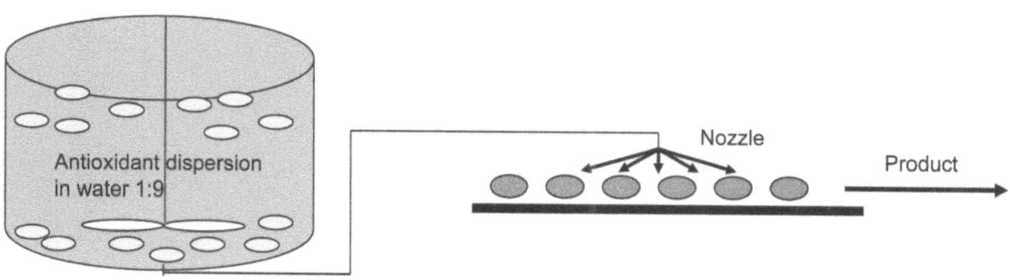

Fig. 13.2. Method of incorporating water-soluble/dispersible antioxidants into a product.

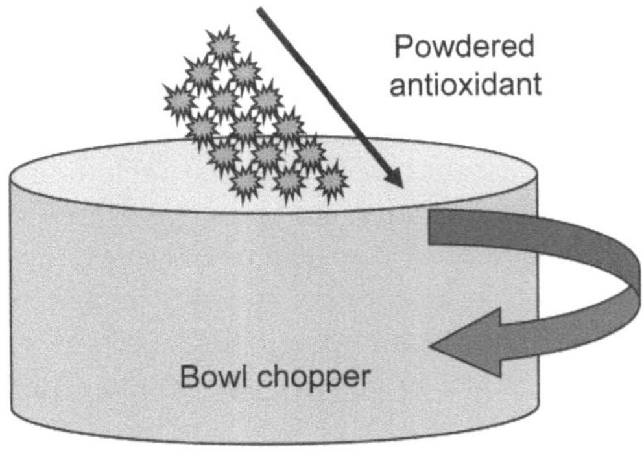

Fig. 13.3. Method of incorporating powdered antioxidants into a product.

Antioxidant Mechanism

Rosemary extract contains several active components that exhibit antioxidant activity. The previously mentioned phenolic diterpenes—comprised of carnosic acid, carnosol, and 12-*O*-methylcarnosic acid—account for nearly all of the antioxidant activity in rosemary extract. Fig. 13.4 illustrates the chemical structures of carnosic acid and carnosol. These diterpenes can be further oxidized with exposure to oxygen, light, or elevated temperatures; however, their antioxidant potency can be maintained. For example, carnosic acid converts to carnosol after oxidation, and carnosol can degrade further to produce other phenolic diterpenes with a γ-lactone structure (Schwarz & Ternes, 1992). The oxidation products so formed are active and can protect edible oils during frying as they carry over into fried foods. This means that rosemary extract can prevent oxidation during frying. Other components of rosemary extract are triterpenes, triterpenic acids, sugars, flavonoids, and other phenolic compounds. Carnosic acid is the primary phenolic diterpene compound found in rosemary leaves (Richheimer et al., 1996). It is a lipophilic antioxidant that scaveng-

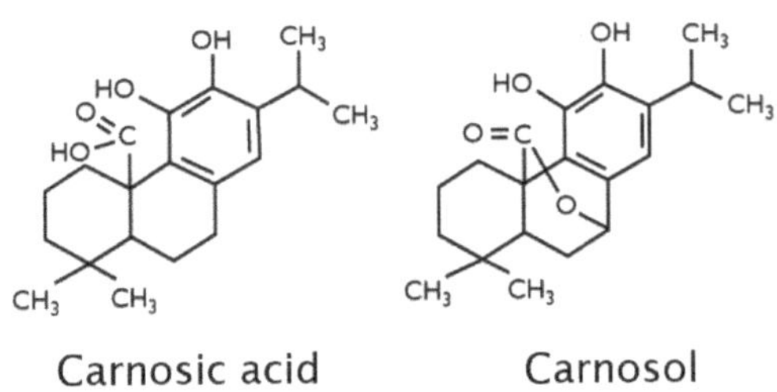

Fig. 13.4. Chemical structures of the carnosic acid and carnosol present in rosemary extract.

es singlet oxygen, hydroxyl radicals, and lipid peroxyl radicals, thus preventing lipid peroxidation and the disruption of biological membranes (Haraguchi et al., 1995). Its radical scavenging activity is caused by the presence of two *o*-phenolic hydroxyl groups found at C_{11} and C_{12} of the molecule (Richheimer et al., 1999).

Earlier studies on rosemary extract, which contains carnosic acid, carnosol, and rosmarinic acid, have shown that the antioxidant activities of these compounds are system-dependent and that their efficacy in different food systems is challenging to predict given the diverse testing systems, protocols, and questionable conditions of oxidation used (Frankel et al., 1996a, 1996b). In bulk vegetable oils (from corn, soybeans, and peanuts) and fish oils, carnosol and carnosic acid function as effective antioxidants (Wanasundara & Shahidi, 2005). It has been hypothesized that the behavior of rosemary phenolic compounds is attributed to partitioning differences in the biphasic system. These hydrophilic antioxidants are oriented in the oil–air interface and protect the bulk oil phase from oxidation (Wanasundara & Shahidi, 2005). However, they are less effective in preventing oxidation at the oil–water interface of oil-in-water emulsions. It has been hypothesized that hydrophilic antioxidants such as rosmarinic acid, gallic acid, and catechins partition more favorably into the aqueous phase and less into the oil phase, which produces less antioxidative protection (Frankel et al., 1996a; Huang et al., 1997; Wanasundara & Shahidi, 2005).

Quantitative Determination of Phenolic Diterpenes

The quality and the price of commercial rosemary extract correlates first with the content of carnosic acid and then with the total content of phenolic diterpenes, including carnosol. Therefore, accurate quantitative determinations of phenolic diterpenes are important, from a commercial standpoint. Several authors have used carnosic acid and carnosol as reference compounds for the quantitative determination of phenolic diterpenes (Schwarz & Ternes, 1992; Richheimer et al., 1996; Thorsen & Hildebrandt, 2003). These tests were carried out by capillary zone electrophoresis (Saenz-Lopez et al., 2002), high-performance liquid chromatography (HPLC) after extraction using an organic solvent (Schwarz & Ternes, 1992; Richheimer et al., 1996), and supercritical fluid extraction (Hidalgo et al, 1998). These studies encountered some practical challenges surrounding the accurate quantification of carnosic acid, carnosol, and other phenolic diterpenes in rosemary extract; however, Thorsen and Hildebrandt (2003) developed a reliable quantification method for phenolic diterpenes. These authors used preparative HPLC chromatography to isolate the carnosic acid from rosemary extract with a purity of 98% or greater. A standard addition experiment, covering a time interval of 21 h, led to the recovery of carnosol and carnosic acid at 100% and 96%, respectively (Thorsen & Hildebrandt, 2003). There were, however, considerable differences in most of the methods of analysis employed, both those described in the literature and those used by suppliers of rosemary extract. Consequently, no standardized international method exists to date.

There are many publications dealing with the determination of phenolic diterpenes, and most of the protocols are based on the reverse-phase HPLC technique coupled with spectrophotometric detection. A study by Hidalgo and colleagues (1998) used supercritical CO_2 extraction followed by reversed-phase HPLC to evaluate variability in the carnosic acid content of rosemary leaves. The content of wild rosemary leaves was found to be highly variable as mature leaves contained smaller amounts of carnosic acid than young leaves. Moreover, carnosic acid content peaked in the summer and was greater in rosemary grown under cultivated conditions. In another study, Richheimer and colleagues (1996) analyzed the phenolic diterpenes present in commercial rosemary extract using HPLC. Carnosic acid was the major phenolic diterpene present in rosemary leaves, with lesser amounts of 12-methoxycarnosic acid and carnosol. In addition to these three compounds, the researchers found other phenolic diterpenes, such as 7-methoxyrosmanol, 7-methoxy-epirosmanol, and rosmanol, in some samples. The latter three compounds seemed to be artifacts produced from carnosic acid by oxidation and cyclization.

Regulatory Status

In the United States, rosemary extract has FDA GRAS status (21 CFR 182.10) for use as a flavoring agent. It can be used in any application in accordance with good manufacturing practice (that is, at levels not to exceed their intended use); however, there is no specific approval for the use of deodorized rosemary extract as an antioxidant. In the European Union (EU), rosemary extract was adopted formally into food-additive legislation in late 2010 as an antioxidant. It is listed as "Extracts of Rosemary" and assigned E 392 as the E number (Official Journal of the European Union, 2010). Directives 2010/67/EU and 2010/69/EU, respectively, established purity criteria and the scope of rosemary extract's use as an antioxidant, thus positioning rosemary extract in the market as a safe, effective, and natural alternative to synthetic antioxidants (Table 13.A). Therefore, rosemary extracts can be applied where there is technological justification for their use, and food companies in the EU region can declare "antioxidant: rosemary extract" on their labels to address the growing consumer demand for natural food products.

Application of Rosemary Extract in Food

Meat, poultry, and seafood products are an important application area for rosemary extract. The products' susceptibility to lipid oxidation can cause a warmed-

Table 13.A. Intended Use of Rosemary Extract (E 392) under Commission Directive 2010/69/EU

Food Category	**Levels of Use (ppm)**
Vegetable oils (excluding virgin oils and olive oils) and fat in which the content of polyunsaturated fatty acids is higher than 15% w/w of the total fatty acid, for their use in non heat-treated food products	30 ppm (sum of carnosol and carnosic acid) on fat basis
Fish oils and algal oil	50 ppm (sum of carnosol and carnosic acid) on fat basis
Lard, beef, poultry, sheep, and porcine fat	50 ppm (sum of carnosol and carnosic acid) on fat basis
Fats and oils for the professional manufacture of heat-treated foodstuffs	50 ppm (sum of carnosol and carnosic acid) on fat basis
Frying oil and frying fat, excluding olive oil and olive pomace oil	50 ppm (sum of carnosol and carnosic acid) on fat basis
Snack foods (based on cereals, potatoes, or starches)	50 ppm (sum of carnosol and carnosic acid) on fat basis
Sauces	100 ppm (sum of carnosol and carnosic acid) on fat basis
Fine bakery wares	200 ppm (sum of carnosol and carnosic acid) on fat basis
Food supplements, as defined in Directive 2002/46/EC	400 ppm (sum of carnosol and carnosic acid)
Dehydrated potatoes, egg products, chewing gum	200 ppm (sum of carnosol and carnosic acid)
Milk powder for vending machines, seasoning and condiments, processed nuts	200 ppm (sum of carnosol and carnosic acid) on fat basis
Dehydrated soups and broths	50 ppm (sum of carnosol and carnosic acid)
Dehydrated meat	150 ppm (sum of carnosol and carnosic acid)
Meat and fish products, excluding dehydrated meat and dried sausage	150 ppm (sum of carnosol and carnosic acid) on fat basis
Dried sausage	100 ppm (sum of carnosol and carnosic acid)
Flavorings	1000 ppm (sum of carnosol and carnosic acid)
Dried milk for manufacturing ice cream	30 ppm (sum of carnosol and carnosic acid)

Adapted from the Official Journal of the European Union (2010).

over flavor (off-flavor and off-odor), discoloration, and protein degradation. Rosemary extract can effectively retard this development. Lipid oxidation is a serious problem in raw meat products, such as turkey breast and refrigerated and frozen minced meat. In cooked products and ready-meals, lipid oxidation is also an increasing concern. These products are susceptible to meat-flavor changes and oxidative rancidity because such products generally contain significant amounts of fat. Hence, the most likely cause of meat-flavor changes is from lipid oxidation. This undesirable effect can be inhibited with rosemary extract. A warmed-over flavor (WOF) has long been recognized as one of the primary causes of quality deterioration in cooked, refrigerated, and pre-cooked frozen meat products (Tims & Watts, 1958). The term "meat flavor deterioration" has been proposed to include the development of undesirable flavors with a concurrent loss of desirable meat-flavor characteristics (Spanier et al., 1988; St. Angelo et al., 1990). WOF is a term used to describe off-flavors and off-odors commonly described as "stale," "cardboard-like," "painty," or "rancid" in meat and poultry products (Vega & Brewer, 1994; Love, 1988). WOF is usually associated with reheated meats that have been refrigerated for 48 hours or less, and WOF can develop in precooked frozen

Table 13.B. Effect of Rosemary Extract on Oxidative Stability (as Measured by Inhibition of TBARS) of Various Meat and Poultry Products during Refrigerated and Frozen Storage

Natural Extract	Food Product	Storage Conditions	Reduction in TBARS (%)
Oleoresin rosemary	Raw restructured beef steaks	−20 °C, 6 months	39
Oleoresin rosemary	Cooked restructured beef steaks	4 °C, 6 days	14
Extracts of rosemary	Frankfurters	4 °C, 18–35 days	0–72
Oleoresin rosemary	Restructured chicken nuggets	4 °C, 6 days	11
Oleoresin rosemary	Restructured chicken nuggets	−20 °C, 6 months	1
Rosemary type W	Cooked beef patties	4 °C, 2 days	56
Rosemary type O	Cooked beef patties	4 °C, 2 days	31
Rosemary oleoresin	Sausages	−18 °C, 20 days	46
Rosemary	Cooked meat balls	4 °C, 5 days	35
Rosemary	Cooked meat balls	5 °C, 10 days	20
Rosemary	Boned turkey meat	3 °C, 13 days	82

Table modified from Madsen and Bertelsen (1995).

meats in a few days or weeks. Table 13.B provides the effect of rosemary extract on the oxidative stability of various meat and poultry products during storage. These experiments determined oxidative stability by locating thiobarbituric acid reactive substances (TBARS) and, in most cases, observed a significant reduction in TBARS. In a study of rosemary extract's antioxidant activity in Atlantic mackerel fish burgers, Ucak and colleagues (2011) found that rosemary extract was responsible for extending the shelf life of fish burgers, as compared to the control with nothing added. In another study, Güntensperger and colleagues (1998) investigated the influence of precooking and the addition of rosemary extract on lipid oxidation during the storage of heat-sterilized ground pork. In fact, they found that adding rosemary extract extended this product's shelf life. Rosemary extract was more effective when added after precooking and just before heat sterilization.

In addition to the meat, poultry, and seafood products described above, other application areas of rosemary extract include vegetable and animal fats and oils, frying oils, shortenings, margarine and spreads, mayonnaise and dressings, sauces, marinades, condiments, baked products, snack foods, cereals, and potato products, among others (Table 13.C). When added to potato flakes, rosemary extract (at 700 ppm) retarded the development of secondary oxidation products, hindered carotenoid loss, and protected lipids and lipid-like materials from oxygen attack. Sensory analysis confirmed the antioxidant effect of rosemary extract, which showed a satisfactory correlation with the analytical figures (Bracco et al., 1981). Another study found that dehydrated potato flakes stabilized by rosemary extract had unacceptable levels of pentane (a secondary degradation product of oxidation) after 2 months (nothing added), 4 months (100 ppm added), and 8 months (500 ppm added) at 20 °C. The pentane results generally correlated well with sensor

Table 13.C. Application Opportunities with Rosemary and Green Tea Extracts

Application	Specific Oxidation Issues
Meat, poultry, and seafood	Highly susceptible to oxidation, resulting in a warmed-over flavor, discoloration, and protein degeneration
Ready meals	Reheating of meat promotes the oxidation process
Bakery products	Products are susceptible to oxidation because of long shelf-life requirements.
Mayonnaise, dressings, soups, and sauces	Large oil–water interface and complex food matrix increase susceptibility to lipid oxidation.
Margarine and fat spreads	Large water–oil interface and complex food matrix increase susceptibility to lipid oxidation.
Shortening	Low oxidative stability, due to trans fatty–acid regulations, increases the need for enhanced antioxidant protection.

evaluation (Löliger, 1991). In a subsequent study, Lalas and Dourtoglou (2003) evaluated the impact of rosemary extract on the stability of soybean oil used for frying potato chips. The extract-containing oil showed greater antioxidant activity and reduced darkening and rancidity of the oil. Potato chips fried in the oil with rosemary extract added were more acceptable than chips fried in oil containing no extract. A similar study by Jaswir and Che Man (1999) found that rosemary extract added to palm olein retarded oil deterioration during deep-fat frying of potato chips.

Green Tea Extract

Green tea, derived from *Camellia sinensis* L., is an evergreen laurel that grows mainly in tropical and subtropical climates, areas with at least 50 inches of rainfall per year. Such climates are particularly prevalent in tropical and temperate regions of Asia, including China, India, Sri Lanka, and Japan. Additionally, several African and South American countries cultivate green tea. It is a small tree or shrub that can grow to 30 feet but is customarily pruned at 2–5 feet and cultivated for its leaves. The leaves are typically dark green, alternate, and oval with serrated edges, growing from 4–15 cm in length and 2–5 cm in breadth. The blossoms are white, fragrant, and appear individually or in clusters. The plant's immature, light-green leaves are typically harvested for tea production, while its mature leaves become deeper green in color. Different leaf ages produce various tea qualities as their chemical compositions are different. Typically, the buds (tips) and the first two to three leaves are harvested for processing. The leaves are hand picked, a process that is generally repeated every one to two weeks. Two main varieties of *C. sinensis* are grown: sinensis (larger-leafed, tall trunk) and assamica (smaller-leafed, bush-type trunk). Green, black, and oolong teas all originate from *C. sinensis;* however, different types of tea are classified according to the degree of fermentation that takes place during processing: green tea being unfermented, oolong tea semi-fermented, and black tea fully fermented (Chu & Juneja, 1997).

Chemistry

The beneficial effects of green tea are attributed to its polyphenolic compounds, predominantly the catechins, which make up 30% of the dry weight of green tea leaves (Graham, 1992). Polyphenolic catechins belong to the family of flavonoids and the subgroup flavan-3-ols, or flavanols. These catechins possess two benzene rings (referred to as the A- and B-rings) and a dihydropyran heterocycle (the C-ring), with a hydroxyl group on carbon 3 (Fig. 13.5). The A-ring is similar to a resorcinol moiety, while the B-ring is similar to a catechol moiety. The molecule has two

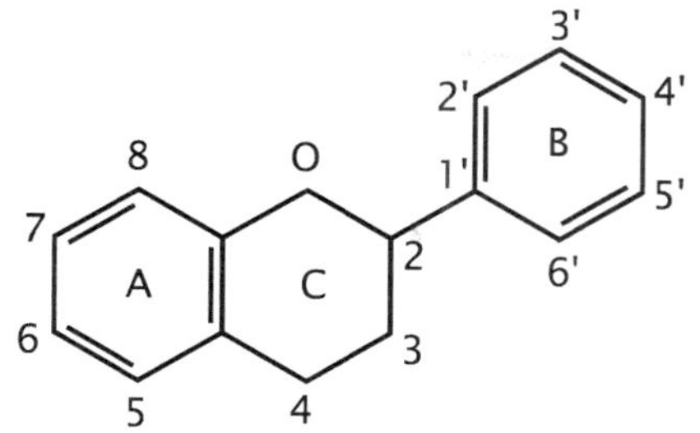

Fig. 13.5. Basic monomeric structure and numbering system for common flavonoids.

chiral centers on carbons 2 and 3; hence, it has four diastereoisomers, with two of the isomers in *trans* configuration and the other two in *cis* configuration. The *trans* and *cis* isomers are referred to as catechin and epicatechin, respectively. Extracts of immature tea leaves are rich in flavanols and their gallic acid derivatives—namely, (+)-catechin, (–)-epicatechin, (+)-gallocatechin, (–)-epicatechin gallate, (–)-epigallocatechin, and (–)-epigallocatechin gallate. In addition, they contain a range of natural green tea flavor components, such as terpenes, oxygenated terpenes, sesquiterpenes, and organic acids. Epicatechin has an ortho-dihydroxyl group in the B-ring at carbons 3' and 4' and a hydroxyl group at carbon 3 on the C-ring (Fig. 13.6). Epigallocatechin differs from epicatechin in that it has a trihydroxyl group at carbons 3', 4', and 5' on the B-ring. Epicatechin gallate differs from epicatechin in its gallate moiety, which is esterified at carbon 3 of the C-ring. However, epigallocatechin gallate has both trihydroxyl groups at carbons 3', 4', and 5' on the B-ring and a gallate moiety esterified at carbon 3 on the C-ring (Fig. 13.6). Yilmaz (2006) has summarized the chemistry and application of green tea, especially in relation to the use its catechins. The relative catechin content of green tea depends on how the leaves are processed before drying. The fermentation and heating of tea leaves during manufacturing can result in the polymerization of monopolyphenolic compounds, such as the catechins, which causes conformational changes and modifies the tea's properties (Cabrera et al., 2006).

Polyphenols are present at significantly higher amounts in green tea than in black or oolong tea (Perumalla & Hettiarachchy, 2011), owing to differences in the processing of leaves after harvest. For green tea, freshly picked tea leaves are steamed at high temperatures and dried to deactivate the polyphenol oxidases, which prevents the oxidation of catechins and maintains the polyphenols in their monomeric forms. In addition to preventing polyphenol degradation, the steaming process also protects against the enzymatic degradation of vitamins. The separation, purification, concentration, and drying processes can produce a highly concentrated green tea catechin extract with low flavor intensity. Black tea, on the other

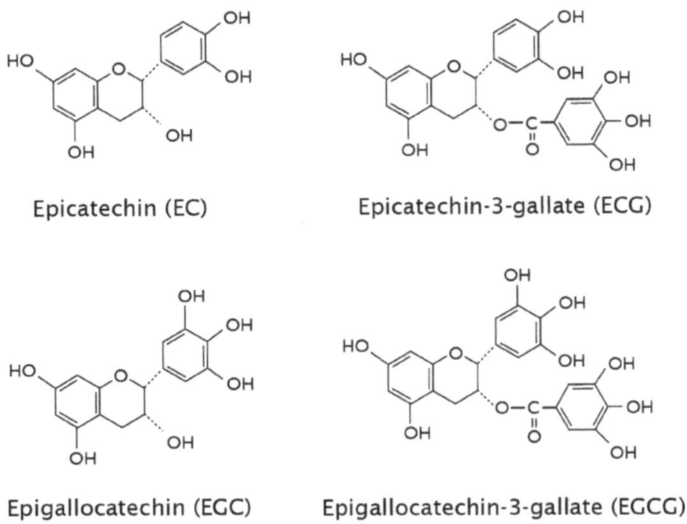

Fig. 13.6. Chemical structures of the active components present in green tea extract.

hand, is produced through the extended fermentation of tea leaves, which results in polymeric compounds, thearubigins, and theaflavins (Zaveri, 2006). Black tea contains mainly gallates of epicatechin. On the contrary, oolong tea, a partially fermented product, contains a mixture of monomeric polyphenols and higher molecular weight theaflavins (Graham, 1992). All three varieties of tea contain significant quantities of caffeine (3–6%), which the different processing methods do not affect (Chu, 1997). Moreover, all three varieties of tea contain (–)-epicatechin, (–)-epigallocatechin, (–)-epicatechin gallate, and (–)-epigallocatechin gallate, but not catechin (Khokhar et al., 1997). Epicatechin gallate and epigallocatechin gallate are the main catechins found in black tea (Obanda et al., 2001). Epigallocatechin gallate is the most abundant catechin in the leaves of green, oolong, and black teas (Graham, 1992). The content of epigallocatechin gallate in green and oolong teas typically ranges from 127 to 550 mg/L, while black teas may contain up to 300 mg/L (Balentine & Paetau-Robinson, 2000).

Antioxidant Mechanism

The antioxidant activity of green tea polyphenols is directly related to the aromatic rings and hydroxyl groups that make up their structure and results from the binding and neutralization of free radicals by the hydroxyl groups. The antioxidant activity of these catechins and their derivatives showed a marked difference depending on the substrate used for evaluation. Green tea catechins, which performed like other hydrophilic antioxidants—such as Trolox (a water-soluble analog of α-tocopherol) and ascorbic acid—have been shown to be active antioxidants in bulk oils but were prooxidants in the corresponding oil-in-water emulsions (Frankel et al., 1994, 1997). In corn-oil triacylglycerols oxidized at 50 °C, epigallocatechin, epigallocatechin gallate, and epicatechin gallate exhibited better antioxidant activity than epicatechin or catechin (Huang & Frankel, 1997). These catechins have been highly effective in retarding the oxidation of polyunsaturated fatty acid–rich marine and vegetable oils (Wanasundara & Shahidi, 1998). In oil-in-water emulsions, all catechins evaluated were prooxidants; however, in the liposomes comprising lecithin, epigallocatechin gallate was the best antioxidant, followed by epicatechin, epigallocatechin, epicatechin gallate, and catechin (Huang & Frankel, 1997). The improved antioxidant activity observed for tea catechins in liposomes compared to emulsions has been explained by the greater affinity that polar catechins have for the polar surface of the lecithin bilayers, which affords better protection (Huang & Frankel, 1997).

Several investigators have speculated on the antioxidant action mechanism of (+)-catechin using some oxidation model studies (Hirose et al., 1990; Koketsu, 1997; Zhu et al., 2000). According to the proposed mechanism, (+)-catechin can scavenge four lipid free radicals per molecule (Hirose et al., 1990; Koketsu, 1997). The antioxidant activity of individual tea polyphenols in different model systems showed a proportional relationship to the number of hydrogen radical donors of catechins. A synergistic effect has been observed among green tea catechins, ascorbic acid, and α-tocopherols (Murakami et al., 2003). Catechin, a monomeric flavanol, is reported to have hydroxyl, peroxyl, superoxide, and DPPH (1,1-diphenyl-2-picrylhydrazyl) radical-scavenging activities (Fukumoto & Mazza, 2000; Bors & Michel, 1999). Moreover, tea catechins have the ability to chelate iron in food model systems (Tang et al., 2002). Nakao and colleagues (1998) found that epicatechin gallate, epicatechin, and catechin have a peroxyl radical-scavenging activity ten times higher than L-ascorbate and beta-carotene when tested on bacteria. Nanjo and colleagues (1996) documented that the DPPH radical-scavenging activities of catechin and epicatechin are less than those of epigallocatechin, epicatechin gallate, and epigallocatechin gallate. Epicatechin is another monomeric flavanol found naturally in green tea. Reports have shown that epicatechin is capable of scavenging hydroxyl, peroxyl, superoxide, and DPPH radicals (Bors & Michel, 1999; Fukumoto & Mazza, 2000; Liu et al., 2000). The peroxyl radical-scavenging activity of epicatechin may be 10 times higher than L-ascorbate or beta-carotene (Nakao et al., 1998).

Green tea polyphenols have been shown to be effective against beta-carotene oxidation. For example, tea catechins have an antidiscoloring effect on beverages and margarine containing beta-carotene (Koketsu, 1997; Unten et al., 1997). It has been suggested that the discoloration of beta-carotene and the oxidation of unsaturated fatty acids progressed by the same mechanism. In particular, carbon-carbon double bonds in beta-carotene and unsaturated fatty acids are attacked by radicals, becoming hydroperoxides. Green tea catechins have the ability to suppress the degradation of double bonds as radical scavengers. It has also been suggested that the hydroxyl group at the 5'-position of the B-ring of the catechin structure contributes to the antidiscoloring effect. Hence, it was speculated that green tea polyphenols delay the degradation of beta-carotene by acting as antioxidants in accordance with the mechanism described above. Among the individual green tea polyphenols examined, gallocatechin gallate, epigallocatechin gallate, epigallocatechin, and gallocatechin had a strong antidiscoloring effect, while epicatechin and catechin showed almost no activity and gallic acid had moderate activity (Unten et al., 1997).

Application of Green Tea Extract in Food

Numerous health claims have made green tea a popular ingredient in dietary supplements and in the growing market for functional foods. Table 13.D reports some benefits of green tea extract. Studies conducted by DuPont Nutrition and Health laboratories have shown that green tea extract significantly retarded oxidation in roasted turkey burgers using a dosage level of 50 to 100 ppm (Figure 13.7). Another study conducted by DuPont Nutrition and Health compared the efficiency of green tea extract with equivalent dosages of butylated hydroxyanisole (BHA). Figure 13.8 demonstrates that a dosage of 200 ppm green tea extract easily provides the

Table 13.D. Benefits of Green Tea Extract

Natural alternative to synthetic antioxidants
Promotes good health
Extended food product shelf-life
Retards development of rancid off-flavors and off-odors
Consistent product quality
Easy to handle and apply
Low cost-in-use
Availability of optimized tailor-made solutions
Brand image protection

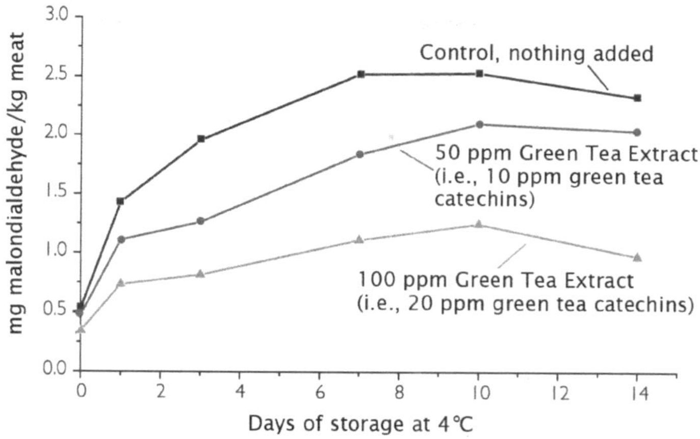

Fig. 13.7. Effect of green tea extract on the inhibition of TBARS in chilled (4 °C), roasted turkey burgers.

same antioxidant ability as 40 ppm BHA in chilled (4 °C), roasted turkey burgers. A sensory evaluation further demonstrated the advantage of using green tea extract compared to BHA. Table 13.E summarizes the most important observations. During internal testing on roasted beef burgers, the addition of 150 ppm green tea extract was sufficient to inhibit hexanal development (a secondary product of oxidation)

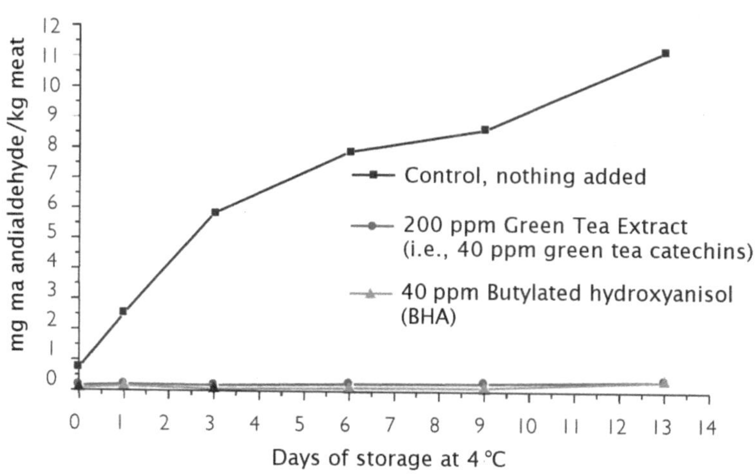

Fig. 13.8. Effect of antioxidants on the inhibition of TBARS in chilled (4 °C) turkey burgers.

Table 13.E. Sensory evaluation results of chilled (4 °C), roasted turkey burgers during storage

Treatment	Sensory Evaluation
Control, nothing added	Strong warmed-over flavor (WOF), resulting in detectable meat-flavor deterioration (MFD) after just 24 hours in storage.
Green tea extract (40 ppm catechins)	The meat flavor was maintained and no signs of warmed-over flavor (WOF) appeared during the 13-day storage period.
BHA (40 ppm)	Slight meat-flavor deterioration (MFD) detected after 6 days in storage.

during 13 days of storage. Figure 13.9 compares the antioxidant ability obtained from 150 ppm green tea extract with that obtained from 375 ppm rosemary extract.

Besides its antioxidant activity and natural flavor attributes, green tea extract can also improve the color stability of fresh meat products. The color of fresh meat depends on myoglobin, which stores oxygen for aerobic metabolism in the muscle. Iron is another important factor in meat color, particularly its chemical state. Oxymyoglobin gives meat its red color, and a brownish color is due to metmyoglobin formed by the oxidation of oxymyoglobin. How green tea extract influences the color cycle in meat products is not well understood. Internal testing of samples have shown that the red color of raw ground beef patties (20% fat) significantly improved with the addition of rosemary extract (1000 ppm) and green tea extract (250 ppm), compared to browning in control samples (data not shown). All sam-

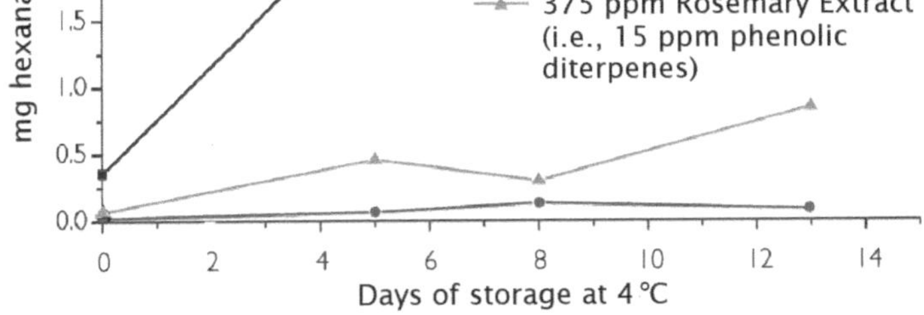

Fig. 13.9. Effect of green tea extract and rosemary extract on the inhibition of hexanal in chilled (4 °C), roasted beef burgers.

ples were stored at 0–2 °C for 22 days and packaged under modified atmosphere (80% oxygen and 20% carbon dioxide) conditions. The color "a" value of meat that was measured using Minolta colorimeter corresponded well with the development of the visible red color (Fig. 13.10). The meat had lost virtually all visible red color when the "a" value reached approximately 12. The control reached this point around day 9, and the treated samples around day 16. This study showed that 1000 ppm rosemary extract and 250 ppm green tea extract protected color well for several days beyond the control sample, and minimal difference existed between treatments. Sensory evaluation of the burgers showed that green tea extract had no flavor impact on the finished burgers, whereas a slight rosemary flavor could be detected in the samples containing rosemary extract (data not shown). In flavor-sensitive meat systems, green tea extract would be a more viable option than rosemary extract as it is just as effective but has less impact on flavor.

Several investigators have demonstrated the antioxidant properties of green tea catechins in various lipid and food model systems (Frankel et al., 1997; Huang & Frankel, 1997; Koketsu & Satoh, 1997; Roedig-Penman & Gordon, 1997; Wanasundara & Shahidi, 1998). When tea catechins were added to noodles and frying oils, they improved the oxidative stability of the fried product and the oil used for frying (Frankel et al., 1997; Koketsu & Satoh, 1997). Wanasundara & Shahidi (1998) investigated the antioxidant and prooxidant effects of green tea extract in marine oils, such as seal blubber oil (SBO) and menhaden oil (MHO). Green tea extract exhibited a prooxidant effect in both oils examined, perhaps due to the catalytic effect of their chlorophyll constituents. Therefore, follow-up experiments

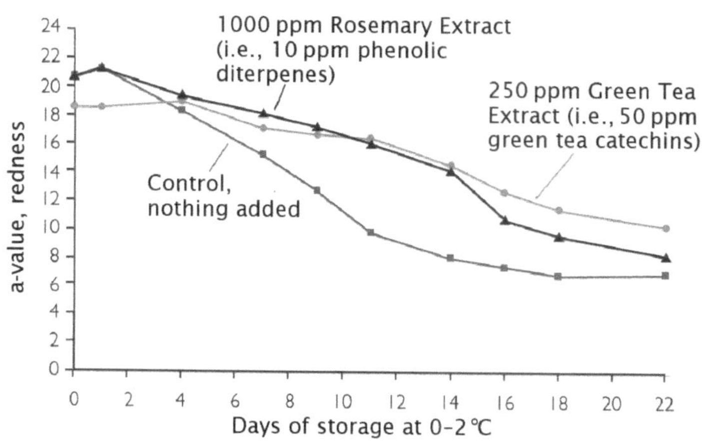

Fig. 13.10. Effect of green tea extract and rosemary extract on the color stability of chilled, MAP packed, raw ground beef patties.

employed a column chromatographic technique to remove chlorophyll from green tea extract. The resultant dechlorophyllized extract was applied at various levels to both SBO and MHO. The antioxidant activity of dechlorophyllized green tea extract was compared with the effects of commonly used antioxidants, such as BHA, butylated hydroxytoluene (BHT), *tert*-butylhydroquinone (TBHQ), and α-tocopherol. Dechlorophyllized green tea extract exhibited excellent antioxidant activity in both oils at ≥200 ppm, and its efficacy was higher than that of BHA (at 200 ppm), BHT (at 200 ppm), and α-tocopherol (at 500 ppm) but less than that of TBHQ (at 200 ppm). Another study examined the antioxidant activity of tea polyphenols in lard and edible oils (Koketsu, 1997; Koketsu & Satoh, 1997). Tea polyphenols clearly suppressed the oxidation of lard, and the effect depended on the concentration of catechins. Furthermore, tea polyphenols had a distinct inhibitory effect on rancidity in soybean oil (Koketsu, 1997). When compared with BHA and tocopherols, green tea catechins had better antioxidant activity in fish oil (Koketsu & Satoh, 1997).

Green tea catechins have been reported to have antibacterial and antiviral properties. Kumudavally and colleagues (2008) examined the feasibility of using green tea extract to extend the shelf life of fresh mutton at ambient storage conditions (25 °C; 85% RH). Their results indicated that green tea extract had a profound bacteriostatic effect on meat spoilage organisms. This finding suggests that the antibacterial components of green tea extract support its practical use for extending the shelf life of fresh mutton for up to four days at ambient conditions without adverse effects on its physical, chemical, and organoleptic properties. Meanwhile, control samples without green tea extract showed signs of spoilage within a day. Green tea catechins also function as inhibitors of key enzymes important for growth and proliferation of microorganisms. Epigallocatechin gallate effectively inhibited protein tyrosine phosphatase (PTPase) activity in *Proveta intermedia* and related species (Okamoto et al., 2003). Studies have also demonstrated that green tea catechins exhibit antibacterial properties against a variety of gram-negative and gram-positive species (Chou et al., 1999). A U.S. patent claims that chewing gum containing tea polyphenols prevents viral infections and inhibits the dissemination of influenza (Hara & Nakayama, 2001).

Conclusion

In recent years, consumers and food manufacturers have opted for products with clean, natural labels. In some cases, natural plant extracts have the advantage of being labeled as spices or natural flavors. Although synthetic antioxidants have had wide use to combat lipid oxidation in the food industry, the current trend is to decrease their usage because of growing consumer health concerns over such chemical additives. Consequently, much emphasis has been given to the use of natural extracts in food. In particular, plant-based extracts, such as those derived from rosemary and green tea, have been used commercially to extend the shelf life of various food and feed products. Today, research focuses on other plant-derived extracts that are part of our modern diet. However, more research is needed to investigate emerging trends, such as the identification, further purification, and incorporation of new natural extracts in food. Hence, the discovery of novel sources of natural extracts is essential to effectively utilize these products as sources for natural antioxidants on a commercial scale.

References

Aldrich, G. Rosemary Extract Acts as Natural Antioxidant. http://www.petfoodindustry.com (2010).

Balentine, D. A.; Paetau-Robinson, I. Tea as a Source of Dietary Antioxidants with a Potential Role in Prevention of Chronic Diseases. Herbs, botanicals, and teas; G. Mazza, B. D. Oomah, Eds.; Technomic Publishing Co., Inc: Lancaster, PA, 2000; 265–287.

Bors, W.; Michel, C. Antioxidant capacity of flavanols and gallate esters: Pulse radiolysis studies. Free Radical Bio. Med. 1999, 27, 1413–1426.

Bracco, U.; Löliger, J.; Viret, J.-L. Production and use of natural antioxidants. J. Am. Oil Chem. Soc., 1981, 58, 686–690.

Cabrera, C.; Artacho, R.; Giménez, R. Beneficial effects of green tea—A review. J. Am. Coll. Nutr. 2006, 25, 79–99.

Chou, C. C.; Lin, L. L.; Chung, K. T. Antimicrobial activity of tea as affected by the degree of fermentation and manufacturing season. Int. J. Food Microbiol. 1999, 48, 125–130.

Chu, D.-C. Green Tea—Its Cultivation, Processing of the Leaves for Drinking Materials, and Kinds of Green Tea. Chemistry and Applications of Green Tea; Juneja, L. R., Chu, D.-C., Kim, M., Eds.; CRC Press: Boca Raton, FL, 1997; 1–11.

Chu, D.-C.; Juneja L. R. General chemical composition of green tea and its infusion. Chemistry and applications of green tea; Yamamoto, T., Juneja, L. R., Chu, D. C., Kim, M. CRC Press: New York, 1997; 13–22.

Frankel, E. N.; Huang, S.-W.; Aeschbach, R. Antioxidant activity of green teas in different lipid systems. J. Am. Oil Chem. Soc. 1997, 74, 1309–1315.

Frankel, E. N.; Huang, S.-W.; Aeschbach, R.; Prior, E. Antioxidant activity of a rosemary extract and its constituents, carnosic acid, carnosol, and rosmarinic acid, in bulk oil and oil-in-water emulsion. J. Agric. Food Chem., 1996a, 44, 131–135.

Frankel, E. N.; Huang, S.-W.; Kanner, J.; German, J. B. Interfacial phenomena in the evaluation of antioxidants: Bulk oils vs. emulsions. J. Agric. Food Chem. 1994, 42, 1054–1059.

Frankel, E. N.; Huang, S.-W.; Prior, E.; Aeschbach, R. Evaluation of antioxidant activity of rosemary extracts, carnosol and carnosic acid in bulk vegetable oils and fish oil and their emulsions. J. Sci. Food Agric. 1996b, 72, 201–208.

Fukumoto, L. R.; Mazza, G. Assessing antioxidant and prooxidant activities of phenolic compounds. J. Agric. Food Chem. 2000, 48, 3597–3604.

Graham, H. N. Green tea composition, consumption, and polyphenol chemistry. Prev. Med. 1992, 21, 334–350.

Güntensperger, B.; Hämmerli-Meier, D. E.; Escher, F. E. Rosemary extract and precooking effects on lipid oxidation in heat-sterilized meat. J. Food Sci. 1998, 63, 955–957.

Hara, Y.; Nakayama, M. Chewing gum and production of the same. U.S. Patent 6,248,346, 2001.

Haraguchi, H.; Saito, T.; Okamura, N.; Yagi, A. Inhibition of lipid peroxidation and superoxide generation by diterpenoids from Rosmarinus officinalis. Planta Medica 1995, 61, 333–336.

Hidalgo, P. J.; Ubera, J. L.; Tena, M. T.; Valcarcel, M. Determination of the carnosic acid content in wild and cultivated Rosmarinus officinalis. J. Agric. Food Chem. 1998, 46, 2624–2627.

Hirose, Y.; Yamamoto, H.; Nakayama, M. Oxidation product of (+)-catechin from lipid peroxidation. Agric. Biol. Chem. 1990, 54, 567–569.

Huang, S.-W.; Frankel, E. N. Antioxidant activity of tea catechins in different lipid systems. J. Agric. Food Chem. 1997, 45, 3033–3038.

Huang, S.-W.; Frankel, E. N.; German, J. B.; Aeschbach, R. Partition of selected antioxidants in corn oil-water model systems. J. Agric. Food Chem. 1997, 45, 1991–1994.

Jaswir, I.; Che Man, Y. B. Use optimization of natural antioxidants in refined, bleached, and deodorized palm olein during repeated deep-fat frying using response surface methodology. J. Am. Oil Chem. Soc. 1999, 76, 341–348.

Khokhar, S.; Venema, D.; Hollman, P. C.; Dekker, M.; Jonge, W. A RP-HPLC method for the determination of tea catechins. Cancer Lett. 1997, 114, 171–172.

Koketsu, M. Antioxidant Activity of Tea Polyphenols. Chemistry and Applications of Green Tea; Yamamoto, T., Juneja, L. R., Chu, D.-C., Kim, M., Eds.; CRC Press: Boca Raton, FL, 1997; 37–50.

Koketsu, M.; Satoh, Y. Antioxidative activity of green tea polyphenols in edible oils. J. Food Lipids 1997, 4, 1–9.

Kumudavally, K. V.; Phanindrakumar, H. S.; Tabassum, A.; Radhakrishna, K.; Bawa, A. S. Green tea—A potential preservative for extending the shelf life of fresh mutton at ambient temperature (25± 2°C). Food Chem. 2008, 107, 426–433.

Lalas, S.; Dourtoglou, V. Use of rosemary extract in preventing oxidation during deep-fat frying of potato chips. J. Am. Oil Chem. Soc. 2003, 80, 579–583.

Liu, Z. Q.; Ma, L. P.; Zhou, B.; Yang, L.; Liu, Z. L. Antioxidative effects of green tea polyphenols on free radical initiated and photosensitized peroxidation of human low density lipoprotein. Chem. Phys. Lipids 2000, 106, 53–63.

Löliger, J. Natural antioxidants. Lipid Technol. 1991, April–June, 58–61.

Love, J. D. Sensory analysis of warmed-over flavor in meat. Food Technol. 1988, 42, 140–143.

Madsen, H. L.; Bertelsen, G. Spices as antioxidants. Trends Food Sci. Technol. 1995, 6, 271–277.

Murakami, M.; Yamaguchi, T.; Takamura, H.; Matoba, T. Effects of ascorbic acid and α-tocopherol on antioxidant activity of polyphenolic compounds. J. Food Sci. 2003, 68, 1622–1625.

Nakao, M.; Takio, S.; Ono, K. Alkyl peroxyl radical scavenging activity of catechins. Phytochem. 1998, 49, 2379–2382.

Nanjo, F.; Goto, K.; Seto, R.; Suzuki, M.; Sakai, M.; Hara, Y. Scavenging effects of tea catechins and their derivatives on 1,1-diphenyl-2-picrylhydrazyl radical. Free Radical Bio. Med. 1996, 21, 895–902.

Obanda, M.; Owuor, P. O.; Mang'oka, R. Changes in the chemical and sensory quality parameters of black tea due to variations of fermentation time and temperature. Food Chem. 2001, 75, 395–404.

Official Journal of the European Union. Commission Directive 2010/69/EU. Amending the Annexes to European Parliament and Council Directive 95/2/EC on food additives other than colours and sweeteners. 2010, L 279/22.

Okamoto, M.; Leung K. P.; Ansai T.; Sugimoto A.; Maeda N. Inhibitory effects of green tea catechins on protein tyrosine phosphatase in *Prevotella intermedia*. Oral Microbiol. Immunol. 2003, 18, 192–195.

Perumalla, A. V. S.; Hettiarachchy, N. S. Green tea and grape seed extracts—Potential applications in food safety and quality. Food Res. Intl. 2011, 44, 827–839.

Richheimer, S. L.; Bailey, D. T.; Bernart, M. W.; Kent, M.; Vininski, J. V.; Anderson, L. D. Antioxidant activity and oxidative degradation of phenolic compounds isolated from rosemary. Recent Res. Dev. Oil Chem. 1999, 3, 45–58.

Richheimer, S. L.; Bemart, M. W.; King, G. A.; Kent, M. C.; Bailey, D. T. Antioxidant activity of lipid-soluble phenolic diterpenes from rosemary. J. Am. Oil Chem. Soc. 1996, 73, 507–514.

Roedig-Penman A.; Gordon, M. H. Antioxidant properties of catechins and green tea extracts in model food emulsions. J. Agric. Food Chem. 1997, 45, 4267–4270.

Saenz-Lopez, R.; Fernandez-Zurbano, P.; Tena, M. T. Capillary electrophoretic separation of phenolic diterpenes from rosemary. J. Chromatogr. 2002, 953, 251–256.

Schwarz, K.; Ternes, W. Antioxidative constituents of rosmarinus officinalis and salvia officinalis. I. Determination of phenolic diterpenes with antioxidative activity amongst tocochromanols using HPLC. Z. Lebensm. Unters. Forsch. 1992, 195, 95–98.

Spanier, A. M.; Edwards, J. F.; Dupuy, H. P. The warmed-over flavor process in beef. A study of meat proteins and peptides. Food Technol. 1988, 42, 110–118.

St. Angelo, A. J.; Crippen, K. L.; Dupuy, H. P.; James, C., Jr. Chemical and sensory studies of antioxidant treated beef. J. Food Sci. 1990, 55, 1501–1505.

Tang, S. Z.; Kerry, J. P.; Sheehan, D., Buckley, D. J. Antioxidative mechanisms of tea catechins in chicken meat systems. Food Chem. 2002, 76, 45–51.

Thorsen, M. A.; Hildebrandt, K. S. Quantitative determination of phenolic diterpenes in rosemary extracts. Aspects of accurate quantification. J. Chromatogr. 2003, 995, 119–125.

Tims, M. J.; Watts, B. M. Protection of cooked meats with phosphates. Food Technol. 1958, 12, 240–243.

Ucak, I.; Ozogul, Y.; Durmus, M. The effects of rosemary extract combination with vacuum packing on the quality changes of Atlantic mackerel fish burgers. Intl. J. Food Sci. Technol. 2011, 46, 1157–1163.

Unten, L.; Koketsu, M.; Kim, M. Antidiscoloring activity of green tea polyphenols on β-carotene. J. Agric. Food Chem. 1997, 45, 2009–2012.

Vega, J. D.; Brewer, M. S. Detectable odor thresholds of selected lipid oxidation compounds at various temperatures in a gelatin model system. J. Food Lipids 1994, 1, 229–245.

Wanasundara, P. K. J. P. D.; Shahidi, F. Science, Technology, and Applications. Bailey's Industrial Oil and Fat Products; Shahidi, F., Ed.; John Wiley & Sons, Inc.: Hoboken, NJ, 2005; 431–489.

Wanasundara, U. N.; Shahidi, F. Antioxidant and pro-oxidant activity of green tea extracts in marine oils. Food Chem. 1998, 63, 335–342.

Yilmaz, Y. Novel uses of catechins in foods. Trends Food Sci. Technol. 2006, 17, 64–71.

Zaveri, N. T. Green tea and its polyphenolic catechins: Medicinal uses in cancer and noncancer applications. Life Sci. 2006, 78, 2073–2080.

Zhu, N.; Sang, S.; Huang, T.; Bai, N.; Yang, C. S.; Ho, C.-T. Antioxidant chemistry of green tea catechins: Oxidation products of (–)-epigallocatechin gallate and (–)-epigallocatechin with peroxidase. J. Food Lipids 2000, 7, 275–282.

CHAPTER 14

Using Natural Plant Extracts to Delay Lipid Oxidation in Foods

Kristen Robbins[1] and William Schroeder[1]

Introduction

Autooxidation of unsaturated fatty acids forms volatile compounds that cause rancidity, which changes the color, flavor, and odor of foods. These changes significantly impact purchasing decisions made by consumers. Currently, consumer preferences trend toward more natural products and ingredients that are perceived as more easily recognizable on labels. Consumer demand is a key driver in the shift from traditional synthetic antioxidants to natural plant extracts, which have antioxidant qualities. It is difficult to meet all consumer demands; however, the goal of reducing oxidation and delivering quality food products can be accomplished using plant-derived extracts from botanicals such as rosemary, green tea, oregano, and sage. Identifying which natural plant extracts deliver the desired product shelf life requires food manufacturers to consider such aspects as regulatory constraints and the extract's flavor threshold and solubility.

Several plant-derived extracts are known for their antioxidant behavior. Two well-known botanicals are rosemary (*Rosmarinus officinalis*) and green tea (*Camellia sinensis*). Rosemary is commonly associated with two antioxidant molecules, carnosic acid and carnosol (see Fig. 14.1), which are typically extracted with non-polar solvents. Carnosic acid and carnosol are free-radical sequestering antioxidant molecules. Another antioxidant molecule associated with rosemary extract is

[1]Kemin Industries, Inc., 2100 Maury St., Des Moines, IA, USA.

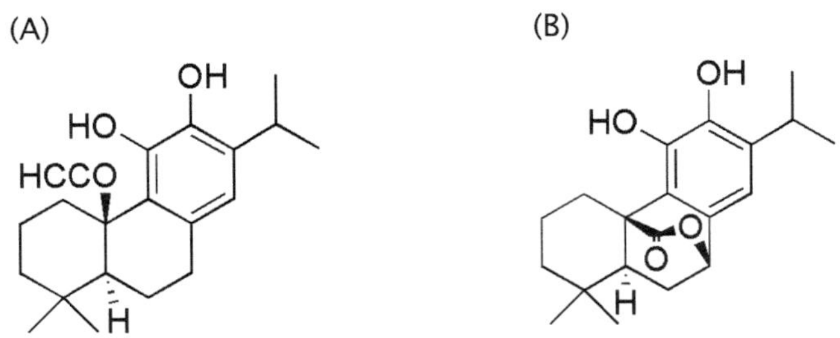

Fig. 14.1. Phenolic diterpenes in rosemary extract. (A) carnosic acid; (B) carnosol.

rosmarinic acid (see Fig. 14.2). Rosemary contains only small amounts of rosmarinic acid; however, rosmarinic acid is found in higher concentrations in lemon balm (*Melissa officinalis*) and spearmint (*Mentha spicata*). The antioxidant activity of green tea extracts is generally attributed to the water-soluble flavan-3-ols collectively known as catechins. Fig. 14.3 depicts several catechins present in green tea extract. These water-soluble compounds differ in their stereochemistry and esterification with gallic acid. Catechin and its epimer, epicatechin (EC), contain the fewest number of phenolic hydrogens available to quench free radicals. Another catechin present in green tea extract is epicatechin gallate, which is the ester of EC and gallic acid. Epigallocatechin (EGC) contains one more phenolic hydroxyl group than EC. The ester of EGC and gallic acid, epigallocatechin gallate (EGCG), is typically the most predominant catechin, and it has the highest theoretical antioxidant activity because of the additional phenolic hydrogens from the gallic acid. The performance of these plant-derived extracts in meat, poultry, and fish will be reviewed later in this chapter.

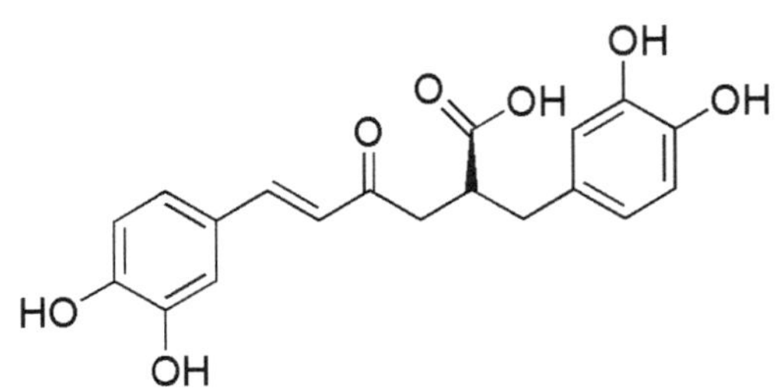

Fig. 14.2. Rosmarinic acid.

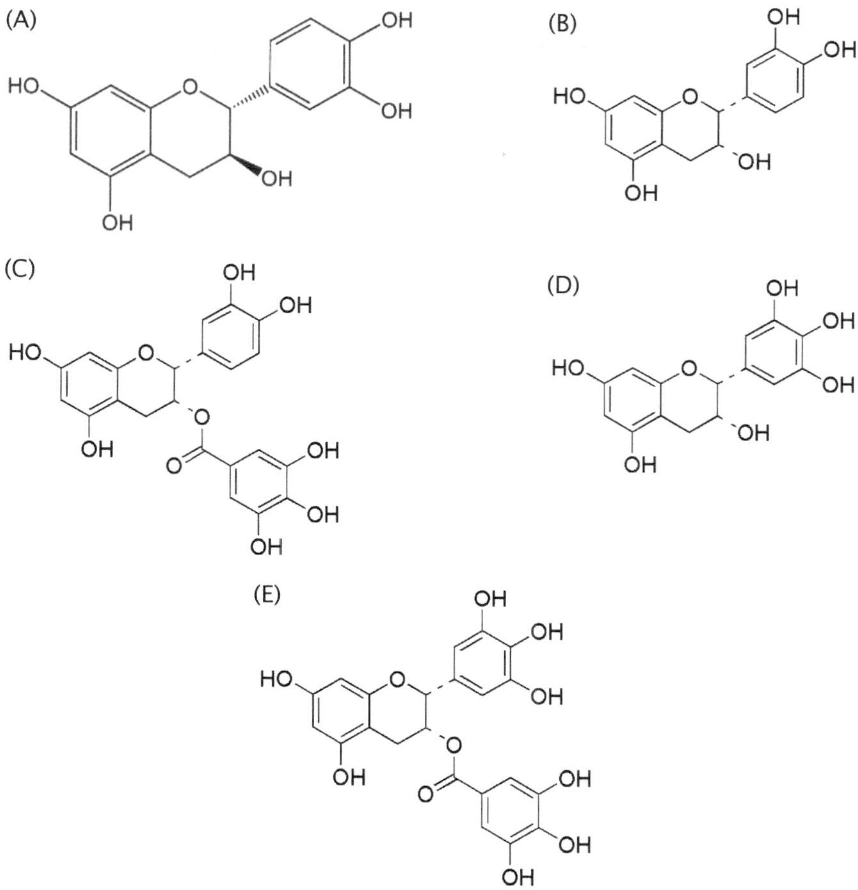

Fig. 14.3 Catechins in green tea extract. (A) catechin; (B) epicatechin; (C) epicatechin gallate; (D) epigallocatechin; (E) epigallocatechin gallate.

Factors Motivating the Shift from Synthetic Antioxidants to Natural Plant Extracts

Decreased Consumer Demand

There has been a global shift away from synthetic antioxidants, such as BHT, BHA, PG, and TBHQ. Consumers want to see easily recognizable ingredients on packaging labels so that they can understand exactly what is in their food. From 2000 to 2010, the global volume of synthetic antioxidants in food products decreased 0.5% (Euromonitor, 2012). This relatively slow reduction occurred primarily because of populous developing countries where consumers are unwilling and unable to pay premiums for natural alternatives. The European Union (EU) was the global leader in the shift toward natural antioxidants, decreasing its synthetic antioxidant use by more than 3% (Euromonitor, 2012). This trend is expected to continue, and it is predicted that global use of synthetic antioxidants will decrease another 0.3% from 2010 to 2015 (Euromonitor, 2012). The most common food categories for synthetic antioxidant use are sweet cookies and biscuits, crackers, instant rice/noodles/pasta, cakes, pastries and sweet goods, oils, cold cereals, gum, pizza, and energy bars. (Mintel GNPD, 2012).

Regulatory

Because of the limitations on synthetic antioxidants in foods (see Tables 14.A and 14.B), there are situations in which a plant-derived extract with antioxidant qualities may be used to supplement a synthetic antioxidant for additional protection. In two studies (Romano et al., 2006, 2009) that used DPPH to measure antioxidant capacity, the combination of BHT or BHA with rosemary extract appeared to not only increase the antioxidant capacity based on the two components, but there was also a synergistic interaction between the natural and synthetic antioxidants. The authors confirmed this finding by measuring the time needed to quench the initial level of DPPH by 50% (EC_{50}) of the separate ingredients and then compared these intervals to the different combined ingredients.

Sustainability

Another driving force in the trend toward plant-derived antioxidants is sustainability. Since 2000, more than 1,000 companies have published reports on their performance under this measure, and sustainability trends are predicted to increase from 2010 to 2015 based on the reassessment of chemical safety for food, beverage, and agricultural production (Frost & Sullivan, 2010). Also, customers increasingly support companies that use sustainable resources, and as the desire to include health claims increases, it drives the demand for natural, sustainable products (Frost & Sullivan, 2010). Natural plant extracts can provide sources for phyto-factories, allowing the sustainable manufacture of fine chemicals. Plants use solar energy to produce a target molecule while removing carbon dioxide from the atmosphere at the same time. After extract production, the spent biomass is returned to the earth so that resources are continuously recycled.

Table 14.A. U.S. FDA Limits and Tolerances for Synthetic Antioxidants in Select Food Products*

Reference	Application	Antioxidant	Limit (percent weight of food unless noted)
21 CFR 182.3169	general	BHA	0.02% singly or in combination of fat or oil portion of food including the essential oil, except where prohibited by Standard of Identity
21 CFR 182.3173		BHT	
21 CFR 172.185		TBHQ	
21 CFR 184.1660		PG	
Specific Foods			
21 CFR 172.615	chewing gum base	BHA, BHT, PG	0.1% singly or in combination
21 CFR 172.5(15	synthetic flavorings	BHA	0.5% of essential (volatile) oil
21 CFR 172.110	active dry yeast	BHA	0.1%
	beverages and desserts prepared from dry mixes		0.0002%
	dry mixes for beverages and desserts		0.009%
	dry diced glazed fruit		0.0032%
21 CFR 172.110 and/or 21 CFR 172.115	dry breakfast cereals	BHA and/or BHT	0.005%
	emulsion stabilizers for shortening		0.02%
	potato granules		0.001%
	potato flakes, sweet potato flakes, and dehydrated potato shreds		0.005%
Standardized Foods			
21 CFR 137.350	enriched parboiled rice	BHT	0.0033%
21 CFR 161.175	frozen raw breaded shrimp	BHA, BHT	0.02% of fat or oil content
21 CFR 166.110	margarine	BHA, BHT, PG	0.02% singly or in combination based on finished product

*Regulations subject to change

FDA (2012).

Table 14.B. USDA-FSIS Limits and Tolerances for Synthetic Antioxidants in Select Meat and Poultry Products*

Meat and Poultry Products			
Reference	**Application**	**Antioxidant**	**Limit (percent weight of food unless noted)**
9 CFR 424.21	dry sausage	BHA,BHT, TBHQ, PG†	0.003% singly, 0.006% in combination, with no antioxidant exceeding 0.003%
	fresh pork and/or beef sausage, brown-and-serve sausage, pre-grilled beef patties, pizza toppings, and meatballs		0.01% singly, 0.02% in combination, with no antioxidant exceeding 0.01%, based on fat content
	dried meats		0.01% singly, or in combination
	rendered animal fat or a combination of such fat and vegetable fat		0.01% singly, 0.02% in combination, with no antioxidant exceeding 0.01%
	various poultry products		0.01% singly, 0.02% in combination, with no antioxidant exceeding 0.01% based on fat content

*Regulations subject to change
†TBHQ and PG cannot be used in combination

FDA (2012).

Use of Natural Plant Extracts in Food

Meat, Poultry, and Fish

The literature has reviewed many studies that demonstrated the ability that plant extracts have to delay oxidation. A non-exhaustive list of plant extract sources and selected applications is available in Table 14.C.

Nuñez de Gonzalez and colleagues (2008a) found that 3% dried plum puree (TBARS = 0.44 mg/kg) was as effective as the combination of 0.01% BHA and 0.01% BHT (fat basis) (TBARS = 0.39 mg/kg) at retarding lipid oxidation in

Table 14.C. Plant-Derived Extracts Used to Delay Lipid Oxidation in Select Muscle Food Applications

Additive	Applications
Grape seed	cooked beef patties[1], pork patties (cooked[1], cooked-frozen[2])
Green tea	fresh pork (sausage[3], patties[4]), chicken patties (fresh[5], cooked[5,6]), beef patties (fresh[5], cooked[5,6])
Oregano	cooked beef patties[1], pork patties (cooked[1], cooked-frozen[2])
Plum	pork sausage[7] (fresh, cooked), cooked roast beef[8]
Rosemary	fresh beef (ground[9], steak[10]), cooked beef (patties[1],meatballs[11]) pork sausage (fresh[3], raw-frozen[12], cooked-frozen[12]), pork patties (cooked[1], cooked-frozen[2]), cooked chicken patties[13], frozen salmon patties[14], frozen trout flakes[15]
Rosemary and green tea	fresh ground pork[4], fresh beef steak[10], cooked chicken patties[1]
α-tocopherol	cooked beef patties[6], cooked chicken patties[6]

[1]Rojas & Brewer, 2007
[2]Sasse et al., 2009
[3]Martinez et al., 2006
[4]Robbins, 2009
[5]Mitsumoto et al., 2005
[6]Tang et al., 2001
[7]Nuñez de Gonzalez et al., 2008a
[8]Nuñez de Gonzalez et al., 2008b
[9]Balentine et al., 2006
[10]Robbins, 2010
[11]Fernandez-Lopez et al., 2005
[12]Sebranek et al., 2005
[13]Robbins, 2011
[14]Tironi et al., 2010
[15]Boyd et al., 1993

cooked pork sausage refrigerated for 28 days. The 3% dried plum puree treatment also maintained low TBARS values for cooked pork sausage frozen for 90 days (TBARS = 0.95 mg/kg), which were similar to those found in the cooked-frozen pork patties treated with the BHA/BHT combination (TBARS = 1.05 mg/kg). The consumer sensory evaluations for overall like/dislike concluded that the pork sausage patties treated with 3% dried plum puree were as well liked as the untreated control and the BHA/BHT treated patties. In another study by Nuñez de Gonzalez et al. (2008b), TBARS for refrigerated-precooked roast beef treated with either 2.5% fresh plum juice concentrate (0.26 mg/kg) or 2.5% dried plum juice concentrate (0.39 mg/kg) had lower ($P < 0.05$) TBARS than an untreated control (0.62 mg/kg) after 21 days of storage. The authors attributed dried plum concentrate's ability to improve meat quality to its high content of phenolic compounds, although the phenolic composition was not specified by the authors.

Table 14.D Visual Color Evaluation During Frozen Storage of Raw Pork Sausage.*

	Days of storage								
Treatment	**1**	**14**	**28**	**42**	**56**	**70**	**84**	**98**	**112**
Control	3.5^{ab}	6.4^{b}	6.3^{bc}	8.7^{c}	11.1d	9.8^{b}	10.0^{b}	9.3^{b}	10.1^{b}
BHA/BHT	5.0^{bc}	6.6^{b}	5.8^{bc}	8.4^{c}	7.8^{c}	9.3^{b}	9.5^{b}	5.6^{a}	10.3^{b}
RE 0.15%	6.3^{c}	6.4^{b}	7.1^{c}	9.0^{c}	5.7^{b}	4.1^{a}	5.0^{a}	7.1^{a}	6.3^{a}
RE 0.25%	2.8^{a}	3.4^{a}	3.7^{a}	3.2^{a}	3.3^{a}	3.5^{a}	4.7^{a}	6.1^{a}	5.3^{a}
SEM	2.7	3.2	2.7	3.0	3.5	3.5	3.4	3.0	3.5

$^{a\text{-}d}$Means within a column with different superscripts are significantly different ($P < 0.05$).
*Pink = 0, Brown = 15. RE = rosemary extract. SEM = standard error of the mean.
Reprinted with permission from Sebranek et al. (2005).

Sebranek and colleagues (2005) found that raw-frozen pork sausage treated with 0.25% (fat basis) rosemary extract had lower ($P < 0.05$) TBARS than sausages treated with a combination of 0.01% BHA and 0.01% BHT (fat basis) after 28 days of a 112-day study. After 42 days of frozen storage, the pork sausage treated with 0.15% rosemary extract had lower ($P < 0.05$) TBARS than the BHA/BHT-treated sausage. A sensory panel evaluation of visual color (Table 14.D) concluded that pork sausage treated with 0.25% rosemary extract browned less ($P < 0.05$) than the untreated and BHA/BHT-treated sausage throughout the study. Furthermore, sausage with 0.15% rosemary extract browned less ($P < 0.05$) than the BHA/BHT–treated sausage from 56 days onward, which confirms that rosemary extract was more effective than the synthetic antioxidant treatment in maintaining the characteristic pink color of raw-frozen pork sausage.

In a study by Boyd and colleagues (1993), cooked, flaked trout was treated with either a combination of TBHQ (0.05%) and ascorbic acid (2%) or rosemary extract (0.25%) and was frozen at –20 °C for 90 days. Although the TBHQ and

Table 14.E Oxidative Deterioration of Cooked Trout Flakes as Noted by Changes in TBA Number

Treatment	**TBA (mg/kg)**
Untreated	1.60^{c}
0.25% rosemary extract	0.74^{b}
0.05% TBHQ-2% AA	0.49

$^{a\text{-}c}$Means within a row not having a common letter are significantly different ($P < 0.05$)
Reprinted with permission from Boyd et al (1993).

Table 14.F Effects of Treatments of Ranked Panel Scores for Aroma

	Aroma Attributes		
Treatment	**Green**	**Fish Oil**	**Chemical**
Untreated	28.8[b]	64.5[a]	29.8[b]
0.25% rosemary extract	58.5[a]	41.6[b]	31.0[b]
0.05% TBHQ-2% AA	33.7[b]	38.8[b]	65.5[a]

[a-b]Means within a row not having a common letter are significantly different ($P < 0.05$)
Reprinted with permission from Boyd et al. (1993). Copyright 1993 Wiley.

ascorbic acid treatment was more ($P < 0.05$) effective than rosemary extract in preventing the formation of TBARS (Table 14.E), rosemary extract was more effective than the untreated control. The sensory results revealed an advantage of the rosemary extract treatment because those samples had less ($P < 0.05$) chemical odor and chemical flavor (see Tables 15.F–G) than the TBHQ-treated fish flakes. The rosemary-extract treatment resulted in a stronger ($P < 0.05$) green flavor and aroma compared with the other samples. The frozen trout treated with TBHQ and ascorbic acid had a more ($P < 0.05$) sour flavor, which was likely caused by the ascorbic acid. Both treatments resulted in less ($P < 0.05$) fish-oil flavor than the untreated trout flakes, indicating that the treatments helped to delay the formation of off-flavors characteristic of oxidized fish.

In frozen, minced sea-salmon (*Pseudopercis semifasciata*) patties, treatment with 0.02% or 0.05% rosemary extract reduced both lipid oxidation and color loss (Tironi et al., 2010). An aqueous dispersion (10% w/w) of rosemary extract was sprayed onto the surface of the minced salmon, and the patties were frozen (–11 ± 2 °C) for nine months.

Patties treated with 0.05% rosemary extract experienced no increase in the TBA value during the first six months of storage. The maximum TBA value was

Table 14.G Effects of Treatments of Ranked Panel Scores for Flavor

	Flavor Attributes			
Treatment	**Green**	**Fish Oil**	**Sour**	**Chemical**
Untreated	28.9[b]	60.6[a]	32.5[b]	34.1[b]
0.25% rosemary extract	55.7[a]	34.9[b]	35.3[b]	34.2[b]
0.05% TBHQ-2% AA	35.2[b]	42.0[b]	59.5[a]	58.1[a]

[a-b]Means within a row not having a common letter are significantly different ($P < 0.05$)
Reprinted with permission from Boyd et al. (1993). Copyright 1993 Wiley.

reached at 7.5 months, and it was half as high as the maximum TBA value achieved by the untreated samples. The TBA values of the 0.02% rosemary extract–treated samples did not increase for the first two months of frozen storage, and after reaching their maximum TBA value at three months, this value was still only half as high as the maximum TBA value of the untreated samples.

Fernandez-Lopez and colleagues (2005) reported that rosemary extract retarded lipid oxidation caused by cooking in Swedish-style beef meatballs, and overall sensory acceptability improved in relation to a control during 12 days of refrigerated storage. The authors compared the effects of untreated meatballs and an oil-soluble rosemary extract (0.1%, w/w), a water-soluble rosemary extract (0.15%), a water and oil–soluble rosemary extract (0.25%), lemon fiber (5%), and orange fiber (5%). The meatballs were stored in the dark under refrigeration (8 ± 1 °C) for 12 days. On Day 1, the TBA values of all treated meatballs were lower ($P < 0.05$) than the control, which suggested that the extracts retarded the lipid oxidation that occurred during the cooking process. On Day 12, the TBA values for all treatments were lower ($P < 0.05$) than those of the control. Furthermore, meatballs that received the three rosemary-extract treatments had the lowest ($P < 0.05$) TBA values at each sampling period. Samples treated with the oil-soluble rosemary extract and the water-soluble rosemary extract did not increase in TBA values ($P > 0.05$) during the storage period, remaining below 1.5 mg/kg. The lemon fiber–treated meatballs had higher ($P < 0.05$) TBA values than the orange fiber–treated meatballs on Day 12, but the TBA values were similar during the earlier portion of the storage period. The sensory panelists found no differences between the treatments for overall acceptability and rancid aroma on Days 1 and 7. As expected, overall acceptability scores inversely correlated with rancid aroma scores and TBA values.

Oregano extract and grape-seed extract are additional examples of water-soluble extracts with demonstrated success delaying lipid oxidation in meats. Oregano extract (0.02%) containing 5–10% rosmarinic acid delayed lipid oxidation as well as 0.01% grape-seed extract (89% proanthocyanidins) in both cooked beef and pork patties that contained 2% salt (Rojas & Brewer, 2007). The ground beef and ground pork patties were treated with oregano extract (0.02%) or grape-seed extract (0.01% and 0.2%) and frozen (−18 °C) for one week. After the week of frozen storage, they were thawed, cooked, and refrigerated for eight days while chemical and sensory tests were used to assess the effects of treatment. The TBARS values of the beef patties treated with either level of grape-seed extract were lower ($P < 0.05$) than the control during the entire eight-day storage period. The relative difference between the TBARS values of the treatments and the control patties increased during storage (Table 14.H) as lipid oxidation intensified during storage. The TBARS values of samples containing oregano extract were lower ($P < 0.05$)

Table 14.H. Percent Reduction in TBARS Values Obtained with Each Treatment Compared to the Untreated Control for Each Species and Storage Time

	Beef		Pork	
Treatment	**Day 6**	**Day 8**	**Day 4**	**Day 6**
0.01% grape-seed extract	29.5	43.4	31.0	31.0
0.02% grape-seed extract	36.5	41.2	45.7	50.8
0.02% oregano extract	23.9	36.0	2.4	30.0

than the control on Days 2, 6, and 8. The sensory panel found no significant differences in the off-odor intensities of the patties during eight days of storage, but when the data were pooled over time, the patties treated with 0.02% grape-seed extract had lower rancid and wet cardboard off-odor scores.

The TBARS values of the pork patties treated with oregano extract were lower ($P < 0.05$) than those of the control on Days 2 and 6, and the patties treated with grape-seed extract remained lower ($P < 0.05$) than the control from Days 2 through 8. The sensory trends were similar to those from the beef study, although no significant differences emerged across the treatments in pork. Sasse and colleagues (2009) found that 0.02% grape-seed extract had similar success delaying lipid oxidation in ground pork patties frozen for six months. Although patties treated with grape-seed extract had lower L^* values than control patties, the sensory results did not indicate that the panelists detected a visual difference as they evaluated the intensity of green, gray, and tan hues for each sample. These studies showed that grape-seed extract, a by-product of grape-juice and wine processing, is a viable option for beef and pork processors seeking a plant-derived extract that is effective at low levels.

Proanthocyanidins, the active compounds in grape-seed extract, are an oligomeric form of flavan-3-ols, a class of compounds that includes catechins. Green tea catechins (0.03%) and 0.03% α-tocopherol both counteracted the prooxidant effects of 1% sodium chloride in cooked beef and chicken patties (Tang and colleagues 2001). Green tea extract had mixed success in other studies. Martinez and colleagues (2006) determined that 50–200 ppm green tea extract (unspecified catechin concentration) reduced ($P < 0.05$) lipid oxidation in fresh pork sausage but failed to delay color loss during refrigerated storage. Mitsumoto et al. (2005) added either 0.02% or 0.04% green tea extract (80.15% catechins) to ground beef strip loin and ground chicken breast, along with 1% sodium chloride. Patties, both raw and cooked, were refrigerated (4 °C) for seven days under fluorescent lighting (700 lx). Both green tea treatments reduced lipid oxidation (TBARS) in cooked ($P < 0.001$) and raw beef ($P < 0.05$) patties. The raw and cooked chicken patties remained relatively resistant to lipid oxidation during storage, so there was no treatment effect. Neither green tea extract treatment affected Hunter L or a values in raw patties, compared with the untreated controls. In the cooked patties pooled across species, the samples treated with green tea extract had lower sensory color scores ($P < 0.05$), *b* values ($P < 0.05$), and *a* values ($P < 0.01$). Focusing on the differences across meat type, the treatments had no effect ($P > 0.10$) on the overall acceptability of cooked beef patties. However, the corresponding scores for the cooked chicken patties with either green tea extract treatment were lower ($P < 0.05$) than those of the untreated patties, likely due to the influence of visual appearance. The authors suggested that discoloration was caused by the green tea binding with the heme iron in myoglobin.

Robbins (2009, 2010) presented a combination of rosemary and green tea extracts as a solution to such discoloration problems in fresh beef and pork. In fresh ground pork, the combination of rosemary and green tea extracts was more effective ($P < 0.05$) in delaying the decline in redness than the green tea extract and the untreated control. The difference became significant starting on Day 3 of a 14-day study (Robbins, 2009).

Maximizing the Performance of Plant-Derived Flavors with Antioxidant Qualities

To achieve the best results, antioxidants should be applied to food during the earliest processing stages, before prooxidants and catalysts have come in contact with food. Balentine and colleagues (2006) added 0.3% of a commercial rosemary extract during four different stages of ground beef processing to determine the impact of addition stage on myoglobin and lipid oxidation during 144 hours of lighted refrigerated storage. The rosemary extract was added at four stages: 12 cm × 12 cm trim pieces, 3-cm^3 cubed pieces, coarse ground meat (0.9525 cm plate), or fine ground meat (0.635 cm plate). The effects of the different extract addition stages are shown in Table 14.I.

The beef had the best ($P < 0.05$) color and the lowest ($P < 0.05$) TBARS values when the rosemary extract was added to the trim and cubed beef before grinding; however, the addition of rosemary extract at any stage improved ($P < 0.05$) the TBARS values over those of the untreated control. Furthermore, the oxymyoglobin content of the beef treated at the trim stage was higher ($P < 0.05$) than that of all other treatments, except the cube treatment. The authors hypothesized that the rosemary extract was more effective when added during the earliest processing

Table 14.1. Effect of Rosemary Extract Treatments Applied to Different Stages of Ground Beef on Mean Values of Color and TBARS through 144 Hours of Simulated Retail Storage

Attribute	Control	Trim	Cube	Coarse	Fine
CIE^{a*}	23.17^{b}	25.68^{a}	24.72^{a}	22.82^{b}	22.15^{b}
CIE^{b*}	21.92^{b}	23.74^{a}	23.48^{a}	22.37^{b}	22.02^{b}
630/580 nm	3.28^{bc}	3.86^{a}	3.54^{ab}	3.10^{c}	3.01^{c}
Sat. index†	32.19^{b}	35.16^{a}	34.33^{a}	32.23^{b}	31.47^{b}
TBARS (mg/kg)	4.75^{a}	0.78^{c}	0.83^{c}	1.15^{c}	2.77^{b}

$^{a\text{-}c}$Means within a row not having a common letter are significantly different ($P < 0.05$)
†Saturation index
Reprinted with permission from Balentine et al. (2006). Copyright 2006 Elsevier.

stages because it had contact with more surface area while being dispersed by the grinding process.

Hurdles Encountered When Using Plant Extracts

Flavor Threshold

Flavors derived from plant extracts effectively control lipid oxidation in food products, but their inherent herbal flavors can limit their application. The flavor threshold of plant extracts will vary based on the level of background flavor in the food. For example, highly seasoned sausage can accommodate higher levels of plant extracts than mild-flavored foods, such as mayonnaise. Plant-extract suppliers often use specialized extraction and processing techniques to create partially deodorized extracts with less flavor impact (Haworth, 2002). In some food products, the flavor imparted by the plant extract enhances the flavor of the food product. Liessens and Dooghe (2011) found that 0.25% rosemary extract contributed a noticeable herbal flavor to cooked chicken patties; however, sensory panelists viewed the herbal flavor as a positive attribute. Using sensory testing to pinpoint the extract's flavor threshold will help reduce the likelihood of flavor contribution. The triangle and difference-from-control tests are well suited for this purpose. The triangle test is appropriate if the sole objective is to determine whether a perceptible difference exists between two samples (Meilgaard et al., 1999). Panelists evaluate two untreated samples and one sample treated with the plant extract, and they choose the one they perceive as the odd sample. If a significant number of panelists are unable to choose correctly, the test should be repeated using higher extract dosages. The difference-from-control test offers a more expedient approach because multiple levels of plant extracts are evaluated in a single test. The results suggest whether a difference exists between the treated samples and a control, and they also indicate the magnitude of the differences (Meilgaard et al., 1999).

Solubility

Plant-based extracts are usually lipophilic, which presents challenges when these products are added to oil-in-water emulsions or aqueous foods. Manufacturing conditions sometimes require adding the extract to the water phase, which makes a hydrophilic rosemary extract containing rosmarinic acid the logical choice. Rosmarinic acid and carnosic acid, the predominant phenolic diterpene in lipophilic rosemary extract, inhibited hydroperoxides and hexanal formation in bulk corn oil (see Table 14.J); however, rosmarinic acid promoted oxidation in emulsified corn oil (Frankel et al., 1996). On the other hand, carnosic acid, carnosol, and rosemary extract (10.3% carnosic acid, 4.4% carnosol) were effective antioxidants in emulsified corn oil. The hypothesis was that the hydrophilic rosemary extract partitioned into the water phase rather than migrating to the oil–water interface, where oxidation was occurring.

Blending the lipophilic rosemary extract with various combinations of agents that modify solubility, such as lecithin, glyceryl monooleate, polyoxyethylene (20) sorbitan monooleate, and propylene glycol, helps to facilitate extract dispersion so that these compounds can be blended with either the oil- or water-phase of emul-

Table 14.J. Inhibition of Hydroperoxide and Hexanal Formation by a Rosemary Extract, Rosmarinic Acid, Carnosic Acid and Carnosol in Stripped Corn Oil and Emulsified Corn Oil Oxidized at 60 °C*

	Hydroperoxide (%)		Hexanal (%)	
Samples	**Bulk Oil (6 days)**	**Emulsion (4 days)**	**Bulk Oil (6 days)**	**Emulsion (4 days)**
Rosmarinic acid, 50 ppm	68	−11	78	−53
Carnosic acid, 50 ppm	70	37	29	83
Carnosol, 50 ppm	6	42	−20	84
Rosemary extract, 250 ppm	64	11	−3	35
Rosemary extract, 500 ppm	81	12	98	28

*Inhibition. Negative values represent prooxidant activity.

sions. Carriers such as maltodextrin, salt, and silicon dioxide are routinely used in powdered extracts, which can be dry-blended into powdered foods or added to liquid foods.

Regulatory Constraints

As mentioned earlier in this chapter and highlighted in Tables 14.A and 14.B, many regulations govern the use of synthetic antioxidants. In the EU, some plant-derived extracts also have usage limits. One of the antioxidants receiving recent attention is rosemary, specifically in regard to its approval as antioxidant E392 and compliance with Directives 2010/67/EU and 2010/69/EU. If a rosemary extract complies with the parameters outlined for extraction solvent, percent actives (carnosic acid + carnosol), residual solvents, and ratio of actives to volatiles outlined in Directive 2010/67/EU, then it is categorized by the EU as an antioxidant (E392). The application of these antioxidant extracts is dictated by Directive 2010/69/EU. For example, in meat and fish products, only 150 ppm (sum of carnosol + carnosic acid) of an E392 antioxidant can be used (based on fat weight), while non-E392 rosemary extract is permitted as a natural flavor with no maximum limit under Regulation (EC) 1334/2008.

In addition to regulations governing natural and synthetic antioxidants worldwide, there are also regulations pertaining to the emulsifiers and carriers used in the antioxidant products themselves. For example, there are regional differences in the regulations regarding the emulsifier polysorbate 80, its level of use,

Table 14.K. FDA and E.U. Regulations for Polysorbate 80 in Select Foods*

Application	Function	21 CFR 172.840	EU 1333/2008
Ice cream	emulsifier	Less than 0.1% total weight of finished product, alone, or in combination with polysorbate 65	0.1%
Yeast	defoamer	less than or equal to 4% of the finished defoamer and less than or equal to 4 ppm in the yeast itself	not permitted
Natural and artificial colors for use in barbecue sauce	surfactant/ wetting agent	less than or equal to 50 ppm of the total weight of barbecue sauce	0.1% in final food

*Regulations subject to change

EU (2008); FDA (2012).

the material in which it can be applied, and its functionality. Table 14.K shows the differences in restrictions among some of the applications of polysorbate 80. When developing a new product, the formulator must consider the regulations surrounding the active molecule, as well as other ingredients used in the formula. Despite the global shift toward plant-derived extracts, there are many factors to consider when using these ingredients, including processing conditions, solubility, carriers, flavor contribution, and regulations surrounding these aspects. As developing countries become more conscious of how chemical manufacturing impacts air, soil, and water quality, there might be increased interest in using natural plant extracts derived from sustainable agricultural sources. However, economic conditions and consumer demand remain important factors that drive the demand for natural solutions, so the shift away from synthetic antioxidants will be expected to progress at a modest pace for a majority of the world (Euromonitor, 2012).

References

Balentine, C. W.; Crandall, P. G.; O'Bryan, C. A.; Duong, D. Q.; Pohlman, F. W. The pre- and post-grinding application of rosemary and its effects on lipid oxidation and color during storage of ground beef. Meat Sci. 2006, 73, 413–421.

BHA. Code of Federal Regulations, Part 172.110, Title 21, http://ecfr.gpoaccess.gov (accessed April 11, 2012).

BHT. Code of Federal Regulations, Part 172.115, Title 21, http://ecfr.gpoaccess.gov (accessed April 11, 2012).

Boyd, L. C.; Green, D. P.; Giesbrecht, F. B.; King, M. F. Inhibition of oxidative rancidity in frozen cooked fish flakes by tert-butylhydroquinone and rosemary extract. J. Sci. Food Agric. 1993, 61, 87–93.

Butylated hydroxyanisole. Code of Federal Regulations, Part 182.3169, 182.3173, Title 21, http://ecfr.gpoaccess.gov (accessed April 11, 2012).

Chewing gum base. Code of Federal Regulations, Part 172.615, Title 21, http://ecfr.gpoaccess.gov (accessed April 11, 2012).

Commission Directive 2010/67/EU. Official Journal of the European Union, Oct 20, 2010.

Commission Directive 2010/69/EU. Official Journal of the European Union, Oct 22, 2010.

Enriched rice. Code of Federal Regulations, Part 137.350, Title 21, http://ecfr.gpoaccess.gov (accessed April 11, 2012).

Frankel, E. N.; Huang, S.W.; Aeschbach, R.; Prior, E. Antioxidant activity of a rosemary extract and its constituents, carnosic acid, carnosol, and rosmarinic acid, in bulk oil and oil-in-water emulsion. J. Agric. Food Chem. 1996, 44, 131–135.

Frost & Sullivan, 2020 Vision Global Food & Beverage Industry Outlook, 9833-88, March 2010.

Frozen raw breaded shrimp. Code of Federal Regulations, Part 161.175, Title 21, http://ecfr.gpoaccess.gov (accessed April 11, 2012).

Haworth, J. E. Method for removing essential oils and antioxidants from extract products of lamiaceae species using rolled film evaporation. U.S. Patent 6,450,935, Sept. 17, 2002.

Ingredients. Euromonitor International. http://www.euromonitor.com (accessed Feb. 24, 2012).

Liessens, M.; Dooghe, W. Kemin Food Technologies, Herentals, Belgium. Unpublished work, 2011.

Margarine. Code of Federal Regulations, Part 166.110, Title 21, http://ecfr.gpoaccess.gov (accessed April 11, 2012).

Martinez, L.; Cilla, I.; Beltran, J. A.; Roncales, P. Antioxidant effect of rosemary, borage, green tea, pu-erh tea and ascorbic acid on fresh pork sausages packaged in a modified atmosphere: Influence of the presence of sodium chloride. J. Sci. Food Agric. 2006, 86, 1298–1307.

Meilgaard, M.; Civille, G. V.; Carr, B. T. Sensory evaluation techniques. CRC Press: New York, 1999.

Mintel. Global new product database. http://www.mintel.com (accessed Feb. 24, 2012).

Mitsumoto, M.; O'Grady, M. N.; Kerry, J. P.; Buckley, D. J. Addition of tea catechins and vitamin C on sensory evaluation, colour and lipid stability during chilled storage in cooked or raw beef and chicken patties. Meat Sci. 2005, 69, 773–779.

Nuñez de Gonzalez, M. T.; Boleman, R. M.; Miller, R. K.; Keeton, J. T.; Rhee, K. S. Antioxidant properties of dried plum ingredients in raw and precooked pork sausage. J. Food Sci. 2008a, 73, H63–H71.

Nuñez de Gonzalez, M. T.; Hafley, B. S.; Boleman, R. M.; Miller, R. K.; Rhee, K. S.; Keeton, J. T. Antioxidant properties of plum concentrates and powder in precooked roast beef to reduce lipid oxidation. Meat Sci. 2008b, 80, 997–1004.

Polysorbate 80. Code of Federal Regulations, Part 172.840, Title 21, http://ecfr.gpoaccess.gov (accessed April 11, 2012).

Propyl gallate. Code of Federal Regulations, Part 184.1660, Title 21, http://ecfr.gpoaccess.gov (accessed April 11, 2012).

Regulation (EC) No 1333/2008 of the European Parliament and of the Council. Official Journal of the European Union, Dec. 16, 2008.

Regulation (EC) No 1334/2008 of the European Parliament and of the Council. Official Journal of the European Union, Dec. 16, 2008.

Robbins, K. Combining rosemary and green tea extract to offset the negative impact green tea imparts on meat color. Presented at the 62nd Reciprocal Meat Conference of the American Meat Science Association, Rogers, AR, June 21–24, 2009; Poster 17.

Robbins, K. Impact of rosemary and green tea extract on the color stability, flavor, and oxidative properties of enhanced beef steaks. Abstracts from the 63rd Reciprocal Meat Conference of the American Meat Science Association, Lubbock, TX, June 20–23, 2010. Meat Sci. 86, 533–576; Poster 90.

Robbins, K. Impact of rosemary and green tea extract on the oxidative and flavor stability of cooked ground chicken patties. Abstracts from the 64th Reciprocal Meat Conference of the American Meat Science Association, Manhattan, KS, June 19–22, 2011. Meat Sci. 89, 325–371; Poster 71.

Rojas, M. C.; Brewer, M. S. Effect of natural antioxidants on oxidative stability of cooked, refrigerated beef and pork. J. Food Sci. 2007, 72, S282–S288.

Romano, C. S.; Abadi, K.; Repetto, M. V.; Vichera, G.; Vojnov, A. A.; Moreno, S. Study of bioactive compounds from plants of *Rosmarinus officinalis L.* with antioxidant activity. Mol. Med. Chem. 2006, 11, 43–46.

Romano, C. S.; Abadi, K.; Repetto, M. V.; Vojnov, A. V.; Moreno, S. Synergistic antioxidant and antibacterial activity of rosemary plus butylated derivatives. Food Chem. 2009, 115, 456–461.

Sasse, A.; Colindres, P.; Brewer, M. S. Effect of natural and synthetic antioxidants on the oxidative stability of cooked, frozen pork patties. J. Food Sci. 2009, 74, S30–S35.

Sebranek, J. G; Sewalt, V. J. H.; Robbins, K. L.; Houser, T. A. Comparison of a natural rosemary extract and BHA/BHT for relative antioxidant effectiveness in pork sausage. Meat Sci. 2005, 69, 289–296.

Synthetic flavoring substances and adjuvants. Code of Federal Regulations, Part 172.515, Title 21, http://ecfr.gpoaccess.gov (accessed April 11, 2012).

Tang, S.; Kerry, J. P.; Sheehan, D.; Buckley, D. J. A comparative study of tea catechins and alpha-tocopherol as antioxidants in cooked beef and chicken meat. Eur. Food Res. Tech. 2001, 213, 286–289.

TBHQ. Code of Federal Regulations, Part 172.185, Title 21, http://ecfr.gpoaccess.gov (accessed April 11, 2012).

Tironi, V. A.; Tomas, M. C.; Anon, M.C. Quality loss during the frozen storage of sea salmon (*Pseudopercis semifasciata*). Effect of rosemary (*Rosmarinus officinalis L.*) extract. Lebsnsm.–Wiss. Technol. 2010, 43, 263–272.

Use of food ingredients and sources of radiation. Code of Federal Regulations, Part 424.21, Title 9, http://ecfr.gpoaccess.gov (accessed April 11, 2012).

CHAPTER

15

Strategies to Prevent Oxidative Deterioration in Oil-in-Water Emulsion Systems

Canola-Based Phenolic Applications

Amy Logan[1] and Peter Fagan[1]

Introduction

The ability to protect oil-in-water emulsion systems from oxidative deterioration has growing importance for the food and nutraceutical industries as new products are entering the market that have both enhanced levels of unsaturated fatty acids (FAs) and reduced levels of saturated and *trans* fats. Canola (*Brassica napus*) is the major oilseed crop grown within Australia and naturally contains high levels of antioxidants, including both α- and γ-tocopherols and phenolic acids (Ratnayake & Daun, 2004; Richards et al., 2008). The predominant phenolic acid found in all *Brassica* oilseeds is sinapic acid (3,5-dimethoxy-4-hydroxycinnamic acid) (Shahidi & Naczk, 1992), and the major mechanism for the antioxidant activity of phenolic acids derives from their ability to scavenge free radicals by donating their phenolic hydrogen atoms. The resulting phenolic radical is resonance stabilized, with various functional groups either increasing or reducing this resonance stabilization (Rice-Evans et al., 1996; Wanasundara et al., 1997). Tocopherols use a similar mechanism to protect against oxidative deterioration (Seppanen et al., 2010; Lampi, 2011).

Preheating canola seeds prior to oil extraction can further improve their inherent oxidative stability within bulk-oil applications (Prior et al., 1991; Veldsink et al., 1999; Szydowska-Czerniak et al., 2010a, 2010b). This is partly because

[1]CSIRO Animal, Food and Health Sciences, Werribee, Victoria, Australia, 3030.

the thermal decarboxylation of sinapic acid forms a potent lipophilic antioxidant, 2,6-dimethoxyvinylphenol, also known as 4-vinyl syringol (4-VS) or canolol (Tsunehiro et al., 2002; Koski et al., 2003; Kuwahara et al., 2004). Details pertaining to this mechanism will be discussed later on in the chapter.

Further, results from the study reported here show that fortification can improve the oxidative stability of 10% water-in-water tuna oil-in-water emulsions systems at levels of 5, 10, and 20% water-in-water total oil phase (seed preheated to 165 °C, 12 min). Moreover, the results indicate that oil from preheated seed can provide protection from oxidative deterioration that is as good as or, in some cases, better than a selection of commonly used antioxidants. Additionally, the preheating process does not appear to reduce oil quality (that is, both FA composition and tocopherol content remain unchanged); however, it does produce a nutty sensory attribute and a deepened color, which may or may not be desirable, depending on the end use. Further details will be discussed later in this chapter.

Oxidation within Oil-in-Water Emulsion Systems

The deterioration of fats and oils is a key issue to the food industry because it not only affects the sensory attributes (for example, taste and odor) but also the nutritional value and safety of the food itself through the production of harmful byproducts from degradation, such as lipid hydroperoxides and reactive oxygen species (Bodamyali et al., 2000; Kawanishi et al., 2002; Sachidanandame et al., 2005; Turner et al., 2006; Laguerre et al., 2007). The rate at which oxidative deterioration takes place depends on storage temperature, the presence of pro- or antioxidants, the fat or oil's chemical characteristics, and the way in which it is dispersed within the food product.

Fats and oils can be emulsified into oil-in-water emulsion systems, where the lipid phase is present as a dispersion of fine oil droplets within a continuous aqueous phase. Under these conditions, the emulsifier is located at the droplet interface, which facilitates emulsification by decreasing interfacial tension (Frankel, 2005). Food products based on oil-in-water formulations include milk, mayonnaise, yogurts, and salad dressings. Rancidity is a major problem for a number of these products, in particular, those formulations high in unsaturated FAs. This is because FA susceptibility to oxidative deterioration is proportionate to the dissociation energies of respective C-H bonds along the chain. A double bond along the FA chain would weaken the C-H bond on the adjacent carbon, allowing easier hydrogen removal. Consequently, the relative rate of oxidation increases as the number of double bonds within a FA chain increases (Holman & Elmer, 1947).

Lipid oxidation within an emulsion is essentially governed by the interfacial dynamics of the system and influenced by interactions between the oil and aqueous phases. This is because the surface-active components within an oil-in-water emulsion system are distributed on the basis of polarity and tend to concentrate at the interface, where they can interact with both phases (Frankel, 2005). In contrast, oxidation reactions within bulk oil systems generally occur at the interface between the oil and air and/or within dissolved oxygen located within the oil interface. As a result, oxidation generally occurs at a faster rate within emulsion systems because they have a larger interfacial surface area. This surface area is influenced by droplet size; however, recent studies (Day et al., 2007; Dimakou et al., 2007) have shown that small changes in droplet size have negligible effects on oxidative stability compared to other factors, such as the thickness and charge of the interfacial layer (Donnelly et al., 1998; McClements & Decker, 2000; Klinkersorn et al., 2005; Let et al., 2005), as well as the metal binding capability of the emulsifier itself (Jacobsen et al., 2001).

Emulsifiers are amphiphilic compounds and orientate themselves at the interface of oil-in-water systems so that the polar and nonpolar groups partition between the two phases. Common emulsifiers used within the food industry include FAs, surface-active proteins, small-molecule surfactants, monoglycerides, and phosopholipids. Oxidative stability can be manipulated by controlling the emulsifier type, interfacial orientation, and concentration due to packing density (Mancuso et al., 1999; Formusco et al., 2002; Frankel, 2005; Klinkersorn et al., 2005). Sodium caseinate is commonly used in the food industry to rapidly establish thick, sterically stabilized layers that protect the oil droplets against flocculation and coalescence. These emulsions tend to resist physical deterioration when compared to those stabilized by globular proteins. This is attributed to the physical characteristics of casein, which, in comparison, tends to be more resistant to heat and, therefore, less likely to undergo conformational changes when temperature is applied (Hunt & Delgleish, 1995; Srinivasan et al., 2002; Surh et al., 2006). Sodium caseinate is also known to protect against autoxidation due to its iron-chelating properties (Hu et al., 2003). Having completely covered the oil-in-water interface, excess protein will remain within the aqueous phase where it increases the viscosity surrounding the droplet and may restrict the penetration and diffusion of radical initiators into the lipid phase (Frankel, 2005). Studies have shown transition metal–based catalysis to be the primary cause of lipid oxidation in food emulsions (Kellerby et al., 2006), and the rate of oxidative deterioration is generally greater within emulsion systems formulated using anionic emulsifiers, compared to those using non-ionic emulsifiers (Richards et al., 2011). The ability of metal ions to degrade lipid hydroperoxides can depend on their location relative to the droplet's interface (Kellerby et al., 2006). In anionic systems, an electrostatic attraction occurs between the negatively charged droplet interface and the positively charged metal ions found in all food systems, such as iron, resulting in metal-catalyzed oxidation (Frankel, 2005). With an affinity for the hydrated layer surrounding the emulsion droplet, metal ions will form complexes through the carboxylic acid group of the free FAs present within the oil. However, the presence of bulky protein molecules and/or surfactants at the emulsion interface may physically hinder these interactions (McClements & Decker, 2000; Kellerby et al., 2006).

The Canola Industry

Canola oil was first developed in Canada from traditional rapeseed and is obtained from the oilseeds of *Brassica napus* (swede rape), *B. rapa* (turnip rape), and, more recently, *B. juncea* (Indian mustard). The word "rape" originates from the Latin word *rapum*, meaning "turnip." Rapeseed was originally cultivated by ancient civilizations in Asia and the Mediterranean and used as a lighting fuel (Colton & Potter, 2012). The species now used for canola production are three diploid species: *B. nigra*, *B. oleracea*, and *B. rapa*. The remaining three species are polyploid, combining two of the diploid species (U, 1935).

The most valuable component of canola is its oil, which comprises approximately 40% of the seed mass; 80% of this oil is concentrated in droplet form within cotyledon cells. The majority of the remaining oil (7–12% of total seed oil) is located within the seed coat and endosperm. In 2010–2011, rapeseed and canola were the third most abundant commercial vegetable oils in the world, after palm and soybean, with global production of 57 million tons of seed and 22 million tons of oil (Mathews, 2012).

Traditional rapeseed oil contained high levels of erucic acid (22:1), approximately 40% of the FA composition, an FA associated with Keshan's disease characterized by fibrotic lesions of the heart. While there is no evidence to indicate that 22:1 is toxic in humans, erucic acid has been observed to cause cardiac lipidosis and necrosis in rats (Larvea et al., 1992). Another constituent of traditional rapeseed is aliphatic glucosinolates (commonly 70 to 120 μmoles/g), which are known to enlarge the thyroid gland and are associated with palatability and nutritional problems in non-ruminant animals (Bell & Belzik, 1965; Fenwick et al., 1982; Salisbury & Wratten, 1999). Due to the negative effects of these constituents, efforts were made by canola breeders to diminish the levels of both aliphatic glucosinolates and 22:1 in specific rapeseed genotypes. The new "double low" genotype was reclassified in 1979 as canola after its Canadian origin. According to the current Australian definition, canola oil must contain no more than 2% 22:1 and less than 30 μmoles/g of aliphatic glucosinolates in the defatted meal (Lamont, 1991; Woods et al., 1991). In Europe, the low-22:1, low glucosinolate rapeseed produced is called oilseed rape.

Canola Development in Australia

Australian environments vary markedly throughout the country. Annual rainfall can range from 325 to 700 mm in areas used for canola cultivation (65–75%, generally between May and October) and temperatures range from a minimum of 4 °C to a maximum of 30 °C. Canola is normally sown between April and June and harvested in November and December (Potter et al., 2012). Canola plants flower between August and September, and the growing season ranges from 150 to 210 days.

The initial focus of canola development was to produce canola with low 22:1, low glucosinolate levels, and increased blackleg resistance (Murray & Brennan, 2012). The first *B. napus* genotypes with canola quality and some blackleg resistance were released in 1980. The cultivation area of canola increased markedly during the 1990s, with around one million hectares or more grown annually since 1999. Canola is now the most abundant oilseed produced in Australia, accounting for more than 70% of total production (Mathews, 2012). In recent years, the Australian canola industry has expanded its focus beyond the production of canola-quality *B. napus*.

Recent trends include interest in the development of *B. juncea* as an alternative canola species for low rainfall areas of Australia (Burton et al., 2003; Salisbury et al., 2004). Beneficial characteristics of *B. juncea* include enhanced seedling vigor, shatter resistance, and enhanced tolerance to drought, higher temperatures, and water stress (Woods et al., 1991; Simbaya et al., 1995; Wright et al., 1996; Niknam & Turner, 1999). In addition, *B. juncea* is more resistant than *B. napus* to blackleg disease (Rimmer & van den Berg, 1992; Ballinger & Salisbury, 1996).

Canola as a Source of Antioxidants

Tocopherols

Tocopherols are effective lipid soluble antioxidants that reduce the rate at which oxidation proceeds by acting as phenolic hydrogen donors and blocking free-radical chain reactions. Four known tocopherol isomers are found in canola oil: α-, β-, γ- and δ-tocopherol, and their relative concentration can vary with seed genetics and environment. γ-Tocopherol (Fig. 15.1D) tends to be the most dominant isomer present within canola oil, ranging from 165–634 ug/g; α-tocopherol (Fig. 15.1C) the second most abundant, ranging between 76–336 ug/g (Richards et al., 2008). The α form has been proven to be most active *in vivo* (Lampi, 2011). This is due to the structure of isomers and their relative ability to donate hydrogen. The presence of an electron-releasing substituent in the *ortho* and/or *para* position to a hydroxyl functional group increases electron density, thereby increasing the stability of the phenoxyl radical (Kamal-Eldin & Appelqvist, 1996). In contrast, the γ-tocopherol homologue is more effective against lipid oxidation *in vitro* (Gottstein & Grosch, 1990; Seppanen et al., 2010). Tocotrienols are similar in structure to the tocopherols and have been shown to protect against oxidative deterioration (Wagner et al., 2001). However, tocotrienols do not naturally occur within canola oil (Aladedunye & Przybylski, 2012).

Hydrophilic Phenolic Content

Brassica oilseed crops are a rich source of phenolics and contain the highest level of phenolics of all oilseed crops, with most of the phenolic components retained in the defatted meal. On a dry weight basis, defatted meal contains 1.5–1.8% phenolics, with most (1.2–1.5%) in an esterified form and some (0.1–0.25%) available as free phenolic acids (Krygier et al., 1982; Shahidi & Naczk, 1992; Szydlowska-Czerniak et al., 2010c). The levels of free phenolic acids in cold, pressed virgin rapeseed oil are approximately 1/1000th of those found in defatted meal (Siger et al., 2008). Sinapic acid (Fig. 15.1A) is the major phenolic acid found in *Brassica* oilseeds, with the majority (71–97%) occurring as the choline ester (sinapine) (Krygier et al., 1982; Shahidi & Naczk, 1992). Recent investigations have identified several new sinapic acid derivatives in canola-seed processing that utilize LC-MS (Harbaum-Piayda et al., 2010; Engels et al., 2012).

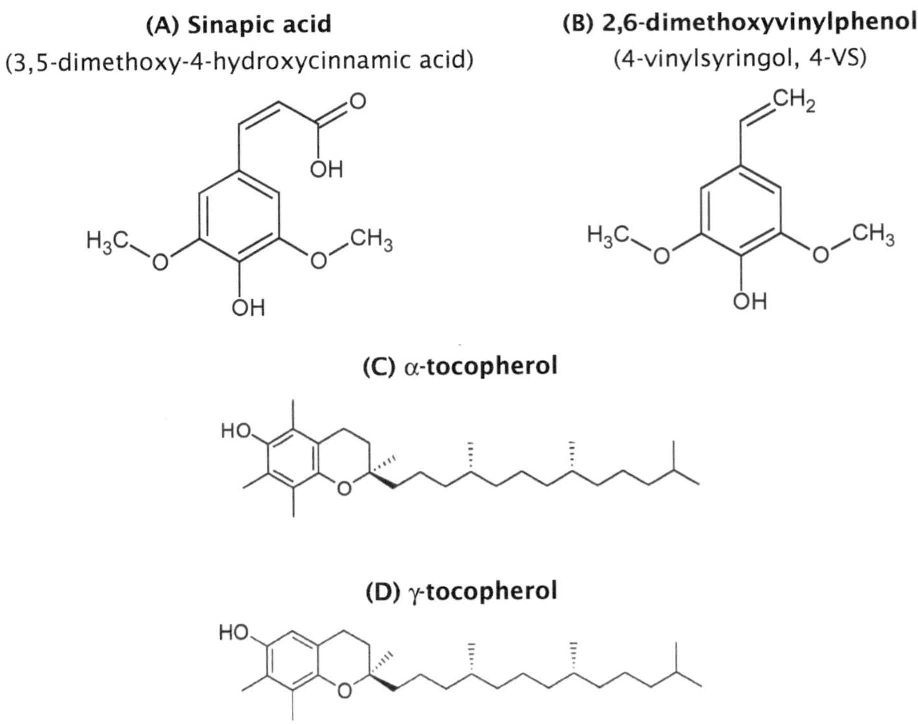

Fig. 15.1. Structures of (A) 3,5-dimethoxy-4-hydroxycinnamic acid (sinapic acid) and its decarboxylation product (B) 2,6-dimethoxyvinylphenol (4-vinyl syringol); (C) α- and (D) γ-tocopherol isomers.

Multiple groups have succeeded in removing some of these phenolic components from the defatted meal through aqueous and ethanolic extractions (Shahidi & Naczk, 1992; Naczk et al., 1998; Xu & Diosady, 2002; Thiyam et al., 2004; Terpinc et al., 2012). These extracts exhibit enhanced antioxidant and antimicrobial activity, although it has been noted that the antioxidant activity of the total extract is greater than that of individual components, which indicates additive or synergistic effects (Nowak et al., 1992; Wanasundara et al., 1994; Wanasundara et al., 1997; Tesaki et al., 1998; Thiyam et al., 2006; Frankel & Finley, 2008; Wakamatsu et al., 2008; Engels et al, 2012). Other groups have focused on developing commercial applications for these extracts in order to mine for valuable byproducts in what was once considered a waste stream of the oilseed industry (Thiyam et al., 2004, 2006; Oskoueian et al., 2011; Terpinc et al., 2012).

Lipophilic Phenolic Content

Over the past 20 years, numerous investigations have reported that canola oil extracted from preheated canola seeds or flour exhibits enhanced oxidative stability (Table 15.A). The focus of many earlier articles was to enhance oxidative stability through thermal inactivation of enzymes utilizing microwave or radio frequencies (Ponne et al., 1996; Oberndorfer & Lücke, 1999; Irfan & Pawelzik, 1999; Valentová et al., 2000, 2002) and/or improve oil recovery as it is expelled. Prior and colleagues (1991) demonstrated that preheating seeds prior to the expelling of oil significantly improved the oxidative stability of crude oil. A relationship was observed between the enhanced stability and an increase in the crude oil's phosphorous content, leading to the conclusion that increased stability was due to an increased concentration of phospholipids.

While investigating the effect of heating on lipase activity in rapeseed, Ponne and colleagues (1996) reported that two distinct lipases appeared to deactivate either below or above 100 °C. Deactivation of the lipases reduced the formation of free FAs during processing and improved oxidative stability. In a related investigation, Veldsink and colleagues (1999) reported that degummed oil extracted from preheated rapeseed had enhanced oxidative stability. There was an almost direct relationship between a preheating temperature above 90 °C and increased oxidative stability. A rapid increase in total phosphorous and total phenols was also observed when preheating temperatures rose above 90 °C. Because the oils were water-degummed, which resulted in the removal of hydratable phospholipids (van Nieuwenhuyzen & Tomás, 2008) prior to the oxidative stability study, it was concluded that the enhanced stability was due to phenolics.

Subsequently, Japanese and Finnish groups independently isolated and characterized a phenolic compound, 4-vinylsyringol (4-VS) (Fig. 15.1B), in the crude oil, which was a potent antioxidant and antimutagen (Tsunehiro et al., 2002;

Table 15.A. Heating Conditions Reported for the Thermal Treatment of Canola Seed or Flour

Heating Method	Temperature (°C)	Reference
Screw-press	120	Prior et al., 1991
	> 100	Koski et al., 2003; Vuorela et al., 2003
Microwave, dielectric, & radio frequency	75–130	Ponne et al., 1996
	100–120	Veldsink et al., 1999
	80–100	Oberndorfer & Lücke, 1999
	120	Irfan & Pawelzik, 1999
	~100	Valentová et al., 2000, 2002
	160	Spielmeyer et al., 2009
	N/A	Azadmard-Damirchi et al., 2010
	N/A	Oskoueian et al., 2011
Oven	165	Kuwahara et al., 2004
	165	Wakamatsu et al., 2005
	145–165	Wijesundera et al., 2008
	165	Vaidya & Choe, 2011
	160	Shrestha et al., 2012
	180	Huidrom, 2012
Pressurized steam	MC 11%, 145	Zacchi et al., 2005, 2008
Oven, Moisture control	MC 11%, 120	Szydlowska-Czerniak et al., 2010a, 2010b

N/A = Temperature data not available

MC = Moisture content

Koski et al., 2003; Vuorela et al., 2003; Kuwahara et al., 2004; Wakamatsu et al., 2005). The Japanese investigated natural phenolics for antiradical behavior, with a focus on antimutagenic activities (Kanazawa et al., 2002), while the Finnish group investigated processing's effect on the oxidative stability of rapeseed oil (Pekkarinen et al., 1998). Both groups attributed 4-VS to the thermal decarboxylation of sinapic acid. Since that time, several other groups have confirmed the formation and structure of 4-VS (Zacchi & Eggers, 2008; Wijesundera et al., 2008; Spielmeyer et al., 2009; Azadmard-Damirchi et al., 2010; Shrestha et al., 2012).

The effect of the canola seed or flour matrix is critical to the decarboxylation of sinapic acid. It has been shown that heating sinapic acid by itself or in water produced very little 4-VS (Speilmeyer et al., 2009). This is not surprising given the requirement for a catalyst, which we will discuss later. The amount of free sinapic acid in canola seed is quite low. Tesakai et al. (1998) reported that ~3% of the total sinapic acid in defatted *B. juncea* meal was free. In a very recent study, Shrestha and colleagues (2012) reported that the formation of 4-VS in heated high erucic mustard seed was highly correlated (> 0.93) with the initial and residual free sinapic acid contents, as well as the loss of total sinapic acid content after alkaline hydrolysis. The correlation was even higher for canola seed flour (0.96), possibly because of the ease with which CO_2 could be released (Shrestha et al., 2012). Based on the 4-VS content and loss of total sinapic acid content, the 4-VS yield was calculated to be between 12.7% and 37.8%. This loss was attributed, at least in part, to the reaction of the vinyl bond in 4-VS with Maillard reaction intermediates (Shrestha et al., 2012).

Several groups have demonstrated that the oxidative stability of refined canola/rapeseed oil is less than that of crude oil. This has been attributed to synergistic effects and the loss of phenolics (including 4-VS) during the refining process (Koski et al., 2003; Zacchi et al., 2008; Speilmeyer et al., 2009). A very recent study found a hexane extract from preheated canola seeds to be a superior antioxidant when compared to isolated 4-VS in a canola oil-in-water emulsion system stabilized with whey protein (Huidrom, 2012).

The content of 4-VS in the oil markedly decreased during each refining step, with none observed after deodorization (Koski et al., 2003; Zacchi et al., 2008). Zacchi and colleagues (2008) noted that an additional component appeared in the high-pressure liquid chromatography HPLC chromatograms after the bleaching step of the refining process. Refined oil obtained from preheated rapeseeds was also noted to have slightly greater oxidative stability than refined oil from seeds that were not preheated. They appear to have had these results because of the pressurized steam heat employed in this work. Continuing these investigations, recent studies have identified the *cis*- and *trans*-diastereomers of the 4-VS dimer in the oils and deodorizer distillate (Harbaum-Piayda et al., 2010). The 4-VS dimer has similar antioxidant activity to 4-VS in methanolic extracts.

Recently, Szydlowska-Czerniak and colleagues (2010a, 2010b) used a response surface methodology to assess the effect of multiple parameters on the oxidative stability of virgin rapeseed oil extracted from seeds that had been pretreated with heat, moisture, and/or enzymes. They found that there is a significant correlation between oxidative stability and both total phosphorous and total phenols after treatment with cellulolytic enzymes. The highest oxidative stability was found for oil pressed from rapeseeds heated at 120 °C with a moisture content of 11% (Szydlowska-Czerniak et al., 2010b).

Enhanced oxidative stability through the thermal pretreatment of canola seed can be attributed to: (1) the deactivation of endogenous enzymes, (2) the release of phospholipids, and (3) the decarboxylation of sinapic acid to produce the neutral, lipophilic 4-VS that will favorably partition into the extracted oil.

The enhanced oxidative stability provided to virgin canola oil by 4-VS can be explained in terms of lipophilicity and structure-activity relationship. The calculated octanol-water partition coefficients (Log P) for sinapic acid (1.26) and 4-VS (1.96) show that 4-VS is considerably more lipophilic than its precursor. Previous studies relating antioxidant activity to phenolic structure have shown that a vinyl group enhances the oxidative stability of a phenol due to its participation in the radical's resonance stabilization (Wanasundara et al., 1997; Wright et al., 2001). Converting the -CH=CH-COOH group in sinapic acid to -CH=CH_2 in 4-VS has a positive influence on the resonance stabilization of the radical and, therefore, antioxidant activity overall. In a recent computational study, Galano and colleagues (2011) found that the reactivity of 4-VS toward peroxyl radicals takes place almost exclusively by H atom transfer from the phenolic functional group, regardless of the environment's polarity and, as such, 4-VS is a very good antioxidant.

The thermal and enzymatic decarboxylation of 4-hydroxycinnamic acids [4-hydroxycinnamic (p-coumaric); 3,4-dihydroxycinnamic (caffeic); 3-methoxy-4-hydroxycinnamic (ferulic), and 3,5-dimethoxy-4-hydroxycinnamic (sinapic) acids] results in the formation of the respective 4-hydroxystyrenes [4-vinylphenol, 3,4-dihydroxystyrene, 3-methoxy-4-vinylphenol (4-vinylguaiacol), and 2,6-dimethoxyvinylphenol (4-vinyl syringol)]. These decarboxylation products and their adducts are present in a diverse range of beverages and foods, including various wines, beers, juices, and berries (Pyysalo et al., 1977; Heresztyn, 1986; Donaghy et al., 1999; Suarez et al., 2007; Vanbeneden et al., 2008; Callemein & Collin, 2010).

The thermal decarboxylation of 4-hydroxycinnamic acids has been studied since the 1940s with various catalysts, including copper/quinoline (Walling & Wolfstirn, 1947), KCl (Cohen & Jones, 1960), amines (Nomura et al., 2005), and sodium acetate (Terpinc et al., 2011). Cohen and Jones (1960) postulated that the decarboxylation reaction proceeds via a *para*-quinone methide intermediate. This mechanism was supported by the rapid decarboxylation of a synthetic analogue (3,5-tert-butyl-4-hydroxycinnamic) (Munteanu, 1991). Recent studies have used microwave heating to provide rapid, highly efficient decarboxylation reactions of substituted 4-hydroxycinnamic acids (Nomura et al., 2005; Bernini et al., 2007). They also demonstrated that the decarboxylation reaction is catalyzed with bases having high pKa values, and proceed via the above reaction mechanism (Nomura et al., 2005). The mechanism for enzymatic decarboxylation also occurs via a *para*-quinone methide intermediate (Hasidoko et al., 1998), although the enzymatic process is stereospecific, with both the cinnamic acid substrate and styrene products in the *trans* (E) configuration (Gramatica et al., 1981).

Several articles have noted that 4-VS would not be suitable as an antioxidant in frying oils (Veldsink et al., 1999; Wakamatsu et al, 2008; Spielmeyer et al., 2009). Veldsink and colleagues (1999) reported that while preheated rapeseed oil displayed enhanced oxidative stability in frying tests, it produced unpleasant odors during frying and more pronounced browning. The odors were characterized as "rubbery," and the browning was attributed to condensation of phenolics. These organoleptic changes were said to be more detrimental than the positive counter effects of enhanced oxidative stability. Spielmeyer and colleagues (2009) noted that 4-VS was thermally unstable, and the content in a rapeseed oil dropped exponentially when the rapeseed oil was exposed to 180 °C (that is, the frying temperature) for 20 minutes.

Various approaches have been developed to exploit the inherent antioxidant, anti-inflammatory, and antimutagenic properties of 4-VS, while addressing its limitations—principally, its loss during refining and its unsuitability in frying oils. Mildola (a Finnish vegetable oil company) has produced a "virgin" rapeseed oil ("Nieto") in partnership with the Finnish research group (Mildola, 2008). The refining process has been significantly modified in order to retain components like phenols and the strong yellow color of the oil. Wakamatsu et al. (2008) employed steam distillation to recover 4-VS from crude canola oil and subsequently added this distillate back to refined oil for use in salad dressings. Meyer et al. (2011) have used short path distillation to recover the 4-VS dimer from deodorizer distillate and proposed its addition to refined oils. However, the separation of the 4-VS dimer from free FAs is problematic, and further investigations continue.

Enhanced Oxidative Stability within Oil-in-Water Emulsion Systems: A Case Study

The Influence of 4-VS on Oxidative Stability of Oil-in-Water Emulsions

The antioxidant properties of canola oil from preheated seed was measured within oil-in-water emulsions (10% water-in-water) stabilized by sodium caseinate (1.75% water-in-water). Accelerated oxidation was performed according to AOCS recommendations (Cg 5-97) at 50 °C. Aliquots (2 mL) of sample were held within 10 mL sealed vials. As described earlier in this chapter, 4-VS is formed through a thermal decarboxylation reaction. More lipophilic than the sinapic acid precursor, 4-VS will partition from the meal into the oil phase upon extraction. In this study, canola seeds were heated at 165 °C for 12 minutes to allow the decarboxylation reaction to occur. It has been shown that preheating the seed in this manner does not affect the FA profile of the extracted oil or the abundance of endogenous tocopherols (Wijesundera et al., 2008). The levels of major FAs, tocopherols, and 4-VS within the respective oil phase blends used in this study are shown in Table 15.B.

Table 15.B. Major Fatty Acid, Tocopherol, and 4-VS Profile within the Oil Phase of Prepared Emulsions ($n = 3$)

	Fatty Acid (%)			
Emulsion Oil Phase	**16:0**	**18:0**	**18:1**	**18:2**
TO + RCO [80:20]	19.30 ± 0.42	5.71 ± 0.05	24.48 ± 0.22	6.00 ± 0.05
TO + PSCO(165) [80:20]	18.99 ± 0.18	5.79 ± 0.01	24.72 ± 0.08	5.87 ± 0.03
TO + PSCO(165) + RCO [80:10:10]	18.88 ± 0.18	5.72 ± 0.01	24.55 ± 0.01	5.93 ± 0.01
TO + PSCO(165) + RCO [80:5:15]	18.72 ± 0.04	5.76 ± 0.02	25.08 ± 0.04	6.17 ± 0.01
TO + PSCO(145) [80:20]	18.67 ± 0.32	5.72 ± 0.02	24.88 ± 0.06	5.89 ± 0.01
TO + CCO [80:20]	18.66 ± 0.18	5.73 ± 0.04	24.97 ± 0.16	5.96 ± 0.04
TO + PSCO(165)-degummed [80:20]	19.03 ± 0.20	5.71 ± 0.01	24.23 ± 0.01	5.77 ± 0.01

RU = relative units based upon peak area per mg/oil; TO = tuna oil; RCO = refined canola oil; PSCO = preheated seed canola oil; CCO = crude canola oil

Identification of 4-VS

The presence of 4-VS within the lipid phase of each sample was confirmed via HPLC analysis. Fig. 15.2 presents a typical elution profile for both 4-VS and the tocopherol isomers contained within a tuna- and canola-oil blend.

Determination of Oxidative Stability

Solid-phase micro extraction (SPME) headspace (HS) analysis, coupled with gas chromatography, was used to measure oxidative deterioration at regular time intervals throughout the experiment. This method involves the insertion of a SPME fiber into the HS of a vial containing the sample in order to monitor the development over time of volatile secondary oxidation products, such as aldehydes and ketones. For a detailed description of this method, see Kataoka et al. (2000) and Pawliszyn (2000). The specific type of volatile present within the sample HS will be specific to the FA profile of the oil phase itself and can be influenced by the conditions of the oxidation trial (specifically, applied heat), type of HS method (that is, SPME may differ from static or dynamic HS analysis), or the type of SPME fiber used (regarding its affinity or sensitivity to specific compounds). The dominant HS volatiles routinely observed within canola oil heated to 50–60 °C are *c,t*-2,4-heptadienal; *t,t*-2,4-heptadienal; propanal; 1-penten-3-ol; 1-penten-3-one; and hexanal. *c,t*-2,4-Heptadienal was the key indicator of lipid oxidation for the study reported here.

			Tocopherol (T) (μg/g oil)			(RU/mg oil)
18:3	EPA	DHA	α-T	γ-T	δ-T	4-VS
2.51 ± 0.02	5.30 ± 0.06	27.31 ± 0.16	498 ± 9	923 ± 50	404 ± 24	Nil
2.99 ± 0.02	5.33 ± 0.04	26.81 ± 0.29	555 ± 1	1067 ± 70	479 ± 34	323 ± 1
2.74 ± 0.01	5.35 ± 0.04	27.36 ± 0.11	572 ± 1	1164 ± 6	529 ± 3	152 ± 4
2 .71 ± 0.01	5.30 ± 0.01	26.98 ± 0.02	567 ± 1	1060 ± 25	476 ± 12	78 ± 1
3.06 ± 0.01	5.31 ± 0.05	27.24 ± 0.32	537 ± 20	1096 ± 45	496 ± 22	53 ± 1
3.05 ± 0.01	5.30 ± 0.03	27.22 ± 0.01	490 ± 7	964 ± 9	430 ± 4	34 ± 2
2.96 ± 0.01	5.34 ± 0.03	27.39 ± 0.24	471 ± 3	924 ± 9	406 ± 4	196 ± 3

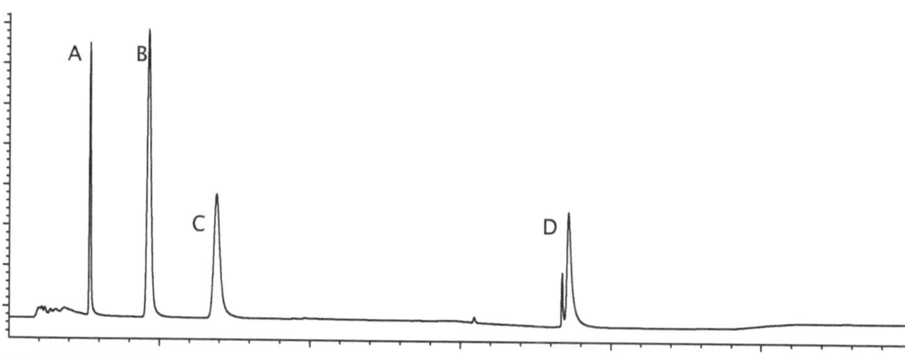

Figure 15.2. Fluorescence chromatograms for the separation of (a) α-tocopherol, (b) γ-tocopherol, (c) δ-tocopherol, and (d) 4-VS in TO + PSCO(165).

Oil dissolved in hexane/IPA (99:1, v/v) and analyzed on an Allsphere Cyano HPLC column [150 mm × 4.6 mm ID (5 μm)] (Grace Davidson Discovery Sciences, Melbourne, Australia) fitted with an Allsphere Cyano guard column. Solvents: (A) Hexane/IPA (99.5:0.5, % v/v); (B) tert-butyl methyl ether; (C) Methanol. Gradient 0–15 min: 100% (A) 15–25 min: 0-20% (B); 25–33 min: 20–100% (B); 33–36 min: 0-100% (C); 36–42 min: 100% (C) 42–45 min: 0–100% B; 45–48 min: 0–100% (A); 48–60 min: 100% (A). The flow rate was constant at 1.0 mL/min, with a total run time of 60 min.

A series of oil-in-water emulsions were formulated and exposed to accelerated oxidative conditions over a period of seven days. The oil phase for the emulsions consisted of:

1. 80% tuna oil, 20% refined canola oil (RCO) (control)
2. 80% tuna oil, 20% canola from preheated seed (PSCO) heated to 165 °C
3. 80% tuna oil, 10% PSCO, 10% RCO
4. 80% tuna oil, 5% PSCO, 15% RCO
5. 80% tuna oil, 20% PSCO heated to 145 °C
6. 80% tuna oil, 20% crude canola oil (extracted from non-preheated seed) (CCO)
7. 80% tuna oil, 20% PSCO heated to 165 °C and degummed in the laboratory

The Influence of 4-VS Concentration on Oxidative Stability (Emulsions 1–4)

Here, the amount of PSCO within the respective oil phase varied from zero (control) to 20%. As such, the levels of 4-VS found naturally within the PSCO oil—hence, within the emulsions systems—varied proportionately (Fig. 15.3). Fig. 15.4 presents the development of *c,t*-2,4-heptadienal over time for these four emulsions systems.

With the exception of the tuna-oil/refined canola–oil control, all emulsions contained oil from preheated seed at varying concentrations. These emulsions were shown to be more stable to oxidative deterioration than the control, as determined via the formation of HS volatiles over time. This indicates that the oil from preheated seed behaves as an antioxidant and protects against oxidative deterioration. We believe that this result occurs because of the 4-VS content, as virtually negligible levels of other phenolic components commonly found within canola are extracted into the oil and instead remain within the meal (Khattab et al., 2010). Further, the FA profile and tocopherol levels within each sample were similar. Under these conditions, the concentration of 4-VS did not appear to have a significant effect so that enhanced stability was achieved when oil from preheated canola seed was used for as little as 5% of the total oil phase (0.05% water-in-water of total emulsion). Further studies need to determine the lowest effective concentration possible to achieve enhanced protection toward oxidative deterioration.

The Influence of Preheating Temperature and Water Degumming (Emulsions 1, 5, and 7)

The concentration of 4-VS within the oil increases with both time and temperature throughout the preheating process. Spielmeyer and colleagues (2009) indicated that maximal amounts of 4-VS are produced when seed is heated for 7.5

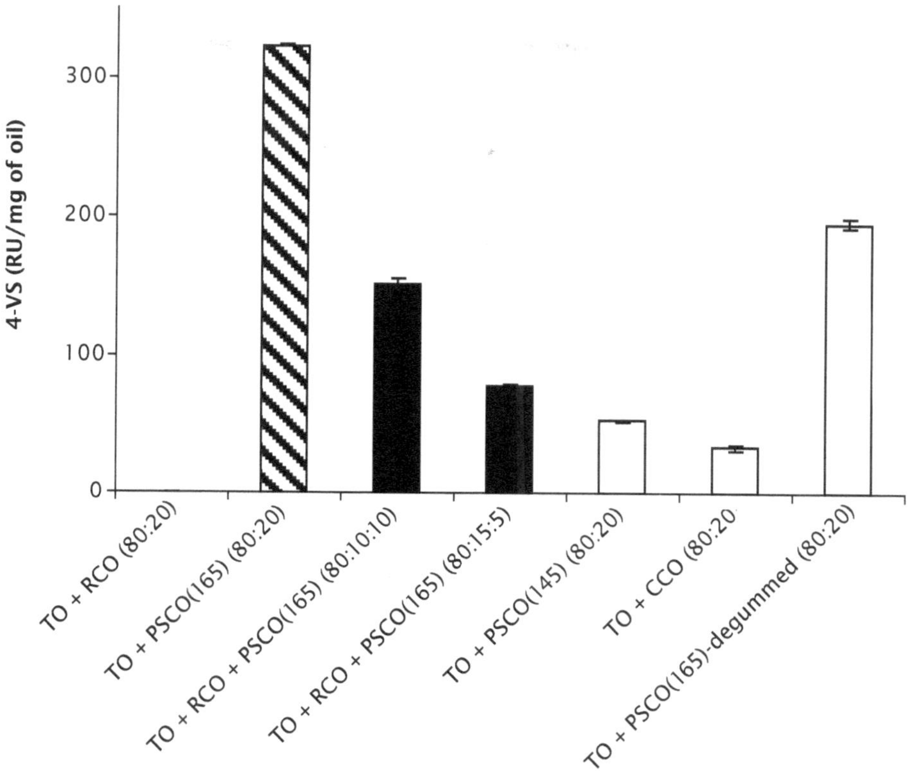

Fig. 15.3. Level of 4-VS within the oil phase for prepared emulsions (*n* = 3).

RU = relative units; TO = tuna oil; RCO = refined canola oil; PSCO = preheated seed canola oil; CCO = crude canola oil.

minutes under microwave-oven conditions (at 1.5 minute intervals, stirring the seed between each interval), reaching a final seed temperature of 160 °C. A study conducted within our laboratory observed levels of 4-VS within seed heated using a conventional oven between 105 and 165 °C for 5–15 minutes. Heating seed under these conditions at 165 °C for 12 minutes creates optimum conditions for 4-VS formation, without degrading or charring the seed. In the current study, we made a comparison between oil from seed heated to 165 °C (with a higher 4-VS concentration) and 145 °C (with a lower concentration). The emulsion containing seed heated to a higher temperature was shown to be only somewhat more stable to oxidative deterioration than that heated to 145 °C (Fig. 15.5). This was not surprising, as the observed effect was essentially that of concentration. As shown

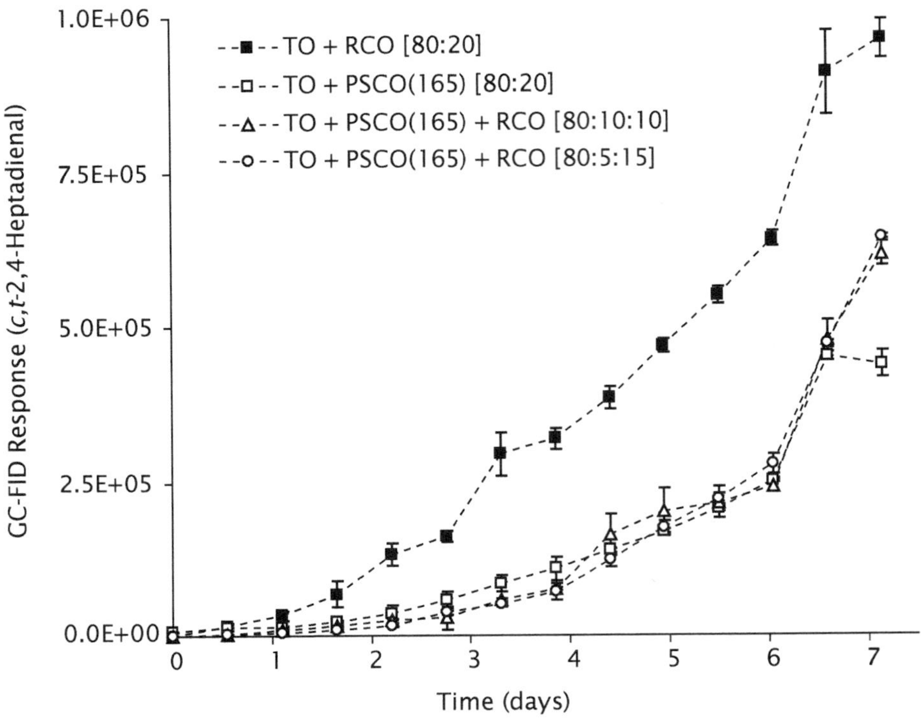

Fig. 15.4. Formation of headspace *c,t*-2,4-heptadienal in 10% water-in-water emulsion systems: A comparison between preheated (165 °C) seed canola oil (PSCO) blends (20, 10, and 5%) and refined canola oil (RCO).

earlier in this chapter, the influence of concentration was minimal whereby the two concentrations used within this study do not appear to influence the oil's antioxidant potential.

The levels of total polar lipids (including phospholipids and glycolipids) reported within crude canola oil vary between ca. 0.5–1 and ca. 4% of the total oil. This is because the amount of polar lipids extracted can vary between extraction methods, for example, whether the oils are expeller pressed or solvent extracted with hexane (Przybylski & Eskin, 1991) or chloroform:methanol (Sosulki et al., 1981). Phospholipids make up a greater portion of the polar lipid fraction, the majority of which is removed during the degumming process.

Phosphatidic acid (PA) is a major component of the canola polar-lipid fraction. It contains a negative charge within the polar head group and is, therefore,

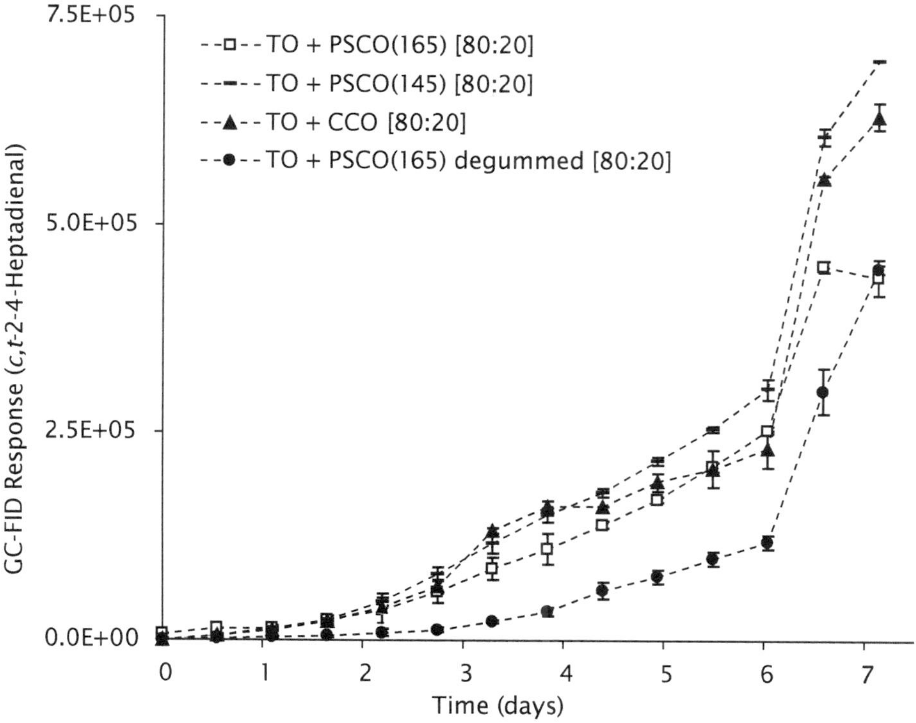

Fig. 15.5. Formation of HS *c,t*-2,4-heptadienal in 10% water-in-water emulsion systems: A comparison between preheated (165 °C & 145 °C) seed canola oil (PSCO), degummed PSCO (165 °C), and unheated seed canola oil (CCO).

anionic. As an intermediary in the biosynthesis of phospholipids, PA can also be generated through the hydrolysis of phosphatidylcholine acid (PC) during processing (Przybylski & Eskin, 1991). Another component of the polar lipid fraction is phosphatidylinositol (PI), an anionic phospholipid. The two remaining major components of the polar lipid fraction of crude canola oil are PC and phosphatidylethanolamine (PE). Both of these phospholipids are uncharged, at neutral pH. As demonstrated in the early chapters of this book, polar lipid components tend to play a larger role in the oxidative stability of emulsion systems, compared with bulk-oil applications, as they are amphiphilic and, therefore, likely to congregate predominantly at the droplet interface and/or partition into the aqueous phase where lipid oxidation takes place. Anionic polar lipids (predominantly, PA and PI) decrease the oxidative stability of oil-in-water emulsion systems by imposing a negative charge at the droplet's surface. The negative charge will attract positive metal ions, such as iron and copper, from within the aqueous phase, and the level of metal-catalyzed oxidation will subsequently increase (Richards et. al., 2011). Once removed or reduced during the degumming process, the interfacial negative charge would subsequently decrease, and the degree of metal-catalyzed oxidation would be reduced. This explains the differences in *c,t*-2,4-heptadienal formation within the degummed and non-degummed emulsions, where the degummed PSCO oil was a more stable system during the autoxidative process (Fig. 15.5). Further, it has been suggested that the removal of phospholipids during oil refining may also result in the removal of metal ions bound to the phospholipids molecules via chelation, increasing the oil's stability (List et al. 1978; Goh et al., 1982; Przybylski & Eskin, 1991). The level of polar components present within the oil may also affect the packing structure of the sodium caseinate interface. Due to the comparatively smaller size of phospholipid compounds, a complex lipid–protein interface would be produced, whereby the phospholipids would fit in between the larger caseinate molecules, producing a more densely packed interfacial structure.

Crude Canola Oil from Non-Preheated Seed (Emulsions 1 and 6)

We extracted oil from both preheated (PSCO) and non-preheated seed (CCO) using a Soxhlet apparatus (petroleum ether, b.p. 40–60 °C). CCO oil was expected to be devoid of 4-VS, similar to the control, as the seed had not been preheated, thereby inducing decarboxylation. However, we detected some 4-VS (Fig. 15.3), which was possibly produced when heat was applied to the crushed seed during the extraction process itself, converting some sinapic acid into 4-VS. As such, the CCO exhibited protection to oxidation similar to that of the PSCO oil (Fig. 15.5), despite the fact that the levels of 4-VS were approximately half that of the 5% oil blend (which was the lowest PSCO blend concentration analyzed within this study) and lower than the concentration of oil from seed preheated to a lower temperature. Again, this reaffirms that 4-VS only needs be present in small amounts to provide enhanced protection within an oil-in-water emulsion system.

How Does PSCO Compare with Other Antioxidants?

In the second part of this study, the antioxidant behavior of PSCO within a 10% water-in-water/oil-in-water emulsion system was compared to a series of other commonly used antioxidants. A series of oil-in-water emulsions were formulated and exposed to accelerated oxidative conditions over a period of eight days. The oil phase for the emulsions consisted of:

1. 80% tuna oil, 20% refined canola oil (RCO) (control)
2. 80% tuna oil, 20% canola from preheated seed (PSCO) heated to 165 °C

3. 80% tuna oil, 20% RCO + 16 mg/kg α-tocopherol
4. 80% tuna oil, 20% RCO + 16 mg/kg α-tocopherol + 16mg/kg sodium dicalcium-EDTA
5. 80% tuna oil, 20% RCO + 16 mg/kg α-tocopherol + 4mg/kg ascorbyl palmitate
6. 80% tuna oil, 20% RCO + 4mg/kg ascorbyl palmitate

Again, levels of *c,t*-2,4-heptadienal were used to measure oxidative deterioration (Fig. 15.6), and RCO without the addition of antioxidants functioned as a control. The most stable formulation was protected by α-tocopherol (an antioxidant mechanism described earlier in this chapter) in combination with EDTA, an effective chelator, binding to metal ions from within the serum phase (Let et al.,

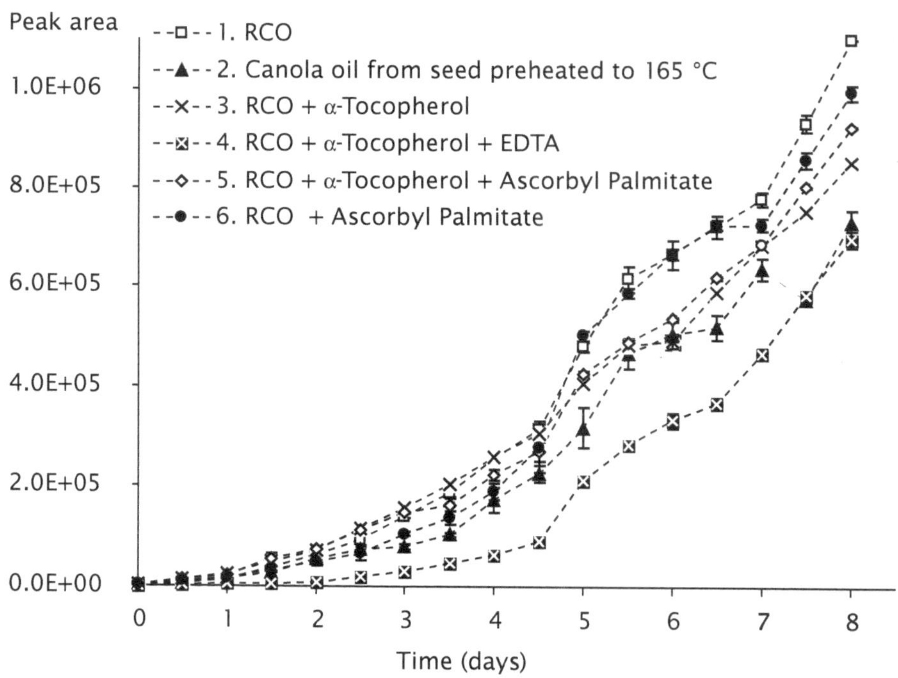

Fig. 15.6. Formation of headspace *c,t*-2,4-heptadienal in 10% water-in-water emulsion systems with an oil phase of 80% TO and 20% of the following:

1. RCO (control)
2. Canola oil from preheated seed (PSCO) heated to 165 °C
3. RCO + 16 mg/kg α-tocopherol
4. RCO + 16 mg/kg α-tocopherol + 16mg/kg sodium dicalcium-EDTA
5. RCO + 16 mg/kg α-tocopherol + 4mg/kg ascorbyl palmitate
6. RCO + 4mg/kg ascorbyl palmitate

2007). On its own and in combination with ascorbyl palmitate, α-tocopherol provided some protection compared to the control; yet, neither was as effective as the combination of α-tocopherol and EDTA. Ascorbyl palmitate did not provide any discernible advantage as it followed the same pattern of volatile formation (that is, oxidative deterioration over time) as the control. In comparison, the emulsion containing PSCO was more stable to oxidative deterioration than all other formulations within this study, with the exception of the α-tocopherol/EDTA combination. Therefore, we demonstrated that PSCO was an effective antioxidant within these systems because it performed better than α-tocopherol, ascorbyl palmitate, and a combination of α-tocopherol/ascorbyl palmitate. This finding could be of interest to the food industry for formulations where canola oil from preheated seed could provide a source of natural antioxidant activity. This is important as trends for cleaner product labeling have produced a push to reduce the occurrence of added antioxidants.

Acknowledgments. The authors would like to acknowledge Dr. Chakra Wijesundera of CSIRO Animal, Food and Health Sciences, Australia, and Associate Professor Matt Golding of Massey University, New Zealand, for invaluable scientific discussion. We would also like to acknowledge our CSIRO Animal, Food and Health Sciences colleagues Claudio Ceccato and Dr. Peter Watkins for analysis.

References

Aladedunye, F. A.; Przybylski, R. Frying performance of canola oil triacylglycerides as affected by vegetable oils minor components. J. Am. Chem. Soc. 2012, 89, 41–53.

Azadmard-Damirchi, S.; Habibi-Nodeh, F.; Hesari, J.; Nemati, M.; Achachlouei, B. F. Effect of pretreatment with microwaves on oxidative stability and nutraceuticals content of oil from rapeseed. Food Chem. 2010, 121 (4), 1211–1215.

Ballinger, D. J.; Salisbury, P. A. Seedling and adult plant evaluation of race variability in Leptosphaeria maculans on Brassica species in Australia. Aust. J. Exp. Agr. 1996, 36 (4), 485–488.

Bell, J. M.; Belzik, R. J. Goitrogenic properties. Rapeseed mean in livestock and poultry—A review; Bowland, J. P.; Clandinin, D. R.; Wetter, L. R.; Eds; Can. Dep. Agr. Publ. 1257: Ottawa, Canada, 1965; 45–60.

Bernini, R.; Mincione, E.; Barontini, M.; Provenzano, G.; Setti, L. Obtaining 4-vinylphenols by decarboxylation of natural 4-hydroxycinnamic acids under microwave irradiation. Tetrahedron 2007, 63, 9663–9667.

Bodamyali, T.; Stevens, C. R.; Blake, D. R.; Winyard, P. G. Reactive oxygen/nitrogen species and acute inflammation: A Physiological Process. Free Radicals and Inflammation; Winyard, P. G.; Blake, D. R.; Evans, C. H.; Eds; Birkhauser: Basel, Switzerland, 2000; 11–19.

Burton, W.; Salisbury, P.; Potts, D. The potential of canola quality *Brassica juncea* as an oilseed crop for Australia. Proceedings of Thirteenth Biennial Australian Research Assembly on Brassicas, Tamworth, New South Wales, Australia, September 8–12, 2003; 62–64.

Burton, W. A.; Ripley, V. L.; Potts, D. A.; Salisbury, P. A. Assessment of genetic diversity in selected breeding lines and cultivars of canola quality *Brassica juncea* and their implications for canola breeding. Euphytica 2004, 136 (2), 181–192.

Callemien, D.; Collin, S. Structure, organoleptic properties, quantification methods, and stability of phenolic compounds in beer—A review. Food Rev. Int. 2010, 26 (1), 1–84.

Cohen, L. A.; Jones, W. M. A study of pH dependence in the decarboxylation of p-hydroxycinnamic acid. J. Am. Chem. Soc. 1960, 82, 1907–1911.

Colton, B.; Potter, T. D. Canola in Australia: The first 30 years: History. http://www.regional.org.au/au/gcirc/canola (accessed Oct. 1, 2012).

Day, L.; Xu, M.; Hoobin, P.; Burgar, I.; Augustin, M. A. Characterization of fish oil emulsions stabilised by sodium caseinate. Food Chem. 2007, 105 (2), 469–479.

Dimakou, C. P.; Kiokias, S. N.; Tsaprouni, I. V.; Oreopoulou, V. I. Effect of processing and storage parameters on the oxidative deterioration of oil-in-water emulsions. Food Biophys. 2007, 2 (1), 38–45.

Donaghy, J. A.; Kelly, P. F.; McKay, A. Conversion of ferulic acid to 4-vinyl guaiacol by yeasts isolated from unpasteurized apple juice. J. Sci. Food Agric. 1999, 79 (3), 453–456.

Donnelly, J. L.; Decker, E. A.; McClements, D. J. Iron-catalyzed oxidation of Menhaden oil as affected by emulsifiers. J. Food Sci. 1998, 63 (6), 997–1000.

Engels, C.; Schieber, A.; Gänzle, M. G. Sinapic acid derivatives in defatted Oriental mustard (*Brassica juncea L.*) seed meal extracts using UHPLC-DAD-ESI-MS n and identification of compounds with antibacterial activity. Eur. Food Res. Technol. 2012, 234 (3), 535–542.

Fenwick, G. R.; Heaney, R. K.; Mullin, W. J.; VanEtten, C. H. Glucosinolates and their breakdown products in food and food plants. CRC Crit. Rev. Food Sci. 1982, 18 (2), 123–201.

Fomuso, L. B.; Corredig, M.; Akoh, C. C. Effect of emulsifier on oxidation properties of fish oil-based structured lipid emulsions. J. Agric. Food Chem. 2002, 50 (10), 2957–2961.

Frankel, E. N. Lipid Oxidation, 2nd ed.; The Oily Press: Bridgewater, England, 2005.

Frankel, E. N.; Finley, J. W. How to standardize the multiplicity of methods to evaluate natural antioxidants. J. Agric. Food Chem. 2008, 56 (13), 4901–4908.

Galano, A.; Francisco-Marquez, M.; Alvarez-Idaboy, J. R. Canolol: A promising chemical agent against oxidative stress. J. Phys. Chem. B 2011, 115 (26), 8590–8596.

Goh, S. H.; Khor, H. T.; Gee, P. T. Phospholipids of palm oil (*Elaeis guineensis*). J. Am. Oil Chem. Soc. 1982, 59 (7), 296–299.

Gottstein, T.; Grosch, W. Model study of different antioxidant properties of α- and γ-tocopherol in fats. Fett 1990, 92 (4), 139–144.

Gramatica, P.; Ranzi, B. M.; Manitto, P. Decarboxylation of cinnamic acids by *Saccharomyces cerevisiae*. Bioorg. Chem. 1981, 10 (1), 14–21.

Harbaum-Piayda, B.; Oehlke, K.; Soennichsen, F. D.; Zacchi, P.; Eggers, R.; Schwarz, K. New polyphenolic compounds in commercial deodistillate and rapeseed oils. Food Chem. 2010, 123 (3), 607–615.

Hashidoko, Y.; Tahara, S. Stereochemically specific proton transfer in decarboxylation of 4-hydroxycinnamic acids by 4-hydroxycinnamate decarboxylase from *Klebsiella oxytoca*. Arch. Biochem. Biophys. 1998, 359 (2), 225–230.

Heresztyn, T. Metabolism of volatile phenolic compounds from hydroxycinnamic acids by Brettanomyces yeast. Arch. Microbiol. 1986, 146 (1), 96–98.

Holman, R. T.; Elmer, O. C. The rates of oxidation of unsaturated fatty avids and esters. J. Am. Oil Chem. Soc. 1947, 24 (4), 127–129.

Hu, M.; McClements, D. J.; Decker, E. A. Lipid oxidation in corn oil-in-water emulsions stabilized by casein, whey protein isolate and soy protein isolate. J. Agric. Food Chem. 2003, 51, 1696–1700.

Huidrom, D. Development of physically stable canola oil-in-water emulsion and evaluation of the effect of endogenous phenolics on the oxidative stability. M.Sc., University of Manitoba, http://hdl.handle.net/1993/8614 2012.

Hunt, J. A.; Dalgleish, D. G. Heat stability of oil-in-water emulsions containing milk proteins: Effect of ionic strength and pH. J. Food Sci. 1995, 60 (5), 1120–1123 & 1131.

Irfan, I.; Pawelzik, E. The effect of rapeseed treatment by microwave and radio-frequency application on oil extraction and oil quality. Part II: Influence on oil quality. Fett/Lipid 1999, 101 (5), 168–171.

Jacobsen, C.; Timm, M.; Meyer, A. S. Oxidation in fish oil enriched mayonnaise: Ascorbic acid and low pH increase oxidative deterioration. J. Agric. Food Chem. 2001, 49 (8), 3947–3956.

Kamal-Eldin, A.; Appelqvist, L. A. The chemistry and antioxidant properties of tocopherols and tocotrienols. Lipids 1996, 31 (7), 671–701.

Kanazawa, A.; Sawa, T.; Akaik, T. Dietary lipid peroxidation products and DNA damage in colon carcinogenesis. Eur. J. Lipid. Sci. Technol. 2002, 104, 439–447.

Kataoka, H.; Lord, H. L.; Pawliszyn, J. Applications of solid-phase microextraction in food analysis. J. Chromatogr. A 2000, 880 (1–2), 35–62.

Kawanishi, S.; Hiraku, Y.; Murata, M.; Oikawa, S. The role of metals in site-specific DNA damage with reference to carcinogenesis. Free Radic. Biol. Med. 2002, 32, 822–832

Kellerby, S. S.; McClements, D. J.; Decker, E. A. Role of proteins in oil-in-water emulsions on the stability of lipid hydroperoxides. J. Agric. Food Chem. 2006, 54 (20), 7879–7884.

Klinkesorn, U.; Sophanodora, P.; Chinachoti, P.; McClements, D. J.; Decker, E. A. Increasing the oxidative stability of liquid and dried tuna oil-in-water emulsions with electrostatic layer-by-layer deposition technology. J. Agric. Food Chem. 2005, 53 (11), 4561–4566.

Koski, A.; Pekkarinen, S.; Hopia, A.; Wahala, K.; Heinonen, M. Processing of rapeseed oil: Effects on sinapic acid derivative content and oxidative stability. Eur. Food Res. Technol. 2003, 217, 110–114.

Krygier, K.; Sosulski, F.; Hogge, L. Free, esterified, and insoluble-bound phenolic acids. 2. Composition of phenolic acids in rapeseed flour and hulls. J. Agric. Food Chem. 1982, 30, 334–336.

Kuwahara, H.; Kanazawa, A.; Wakamatu, D.; Morimura, S.; Kida, K.; Akaike, T.; Maeda, H. Antioxidative and antimutagenic activities of 4-vinyl-2,6-dimethoxyphenol (Canolol) isolated from canola oil. J. Agric. Food Chem. 2004, 52 (14), 4380–4387.

Laguerre, M.; Lecomte, J.; Villeneuve, P. Evaluation of the ability of antioxidants to counteract lipid oxidation: Existing methods, new trends and challenges. Prog. Lipid Res. 2007, 46, 244–282.

Lamont, J. A History of Canola Production in New South Wales; The Canola Association of Australia Inc.: Junee, NSW, 1991; 6–7.

Lampi, A.-M. The AOCS lipid library: Analysis of tocopherols and tocotrienols by HPLC. http://lipidlibrary.aocs.org/topics/tocopherols/index.htm (accessed Sept 21, 2012).

Larvea, M. D.; Jiang, Y. F.; Xu, G. L.; Lombeck, I. Fatty acid composition of blood lipids on Chinese children consuming high erucic acid rapeseed oil. Ann. Nutri. Metabolism 1992, 36 (5), 273–278.

Let, M. B.; Jacobsen, C.; Pham, K. A.; Meyer, A. S. Protection against oxidation of fish-oil-enriched milk emulsions through addition of rapeseed oil or antioxidants. J. Agric. Food Chem. 2005, 53 (13), 5429–5437.

List, G. R.; Evans, C. D.; Black, L. T.; Mounts, T. L. Removal of phosphorus and iron by commercial degumming of soybean oil. J. Am. Oil Chem. Soc. 1978, 55 (2), 275–276.

Mancuso, J. R.; McClements, D. J.; Decker, E. A. The effects of surfactant type, pH, and chelators on the oxidation of salmon oil-in-water emulsions. J. Agric. Food Chem. 1999, 47 (10), 4112–4116.

Mathews, L. Oilseeds industry review. Commodities: Agri updates. http://www.commbank.com.au/corporate/research/publications/commodities/agricultural-insights/201010-Agri_Updates_Oilseeds.pdf (accessed Oct. 3, 2012).

McClements, D. J.; Decker, E. A. Lipid oxidation in oil-in-water emulsions: Impact of molecular environment on chemical reactions in heterogeneous food systems. J. Food Sci. 2000, 65 (8), 1270–1282.

Meyer, F.; Eggers, R.; Oehlke, K.; Harbaum-Piayda, B.; Schwarz, K.; Siddiqi, M. A. Application of short path distillation for recovery of polyphenols from deodorizer distillate. Eur. J. Lipid Sci. Technol. 2011, 113 (11), 1363–1374.

Mildola Oy, Neito Specialty Rapeseed Oil. Annual Report 2005: Midola. http://www.lannen.fi/en/annual_report_2005/business_divisions/page.asp?DocumentID=31 (accessed Aug. 11, 2008).

Munteanu, D.; Csunderlik, C.; Tincul, L. Synthesis of the monomeric antioxidant 3,5-di-tert-butyl-4-hydroxy-styrene by the thermal decomposition of trans-3,5-di-tert-butyl-4-hydroxycinnamic acid. J. Therm. Anal. 1991, 37, 411–426.

Murray, G. M.; Brennan, J. P. The current and potential costs from diseases of oilseed crops in Australia; GRDC Report CER00002; May 2012.

Naczk, M.; Amarowicz, R.; Sullivan, A.; Shahidi, F. Current research developments on polyphenolics of rapeseed/canola: A review. Food Chem. 1998, 62 (4), 489–502.

Nawar, W. W. Lipids. Food chemistry, 3rd ed.; Fennema, R. R., Ed; Marcel Dekker: New York, 1996; 225–319.

Niknam, S. R.; Turner, D. W. A Single Drought Event, at Different Stages of Development Had Different Effects on the Final Yield of *Brassica Napus* cv Monty and *B. Juncea* Line 397-23-2-3-3; Agriculture Western Australia: Northam, WA, 1999; 14–15.

Nomura, E.; Hosoda, A.; Mori, H.; Taniguchi, H. Rapid base-catalyzed decarboxylation and amide-forming reaction of substituted cinnamic acids *via* microwave heating. Green Chem. 2005, 7, 863–866.

Nowak, H.; Kujawa, K.; Zadernowski, R.; Roczniak, B.; Kozowska, H. Antioxidative and bactericidal properties of phenolic compounds in rapeseeds. Fett 1992, 94 (4), 149–152.

Oberndorfer, C.; Lücke, W. The effect of rapeseed treatment by microwave and radio-frequency application on oil extraction and oil quality. Part I: Influence on mechanical oil extraction. Lipid-Fett 1999, 101 (5), 164–167.

Oskoueian, E.; Abdullah, N.; Hendra, R.; Karimi, E. Bioactive compounds, antioxidant, xanthine oxidase inhibitory, tyrosinase inhibitory and anti-inflammatory activities of selected agro-industrial by-products. Int. J. Mol. Sci. 2011, 12 (12), 8610–8625.

Pawliszyn, J. Theory of solid-phase microextraction. J. Chromatogr. Sci. 2000, 38 (7), 270–278.

Pekkarinen, S. S.; Stockmann, H.; Schwarz, K.; Heinonen, M.; Hopia, A. I. Antioxidant activity and partitioning of phenolic acids in bulk and emulsified methyl linoleate. J. Agric. Food Chem. 1999, 47, 3036–3043.

Ponne, C. T.; Moller, A. C.; Tijskens, L. M. M.; Bartels, P. V.; Meijer, M. M. T. Influence of microwave and steam heating on lipase activity and microstructure of rapeseed (*Brassica napus*). J. Agr. Food Chem. 1996, 44 (9), 2818–2824.

Potter, T.; Marcroft, S.; Walton, G.; Parker, P. Canola in Australia: The first 30 years: Climate and Soil. http://www.regional.org.au/au/gcirc/canola (accessed Oct. 1, 2012).

Prior, E. M.; Vadke, V. S.; Sosulski, F. W. Effect of heat treatments on canola press oils. II. Oxidative stability. J. Am. Oil Chem. Soc. 1991, 68 (6), 407–411.

Przybylski, R.; Eskin, N. A. M. Phospholipid composition of canola oils during the early stages of processing as measured by TLC with flame ionization detector. J. Am. Oil Chem. Soc. 1991, 68 (4), 241–245.

Pyysalo, T.; Torkkeli, H.; Honkanen, E. The thermal decarboxylation of some substituted cinnamic acids. Lebensm. Wiss. Technol. 1977, 10 (3), 145–147.

Ratnayake, W. M. N.; Daun, J. K. Chemical Composition of Canola and Rapeseed Oils. Rapeseed and Canola Oil: Production, Processing, Properties and Uses; Gunstone, F. D. Ed.; Blackwell Publishing: Oxford, UK, 2004; 37–78.

Rice-Evans, C. A.; Miller, N. J.; Paganga, G. Structure-antioxidant activity relationships of flavonoids and phenolic acids. Free Radic. Biol. Med. 1996, 20 (7), 933–956.

Richards, A.; Golding, M.; Wijesundera, C.; Lundin, L. The influence of secondary emulsifiers on lipid oxidation within sodium caseinate-stabilized oil-in-water emulsions. J. Am. Oil Chem. Soc. 2011, 88 (1), 65–73.

Richards, A.; Wijesundera, C.; Salisbury, P. Genotype and growing environment effects on the tocopherols and fatty acids of *Brassica napus* and *B-Juncea*. J. Am. Oil Chem. Soc. 2008, 85 (2), 159–168.

Rimmer, S. R.; van den Berg, C. G. J. Resistance of oilseed Brassica ssp. to blackleg caused by Leptosphaeria maculans. Can. J. Plant Path. 1992, 14, 56–66.

Sachidanandame, K.; Fagan, S. C.; Ergul, A. Oxidative stress and cardiovascular disease: Antioxidants and unresolved issues. Cardiovasc. Drug Rev. 2005, 23, 115–132.

Salisbury, P.; Wratten, N. Canola in Australia: The first 30 years: Brassica napus breeding. http://www.regional.org.au/au/gcirc/canola (accessed Oct. 1, 2012).

Salisbury, P.; Norton, R.; Burton, W. Canola quality *B. juncea* for Australia. Fourth Int. Crop Sci. Congress, Brisbane, Australia, 2004.

Seppanen, C. M.; Song, Q.; Csallany, S. A. The antioxidant functions of tocopherol and tocotrienol homologues in oils, fats, and food systems. J. Am. Oil Chem. Soc. 2010, 87 (5), 469–481.

Shahidi, F.; Naczk, M. An overview of the phenolics of canola and rapeseed: Chemical, sensory and nutritional significance. J. Am. Oil Chem. Soc. 1992, 69 (9), 917–924.

Shrestha, K.; Stevens, C. V.; De Meulenaer, B. Isolation and identification of a potent radical scavenger (Canolol) from roasted high erucic mustard seed oil from Nepal and its formation during foasting. J. Agric. Food Chem. 2012, 60 (30), 7506-7512.

Siger, A.; Nogala-Kalucka, M.; Lampart-Szczapa, E. The content and antioxidant activity of phenolic compounds in cold-pressed plant oils. J. Food Lipids 2008, 15 (2), 137–149.

Simbaya, J.; Slominski, B. A.; Rakow, G.; Campbell, L. D.; Downey, R. K.; Bell, J. M. Quality characteristics of yellow-seeded *Brassica* seed meals: Protein, carbohydrate, and dietary fiber components. J. Agric. Food Chem. 1995, 43 (8), 2062–2066.

Sosulski, F.; Zadernowski, R.; Babuchowski, K., Composition of polar lipids in rapeseed. J. Am. Oil Chem. Soc. 1981, 58 (4), 561–564.

Spielmeyer, A.; Wagner, A.; Jahreis, G. Influence of thermal treatment of rapeseed on the canolol content. Food Chem. 2009, 112 (4), 944–948.

Srinivasan, M.; Singh, H.; Munro, P. A. Formation and stability of sodium caseinate emulsions: Influence of retorting (121 °C for 15 min.) before or after emulsification. Food Hydrocoll. 2002, 16 (2), 153–160.

Suarez, R.; Suarez-Lepe, J. A.; Morata, A.; Calderon, F. The production of ethylphenols in wine by yeasts of the genera Brettanomyces and Dekkera: A review. Food Chem. 2007, 102 (1), 10–21.

Surh, J.; Decker, E. A.; McClements, D. J. Influence of pH and pectin type on properties and stability of sodium-caseinate stabilized oil-in-water emulsions. Food Hydrocoll. 2006, 20 (5), 607–618.

Szydlowska-Czerniak, A.; Karlovits, G.; Hellner, G.; Dianoczki, C.; Szlyk, E. Effect of enzymatic and hydrothermal treatments of rapeseeds on quality of the pressed rapeseed oils. Part I: Antioxidant capacity and antioxidant content. Process Biochem. 2010a, 45 (1), 7–17.

Szydlowska-Czerniak, A.; Karlovits, G.; Hellner, G.; Dianoczki, C.; Szlyk, E. Effect of enzymatic and hydrothermal treatments of rapeseeds on quality of the pressed rapeseed oils: Part II. Oil yield and oxidative stability. Process Biochem. 2010b, 45 (2), 247–258.

Szydlowska-Czerniak, A.; Trokowski, K.; Karlovits, G.; Szlyk, E. Determination of antioxidant capacity, phenolic acids, and fatty acid composition of rapeseed varieties. J. Agric. Food Chem. 2010c, 58 (13), 7502–7509.

Terpinc, P.; Čeh, B.; Ulrih, N. P.; Abramovič, H. Studies of the correlation between antioxidant properties and the total phenolic content of different oil cake extracts. Ind. Crop Prod. 2012, 39, 210–217.

Terpinc, P.; Polak, T.; Segatin, N.; Hanzlowsky, A.; Ulrih, N. P.; Abramovic, H. Antioxidant properties of 4-vinyl derivatives of hydroxycinnamic acids. Food Chem. 2011, 128 (1), 62–69.

Tesaki, S.; Tanabe, S.; Ono, H.; Fukushi, E.; Kawabata, J.; Watanabe, M. 4-Hydroxy-3-nitrophenylacetic and sinapic acids as antibacterial compounds from mustard seeds. Biosci. Biotechnol. Biochem. 1998, 62 (5), 998–1000.

Thiyam, U.; Kuhlmann, A.; Stöckmann, H.; Schwarz, K. Prospects of rapeseed oil by-products with respect to antisoxidative potential. C. R. Chimie 2004, 7 (6), 611–616.

Thiyam, U.; Stöckmann, H.; ZumFelde, T.; Schwarz, K. Antioxidative effect of the main sinapic acid derivatives from rapeseed and mustard oil by-products. Eur. J. Lipid Sci. Technol. 2006, 108, 239–248.

Tsunehiro, J.; Yasuda, F.; Wakamatsu, D.; Nakai, C. Isolation and characterization of lipid-radical scavenging component in crude canola oil. Third International Conference and Exhibition on Nutraceuticals and Functional Foods: From Laboratory to the Real World and the Marketplace; American Oil Chemists' Society: Champaign, IL, 2002; San Diego, CA, 2002.

Turner, R.; McLean, C. H.; Silvers, K. M. Are the health benefits of fish oils limited by products of oxidation? Nutr. Res. Rev. 2006, 19, 53–62.

U N., Genomic analysis in *Brassica* with special reference to the experimental formation of *B. napus* and peculiar mode of fertilization. Jap. J. Bot. 1935, 7, 389–452.

Vaidya, B.; Choe, E. Effects of seed roasting on tocopherols, carotenoids, and oxidation in mustard seed oil during heating. J. Am. Oil Chem. Soc. 2011, 88 (1), 83–90.

Valentová, O.; Novotná, Z.; Svoboda, Z.; Pejchar, P.; Káš, J. Influence of microwave treatment on the quality of rapeseed oil (Letter to the Editor). J. Am. Oil Chem. Soc. 2002, 79 (12), 1271–1272.

Valentová, O.; Novotná, Z.; Svoboda, Z.; Schwarz, W.; Káš, J. Microwave heating and γ-radiation treatment of rape seed (*Brassica napus*). J. Food Lipids 2000, 7, 237–245.

van Nieuwenhuyzen, W.; Tomás, M. C. Update on vegetable lecithin and phospholipid technologies. Eur. J. Lipid Sci. Technol. 2008, 110 (5), 472–486.

Veldsink, J. W.; Muuse, B. G.; Meijer, M. M. T.; Cuperus, F. P.; van de Sande, R. L. K. M.; van Putte, K. P. A. M. Heat pretreatment of oilseeds: Effect on oil quality. Lipid-Fett 1999, 101 (7), 244–248.

Vuorela, S.; Meyer, A. S.; Heinonen, M. Quantitative analysis of the main phenolics in rapeseed meal and oils processed differently using enzymatic hydrolysis and HPLC. Eur. Food Res. Technol. 2003, 217, 517–523.

Wagner, K.-H.; Wotruba, F.; Elmadfa, I. Antioxidant potential of tocotrienols and tocopherols in coconut far at different oxidation temperatures. Eur. J. Lipid Sci. Technol. 2001, 103, 746–751.

Wakamatsu, D.; Morimura, S.; Imai, T.; Tang, Y.; Maeda, H.; Kida, K. Production of canola oil showing radical scavenging activity based on a high Canolol concentration. J. Jpn. Soc. Food Sci. Technol. 2008, 55 (5), 233–238.

Wakamatsu, D.; Morimura, S.; Sawa, T.; Kida, K.; Nakai, C.; Maeda, H. Isolation, identification, and structure of a potent alkyl-peroxyl radical scavenger in crude canola oil, canolol. Biosci. Biotechnol. Biochem. 2005, 69 (8), 1568–1574.

Walling, C.; Wolfstirn, K. B., Substituted styrenes. I. The decarboxylation of substituted cinnamic acids. J. Am. Chem. Soc. 1947, 69 (4), 852–854.

Wanasundara, P. K. J. P. D.; Shahidi, F.; Shukla, V. K. S. Endogenous antioxidants from oilseeds and edible oils. Food Rev. Int. 1997, 13 (2), 225–292.

Wanasundara, U.; Amarowicz, R.; Shahidi, F. Isolation and identification of an antioxidative component in canola meal. J. Agric. Food Chem. 1994, 42 (6), 1285–1290.

Wijesundera, C.; Ceccato, C.; Fagan, P.; Shen, Z. Seed roasting improves the oxidative stability of canola (*B. napus*) and mustard (*B. juncea*) seed oils. Eur. J. Lipid Sci. Technol. 2008, 110, 360–367.

Woods, D. L.; Capcara, J. J.; Downey, R. K. The potential of mustard [*Brassica juncea* (L.) Coss] as an edible oil crop on the Canadian prairies. Can. J. Plant Sci. 1991, 71 (1), 195–198.

Wright, J. S.; Johnson, E. R.; DiLabio, G. A. Predicting the activity of phenolic antioxidants: Theoretical method, analysis of substituent effects, and application to major families of antioxidants. J. Am. Chem. Soc. 2001, 123, 1173–1183.

Wright, P. R.; Morgan, J. M.; Jessop, R. S. Comparative adaptation of canola (*Brassica napus*) and Indian mustard (*B-juncea*) to soil water deficits: Plant water relations and growth. Field Crops Res. 1996, 49 (1), 51–64.

Xu, L.; Diosady, L. L. Removal of phenolic compounds in the production of high-quality canola protein isolates. Food Res. Int. 2002, 35, 23–30.

Zacchi, P.; Eggers, R. High-temperature pre-conditioning of rapeseed: A polyphenol-enriched oil and the effect of refining. Eur. J. Lipid Sci. Technol. 2008, 110, 111–119.

Zacchi, P.; Eggers, R.; Jaeger, P.; Ambrogi, A. High temperature–short time preconditioning of rapeseed: A polyphenol-enriched oil and the effect of degumming. Second Mercosur Congress on Chemical Engineering, Village Rio das Pedras, Club Med, Rio de Janeiro, August 14–18, 2005.

Editors and Contributors

Editors

Amy Logan (nee Richards) has over 10 years of experience as a lipid chemist, and is currently working as a research scientist within CSIRO Animal, Food and Health Sciences in Werribee, Australia. With a Ph.D. from the University of Melbourne (School of Agriculture and Food Systems), Amy has worked alongside the Australian Canola Industry studying the effect of genotype and environment on chemical composition and the influence of oxidative stability within Brassica oils. This research was the basis of her 2006 AOCS Honored Student Award. Her current interests include emulsion/colloidal science, lipid oxidative stability, and structure-function relationships within food matrices. She is currently involved in studies concerning the functional properties of dairy products based on the composition and architecture of the milkfat globule. Awarded with a 2008 Victorian Fellowship and a recipient of the 2010–2011 Australian Academy of Science "Scientific Visits to North America" program, she is an active member of the AOCS including roles within both the Australasian Section and the Lipid Oxidation and Quality division.

Uwe Nienaber is an Associate Principal Scientist in the Analytical Sciences Department at Kraft Foods in Glenview, Illinois, where he is currently leading the Physical Sciences group. He earned both M.S. and Ph.D. degrees in Food Chemistry from the University of Münster in Germany. His main interests are in the areas of analytical testing, lipid oxidation, water relations in foods and shelf life. Prior to joining Kraft Foods in 2001, he was a Research Associate at the Food Science and Technology Department at Ohio State University, conducting research on high-pressure processing of foods. From 1994 to 1997 he worked as R&D Manager at Nutrifood Indonesia with responsibilities for product development and analytical testing. He has been a member of the American Oil Chemists' Society since 2008. He is also a professional member of IFT since 1997 and has served the Chicago Section as treasurer and chair.

Xiangqing (Shawn) Pan is a research scientist at DuPont Nutrition and Health (formally Solae, LLC), and his research interests are in the areas of analytical sciences, lipid oxidation and antioxidants, phospholipids, protein and carbohydrate chemistry. Shawn earned his Ph.D. in the area of Food Chemistry with focus on seafood lipid oxidation and antioxidants from Tokyo University of Marine Science and Technology at Japan, where his study was financially supported by Japanese Government (Monbukagakusho), and his graduate program research in Japan was the basis of his 2005 AOCS Honored Student Award. Prior to joining Solae, LLC, Shawn was a postdoctoral research associate from 2005 to 2007 at Donald Danforth Plant Science Center/University of Missouri-St. Louis in the area of plant biochemistry with focus on lipid metabolomics, lipidomics and proteomics. Shawn has 25 publications in the prestigious journals of the field such as Nature Protocols, Plant Cell, Plant Journal, and Phytochemistry.

As an active AOCS member, Shawn actively involved in AOCS LOQ division activities during last five years. He is an active LOQ extended committee member since 2008 and served in different roles in LOQ division such as LOQ Frankel Best Paper Award Committee Chair, Best Student Poster Award Committee Chair and AOCS Honored Student Award Committee Chair as well as AOCS Young Scientist Award Committee member. Shawn also served as AOCS Annual Meeting Technique program Session Chairs since 2010.

Contributors

Huynh Nguyen Duy Bao was born in 1972 in Vietnam. He obtained his B.S. in 1995 and his Master of Engineering in 2003 at the University of Fisheries, Vietnam. In 2004, he was awarded a scholarship by the United Nations University to study the Postgraduate Fisheries Training for Professionals in Iceland. He received his Ph.D. in Applied Marine Biosciences from the Tokyo University of Marine Science and Technology (TUMSAT) in 2009. He was a postdoctoral research fellow at TUMSAT from 2009–2010, researching the bioactive compounds in edible mushrooms. He has been a lecturer at the Nha Trang University since 1996. His major research interests are bioactive natural products and better utilization of by-products from the seafood industry.

Christelle Bayrasay received her bachelor's degree in biochemistry and molecular biology at Université des sciences Touluse III in Toulouse (France) in the field of lipids, proteins and sugars in cell metabolism, and then her Master's degree in chemistry at Université des sciences Montpellier II in Montpellier, France. Her research included searching for a way to protect vitamin C against oxidation with a polymer chitosan, and focus on drug discovery.

She is currently conducting a Ph.D. at CIRAD (International Research Centre for Agronomy and Development) in Montpellier, France. Her interests include the oxidative stress of cells and in the discovery of new drugs against oxidative damages.

Donald R. Berdahl is Chief Technical Officer of Kalsec, Inc., a leading producer of spice and herb extracts providing natural flavors, natural colors, oxidation management systems and modified hop extracts to the food and brewing industries. Don received his B.S. in Chemistry at the University of Wyoming and his M.S., M.Phil. and Ph.D. degrees in Chemistry at Yale University, studying singlet molecular oxygen and photooxidation. Don was a research scientist at GE's Corporate Research and Development Center in Schenectady, NY for 13 years, working in the areas of polymer stabilization, polymer synthesis and environmental chemistry. This work is described in several papers and 13 U.S. patents. Don is a member of the American Chemical Society and the American Oil Chemists Society and is a past chair of the Lipid Oxidation and Quality Division of the AOCS.

Claire Berton is a post-doctoral scholar at The Pennsylvania State University, Department of Food Science, USA. She carries out a research project to explore the use of solid lipid nanoemulsions as delivery systems of lipophilic ingredients in foods. She graduated with a Ph.D. in Food Science at the University of Nantes and INRA (French National Institute for Agricultural Research) UR1268 Biopolymères Interactions Assemblages research unit in 2011. During her Ph.D. project, she studied the influence of the interfacial layer on lipid oxidation in emulsions. She is interested in the relationships between the structure of complex, multiphase food systems, their reactivity and their functionalities.

Bingcan Chen is currently a Post-doctoral Research Associate at University of Massachusetts Amherst Food Science. He majored in Food Chemistry while obtaining his Doctor of Philosophy from University of Massachusetts Amherst, MA, in 2012. Dr. Chen has been actively working on lipid oxidation mechanisms and the strategies to prevent oxidation in various food systems including bulk oil and food dispersions. His research has been recognized by awards from the Institute of Food Technologists, the American Oil Chemists' Society and the Agriculture and Food Chemistry Division of the American Chemical Society. Dr. Chen has served in a number of positions within the American Oil Chemists' Society, including Student Representative of the Lipid oxidation and Quality section and advisory board for Inform magazine.

Ling-Zhi Cheong is a post-doctoral research fellow in the Department of Engineering, Aarhus University. She obtained her Ph.D. in enzyme and food technology from Universiti Putra Malaysia, Malaysia in 2008. Her research area of interests includes enzyme technology, food and lipid technology, food functionality, food colloids and encapsulation. Currently, she is actively researching surface active lipids with unique self-assembling property for use in encapsulation of sensitive compounds. She has more than 20 publications in related areas.

Eric Decker is currently a Professor and the Department Head of UMass Food Science. Dr. Decker has been actively conducting research to characterize mechanisms by which lipids and antioxidants oxidize in foods and the health implications of antioxidants and lipid oxidation products. Dr. Decker has over 300 publications and has been named as one of the Most Highly Cited Scientists in Agriculture. He has served on numerous committees including work with the FDA and the Institute of Medicine. His research has been recognized by awards from the Institute of Food Technologists, the American Oil Chemists' Society, the Agriculture and Food Chemistry Division of the American Chemical Society and the International Life Science Institute.

Ryan Elias is the Frederik Sr. and Faith E. Rasmussen professor of food science and an assistant professor of food chemistry in the Department of Food Science at the Pennsylvania State University. His research group studies free radical mechanisms in lipid foods. He holds a Ph.D. in food science from the University of Massachusetts—Amherst and completed his postdoctoral training in the Department of Viticulture and Enology at the University of California—Davis, before coming to Penn State in 2008. Ryan has been an active member of AOCS since 2003 and is the incoming 2013 chair of IFT's Food Chemistry Division. Ryan was the recipient of AOCS's Honored Student Award in 2005, and the American Chemical Society's Young Scientist Award in 2012. He teaches undergraduate courses in food analysis and plant food processing at Penn State.

Peter Fagan has over 20 years of experience as a separation scientist and analytical chemist in a variety of fields (environmental monitoring, minerals processing and food science). For the past 12 years, Peter has worked within CSIRO and been extensively involved in research activities that have encompassed new bioactive delivery systems, novel lipid bioactives and new food oils. This has included extensive HPLC method development and analysis of lipids, oxidation products, lipidophilic components and bioactives.

Mia Falkeborg received a bachelor's degree in Biotechnology from Aarhus University, Denmark in 2009, after which she was employed as a research assistant at the Department of Molecular Biology at Aarhus University. In 2011 she received her master's degree in Biotechnology from Aarhus University. During her master studies she had a visiting stay at Wilmar Global R&D Center in Shanghai, China. She is now a Ph.D. student at the Department of Engineering, Aarhus University, under supervision of Associate Professor Zheng Guo. Her thesis is focused on development of innovative emulsifiers with potential applications in the area of microencapsulation of sensitive compounds.

Claude Genot is research director at INRA (French National Institute for Agricultural Research), UR1268 Biopolymères Interactions Assemblages Research Unit in Nantes. She obtained her doctor-engineer diploma in 1982 after a Food Science and Technology engineer diploma from ENSIA (Massy, France; now AgroParisTech) and her ability to manage PH.d. students at the University of Nantes in 2002. She is interested in better understanding how and how much the lipid organization and their interactions within food matrices will intervene in their end-use properties. In recent years, she focused her interest on oxidation in model emulsions and emulsified food systems with recent developments on the oxidative fate of lipids in simulated digestive conditions.

Zheng Guo is an Associate Professor at the Department of Engineering, Aarhus University, Denmark. He received his Ph.D. in Chemical Engineering in 2004 at Tianjin University, China. From 2004 to 2008, he worked as a research Assistant Professor at the Technical University of Denmark. In 2011, he was sponsored by Novartis as a visiting scholar at Massachusetts Institute of Technology. His research area involves lipid processing technology, lipid enzymology and reactions, lipid materials and lipid microbiology.

Masashi Hosokawa is an Associate Professor of Biofunctional Material Chemistry, Faculty of Fisheries Sciences, Hokkaido University. He graduated from Hokkaido University, Japan, in 1990. He took Ph.D. from Hokkaido University in 1998. He has worked in the field of Lipid Biochemistry and Molecular Nutrition.

Dejian Huang obtained his B.Sc. degree from Fujian Teachers University and M.Sc. degree from the Chinese Academy of Sciences. In 1990, he joined Chengdu Institute of Organic chemistry, Chinese Academy of Sciences, as research assistant specialized in organometallic chemistry. Four years later, he moved to Indiana University at Bloomington, Indiana, USA, to study under the guidance of Professor Kenneth G. Caulton. He earned his Ph.D. in 1999, and worked on nanocomposite materials for asymmetric catalysis as postdoctoral associate at Professor Jackie Y. Ying's group at Massachusetts Institute of Technology until 2001. He was then appointed as a principal research scientist in charge of R & D on analysis of dietary antioxidants and free radicals at Brunswick Laboratories at Massachusetts. He has been a faculty member at Food Science and Technology Programme of National University of Singapore since 2004. His research interests include anti-aging medicine, bioactive constituents in edible plants, Chinese herbal medicine, dietary antioxidants and oxidative stress, functional foods for controlling hyperglycemia, luminescent molecular and nanoprobes for detection of small molecules (NO, NO_2, H_2S, ClO^-, etc) of biological importance.

Charlotte Jacobsen is professor in lipids and oxidation in foods. She leads the Lipids and Oxidation group at the National Food Institute, Technical University of Denmark. She obtained her M.Sc. in 1990 and her Ph.D. in 1999, both from the Technical University of Denmark (DTU). During the last 20 years she has carried out research in the area of oxidative stability of foods enriched with functional lipids including omega-3 enriched foods. She is internationally renowned for her research in this area and she has received several awards for her work including the Danish Danisco prize 2003, and the French La Médaille Chevreul 2010 awarded by Association Francaise pour l'étude des Corps Gras, as well as two best paper awards from the American Oil Chemists' Society. She was appointed by EFSA as an expert in the Fish Oil working group under the Biohazard Panel to evaluate the potential hazard associated with human intake of refined fish oil. Her publication list includes more than 100 peer-reviewed manuscripts and book chapters.

Derya Kahveci is an Assistant Professor at the Department of Food Engineering, Yeditepe University, Istanbul, Turkey. She received her Ph.D. from the Graduate School of Engineering at Aarhus University, Denmark, in 2011. Her thesis project aimed at development of enzymatic reactions to produce oil enriched in omega-3 polyunsaturated fatty acids from salmon processing by-products. Her research is focused on enzymatic modification of food lipids for improved functional properties as well as modification of lipases towards altered selectivity.

Mickaël Laguerre has been working in the field of lipid oxidation and antioxidant chemistry for seven years. He received his Ph.D. degree from the University of Montpellier in 2008 in Dr Villeneuve's group at CIRAD, where he developed a new theoretical background now referred to as the "cut-off theory." Mickaël received the AOCS Honored Student Award in 2008 and the Ph.D. Thesis Prize of the Lipid Section of the French Society for Biochemistry and Molecular Biology (GERLI) in 2009. After his Ph.D., he joined Dr. Decker's group at the University of Massachusetts, Amherst, for a post-doctoral stay aimed at studying the ability of polyphenols to protect bulk oil from oxidation. He later returned to France to work at the Science for Oenology Reseach Team (INRA) for the development of a novel spectrophotometric assay to determine the encapsulability and inter-micellar exchange rate of polyphenols in surfactant micelles. Mickaël now holds a researcher position focused on the study of the antioxidant activity of lipophilized molecules in the Joint Research Unit IATE in Montpellier, France.

Jérôme Lecomte has a Ph.D. in molecular chemistry at the National Graduate School of Chemistry, Montpellier, France. He started his career studying biomass conversion into valuable compounds (furans derivatives, sugars, biodiesel) through heterogeneous catalysis (zeolites, hydrotalcites). He worked a few years in a private analytical laboratory at Eurofins, and before joining CIRAD (French Agricultural Research Centre for International Development) in 2001, where he worked as manager of the Cotton Chemistry laboratory. There, he worked on processing cottonseed proteins into biomaterials and on developing seed coatings with controlled release properties. In 2006 he joined the team of Dr. Pierre Villeneuve and focused his work on the chemical and bioprocessing of lipids and biomolecules for food and non-food applications. He is currently working on topics such as restructuring of oils and fats triacylglycerols, hydrophobation and evaluation of bioactives compounds (antioxidants), and synthesis of polymer precursors by functionalization of lipids and phenolics (allylation, epoxydation). Lecomte is interested in the screening of the tropical biomass for new sources of enzymes, lipids and bioactive molecules.

David Julian McClements is a Professor at the Department of Food Science at the University of Massachusetts. He specializes in the areas of food biopolymers and colloids, particular ly on the development of food-based structured delivery systems for bioactive components. Dr McClements received his Ph.D. in Food Science (1989) at the University of Leeds (United Kingdom). He then did Post-Doctoral Research at the University of Leeds, University of California (Davis) and University College Cork (Ireland). He is the sole author of the first and second editions of "Food Emulsions: Principles, Practice and Techniques," co-author of "Advances in Food Colloids" with Professor Eric Dickinson, and coeditor of "Developments in Acoustics and Ultrasonics," "Understanding and Controlling the Microstructure of Complex Foods," "Designing Functional Foods" "Oxidation in Foods and Beverages (Volumes 1 and 2)" and "Encapsulation and Delivery Systems for Food Ingredients and Nutraceuticals." He has published over 470 scientific articles in peer-reviewed journals. McClements has received awards from the American Chemical Society, American Oil Chemists' Society, Society of Chemical Industry (U.K.), Institute of Food Technologists, and University of Massachusetts, in recognition of his scientific achievements. His research has been funded by grants from the United States Department of Agriculture, National Science Foundation, U.S. Department of Commerce, Dairy Management Incorporated, and the food industry. He is a member of the editorial boards of a number of journals, and has organized workshops, symposia and conferences in the field of food colloids, food emulsions, and delivery systems.

Kazuo Miyashita is a Professor of Bio-functional Material Chemistry, Faculty of Fisheries Sciences, Hokkaido University. He graduated from Tohoku University, Japan, in 1979. He received his Ph.D. from Tohoku University in 1985. He worked for Hokkaido University as an Instructor from 1985 to 1995 and as an Associate Professor from 1995 to 2000, when he became a Professor. He has worked in the field of lipid oxidation, and in recent years he also focused on lipid soluble nutrients such as fucoxanthin and astaxanthin, which have great impact on human health. He is a board member of several Japanese Scientific Societies including Japan Oil Chemists' Society and a member of Science Council of Japan. He currently serves as a chief editor of Journal of Oleo Science, an official journal of the Japan Oil Chemists' Society.

Reiko Nagasaka is an assistant professor in the Department of Food Science and Technology at the Graduate School of Marine Science and Technology, where she has been a faculty member since 2011. In her current position, she teaches courses in food chemistry and functional nutrition. She majored and obtained her Ph.D. in Fisheries Science from Tokyo University of Fisheries, Japan. She received the 2006 young scientist award of International Union of Biochemistry and Molecular Biology (IUBMB) and the 2009 young scientist award of Federation of Asian and Oceanian Biochemists and Molecular Biologists (FAOBMB). In 2012, Reiko was a visiting professor for one year at the Nutrition Division, Health Canada. Her primary research areas of interest include functional nutrition, energy metabolism of teleost, and cell signaling pathways.

Roger Nahas is Director of R&D, Antioxidants, Kalsec® Inc. Roger Nahas joined Kalsec in October 2007 as a Lead Scientist. Currently he is in charge of antioxidant research and applications. He is also managing several commercial aspects of the antioxidant product line, on an interim basis. His research interests and commercial activities are focused on finding natural solutions to oxidation problems in today's food systems. Nahas is the author of a number of patents, book chapters, and articles in the area of natural antioxidants and management of oxidation in foods and beverages, and has been an invited speaker to several national and international conferences. He holds a Bachelor of Science from the Lebanese University in Beirut, Lebanon, a Master of Science from MAICh Institute in Chania, Greece, and a Doctorate from the University of Missouri in Columbia, MO, United States.

The Han Nguyen is a candidate for Ph.D. in Food Science and Technology at the Tokyo University of Marine Science and Technology, Japan. His Ph.D. study deals with development of the analysis and evaluation of antioxidative properties as well as the application of a natural antioxidant, ergothioneine, to food and biological materials. He graduated with B.S. from Department of Aquatic Products Processing, Faculty of Food Science at the Nha Trang University, Vietnam in 2005 and started his career as a lecturer for more than 6 years in the same university. He received his Master's Degree in Applied Marine Biotechnology and Engineering in 2010 from the Gangneung-Wonju National University, Korea where he focused on the isolation and identification of certain bioactive compounds from marine resources. He has published several book chapter and peer-reviewed research articles in international journals.

Toshiaki Ohshima is a professor of Faculty of Marine Science at Tokyo University of Marine Science and Technology. He joined the Tokyo University of Fisheries in Food Processing and Preservation Laboratory as an Assistant Professor in 1980 and received his Ph.D. in Agricultural Chemistry from Kyushu University in 1984. After Post-Doctoral Fellow experience at Canadian Institute of Fisheries Technology in Halifax, Canada in 1985 and working at UC Davis as a visiting Researcher in 1994, he was stipend as Professor in Department of Food Science and Technology, Tokyo University of Marine Science and Technology in 2006. His scientific background is in Food Chemistry and Nutrition focused on not only marine products but also general agricultural as well as forestry products in terms of application to food functionality in human health.

Worawan Panpipat received a B.S. in Food Technology from Suranaree University of Technology, Nakhon Ratchasima, Thailand in 2001. She pursued her master's degree in Food Technology at Suranaree University of Technology in 2005 with the scholarship from 2002. Upon her graduation, she started working as a lecturer at School of Agricultural Technology, Walailak University, Nakhon Si Thammarat, Thailand. In 2010, she got a scholarship from the Thai government to pursue her Ph.D. She is now a Ph.D. candidate at the Department of Engineering, Aarhus University, Denmark under the supervision of Prof. Xuebing Xu. Her thesis is focused on biosynthesis of broad spectrum of sterol derivatives and physical characterization of their multiple length scale micelling behavior.

Kristen Robbins is the Assistant Manager of Research and Development for Kemin Food Technologies, located at the worldwide headquarters of Kemin Industries, Inc. in Des Moines, Iowa, USA. She has been with Kemin for 10 years. Her responsibilities include the development of new blends of antioxidants and natural extracts to delay lipid oxidation in fresh and processed meats, salad dressings, nuts, and oils. She is a member of the American Meat Science Association and the Institute of Food Technologists.

Marie-Hélène Ropers is researcher at INRA (French National Institute for Agricultural Research), in the research unit Biopolymers Interactions and Assembly (BIA). After getting her Ph.D. in Physical Chemistry in 2000 (University of Nancy, France), she spent two years at the Max-Planck Institute for Colloids and Interfaces (Germany). Since 2002, she works in the research unit BIA. Her objective is to understand the self-assembly of biomolecules, their organization at interfaces and their impact within food matrices. For the last few years, she focused her interest on the possible relationships between the properties of interfaces and lipid oxidation in model emulsions.

Namal Senanayake graduated from University of Peradeniya in Sri Lanka with a B.S. in Agriculture in 1991. After obtaining his Bachelor's degree, he accepted a position of Lecturer at University of Peradeniya in Sri Lanka from 1991 to 1993. He earned a M.Sc. in Food Science in 1995 and a Ph.D. in Food Science in 2001 at the Memorial University of Newfoundland in Canada. After obtaining his doctorate, he was awarded a Post Doctoral Fellowship at Memorial University of Newfoundland from June to August 2001. In October 2001, he accepted a position as a Research Scientist at Martek Biosciences Corporation (Winchester, Kentucky, USA), focusing on the downstream processing aspects and quality improvements on nutritional oils, specifically omega-3 oils derived from microalgae. From 2005 to 2009, he was a Senior Research Scientist at Martek Biosciences Corporation. He is the author of 22 scientific publications and 12 book chapters, and has also presented 29 papers at various scientific meetings. His research interests are in the areas of omega-3 oils, lipid oxidation, antioxidants, specialty oil refining, structured and modified lipids, oxidative stability of oils and phospholipids. He is currently employed as an Application Scientist at DuPont Nutrition & Health (formerly Danisco USA Inc.) in New Century, Kansas, USA. He is a member of the American Oil Chemists' Society, Institute of Food Technologists, and Society of Sensory Professionals.

Karen Schaich has focused her career on oxidative degradation of foods and biological systems. After receiving her Sc.D. in Food Chemistry from the Massachusetts Institute of Technology, she spent 14 years in the Medical Department at Brookhaven National Laboratory studying free radical reactions in toxicology. She then joined the faculty in Food Science

at Rutgers University and redirected her research to lipid oxidation in foods. Her current research encompasses mechanisms of lipid oxidation and protein co-oxidation, electron spin resonance analyses of free radicals to foods, development of improved assays for quantitating lipid oxidation, isolation and application of natural antioxidants from unused biological materials, and development and standardization of antioxidant activity assays. She teaches an undergraduate course and laboratory in Principles of Food Science and graduate courses in Food Chemistry and Lipid Chemistry.

William Schroeder has been Director of Research and Development at Kemin Food Technologies since April 2010. He is responsible for development of new antioxidants and antimicrobials, with a focus on natural sources of both. Schroeder is responsible for research and development, customer laboratory services, and QA/QC (all worldwide) including locations in Europe, United States, South America, Asia Pacific, China, and India.

Schroeder received his Ph.D. from Iowa State University in Chemical Engineering in 2001. He had the focal point of his work on the reaction pathway of selective oxidation of 1,3-butadiene over VMoO catalysts. While in graduate school he co-invented a continuous process for freezing ice cream using liquid nitrogen. After being granted a patent for the ice cream process, he spent the next 10 years executive at a company founded on the ice cream technology, with several franchises running across the country today.

Ann-Dorit Moltke Sørensen, is working as a post doc in the Lipids and Oxidation group at the National Food Institute, Technical University of Denmark. She obtained her M.Sc. in Food Science and Technology in 2005 and her Ph.D. in 2010 from the Technical University of Denmark. Since 2005 she has worked in the area of lipid oxidation in foods with main focus on antioxidant's efficacy in emulsified systems. She has received a Young Scientist award by the Nordic Lipid Forum and an European Student Travel award by the EAOCS board. Moreover, she received the Edwin Frankel Best Paper award 2012 in Lipid Oxidation and Quality.

Pierre Villeneuve is directing the Lipid Research Group at CIRAD, UMR IATE in Montpellier, France. He obtained a Ph.D. in organic chemistry from University of Montpellier in 1995. His Doctorate Thesis was carried out at CIRAD under the supervision of Prof. Jean Graille on the chiral synthesis of mixed triacylglycerols and study of lipase specificities. He then moved to Wyndmoor, Pennsylvania, USA as a Post Doctoral Fellow at USDA, Hides Lipids and Wool Research Unit. Here he worked in the Group of Dr. Thomas A. Foglia on enzymatic restructuration of oils and fats, production of low fat substitutes and enzymatic polymerization of glycerol derivatives. In 1999 he joined the Danone Company R&D in Paris where he worked as a fats and oils expert on lipid oxidation in complex food matrices. In 2002, he obtained a research group leader position at CIRAD (Montpellier, France) where his research interests deal with biocatalysis applied to oils and fats transformation, synthesis of novel antioxidant molecules and lipid oxidation in heterogeneous systems. He is also in various Research Societies, and is currently serving as the Président of Société Française pour l'Etude des Lipides (SFEL), and on the governing board of EuroFedLipid.

Ik Chian Wong obtained her Bachelor of Science (Hons) at the University of Melbourne, Australia. She is currently pursuing her Ph.D. at the National University of Singapore under the supervision of Associtae Professor Dejian Huang. Her research interests encompass various high throughput-screening methodologies for antioxidants and α-amylase and α-glucosidase inhibitory activities and their applications on discovering novel bioactive constituents from tropical medicinal plants.

Xuebing Xu is a Professor of Agro-Biotechnonology Science, Department of Engineering, Aarhus University. He obtained his Ph.D. in Chemical Engineering from the Technical University of Denmark in 2000. He has more than 180 publications in lipid/enzyme technology, food functionality, industrial biocatalysts, and others.

Lisa Zhou received her Ph.D. in 2012 from the Department of Food Science at The Pennsylvania State University in University Park, PA. Her research focuses on the metal-catalyzed oxidation of phenolics and their effect on lipid oxidation in complex foods. She has presented her findings at national meetings and received the Analytical Division Award from the American Oil Chemists' Society in 2012. She plans to pursue an industry career focusing on research in compound encapsulation and delivery after she completes her graduate studies.

Index

Page references followed by f indicate figures and t indicate tables.

A

AAPH. *See* 2,2'-azobis(2-amidino-propane) dihydrochloride (AAPH)

Abeta-induced cytotoxicity, and ergothioneine, 388

absorbance wavelengths
- concentration effects, 77f
- conjugated dienes, 328
- conjugated trienes, 329, 330f
- DPPH and colored test compounds, 402
- fluorescence. *See* fluorescence
- molecular structure effect, 75f
- phosphorescence probes, 331, 331–333f
- spectral overlap, 69
- spectrophotometry. *See* spectrophotometry
- TAG oxidative stability, 166f

ABTS. *See* 2,2'-azino bis(3-ethylbenzothiazoline-6-sulfonic acid) diammonium salt (ABTS)

accelerated shelf-life studies. *See* shelf life, of foods

accelerated solvent extraction (ASE), 67–68, 68t, 69f

acceptable daily intake (ADI) levels, 399–400

acidolysis, enzymatic, 218f, 219–220

activated oxygen, 348

active dry yeast
- polysorbate 80 FDA/EU regulatory limits, 453t
- synthetic antioxidant FDA limits, 443t

additions
- alkoxyl radicals, 23–24, 39t, 40
- facilitation, 39t, 40
- peroxyl radicals, 17–18, 39t, 40

air. *See also* oxygen
- extraction oxidation artifacts, 61

aldehydes
- from alkoxyl scission, 24, 42
- amino acid covalent adducts, 195
- aquatic product deterioration, 353, 357
- browning of foods, 43
- DNPH complexation to HPLC separation, 90, 91f
- as lipid oxidation product, 2, 4f, 6f, 43, 81f, 468
- as protein cooxidation marker, 101t
- protein crosslinking by, 43
- as rancidity source, 43

Alde-Safe™ assay, 70

algae, blue-green (cyanobacteria)
- ergothioneine source, 386t, 387
- lipid class separation, 92

alkaline deacidification process, 231–232

alkoxyl radicals (LO•)
- additions, 23–24, 39t, 40
- cyclization, 22–23, 39–40, 39t
- in foods, 25, 135
- H abstraction, 2, 5, 20–21, 38–39, 39t
- peroxyl radical disproportionation, 18–19
- as propogator, 155, 346, 351, 374
- rates of reaction vs. peroxyl, 19, 20, 38, 39t
- scission, 24–34, 26–31f, 39t, 40

alkyl chain length. *See* cut-off theory; polar paradox

alkyl radicals (L•)
- alternate reactions of, 7–8
- from β-elimination of oxygen, 9
- initiation, 4f, 6f, 155, 346–347, 351
- termination reactions, 8

α-scission. *See also* scission
- α or β, 24–25, 32
- definition, 24

α-tocopherol. *See also* tocopherols
- antioxidant activity in food systems, 299t, 445t, 449
- antioxidant dietary supplementation, 366, 368
- as antioxidant *in vivo,* 461
- BODIPY fluorescence, 336
- in canola oil, 457, 461, 469f
- concentration of, 328
- vs. EDTA, 250
- endogenous and ergothioneine, 372
- vs. green tea extract, 434
- homologs, in membrane systems, 266, 286

α-tocopherol *(continued)*
intermembrane mobility, 285
lipophilic vs. hydrophilic antioxidants, 246
liposome penetration, 286
membranes and internalization theory, 278–279
micelle charge and, 157–158, 158f
vs. mushrooms, 362, 365f
phospholipids and, 144–145
polarity and surface activity, 250, 251f
vs. preheated canola seed oil, 475–476, 475f
reduction potential of, 145, 244
structure of, 298f, 462f
structured lipids, 221–222
synergism with catechins, ascorbic acid, 429
vs. Trolox, 146, 146f, 161, 248, 257, 274
amphiphilicity
amphiphilic antioxidants, 235
antioxidants and amphiphiles, 263
antioxidants into association colloids, 280
emulsifiers, 459
fatty acids, 161
lipase helix peptide sequence, 212
lipophilized phenolics, 235
minor oil components, 256
polar lipid components, 473
resorcinolic lipids, 280
analysis of lipid oxidation, 53–111
assay accuracy, 59–60, 68–76, 73t, 75f
assays, standard, 54, 55–57t, 60
chemical and physical combined, 84–89
chemometrics. *See* chemometrics
closed vs. open samples, 94
cooxidation product analysis, 58, 95–98, 97f, 99–101t
extraction oxidation artifacts, 60–68
fundamental requirements, 53–58
multiple pathways challenge, 41, 57
multiple product analysis, 78–84, 110–111
as reproducible and accurate, 54, 59–60
specific product tracking, 57, 90–94
storage skewing, 61, 61t, 84
"thinking chemistry," 54, 60, 68
anhydrous milk fat, 224t, 232
anthocyanins, 309, 402
antioxidant activity mechanisms, 323–339. *See also* antioxidant evaluation
autoxidation cause and prevention, 324–325, 324f
EGCG, 302
electron transfer, 5
emulsions, 134–135
phenolics, 299–300, 300f
antioxidant evaluation, 323–339
antioxidant activity–guided fractionation, 402, 405
autoxidation cause and prevention, 324–325, 324f
bulk oil ORAC assay, 331, 331–333f, 334
conjugated autoxidizable triene (CAT) assay, 329, 330f
electron transfer (ET) reaction-based redox titration, 325–326
fluorescence, BODIPY-based, 335–337, 336f, 337f
fluorescence, organophosphine quencher, 337–339, 338f
HPLC–DPPH assay
correlation with spectrophotometry, 403, 405
for ergothioneine, 405–406, 406f, 407f
HPLC–DPPH-UV-NMR, 404
online HPLC–DPPH, 402–405, 404f
as rapid antioxidant screening, 383
hydrogen atom transfer (HAT) reaction-based, 325–326
limitations of assays, 326–328
lipid peroxidation–based, 328–329
microemulsion ORAC assay, 334–335
review of methods, 328
antioxidants. *See also* antioxidant activity mechanisms; antioxidant evaluation; cut-off theory; diffusibility; polar paradox
ascorbic acid as, 235, 382
ascorbyl palmitate as, 235
tert-butylhydroxyquinone as, 222
caffeic acid as, 235
canola oil as source of, 457–458, 461–467, 462f, 464t
carnosic acid as, 421
concentration and polar paradox, 244–245, 244t, 246, 257
cooxidation effect, 95

definition, 352, 381
dietary supplementation of
ergothioneine and oxidative stress *in vivo,* 400–401
ergothioneine bioavailability, 396
meat discoloration prevention, 366, 368, 369–371f, 372, 401
DTAB as, 136–137
E. coli for EPA, 161
EDTA as, 183, 228t, 229, 230, 235, 249
electron transfer as, 5
emulsion droplet charge, 137, 139
gallic acid as, 235, 299t
green tea as. *See* green tea extract
herbs as, 235, 382, 419f
lactoferin as, 235
as lipid oxidation retardants, 130–131
micelle charge and, 157–158, 158f
minor components as, 141, 231
for modified lipid stability, 235–236
mushrooms as. *See* mushrooms as antioxidants
natural. *See* natural plant extracts
peptides as, 235
phenolics as, 183, 231, 235, 263, 297, 298, 299t
phospholipids as, 141, 144–145, 197, 201, 352
polysaccharides in emulsions, 142–143
preheated canola seed oil as, 458, 463, 466, 467, 474–476, 475f
preventative, 249
as prooxidants, 267, 327
propyl gallate as, 235
proteins in emulsions, 141–142, 169–170, 182
proteins in lipid dispersons, 311
quercetin as, 226t, 227, 299t
regulation of, 362, 382, 442, 443–444t
rosemary as. *See* rosemary
spectral interference, 69
spices as, 235, 382, 419f
sugar alcohols as, 143
synthetic
BHA. *See* butylated hydroxyanisole (BHA)
BHT. *See* butylated hydroxytoluene (BHT)
decreased use of, 442
regulation of, 362, 382, 442, 443–444t
TBHQ. *See* TBHQ (*tert*-butylhydroquinone)
toxicity of, 382
tocopherols as, 221–222, 231, 235, 299t, 352, 354f, 366, 382, 445t, 449, 461
Trolox vs. α-tocopherol, 146, 146f, 161
ubiquinols as, 352
4-VS (canolol) as, 458, 463, 466, 467, 470, 474
whey protein as, 141, 142, 169, 170, 184, 227, 234
aquatic food products, 345–375
ASE and fish tissue, 67
discoloration of fish meat
mushrooms, mechanisms underlying, 372–374, 373f
mushrooms for prevention, 362–372, 364f, 365f, 367f, 369–371f, 390, 401
from oxidation, 351, 353, 354f, 356, 356f
rosemary for prevention, 424
ergothioneine content of, 386t
factors affecting lipid oxidation, 347–353
fermentation products, HPLC–DPPH, 403–404, 404f
fishy odor, 351, 354f
heme protein as prooxidant, 350–351, 374
inhibitors of oxidation, 352
lipid oxidation reactions, 346–347
lipids of fish muscle, 347–348
lipoxygenase initiation, 351
mackerel. *See* mackerel meat
nutrient loss, 357, 358–359t
oxidative deterioration of, 348, 349, 353–357, 354–356f
phospholipids as endogenous antioxidants, 352
postmortem changes in, 345–346, 347, 348, 349, 366
prooxidants in, 349–352, 350f, 374
rosemary EU regulatory limits, 423t
rosemary extract as antioxidant, 424
salmon meat, 366, 368, 447–448
shrimp meats, 390, 399–400, 443t
texture changes, 354f, 356–357
trout. *See* trout meat

aquatic food products *(continued)*
tuna. *See* tuna meat
yellowtail. *See* yellowtail meat
aqueous lipid oxidation
aqueous micelles, 183
chelation in aqueous phase, 182f, 183
droplet size, 131–132, 132f, 136, 164–167, 165f, 168f
emulsifiers in aqueous phase, 180
free fatty acids as lipid model, 161
free-radical scavengers in, 182f, 183
interfacial oxidant affinity, 157–160, 158f, 159f, 245
micelles and oil-in-water emulsions as models, 155–157, 156f
substrate effects, 160–164
arachidonic acid (AA)
lipase-catalyzed acidolysis of capric acid, 221
lipoxygenase generating, 351
position of H abstraction, 13t, 16
solvents and H abstraction, 20t
argon (Ar)
for assay oxygen independence, 70
extraction oxidation artifacts, 60–61
solubility, 61, 62f, 63t
ultrasound extraction, 66
ascorbic acid
acyl ascorbates vs. acid and esters, 271
amphiphilicity of long-chain ascorbates, 280
as antioxidant in food systems, 235, 382
ascorbate and tocopherols, 300
esters vs. monostearate, 269
lipophilic vs. hydrophilic antioxidants, 246, 264, 267
liposome membrane mobility, 285
metal-catalyzed oxidation, 306
vs. mushrooms, 362, 365f
vs. phenolic compounds, 374
vs. phenolics, 298
vs. rosemary, 446–447
supramolecular aggregates, 281–282
synergism reaction with catechins, α-tocopherol, 429
ascorbyl palmitate
as antioxidant in food systems, 235
lipophilic vs. hydrophilic antioxidants, 246, 264, 267
vs. preheated canola seed oil, 475–476, 475f
ASE. *See* accelerated solvent extraction (ASE)
aspasomes. *See* liposomes
assays for lipid oxidation products. *See also* analysis of lipid oxidation; antioxidant evaluation; *specific assays*
absorbance wavelength. *See* absorbance wavelengths
ABTS, 253, 254, 299, 325
accuracy, 59–60, 68–76, 73t, 75f
detection ranges and product concentrations, 76–78, 77f, 77t, 79f
DPPH, 223t, 254, 299, 325
fluorescence, of protein, 191–192, 193f, 195, 326
oxygen electrode, 328, 329
sensitivity of, 83
spectrophotometry
correlation with HPLC–DPPH, 403
cut-off theory, 278, 283
of ergothioneine, 391–392, 394–395, 401–402, 405
of unadsorbed proteins, 186
standardized, 54, 55–57t, 60
stoichiometry of, 59, 71–73, 73t
association colloids. *See also* micelles
from amphiphilic molecules, 256, 280
in bulk oils, 144–147, 146f, 256, 257f
autoxidation. *See also* antioxidant evaluation
antioxidation-reduction equivalence, 325
cause and prevention of, 324–325, 324f
lipid hydroperoxides, as key intermediate, 2, 4f, 6f, 34, 155, 346–347
oxyMb, inhibition of, 362, 363f, 373f
vs. photosensitized oxidation, 22
2,2'-azino bis(3-ethylbenzothiazoline-6-sulfonic acid) diammonium salt (ABTS)
antioxidant evaluation, 325, 401
measuring free-radical scavenging, 253, 254
phenolic antioxidant activity, 299
2,2'-azobis(2-amidino-propane) dihydrochloride (AAPH)
conjugated autoxidizable triene (CAT) assay, 329
hydroperoxide quantification, 338
microemulsions and ORAC assay, 334
vs. superoxide, 302

B

bacteria
canola oil phenolics, antimicrobial, 463
ergothioneine biosynthesis, 368, 382, 387, 397
fatty acid methyl ester measurement, 89
green tea antibacterial activity, 434
growth of vs. rancid development, 346
as microbial lipase source, 212, 215t, 216
microbial lipid extractions, 65, 67
batch deodorization, modified lipids, 230
batch production, enzymatic modification, 233
BCN (β-casein). *See* caseins
beef. *See also* meat products
antioxidant dietary supplementation, 396, 396f, 399
ergothioneine content of, 386t
minced patties and phenolics, 299f
mushroom as antioxidant, 362, 390–391
natural plant extract uses, 444–450, 445t, 449t, 450, 451t
rosemary EU regulatory limits, 423t
storage and green tea extract, 432–433, 432f, 432t, 433f
storage and rosemary extract, 424–425, 424t, 432–433, 432f, 433f
synthetic antioxidant USDA limits, 444t
tallow, 89, 232, 423t, 444t
beer, ergothioneine content, 387t
Berzelius, Jöns Jacob, 243
beta-carotene
discoloration, 430
green tea polyphenols as antioxidants, 429, 430
β-casein (BCN). *See* caseins
β-lactoglobulin (BLG)
as antioxidant in emulsions, 141, 170, 234
cross-linking and iron decomposition, 139
denatured, partially aggregated, 200–201
designed emulsions for study, 186, 187
direct transesterification in aqueous phase, 187
EGCG prooxidant effect, 311, 312f
electrostatic and steric repulsions, 180
interfacial thickness and homogeneity, 197
phospholipid/BLG mixed interface, 197, 201–203, 201f
protein modification study, 192, 193f, 194f, 195
vs. surfactant in emulsion oxidation, 188, 189f, 190
β-scission. *See also* scission
α or β, 24–25, 32
definition, 24
facilitation, 39, 40
beverages
beer ergothioneine content, 387t
green tea polyphenols and beta-carotene, 430
synthetic antioxidant FDA limits, 443t
BHT. *See* butylated hydroxytoluene (BHT)
bigeye tuna. *See* tuna meat
BLG. *See* β-lactoglobulin (BLG)
blue-green algae (cyanobacteria)
ergothioneine source, 386t, 387
lipid class separation, 92
BODIPY-based fluorescence, 335–337, 336f, 337f
bond dissociation energy (BDE)
hydroperoxide H vs. allylic H, 21
phenolic antioxidant capacity, 263
polar paradox and, 264
temperature and, 15, 35, 39
bovine serum albumin (BSA)
vs. casein, 137
designed emulsions for study, 186
electrostatic and steric repulsions, 180
protein modification, 192
vs. surfactants, 187–191, 189f, 190f
Brij
charge, 137, 159, 159f
chlorigenic acid and internalization hypothesis, 277f
head- and tail-group sizes, 138, 169, 183
hydrophilic stabilization, 276
micelles and alkyl chain length, 278
micelles in continuous phase, 143–144
rosmarinate esters and cut-off effect, 286
browning, of food
aldehyde contribution, 43
cooxidations and, 96
natural plant extracts, 432
4-VS in frying oils, 467

bulk oils
antioxidant polar paradox, 140, 243–244, 256–257
association colloids, 144–147, 146f, 256, 257f
EDTA and, 249–250
vs. emulsions, 134–135, 458–459
fish oil and phenolic prooxidant activity, 305t
green tea extract as antioxidant, 429
lipophilic vs. hydrophilic antioxidants, 246–247
as low surface-to-volume ratio, 243
oxidation as surface effect, 243, 246, 459
phenolics as antioxidants, 298, 299t, 300
preheating canola seeds, 457
rosemary antioxidants, 247, 421, 452
butylated hydroxyanisole (BHA)
FDA limits and tolerances, 443–444t
vs. natural plant extracts, 430–431, 431f, 432t, 434, 444–445
polar paradox, 248–250
potential toxicity, 297
synergism with rosemary extract, 442
as synthetic, 297, 328
butylated hydroxytoluene (BHT)
FDA limits and tolerances, 443–444t
vs. natural plant extracts, 434, 444–445
polar paradox, 160, 248, 264
potential toxicity, 297
synergism with rosemary extract, 442
as synthetic, 297, 328

C
caffeic acid, 235
calcium disodium ethylenediaminetetraacetic acid, 249
cancer prevention, 298
canola oil, 457–476. *See also* canola seeds, preheated
Australian development, 460–461
canola industry, 460
chemometrics for quality, 106f
fat emulsions, oxidative stability, 228t
hydrophilic phenolic content, 462–463, 462f
lipophilic phenolic content, 463–467, 464t
phospholipids in crude, 472–473
plastic fats, oxidative stability, 226t
polar lipid levels in crude, 472
vs. preheated seed canola oil, 468t, 470, 471f, 472f, 474, 476
production levels, 460
rapeseed source, 460
short-path distillation, 231
tocopherols, 457, 461
canola seeds, preheated
flavor of, 458
heating conditions, 463, 464t, 466, 470–471
oxidative deterioration protection by
vs. other antioxidants, 458, 474–476, 475f
4-VS and oil-in-water emulsions, 470
4-VS as antioxidant, 458, 463, 466, 467, 474
oxidative stability enhancement, 465
phospholipids increased, 463
4-VS case study, 467–476
methods, 467–470, 468t, 469f
oxidative deterioration, 470, 472f
preheating temperature, 470–471
tuna emulsion fortification, 458, 470
4-VS concentration in emulsion, 470, 471–472, 471f, 474
water degumming, 472–474
4-VS production, 457–458, 464–465, 464t
capillary zone electrophoresis, 421
capric acid
fat emulsions, oxidative stability, 228t
lipase-catalyzed acidolysis, 221
caprylic acid
enzymatically enhanced, 231
fat emulsions, oxidative stability, 228t
oxidative stability vs. stearic/oleic, 221
carbon dioxide (CO_2)
solubility, 61
supercritical fluid extraction (SFE), 66–67, 422
carbonyls
aldehydes and foods, 43
from alkoxyl scission, 24, 42
assays commonly used, 56t
DNPH complexation to HPLC separation, 90, 91f
protein carbonyls, 191–192, 194f, 195, 196f
as protein cooxidation marker, 100–101t

carcinogenicity
epoxides, 81–82
ingestion of lipid oxidation products, 82, 177
multiple pathways challenge, 43
of synthetic antioxidants, 382
cardiac diseases
erucic acid causing, 460
oxidative stability and, 130
Carica papaya lipase (CPL), 213t, 216
carnosic acid
bulk vs. emulsion oxidation, 452, 452t
HPLC–DPPH assay, 403
HPLC–DPPH-UV-NMR, 404
phenolic diterpene quantification, 421–422
polar paradox antioxidant, 247–248, 247f, 248t
in rosemary extract, 404, 418, 420–421, 439
structure of, 249f, 420f, 440f
carnosol
bulk vs. emulsion oxidation, 452, 452t
phenolic diterpene quantification, 421
polar paradox antioxidant, 247, 248t
in rosemary extract, 404, 418, 420–421, 439
structure of, 249f, 420f, 440f
carotenoids
absorbance wavelength, 328–329
antioxidant dietary supplementation, 368
colorimetric DPPH and, 402
as endogenous antioxidant in fish, 352
as natural food antioxidant, 382
singlet oxygen defense, 327–328
caseins. *See also* sodium caseinate
as antioxidant in emulsions, 141, 169, 184
cross-linked on interface, 139
designed emulsions for study, 186
direct transesterification in aqueous phase, 187
electrostatic repulsion, 180
heat resistance and conformation, 459
as metal chelators, 183
vs. other proteins, oxidative stability, 137
peroxide scavenging, 311
phospholipid/BCN mixed interface, 197, 201–203, 201f, 202f
protein modification, 192, 195
vs. surfactants for stabilized emulsions, 187–191, 189f, 190f
castor-bean lipase, 213t, 216
catechins
as antioxidants in food systems, 433–434, 449
content and green tea processing, 427
green tea antioxidant mechanism, 428–430, 440, 441f
in green tea extract chemistry, 426–428, 427f, 428f
oxidation potential, 302
prooxidant activity, 304, 305t, 429
protein-rich foods and, 311
structure of, 440, 441f
synergism with ascorbic acid, α-tocopherol, 429
catechol
antioxidant activity, 301, 301f
metal-catalyized oxidation, 306
structure of, 298f
Cayman kit, 55t, 77t
cell membranes. *See* membrane systems
chain length. *See* critical chain length; cut-off theory; polar paradox
chain reaction. *See* free-radical chain reaction
charge. *See* electrical charge
chelation
in aqueous phase of emulsions, 182f, 183
chelation potential and antixodant activity, 248–250
by EDTA, 183, 249
by emulsifiers, 139, 183, 459
lipid oxidation control in foods, 130
metal redox cycling disruption, 310, 310f, 312–313
by peptides, 235
by phenolics, 299–300, 300f, 304, 374
by phospholipids, 141, 474
as prooxidant behavior, 250
by proteins, 141, 169–170, 196f, 459
by quercetin, 304
by sodium caseinate, 459
by tea catechins, food model systems, 429
by xanthan gum, 142
chemical interesterification, 211
chemistry for analysis. *See* analysis of lipid oxidation
chemokines, 388

chemometrics
  calibration sets for, 104
  for complex lipid data, 58
  data pattern identification, 98, 102–110, 103t, 105f, 106f
  definition, 98
  to differentiate products, 85, 88
  references for, 102
  with spectral/chromatograph assays, 7, 85, 90, 94–95, 102
chewing gum
  rosemary, EU regulatory limits, 423t
  synthetic antioxidant FDA limits, 443t
  with tea polyphenols, 434
chicken
  chemometrics for breast differentiation, 107
  ergothioneine content of, 386t
  natural plant extract uses, 445t, 449–450
  phenolics as antioxidants, 299t
  poultry meat and ASE, 67
  rosemary EU regulatory limits, 423t
  storage and rosemary extract, 424–425, 424t
  synthetic antioxidant USDA limits, 444t
chlorogenic acid
  antioxidant activity in food systems, 299t
  ester alkyl chain variation
    cut-off effect, 255, 267–268, 268f, 271, 271f
    phenolic conjugation, 303
    polarity and antioxidant activity, 160
  fatty acid–esterified vs. free, 303
  internalization hypothesis and esters, 277, 277f
  self-aggregation hypothesis and esters, 282, 288
  surfactant concentration, 283
Citrem, proteins vs. surfactants, 187, 189f, 190
citric acid, as metal chelator, 183
CMC. *See* critical micelle concentration (CMC)
coalescence
  proteins preventing, 169, 180
  stabilizers limiting, 131, 178, 180
cocoa butter, 217, 219
cocoa polyphenols, 298
coconut oil
  fat emulsions, oxidative stability, 228t
  plastic fats, oxidative stability, 226t
  for *trans*-free fats, 225
cod liver oil
  hydroxytyrosol fatty esters, 269
  mushrooms and oxidation, 362, 364f, 390
cold injection, 94
color
  additives, polysorbate 80 regulations, 453t
  beta-carotene discoloration, 430
  canola oil from preheated seeds, 458
  color loss, flavanols and anthocyanins, 309
  colorimteric DPPH, 402
  ET-based titration, 326
  fish meat discoloration
    mushrooms, mechanisms underlying, 372–374, 373f
    mushrooms for prevention, 362–372, 364f, 365f, 367f, 369–371f, 390–391, 401
    from oxidation, 351, 353, 354f, 356, 356f
    rosemary for prevention, 424
  fresh meat color stability, 432–433, 449–450
  phenolic polymerization, 309
  pork and natural vs. synthetic antioxidants, 446, 446t
  spectrophometric analysis, ergothioneine, 391–392
colorimetric DPPH assay, for ergothioneine, 401–402, 401f, 405
column chromatography, dimers and polymers, 83
concentration
  α-tocopherol, 328
  detection ranges and product concentrations, 76–78, 77f, 77t, 79f
  dispersion saturation, 262
  emulsifier and phenolic partitioning, 304
  ferrithiocyanate assay, 78, 83
  mayonnaise emulsifier concentrations, 168
  phenolic oxidation, 306
  phenolics as prooxidants, 304–305
  polar paradox and, 244–245, 244t, 246
  reverse micelles and antioxidant concentration, 257
  SDS and self-diffusivity, 289
  self-aggregation, 281
  surfactants and emulsifier packing, 168

surfactants and self-aggregation hypothesis, 283
4-VS in oil-in-water emulsion, 470, 471–472, 471f, 474
xylenol orange assay, 78, 83
conjugated autoxidizable triene (CAT) assay, 329, 330f
conjugated dienes (CDs)
accelerated solvent extraction (ASE), 67–68, 69f
as alternate pathway product, 7
assays commonly used, 55t
effect of pH on formation, 190, 190f
as first stable products, 34
as lipid oxidation marker
as ELSD/FID supplement, 93
vs. hexanal, conflicting results, 245
lipid peroxidation, 328
lipophilic vs. hydrophilic antioxidants, 246, 248t
liposomes and phenolipid antioxidants, 270
pH and formation of CDs, 190, 190f
proteins vs. surfactants, 188
solvent mixtures, 198, 198f
spectral overlap, 69
containers, glass vs. plastic, 61, 61t
cooxidation product analysis, 58, 95–98, 97f, 99–101t
copper (Cu)
addition reactions of peroxyl radicals, 18
biological system oxidation, 350
catechins as chelators, 307–308
chelation of, 312
copper-catalyzed human LDL oxidation, 300
cyclization of alkoxyl radicals, 22
EDTA and bulk oil, 250
polar lipids and, 473–474
removal from foods, 310
corn oil
accelerated solvent extraction (ASE), 67–68, 69f
α-tocopherol vs. Trolox, 248
green tea extract antioxidant mechanism, 429
phenolic prooxidant activity, 305t
rosemary as bulk oil antioxidant, 421, 452
cottonseed oil
chemometrics for differentiation, 108f
gallate esters and polar paradox, 244–245, 244t
critical chain length
antioxidant location and, 273, 280
development of, 267–268, 268f
hydroxytyrosol and γCMC, 281
in living cells, 271, 271f
location-based antioxidant distribution, 274f
in membrane systems, 269–270
octyl ester in oil-in-water emulsions, 270
self-aggregation hypothesis, 272, 280, 288
critical micelle concentration (CMC)
emulsion surfactant micelles, 143
phenolipids, 280–281
polar paradox cut-off effect, 255
resorcinolic lipids, 280
self-diffusion coefficient of micelles, 288–289
surface tension at CMC (γCMC), 281
cryotrapping, 94
cut-off theory, 261–291. *See also* polar paradox
critical chain length. *See* critical chain length
cut-off effect, 254–255, 263, 267–272, 268f, 270f, 271f
diffusibility and self-aggregation, 288–289
diffusivity and hydrophobicity, 285–288, 287f
internalization hypothesis, 272, 274f, 275–279, 277f
localization at interface, 273–275, 274f
localization in aqueous phase, 279–284
localization in lipid phase, 275–279, 277f, 279f
polar paradox, experimental contradictions, 265–267
polar paradox, nonlinearity of results, 264–265, 265f
reduced mobility hypothesis, 272–273, 280, 284, 285–288, 287f, 288
self-aggregation hypothesis, 272–273, 274f, 280–284, 288–289
cyanidin
antioxidant activity in lipid systems, 299t
phenolic combinations, 300
structure of, 298f

cyanobacteria (blue-green algae)
ergothioneine source, 386t, 387
lipid class separation, 92
cyclization
alkoxyl radicals, 22–23, 39–40, 39t
facilitation, 39–40, 39t
peroxyl radicals, 14, 15–17, 39–40, 39t
cytotoxicity
epoxides, 82
ergothioneine inhibition of, 388

D
DART–MS. *See* direct analysis in real time–mass spectrometry (DART–MS)
data mining. *See* chemometrics
database of knowledge
libraries of spectral analysis, 85, 88, 89
multiple product analysis, 82
dechlorophyllized green tea extract, 434
degumming
emulsions, 468t, 470, 471f, 472–474, 473f
oil from preheated rapeseed, 463
delivery systems
of enzymatically processed oils and fats, 234
green tea extract application, 430–434, 430t, 431f, 432f, 432t, 433f
for rosemary, 419, 419f, 420f
deodorization
of canola oil and 4-VS content, 465, 467
of enzymatically modified lipids, 230–231
of natural plant extracts, 382, 419f, 451
depletion flocculation
polysaccharides in emulsions, 143
protein droplet–droplet bridging, 180
desorption methods, sensitivity of, 83
DHA (docosahexaenoic acid)
conformation, 161, 162, 164f
as fish lipid PUFA, 345
H abstraction, 13t, 16, 161
vs. linoleic acid, oxidative stability, 167, 168f
lipase-catalyzed acidolysis, 221
lipoxygenase generating, 351
monohydroperoxide (HP) formation, 161–162, 163f
mushroom extracts as food antioxidants, 390
4-VS and oil-in-water emulsions, 469f
diacylglycerols (DAGs), 219–220
dietary supplementation
cyanobacterial human supplements, 387
ergothioneine and oxidative stress *in vivo,* 400–401
ergothioneine bioavailability, 396
meat discoloration prevention, 366, 368, 369–371f, 372, 401
diethylcarbamate (DETC) complexation of epoxides, 90, 91f
diffusibility
aggregates, in lipid phases, 289
entropy driving diffusion, 275, 276
food system phase distribution, 328
hydrophobicity and, 284, 285–288, 287f
as key characteristic, 262
micellization and, 280
reduced mobility hypothesis, 272–273, 284
self-aggregation and, 288–289
self-diffusion coefficient of micelles, 288–289
unadsorbed emulsifiers affecting, 459, 460
dilauroyl phosphatidylcholine (DLPC), 201–203, 201f, 202f
dimers
as alternate reaction product, 2, 6f, 43, 81f
assays commonly used, 57t
peroxyl radicals, 17–18
toxicity of, 43, 82
dinitrophenylhydrazine (DNPH) complexation of carbonyls, 90, 91f
direct analysis in real time–mass spectrometry (DART–MS), 86, 87–89, 87f, 88f
discoloration. *See* color
dismutation
facilitation, 39t, 40
of peroxyl radicals, 19, 39t, 40
dispersions. *See* lipid dispersions
distillation. *See* short-path distillation (SPD)
dittany, 382
dityrosine, protein aggregation, 195
DNA
epoxides vs. aldehydes, 43
ergothioneine damage inhibition, 388, 389
quinone damage, 309
DOPC (phosphatidycholine with oleic acid), 145–147, 146f
DPA (docosapentaenoic acid), 221

DPPH (2,2-diphenyl-1-picrylhydrazyl)
antioxidant evaluation, 325, 383, 401, 401f, 402–405, 404f, 442
catechin radical scavenging, 429
ergothioneine, colorimetric DPPH assay, 401–402, 401f, 405
ergothioneine, radical scavenging, 360t, 383, 388, 405–406, 406f, 407f
as oxidative status marker, 223t, 254
phenolic antioxidant activity, 299
drip loss, 356
droplet size
designed emulsions for study, 186
droplet surface area, 136, 178, 181, 182f, 459
emulsion oxidation and, 164–167, 165f, 168f, 178, 181–182, 182f
food emulsion dispersed phase, 131–132, 132f, 178
microemulsions and ORAC assay, 334
multiple emulsions, 133
nanoemulsions, 132
range of sizes, 131, 178
DTAB (dodecyltrimethylammonium bromide)
as antioxidant, 136–137
electrical charge, 136–137, 159, 159f
vs. Tween 20 and SDS, 159, 159f
dynamic headspace analysis, 86–87, 468

E
E. coli, as EPA antioxidant, 161
EC. *See* epicatechin (EC)
ECG. *See* epicatechin-3-gallate (ECG)
edible mushrooms. *See* mushrooms as antioxidants
EDTA (ethylenediaminetetraacetic acid)
as antioxidant in food systems, 235, 249
fat emulsion stability, 228t, 229
functional foods with SLs, 230
as metal chelator, 183, 249, 313
polar paradox, 249–250
vs. preheated canola seed oil, 475–476, 475f
prooxidant behavior, 313
in protein modification study, 192, 194f
protein/phospholipid interfacial layer, 201, 201f
SOD and metal-catalyzed oxidation, 307
in solvent mixtures study, 198f
EGC. *See* epigallocatechin (EGC)
EGCG. *See* epigallocatechin gallate (EGCG)
eggs
egg yolk as emulsifier, 267
egg yolk liposomes, 269
ergothioneine content of, 386t
eicosapentaenoic acid (EPA)
conformation, 162
E. coli as antioxidant, 161
as fish lipid PUFA, 345
mushroom extracts as food antioxidants, 390
position of H abstraction, 13t
4-VS and oil-in-water emulsions, 469f
electrical charge. *See also* polar paradox
antioxidant activity and, 157–160, 158f, 159f
dispersion oxidative stability, 262
emulsifiers, 179–180, 182, 234, 459
emulsion droplets, 132, 137, 139, 190
emulsion interfacial layer, 182, 182f, 459
emulsion interfacial metals, 159, 159f, 182, 182f
micelles, 157–158, 158f
oil-in-water emulsions and oxidative stability, 134f, 136–137, 140–141, 189f, 190
proteins and pH, 141
scission and, 34
electrochemical detection (ECD), 93
electron transfer (ET)
as antioxidant mechanism, 5, 301, 301f
ET reaction-based redox titration, 325–326
photo-induced electron transfer (PeT), 336–337
electrophoresis
capillary zone electrophoresis, 421
SDS-PAGE, 191, 194f, 195
emulsifiers
anionic vs. non-ionic, 459
in aqueous phase, 180
charge effects, 179–180, 182, 234, 459
chelation by, 139, 183, 459
concentration and phenolic partitioning, 304
as emulsion stabilizers, 178, 459–460
hydrophile-lipophile balance (HLB), 169
oxidative stability effects, 139, 167–170
phospholipid emulsifiers and polar paradox, 266
polysorbate 80, regulation of, 453–454, 453t

emulsifiers *(continued)*
  protein double role in oxidation, 191–195, 193f, 194f, 196f
  surface load, definition, 179
  unadsorbed
    concentrations of, 183–184, 184f
    emulsion designed for study, 185–187, 185f
    minimal vs. excess, 195, 196f
    physical hinderance of interaction, 459–460
    protein double role, 191–192
emulsions. *See also* interfacial layer, oil-in-water emulsions
  antioxidant polar paradox, 140
  vs. bulk oils, 134–135, 458–459
  canola-based oxidation prevention. *See* canola oil; canola seeds, preheated
  definition, 131, 178
  degummed, 472–474
  as delivery system, 234
  designing for interface study, 185–187, 185f
  droplet aqueous structure, 156, 156f
  droplet size. *See* droplet size
  droplet traits, 131, 132, 132f, 133–134, 134f
  droplets and free fatty acids, 140
  droplets and lipid oxidation, 135–140
  droplets and phospholipids, 141
  emulsifiers and oxidative stability, 139, 167–170
  fat emulsions, 227, 228t, 229
  green tea catechin prooxidant behavior, 304, 429
  interface thickness, 138, 179, 182f, 183, 197, 234, 459
  interfacial oxidative affinity, 157–160, 158f, 159f, 245
  layer-by-layer (LbL) emulsions, 138
  lipophilic vs. hydrophilic antioxidants, 246–247
  metals as prooxidants, 134–135
  metals in, 135, 136, 159, 159f, 170, 182, 182f, 249
  microchannel (MC) emulsification, 165, 166f
  microemulsions and ORAC assay, 334–335
  molecular conformation of, 161, 162, 164, 164f
  multiple emulsions, 132–133, 133f
  mushrooms as antioxidants. *See* mushrooms as antioxidants
  nanoemulsions, 132
  oil-in-water. *See* oil-in-water emulsions
  oil-in-water vs. water-in-oil, 227, 228t
  oxidation, main findings, 134–135, 181–184, 182f, 184f, 458–460
  phase overview, 131, 273
  phenolics as prooxidants, 304–305, 305t, 306f
  photosensitizers, 144
  polysaccharides in, 142–143, 178
  prooxidants in, 134–135, 136, 142–143
  protein double role in oxidation, 191–195, 193f, 194f, 196f
  protein vs. surfactant, 183–184, 187–191, 189f, 190f, 197
  protein/phospholipid interfacial layers, 201–203, 201f, 202f
  proteins as antioxidants, 141–142
  riboflavin and light, 144
  rosemary extract as antioxidant, 247, 421, 452
  substrate and aqueous lipid oxidation, 160–164, 163f, 164f
  surface area of, 135, 136, 178, 181, 182f, 459
  surfactant concentration, 283
  surfactant micelles, 143–144
  surfactant mixtures, 197–200, 198f, 199f
  texture modifiers as stabilizers, 178
  triacylglycerols (TAGs) in, 160
  4-VS case study, 467–476
    methods, 467–470, 468t, 469f
    oxidative deterioration, 470, 472f
    preheating temperature, 470–471
    tuna emulsion fortification, 458, 470
    4-VS concentration in emulsion, 470, 471–472, 471f, 474
    water degumming, 472–474
  ζ-potential, 134
encapsulated oil, as delivery system, 234
entropy, 275, 276–277
enzymatically processed oils and fats, 211–236
  advantages of, 211–212, 216
  enzymes for, 212–216

factors influencing oxidative stability, 230–234
improvement of oxidative stability, 234–236
interesterified lipids, 222–229, 223–224t, 226t, 228t
main reaction routes, 217–220, 218–219f
plastic fats, 225, 226t, 227
processing conditions, 233–234
reasons for, 217
short-path distillation, 230–231, 232–233
structured lipids (SLs)
functional foods containing, 229–230
interesterified lipids, 222–229, 223–224t, 226t, 228t
MLM-type SLs, 217, 220–222
EPA. *See* eicosapentaenoic acid (EPA)
epicatechin (EC)
oxidation potential, 302
prooxidant activity in lipid systems, 305t
structure of, 440, 441f
epicatechin-3-gallate (ECG)
oxidation potential, 302
pH and oxidation potential, 302
prooxidant behavior, 304, 305t
quinones and protein aggregates, 309
structure of, 441f
epidioxides
as alternate reaction product, 5, 6f, 81f
facilitation, 39–40, 39t
peroxyl radical cyclization to, 14, 15–17, 39–40, 39t
epigallocatechin (EGC)
oxidation potential, 302
prooxidant behavior, 305t
proteins and, 311
structure of, 440, 441f
epigallocatechin gallate (EGCG)
antioxidant mechanisms, 302
β-lactoglobulin and, 311, 312f
health benefits, 298
liver toxicity, 299
metal-catalyized oxidation, 306
oxidation H_2O_2, 308
oxidation potential, 302
prooxidant behavior, 304, 305t
proteins and, 311
quinones and protein aggregates, 309
epoxides
alkoxyl radical cyclization to, 22–23, 39–40, 39t
as alternate reaction product, 3, 5, 6f, 43, 80–82, 81f
assays commonly used, 56t
detection of, 23, 41, 90, 91f
as dominant products, 22, 82
facilitation, 39–40, 39t
as protein cooxidation marker, 99t
protein rapid reaction, 23, 43, 81–82
toxicity of, 43, 81–82
ergothioneine, 359–375, 381–408
analysis by chromatography, 391, 392–397, 394–396f
analysis by spectrophotometry, 391–392, 394–395, 401–402, 405
animal tissue accumulation, 372, 382, 383–384, 386t, 397–400
antioxidant dietary supplementation, 372, 400–401
antioxidant properties, 359, 362, 382–383, 387–391, 389f
chelation by, 387
detection in biological samples, 383
DPPH, radical scavenging, 360t, 383, 388, 405–406, 406f, 407f
HPLC–DPPH, 383, 402–406, 404f, 406f, 407f
mushrooms
distribution in, 359, 360t, 361f, 361t, 362, 384, 385–386t
meat discoloration prevention, 362–372, 364f, 365f, 367f, 369–371f, 401
mechanisms underlying, 372–374, 373, 373f
oxidative stress and, 372, 383, 388–389, 397, 398, 400–401
as patented, 391
as radical scavenger, 372, 373, 383, 387–388
sources of
in animal tissue, 372, 382
increasing level of, 384
mushrooms, 359, 360t, 361f, 361t, 362, 384, 385–386t, 387
mycobacteria, 387

ergothioneine *(continued)*
sources of *(continued)*
other fungi, 382, 387
plants, 384, 386–387t
structure of, 382f
synthesis of, 384
toxicological studies, 399–400
ESH. *See* ergothioneine
esterases, 212
esterification, enzymatic, 218–219f, 219
ET. *See* electron transfer (ET)
European Union (EU)
polysorbate 80 regulations, 453t
rosemary regulatory status, 422, 423t, 453
synthetic antioxidant usage, 442
evaporative light scattering detection (ELSD)
with electrochemical detection, 93
with HPLC, 92
extraction
methods to limit artifacts, 65–68, 68t
oxidation artifacts, 60–64

F
fat emulsions, oxidative stability, 227, 228t, 229
FDA (U.S. Food and Drug Administration)
food-grade EDTA, 249
polysorbate 80 regulations, 453t
rosemary extract as GRAS, 422
synthetic antioxidant limits and tolerances, 443–444t
trans-fat labeling, 130
Fenton reaction
EDTA preventing, 249
hydroxyl radicals, 304, 308, 350
oxidative damage *in vivo,* 400
fermentation
for ergothioneine content improvement, 384
fish product fermentation, HPLC–DPPH, 403–404, 404f
green, black, and oolong teas, 426, 427, 428
ferric reducing antioxidant power (FRAP) test
antioxidant evaluation, 325
structure–activity studies, 253, 254
ferrithiocyanate assay
detection range for hydroperoxide, 76, 77t, 78, 79f, 83
molecular structure, 74, 75f
oxygen interference, 70
product concentration effects, 78, 83
sensitivity of, 83
solvent stability, 64
spectral overlap, 69
as standard assay, 55t
stoichiometry, 71–72, 73t, 83
filtration
long-chain phenolipids in aqueous phase, 282–283
unadsorbed emulsifier quantification, 185f, 186
fish meat, 345–375
ASE and fish tissue, 67
discoloration of fish meat
mushrooms, mechanisms underlying, 372–374, 373f
mushrooms for prevention, 362–372, 364f, 365f, 367f, 369–371f, 390–391, 401
from oxidation, 351, 353, 354f, 356, 356f
rosemary for prevention, 424
ergothioneine content of, 386t
factors affecting lipid oxidation, 347–353
fermentation products, HPLC–DPPH, 403–404, 404f
fishy odor, 351, 354f
heme protein as prooxidant, 350–351, 374
inhibitors of oxidation, 352
lipid oxidation reactions, 346–347
lipids of, 347–348
lipoxygenase initiation, 351
mackerel. *See* mackerel meat
nutrient loss, 357, 358–359t
oxidative deterioration of, 348, 349, 353–357, 354–356f
phospholipids as endogenous antioxidants, 352
postmortem changes in, 345–346, 347, 348, 349, 366
prooxidants in, 349–352, 350f, 374
rosemary EU regulatory limits, 423t
rosemary extract as antioxidant, 424
salmon meat, 366, 368, 447–448
shrimp meats, 390, 399–400, 443t
texture changes, 354f, 356–357
trout. *See* trout meat
tuna. *See* tuna meat
yellowtail. *See* yellowtail meat

fish oil
  EPA and DHA PUFAs, 345
  fish oil–enriched milk emulsions, 200, 235, 267
  fish pate microencapsulation, 234
  green tea extract antioxidant mechanism, 429, 434
  interesterified SLs and HMFS, oxidative stability, 223t
  internalization hyphothesis and esters, 275
  mayonnaise and antioxidants, 250, 267
  menhaden oil. *See* menhaden oil
  microchannel emulsification, 166f
  pH and charge, 159
  phenolic prooxidant activity, 305t
  phenolics as antioxidants, 299t
  rosemary as bulk oil antioxidant, 421
  rosemary EU regulatory limits, 423t
  salad dressings and phenolics, 299t
  short-path distillation, 231
  SL oxidative stability, 221, 222
  TAG emulsion droplet size, 167, 168f
  TAG oxidative stability, 166f
flame ioization detection (FID)
  gas chromatography (GC) with, 472f, 473f
  HPLC with, 92
flavonoids
  chelation by, 304
  flavanols and color changes, 309
  green tea antioxidant mechanism, 428–430
  in green tea extract chemistry, 426–428, 427f, 428f
flavor. *See also* flavoring agents; off-flavors; rancidity
  antioxidant impact on, 362
  canola seeds, preheated, 458
  chemometrics for flavor components, 102, 107
  green tea vs. rosemary antioxidants, 433
  mayonnaise emulsifier concentrations, 168
  meat flavor deterioration (MFD), 424, 432t
  as natural plant extract challenge, 382, 451–452
  rosemary extract de-flavored, 382
  rosemary vs. green tea antioxidants, 433
  toxic products without, 81–82
  volatiles important for, 129–130
  warmed-over flavor (WOF), 424–425, 432t
"flavor scalping," 83, 86
flavoring agents
  rosemary as FDA GRAS, 422
  rosemary EU regulatory limits, 423t
  synthetic antioxidant FDA limits, 443t
flax oil
  β-lactoglobulin and EGCG, 311, 312f
  phenolic prooxidant activity, 305t
flocculation
  by polysaccharides in emulsions, 142–143
  protein droplet–droplet bridging, 180
  stabilizers limiting, 131, 178, 180, 459
fluidized-bed reactor, 233
fluorescence
  BODIPY fluorophore (B-TOH), 336–337, 337f
  BODIPY-based, 335–337, 336f, 337f
  correlation of fluorescence and peroxidation, 336, 338
  fluorescence chromatograms, canola oil, 469f
  interference of, 339
  organophosphine as quencher, 337–339, 338f
  of protein, 191–192, 193f, 195, 326
food matrices, 129–147
  antioxidant polarity and, 248
  bulk oils. *See* bulk oils
  emulsions. *See* emulsions
  matrix binding of volatiles, 83, 86
  methods to increase oxidative stability, 130–131
foods
  accelerated shelf life studies. *See* shelf life, of foods
  alkoxyl radical scission reactions, 25
  antioxidant evaluation challenges, 328, 421
  aquatic food products. *See* aquatic food products
  browning of. *See* browning, of food
  chemometrics for quality, 107, 108f, 109f
  cooxidation, 97–98
  delivery systems. *See* delivery systems
  EDTA, food-grade, 249
  free fatty acids as lipid model, 161
  functional foods. *See* functional foods
  green tea extract application, 430–434, 430t, 431f, 432f, 432t, 433f

foods *(continued)*
lipids in dispersion, 131
matrices of. *See* food matrices
matrix binding of volatiles, 83, 86
medium-chain FFAs for energy, 220
metals as catalysts, 22–23, 35
methods to increase oxidative stability, 130–131
nutritional loss, 357, 382
odor and safety, 43
phenolic stabilization, 312–313
phenolics as antioxidants, 297–298, 299t
phenolics as bioactives, 298–299
phenolics as prooxidants, 304–305, 305t
purge and trap/thermal desorption to GC, 87
rancidity. *See* rancidity
reactions render inedible, 53
regulation of. *See* regulation
rosemary application, 422–426, 423–425t
safety and odor, 43, 81–82
shelf life. *See* shelf life, of foods
formulated foods. *See* functional foods
fortification
emulsions and preheated canola seeds, 458
functional foods with SLs, 229–230
Fourier transform infrared spectroscopy (FT-IR)
chemometrics for, 102–107
product detection, 84–85
Fourier transform near infrared spectroscopy (FT-NIR)
chemometrics for, 102–107, 105f, 106f, 108f
product detection, 84–85
fractionation, antioxidant activity–guided, 402, 405
FRAP. *See* ferric reducing antioxidant power (FRAP) test
free fatty acids (FFAs)
bulk oil oxidation acceleration, 140–141
fish postmortem, 348
medium-chain FFAs for energy, 220
micelles, 161, 165
quantitation of, 85
removal in oil refining, 140, 144
supercritical fluid extraction (SFE), 66
vs. TAGs in aqueous dispersion, 161
free-radical chain reaction
aquatic food products, 346–347
classical, 2–3, 4f, 11
integrated, 3, 5, 6f, 7, 40–43, 81f, 111
typical products analyzed, 80, 81f
free-radical scavengers
ABTS measuring scavenging, 253, 401
amino acid residues as, 201
amylene as, 64
in aqueous phase of emulsions, 182f, 183
autoxidation prevention, 325
carnosic acid as, 248, 420–421
DPPH in mushrooms, 360t
DPPH measuring scavenging, 401–405
ergothioneine as, 372, 373
food antioxidants as, 130–131
peptides as, 235
peroxide quenching, 310, 310f, 311–312, 312f
phenolics as, 300, 300f, 301–304, 301f
phospholipid reverse micelles and, 145–146, 146f
phospholipids and α-tocopherol, 144–145
polar paradox and, 139–140
polysaccharides as, 142–143
proteins as, 137, 141–142, 169–170, 183
steric hindrance, 302, 304
frying
oil anisidine analyses, 70
rosemary extract, 420, 423t, 425, 426
tea catechins, 433
temperature and scission, 34
4-VS, 466–467
FT-IR. *See* Fourier transform infrared spectroscopy (FT-IR)
functional foods
ergothioneine enriched, 408
green tea extract, 430, 430t
phenolics as bioactives, 297
structured lipids (SLs), 229–230
fungi. *See also* mushrooms as antioxidants
ergothioneine sources
mushrooms, 359, 360t, 361f, 361t, 362, 384, 385–386t, 387
other fungi, 382, 387
microbial lipase source, 213–214t, 216

G
gallic acid
alkyl ester diffusion, 285–286
as antioxidant in food systems, 235, 299t
gallate esters and polar paradox, 244–245, 244t, 264–266, 265f
gallates into micelles, 281
lipophilic vs. hydrophilic antioxidants, 267
micelles and alkyl chain length, 278
prooxidant behavior, 304, 305t
structure of, 298f
γ CMC (surface tension at CMC), 281
γ-tocopherol
as antioxidant *in vitro,* 461
in canola oil, 457, 461, 469f
salmon and dietary antioxidants, 366
SPD effects on, 231
structure of, 462f
gas chromatography (GC)
chemometrics for, 58, 94–95, 102, 107, 110
cyclohexene interference, 64
with flame ioization detection (FID), 472f, 473f
hexanal cycling, 37f
issues with, 85–86
with mass spectrometric detection, 7, 84
purge and trap/thermal desorption, 86–87
quantitation, 86
as standard assay, 56t, 84
4-VS and oil-in-water emulsion, 468, 472f, 473f
gas sparging, 60–61, 66, 70
generally recognized as safe (GRAS)
proteins as antioxidants, 141
rosemary as, 422
genotoxicity
epoxides, 82
quinones, 309
glutathione
as antioxidant, ergothioneine vs., 373, 388, 389f
ergothioneine conservation of, 372
fish meat in ice storage, 357
flavor development, 107
peroxide quenching, 311
glutathione peroxidase
as endogenous antioxidant, 352
flavor development, 107
glycerol
esterification of, 217, 218–219f, 219
fat emulsions, oxidative stability, 228t
transesterification of, 219–220, 219f
glycerolysis, enzymatic, 219–220, 219f
grains and grain products, ergothioneine content, 382, 386–387t
grape pomace extract, prooxidant activity, 305t
grape seed extract
antioxidant activity in food systems, 299t
as antioxidant in meat products, 445t, 448–449, 449t
GRAS. *See* generally recognized as safe (GRAS)
green tea extract, 426–435
as antimicrobial, 434
as antioxidant in food systems, 299t, 439, 445t, 449–450
antioxidant mechanism, 428–430, 440
beta-carotene antidiscoloration, 430
vs. butylated hydroxyanisole (BHA), 430–431, 431f, 432t
catechin content and leaf processing, 427
chemistry, 426–428, 427f, 428f
dechlorophyllized, 434
food application, 430–434, 430t, 431f, 432f, 432t, 433f
fresh meat color stability, 432–433
green tea plant characteristics, 426
marine oils and, 433–434
pH and temperature, 314
prooxidant in o/w emulsions, 304, 305t, 429
vs. rosemary, 432–433, 432f, 433f

H
HBr titration
molecular structure, 74, 83
oxygen interference, 70
sensitivity of, 83
as standard assay, 56t
headspace
oxygen exclusion, 310
samples, 70, 86
solvents, 64
headspace analyses
chemometrics for flavor components, 107

headspace analyses *(continued)*
dynamic headspace, 86–87, 468
preheated seed canola oil, 474–475, 475f
sensitivity of, 85–86
static headspace, 56t, 83, 85–86
4-VS and o/w emulsion, 468, 470, 472, 472f, 473f
health advantages. *See also* toxicity
ergothioneine, 383
functional foods, 229–230
green tea extract, 430t
infant formulas with PUFAs, 222
MLM-type structured lipids, 220
omega-3 fatty acids, 170, 222, 345
phenolics as bioactives, 297, 298
heart diseases
erucic acid causing, 460
oxidative stability and, 130
heat. *See* temperature
hemoglobin (Hb), as prooxidant, 350–351, 374
herbs
as antioxidants in food systems, 235, 382
deodorization of, 382, 419f
HPLC–DPPH-UV-NMR, 404
reverse-phase HPLC online, 404–405
heterolytic reactions, 35
hexanal
amino acid covalent adducts, 195
cycling of, 36, 37f
as expected/monitored product
among few monitored products, 41, 59, 80, 95
beef burgers and green tea extract, 431–432, 432f
vs. conjugated dienes, conflicting results, 245
lipophilic vs. hydrophilic antioxidants, 246, 248t
oleic acid and o/w emulsions, 140
vs. pentane from β-scission, 32
phosphatidylcholine in emulsion, 141
plant-based extracts and bulk vs. emulsion, 452, 452t
polar headgroup size of Brij, 183
rosmarinate ester alkyl chain length, 268
as minor product, 32, 41, 57
higher polyunsaturated fatty acids (HPUFAs)
cyclization, 16, 20–21
H abstraction, 20, 39
scission, 25, 26–31f
HMFS. *See* human milk fat substitutes (HMFS)
HO•. *See* hydroxyl radicals (HO•)
homolytic decomposition, 35
HPLC (high-performance liquid chromatography)
chemometrics for, 107, 110
DART–MS facilitating, 88
with electrochemical detection (ECD), 393
epoxide detection, 83
ergothioneine detection, 383, 392–396, 394f, 395f, 405–406, 406f, 407f
with evaporative light scattering detection (ELSD), 92
filtration prior to, 283
with flame ioization detection (FID), 92
hexanal cycling, 37f
HPLC–DPPH
correlation with spectrophotometry, 403, 405
for ergothioneine, 405–406, 406f, 407f
HPLC–DPPH-UV-NMR, 404
online HPLC–DPPH, 402–405, 404f
as rapid antioxidant screening, 383
limitations of, 92
lipid class separations, 90–94
with mass spectrometry (HPLC-MS), 7, 56t, 84, 391, 397
phenolic diterpene quantification, 421–422
post-column reaction
HPLC post-column derivation, 393–396, 394f, 395f
HPLC–DPPH, 383, 402–406, 404f
hydroperoxide detection, 93
as standard assay, 55t
product detection, 83, 84, 90
reversed phase, 91, 92–93, 404–405, 422
as standard assay, 55–57t
stationary phase, 92
with UV-Vis detector, 392–393
4-VS presence, 468, 469f
human LDL, 300
human milk fat substitutes (HMFS)
as interesterified lipids, 222
oxidative stability of, 223t
hydrated matrices, 177
hydrocolloids, as free-radical scavengers, 183

hydrogen (H) abstraction
addition of alkoxyl vs., 23, 24t
alkoxyl radical reactions, 20–21, 23, 24t, 33t
β-scission vs., 33, 33t
facilitation, 38–39, 39t
oxygen radical absorbance capacity (ORAC), 299, 300
peroxyl radical reactions, 11–15, 347
phenolic antioxidant activity, 299
position of, 11–13, 13t
as rate-limiting step, 155
solvents and, 20, 20t, 33, 33t
traditional explanations, 2
hydrogen peroxide (H_2O_2)
ergothioneine and hydroxyl radical production, 388
ergothioneine and metMb production, 372, 373, 391
formation of, 308, 348
membranes and electron transport, 348
reactive oxygen species generation, 308
hydrogen radicals (H•). *See* hydrogen (H) abstraction
hydroperoxides. *See* lipid hydroperoxides (LOOH)
hydrophile-lipophile balance (HLB)
acyl size of emulsifiers, 169
inhibition of lipid oxidation and, 256
Span 20, 197
hydrophilicity
phenolics as hydrophilic, 276, 421
rosemary extract containing rosmarinic acid, 452
tocopherols and methyl substitution, 244
hydrophobic contact, 285
hydrophobicity, 261–291. *See also* cut-off theory; polar paradox
diffusivity and, 284, 285–288, 287f
γ-CMC increase, 281
increasing chain-length, in emulsions, 303
partitioning and, 274
phenolics and scavenging, 303
hydroxy lipids. *See* lipid alcohols
hydroxyl radicals (HO•)
carnosic acid as scavenger, 420–421
catechin radical scavenging, 429
ergothioneine and, 387–388, 389f
Fenton reaction, 304, 308, 350
H abstraction by, 12
hydroperoxide production, 36, 155
kinetics boosting, 36
peroxide quenching, 310, 311–312
phenolic prooxidant reaction mechanism, 306f, 310
propagation, 4f, 6f, 155, 346, 348
as reactive oxygen species, 308–309
ultrasound-assisted extractions, 66
hydroxytyrosol
alkyl hydroxytyrosol ethers, 253, 254f
CMCs for alkyl esters, 281
cut-off effect and living cells, 271
cut-off effect and reduced mobility, 286
internalization hypothesis and esters, 275
o/w emulsion and esters, 160, 269
olive oil and DART–MS, 89
olive oil and Rancimat, 251–252, 252f, 253
surfactant effectiveness of, 256, 256f

I
ice cream
polysorbate 80 FDA/EU regulatory limits, 453t
rosemary EU regulatory limits, 423t
inert gases
for assay oxygen independence, 70
extraction oxidation artifacts, 60–61
solubility, 61, 62f, 63t
infrared spectroscopy (IR). *See* Fourier transform infrared spectroscopy (FT-IR)
initiation of lipid oxidation
alkyl radicals, 4f, 6f, 155, 346–347, 351
alternate reactions, 2
polar paradox, 248–249
postharvest storage, 347
reactive oxygen species, 325
stage identification, 1
injecting cold, 94
interesterification
chemical interesterification, 211
enzymatic interesterification, 217, 218f
interesterified lipids, 222–229, 223–224t, 226t, 228t
interfacial layer, dispersions
hydrophobicity and partitioning, 274, 276
lipid dispersion engineering, 262
as nonpolar surfactants, 276

interfacial layer, dispersions (continued)
oxidation site, 275, 303
thickness, 262
interfacial layer, o/w emulsions, 177–204
compactness and homogeneity, 197
definition, 178
denatured, partially aggregated BLG, 200–201
designing emulsions for study, 185–187, 185f
hydrophobicity and partitioning, 274, 276
key concepts about, 178–181
as nonpolar surfactants, 276
oxidation, main findings, 181–184, 182f, 184f, 458–460
oxidation site, 275
oxidative affinity, 157–160, 158f, 159f, 245, 459
protein double role in oxidation, 191–195, 193f, 194f, 196f
protein vs. surfactant, 183–184, 187–191, 189f, 190f, 197
protein/phospholipid interfacial layers, 201–203, 201f, 202f
rosemary extracts as antioxidants, 421
surfactant mixtures, 197–200, 198f, 199f
thickness, 138, 179, 182f, 183, 197, 234, 459
interleukin-8 (IL8), 388
internalization hypothesis, 272, 274f, 275–279, 277f
iodometric titration
detection range for hydroperoxide, 77t
molecular structure blindness, 74
oxygen interference, 70
as standard assay, 55t
stoichiometry constant, 73
iron (Fe)
biological system oxidation, 349–351, 350f
chelation by catechins, 307–308
chelation by sodium caseinate, 459
chelation strategies, 312–313
cyclization of alkoxyl radicals, 22, 35
EDTA binding, 250
as emulsion prooxidant, 135, 138, 139, 143
hydroperoxide reduction stoichiometry, 71–72, 73t
mayonnaise antioxidant efficacy, 267
peptide iron-reducing, 235
phenolics binding, 304
polar lipids and, 473–474
removal from foods, 310
Trolox and α-tocopherol depletion, 146–147
Israelavhvili's packing parameter, 281

K
ketones
headspace analysis, 4-VS case study, 468
as lipid oxidation products, 2, 4f, 6f
rancidity in postharvest aquatic foods, 353
kinetics of lipid oxidation
free-radical scavenging, 139
hydroxyl radicals boosting, 36
multiple pathways challenge, 41–42, 42f
ORAC assay, 326
purge and trap/thermal desorption to GC, 86–87
rate-limiting by H abstraction, 155
rates of reaction, alkoxyl vs. peroxyl, 19, 20, 38, 39t
koji fermentation, HPLC–DPPH assay, 403

L
lactoferin, 235
lamb
ergothioneine content of, 386t
fat, rosemary EU regulatory limits, 423t
fresh mutton and green tea extract, 434
lard
chemometrics for differentiation, 108–109f
DART–MS, tallow differentiation, 89
gas solubility, 63t
for HMFS, 222
interesterified SLs and HMFS, oxidative stability, 223t
rosemary EU regulatory limits, 423t
synthetic antioxidant USDA limits, 444t
tea polyphenols as antioxidants, 434
lavender, 382
layer-by-layer (LbL) emulsions, 138, 138f
LDL. *See* low-density lipoprotein (LDL)
lemon balm, rosmarinic acid in, 440
light
carotenoids and photosynthesis, 327–328
extraction oxidation artifacts, 61

homolytic decomposition, 35–36
hydroperoxide quantification, 338
riboflavin, as prooxidant in, 144
solvent mixtures study, 198f
linoleic acid
antioxidant polarity, 248
conformation, 162, 164f
C13 vs. C9 products, 94
cyclization, 16
vs. DHA, oxidative stability, 162, 167, 168f
H abstraction and solvents, 20t
H abstraction position, 13t, 41
H abstraction vs. β-elimination, 9, 11
hexanal production, 41
human milk TAG, 222
micelle charge, 157–158, 158f
micelles and Trolox vs. α-tocopherol, 161
peroxidation primary products, 328
scission, 10f, 28–29f, 32, 42
linolenic acid
vs. EPA and DHA, oxidative stability, 162
H abstraction and solvents, 20t
H abstraction position, 13t, 16
oxidation susceptibility, 130
scission, 30–31f
lipases
amphiphilic helix peptide lid, 212
characteristics of, 212, 213–215t
fish postmortem FFA hydrolysis, 348
heating rapeseed, 463
interesterified SLs and HMFS, 223–224t
mammal lipases, 213t, 216
microbial lipases preferred, 216
plant lipases, 212, 213t, 216
lipid alcohols
assays commonly used, 56t
as products of lipid oxidation, 3, 20
lipid bilayer. *See* interfacial layer, dispersions
lipid dispersions, 261–291
definition, 261
lipid dispersion engineering, 262
phenolic prooxidant activity, 304–305, 305t, 306f
phenolics as antioxidants, 298
polar paradox experimental contradictions, 265–267
protein antioxidant activity, 311
surface area effects, 262
lipid free radicals (L•)
antioxidant hydrogen donation, 155, 352
propagation, 4f, 6f, 155, 346
lipid hydroperoxides (LOOH)
accelerated solvent extraction (ASE), 67–68, 69f
alternate reactions of, 34–38
assays commonly used, 54, 55t, 76, 83
batch deodorization, 230
beta-carotene discoloration, 430
β-elimination of oxygen, 9, 10f, 11
C9 and C13 formation, 94
as expected/monitored product
alternate reactions, 2, 3, 5, 7, 8, 41, 43, 57, 59, 80, 89, 90
antioxidant dietary supplementation, 369f, 371f
chemometrics, 85, 106f, 107, 109f
cornerstone of lipid oxidation analysis, 11
DHA oxidations, 159, 163f
electrochemical detection, 93
emulsion droplet interfacial layers, 138–139
liposome membrane mobility, 287
molecular structure, 74, 75f
mushroom extract, 364f, 365f, 367f
o/w emulsion oxidation, 179, 196f
oleic acid and, 140, 145
oxidative deterioration in aquatic foods, 354f, 355f
plant-based extracts and bulk vs. emulsion, 452, 452t
polar paradox, 245, 246–247, 248t, 250, 265–266
product concentrations and detection ranges, 76, 77f, 77t, 78, 79f, 83
as protein cooxidation marker, 99t
rosmarinate ester alkyl chain length, 268
extraction oxidation artifacts, 60, 66
ferrous iron reduction stoichiometry, 71–72, 73t, 83
H abstraction, 2, 11–13, 15–16, 21
iodiometric titration of, 74
as key intermediate, 2, 4f, 6f, 155, 346–347
micelles and, 143, 145
oxygen interference in assays, 70
peroxide quenching, 310, 311–312
phenolic prooxidant reaction mechanism, 306f

lipid hydroperoxides (LOOH) *(continued)*
phospholipids and, 141
quantification of, 338
by singlet oxygen, 144
solvent polarity and cyclization, 22, 23t
transition metal reactions, 135, 136, 159, 351, 374
by triplet oxygen, 349
lipid oxidation
antioxidant evaluation. *See* antioxidant evaluation
antioxidation-reduction equivalence, 325
causes and prevention of, 324–325, 324f
definition, 129
free-radical chain reaction. *See* free-radical chain reaction
hydroperoxides as cornerstone, 11, 34–35
mechanisms of. *See* mechanisms of lipid oxidation
multiple pathways vs. single chain, 2–7, 4f, 6f, 11, 40–43
references for overview, 1
scheme of, 329f
lipophilicity
of antioxidants and entropy, 275
carnosic acid, 420
linearity of, 246
plant-based extracts as, 452
reduced mobility hypothesis, 285
tocopherols and methyl substitution, 244
4-VS (canolol), 458
lipoprotein. *See* low-density lipoprotein (LDL)
liposomes
α-tocopherol penetration, 286
ascorbate and tocopherols, 300
ascorbic acid supramolecular aggregates, 282
conjugated dienes and phenolipid antioxidants, 270
cut-off effect, 269–270, 270f, 271, 286
as delivery system, 261
DHA conformation, 161–162, 163f, 165
egg yolk liposomes, 269
ester alkyl chain lengths, 140
formation of, 280
green tea antioxidant mechanism, 429
internalization hypothesis, 272, 276, 278–279, 279f
intra-/intermembrane mobility, 285–286
phenolics as antioxidants, 299t
self-aggregation, 281
size-dependent internalization, 289
soybean liposomes, 269, 278–279, 287–288, 287f
lipoxygenases
collection of, 36
as endogenous prooxidant, 349
enzymatic initiation of oxidation, 351
oxidative stress or inflammation, 325
liquid chromatography–mass spectrometry (LC-MS), 396–397
LO•. *See* alkoxyl radicals (LO•)
localization, of antioxidants. *See* cut-off theory; partitioning
LOO•. *See* peroxyl radicals (LOO•)
low surface-to-volume (LSV) ratio, 243
low-density lipoprotein (LDL)
autoxidation, 325
lipid peroxide quantification, 338
lipoxygenase oxidation, 351
phenolic combination antioxidants, 300
protein cooxidation, 99t

M

mackerel meat. *See also* fish meat
antioxidant dietary supplementation, 371f
ice storage and antioxidant decline, 357
ice storage and hydroperoxides, 355f
ice storage and PUFAs, 357, 358–359t
lipids of, 347
lipoxygenases in, 351
minced and phenolic antioxidants, 299t
rosemary extract as antioxidant, 425
mammal lipases, 213t, 216
manual extraction vs. ASE, 67–68, 68t, 69f
margarines
enzymatic transesterification, 225
fat emulsions, oxidative stability, 228t
green tea polyphenols and beta-carotene, 430
plastic fats, oxidative stability, 227, 229
synthetic antioxidant FDA limits, 443t
trans-free, 225
as water-in-oil emulsion, 227
marjoram, 382

mass spectrometry
with chromatography, 7, 56t, 84, 391, 396–397
direct analysis in real time–mass spectrometry (DART–MS), 86, 87–89, 87f, 88f
sample sizes, 86, 87
time of flight spectrometers, 87
verification of oxidation products, 94
matrix binding of volatiles, 83, 86
mayonnaise
antioxidant activity and lipophilicity, 267
EDTA as antioxidant, 249, 250
emulsifier concentrations, 168
emulsion droplet size, 136
as o/w emulsion, 458
with SLs, oxidative stability, 229, 230
tocopherols as antioxidants, 235
meat flavor deterioration (MFD), 424, 432t
meat products
beef. *See* beef
chicken. *See* chicken
discoloration, fish meat
mushrooms, mechanisms underlying, 372–374, 373f
mushrooms for prevention, 362–372, 364f, 365f, 367f, 369–371f, 390–391, 401
from oxidation, 351, 353, 354f, 356, 356f
rosemary for prevention, 424
discoloration, other meats
fresh meat color stability, 432–433, 449–450
pork and natural vs. synthetic antioxidants, 446, 446t
discoloration, prevention
dietary antioxidants, 366, 368, 369–371f, 372, 399, 401
green tea, 432–433, 450
mushrooms, 362–372, 364f, 365f, 367f, 369–371f, 390–391, 401
rosemary, 424
ergothioneine content, 386t
extraction of, 65, 67
fish. *See* fish meat
lamb, 386t, 423t, 434
meat flavor deterioration (MFD), 424, 432t
mushrooms as antioxidant. *See* mushrooms as antioxidants
natural plant extract uses, 311, 444–450, 445–447t, 449t
as o/w emulsions, 177
phenolics as antioxidants, 298, 299t
pork. *See* pork
postharvest storage and initiation reactions, 347
rat muscle cells, cut-off effect, 271, 286
rosemary EU regulatory limits, 423t
rosemary extract as antioxidant, 424–425, 424t
synthetic antioxidant USDA limits, 444t
turkey. *See* turkey
mechanisms of lipid oxidation, 1–44. *See also* antioxidant evaluation
alkoxyl radicals, 19–34. *See also* alkoxyl radicals (LO•)
alkyl radicals, 7–8. *See also* alkyl radicals (L•)
causes and prevention of, 324–325, 324f
classical chain vs. alternate reactions, 1–7. *See also* free-radical chain reaction
conditions shifting pathways, 38–40
hydroperoxides as cornerstone, 11, 34–35
lipid hydroperoxides, 34–38. *See also* lipid hydroperoxides (LOOH)
multiple pathways vs. single chain, 2–7, 4f, 6f, 11, 40–43
peroxidation scheme, 329f
peroxyl radicals, 8–19. *See also* peroxyl radicals (LOO•)
membrane systems
antioxidant dietary supplementation, 366
carnosic acid as antioxidant, 421
cut-off effect in, 269–270, 270f
imaging oxidative events, 335
internalization hypothesis, 278–279, 279f, 289
lipid peroxide quantification, 338
polar lipids in, 347–348
polar paradox experimental contradictions, 266
reduced mobility hypothesis, 286–288, 287f
storage temperature and cell membranes, 352

menhaden oil. *See also* fish oil
chlorogenic acid vs. esterified, 255, 303
emulsion interface cross-linking, 139
fat emulsions, oxidative stability, 228t
green tea extract as antioxidant, 434
green tea extract as prooxidant, 433
phenolics as antioxidants, 299t
polarity and surface activity, 250
metals
biological system oxidation, 349–351, 350f
charge of oil droplet and, 159, 159f
chelation. *See* chelation
in classic chain reaction, 3, 4f
cyclization of alkoxyl radicals, 22
EDTA effects, 249
electron transfer as antioxidant, 5
in emulsions, 135, 136, 159, 159f, 170, 182, 182f, 249
hydroperoxide decomposition, 35
hydroperoxide reduction stoichiometry, 71–72
metal redox cycling disruption, 310, 310f, 312–313
pH and properties, 308
phenolic oxidation, 305–308, 306f, 309–314, 310f, 312f
polar lipids and, 473–474
Trolox and α-tocopherol depletion, 146–147
micelles. *See also* association colloids
alkyl chain and diffusion, 286
alkyl chain and polar head depth, 278
aqueous micelles, 183, 280–284
ascorbic acid supramolecular aggregates, 281
Brij micelles in continuous phase, 143–144
charge and antioxidants, 157–158, 158f
conformation of lipids, 162, 164, 164f
critical micelle concentration. *See* critical micelle concentration (CMC)
DHA micelles, 161–164, 164f, 167
emulsion surfactant micelles, 143–144
entropy-driven antioxidant diffusion, 276
fatty acid anions in emulsions, 161
interface aqueous structure, 156, 156f
linoleic acid micelles, 161–164, 164f, 167, 248
long-chain lipophilized phenolics, 235
reverse micelles. *See* reverse micelles
self-aggregation of antioxidants, 280–281, 288–289
self-diffusion coefficient, 288–289
size of, 167
microbes
bacteria. *See* bacteria
green tea antiviral properties, 434
microbial lipases, 216
microbial lipid extractions, 65, 67
microemulsions
conformation at interface, 162
ester solubilization, 161
ORAC assay and, 334–335
substrate concentration vs. emulsifier, 167
microwave
canola seed preheating, 463, 464t, 466, 471
extractions, 65–66
milk drinks
fish oil–enriched milk emulsions, 200, 235
milk as o/w emulsion, 458
with SLs, oxidative stability, 228t, 230
milk fat, anhydrous, 224t, 232
minor components
definition, 230–231
minor oil components, 140–141
polar antioxidants and, 256–257, 257f
short-path distillation, 230–231
MLM-type structured lipids, 217, 220–222
modification of fats and oils. *See* enzymatically processed oils and fats
monoacylglycerols (MAGs), 219–220
monohydroperoxide (HP; LOOH)
DHA oxidation, 161–162, 163f
from peroxyl radical H abstraction, 155
monolauryl glycerol (MLG), 198–200, 198f, 199f
multiple emulsions, 132–133, 133f
muscle. *See* meat products
mushrooms as antioxidants, 357–375. *See also* ergothioneine
antioxidant dietary supplementation, 372, 400–401
antioxidative properties, 359, 360t, 361f, 361t, 362
fish meat discoloration prevention, 362–372, 364f, 365f, 367f, 369–371f, 390–391, 401

HPLC–DPPH for ergothioneine, 383, 392–396, 394f, 395f, 405–406, 406f, 407f
mechanisms underlying, 372–374, 373f
mutagenicity
epoxides, 82
4-vinylsyringol as animutagen, 463–464, 467
mutton
ergothioneine content of, 386t
fat, rosemary EU regulatory limits, 423t
fresh mutton and green tea extract, 434
mycobacteria, as ergothioneine source, 387
myoglobin (Mb)
discoloration of fish meat
mushrooms, mechanisms underlying, 372–374, 373f
mushrooms for prevention, 362–372, 364f, 365f, 367f, 369–371f, 390–391
from oxidation, 353, 354f, 356, 356f
fresh meat color, 432–433
as prooxidant, 350–351, 374
myricetin, 300, 304

N
nanoemulsions, 132
nanotubes, from self-aggregation, 282
natural plant extracts, 439–454
challenges of, 451–454, 452t, 453t
commonly used, 382
ergothioneine. *See* ergothioneine
factors driving shift toward, 441–442, 443–444t
green tea extract. *See* green tea extract
maximizing performance, 450, 451t
meat, poultry, fish uses, 444–450, 445t, 446t, 447t, 449t
phenolic extracts. *See also* phenolics
antioxidant mechanisms, 299–304, 300f, 301f
combinations of phenolics, 300
as nonsynthetic antioxidants, 297–298, 299t
prooxidant mechanisms, 304–309, 305t, 306f
prooxidant minimization, 309–314, 310f, 312f
structure, 297, 298f
phyto-factories, 442
rosemary. *See* rosemary
navel oranges, phenolic combinations, 300
near infrared spectroscopy (NIR). *See* Fourier transform near infrared spectroscopy (FT-NIR)
neat oil, as delivery system, 234
nitrobenzyl pyridine (NBP) assays
molecular structure, 74, 75f, 83
oxygen presence, 70
sensitivity of, 83
temperature-time issues, 83
nitrogen (N_2)
extraction oxidation artifacts, 61
solubility, 61, 62f, 63t
NMR (nuclear magnetic resonance)
for molecular conformation, 161, 162, 164, 164f
as standard assay, 55–56t
normal phase HPLC. *See* HPLC (high-performance liquid chromatography)

O
odor. *See* off-odors
off-flavors. *See also* rancidity
as first quality loss, 222, 345
green tea extract retarding, 430t
infant formula with SLs, 222
mayonnaise emulsifier concentrations, 168
oxidative deterioration in aquatic foods, 353
rosemary preventing, 424, 446–447, 447t
off-odors
driving product analysis, 80
fishy odor, 351, 354f
infant formula with SLs, 222
natural plant extracts retarding, 424, 430t, 449
safety of foods and, 43, 81–82
scission reactions, 3, 43
4-VS in frying oils, 466–467
oil-in-water (o/w) emulsions. *See also* interfacial layer, o/w emulsions
antioxidant polar paradox, 140
vs. bulk oils, 134–135, 458–459
canola-based oxidation prevention. *See* canola oil; canola seeds, preheated

oil-in-water (o/w) emulsions *(continued)*
charge and oxidative stability, 134f, 136–137, 140–141, 189f, 190
definition, 131, 178
degummed, 472–474
designing for interface study, 185–187, 185f
droplet aqueous structure, 156, 156f
droplet size. *See* droplet size
emulsifiers and oxidative stability, 139, 167–170
green tea catechin prooxidant behavior, 304, 429
interface thickness, 138, 179, 182f, 183, 197, 234, 459
interfacial oxidative affinity, 157–160, 158f, 159f, 245
lipophilic vs. hydrophilic antioxidants, 246–247
metals as prooxidants, 134–135
metals in, 135, 136, 159, 159f, 170, 182, 182f, 249
microemulsions and ORAC assay, 334–335
molecular conformation of, 161, 162, 164, 164f
mushrooms as antioxidants. *See* mushrooms as antioxidants
oxidation, main findings, 134–135, 181–184, 182f, 184f, 458–460
phase overview, 131, 273
phenolics as antioxidants, 299t
phenolics as prooxidants, 304–305, 305t, 306f
photosensitizers, 144
polysaccharides in, 142–143, 178
prooxidants in, 134–135, 136, 142–143
protein double role in oxidation, 191–195, 193f, 194f, 196f
protein vs. surfactant, 183–184, 187–191, 189f, 190f, 197
protein/phospholipid interfacial layers, 201–203, 201f, 202f
proteins as antioxidants, 141–142
riboflavin and light, 144
rosemary extract as antioxidant, 247, 421, 452
substrate and aqueous lipid oxidation, 160–164, 163f, 164f
surface area of, 135, 136, 178, 181, 182f, 459
surfactant concentration, 283
surfactant micelles, 143–144
surfactant mixtures, 197–200, 198f, 199f
texture modifiers as stabilizers, 178
triacylglycerols (TAGs) in, 160
4-VS case study, 467–476
methods, 467–470, 468t, 469f
oxidative deterioration, 470, 472f
preheating temperature, 470–471
tuna emulsion fortification, 458, 470
4-VS concentration in emulsion, 470, 471–472, 471f, 474
water degumming, 472–474
vs. water-in-oil emulsions, oxidation, 227
ζ-potential, 134
oilseed rape, 460. *See also* canola oil
oleic acid
emulsion oxidation by, 140
fat emulsions, oxidative stability, 228t
human milk TAG, 222
oxidative stability vs. stearic/capric, 221
phosphatidycholine with (DOPC), 145–147, 146f
position of H abstraction, 13t
reverse micelles and, 145
scission, 26–27f
solvents and H abstraction, 20t
olive oil
chemometrics for quality, 107
DART–MS analyses for quality, 89
FT-IR to classify, 104
gas solubilities, 63t
interesterified SLs, oxidative stability, 223–224t
Rancimat stabilization, 251–252, 252f
olive-leaf phenolics, 329
omega-3 fatty acids. *See also* structured lipids (SLs)
as food additive, 217, 220, 222, 345
oxidative stability of, 223t
ORAC. *See* oxygen radical absorbance capacity (ORAC)
oregano, as food system antioxidant, 235, 382, 445t, 448–449, 449t
Oswald ripening, 131

oxidation potential
antioxidant activity and, 139, 302
tocopherol methyl substitution, 244
oxidative induction time (OIT), 223–224t, 225, 226t, 231
oxidative stress. *See also* antioxidant evaluation
chronic diseases and, 323
ergothioneine deficiency, 398
ergothioneine disease prevention, 383, 388–389
ergothioneine *in vivo,* 400–401
ergothioneine production, 372, 397
in foods vs. assays, 327
oxygen (O_2)
activated oxygen, 348
alkyl radical reactions, 7–8, 9, 155
assay independence of, 70
consumption analysis
droplet size and oxidative stability, 166f
for fresh products, 54
IR and NIR for, 85
lipid peroxidation, 328
mushrooms and hydroperoxide, 364f
phosphorescence probes, 331, 331–333f
protein fluorescence relationship, 191–192, 193f, 195
protein/phospholipid interfacial layer, 201f, 202
proteins vs. surfactants, 188, 189f
surfactant mixtures, 198, 198f
consumption inhibition, 312
in definition of lipid oxidation, 129, 348
exclusion to minimize prooxidants, 310, 313–314
extraction oxidation artifacts, 60–61, 64
fish postmortem oxidative deterioration, 348, 349, 353–357, 354–356f, 366
food system antioxidant oxidation, 327
interface packing density, 168
interfacial oxidant affinity, 158f
permeability of plastics, 61, 61t
peroxyl β-elimination of, 9–11, 10f
peroxyl disproportionation producing, 19
polysaccharides as antioxidants, 142
as reaction condition, 39–40, 72
reactive oxygen species. *See* reactive oxygen species (ROS)
sample incubation, 94
scavengers. *See* free-radical scavengers
singlet oxygen as prooxidant, 144, 325, 327, 348–349, 349f, 421
solubility, 61, 62f, 63t
triplet oxygen and FFA oxidation, 348, 349
triplet oxygen and phenolic oxidation, 307
oxygen electrode, 328, 329
oxygen radical absorbance capacity (ORAC)
bulk oil, 331, 331–333f, 334
as hydrogen atom transfer (HAT) assay, 326
microemulsions, 334–335
phenolic antioxidant capacity, 299, 300
radical scavenging activity, 303
oxymyoglobin (oxyMb)
autoxidation inhibition, 362, 363f, 373f, 391
autoxidation of, 338
rosemary as beef antioxidant, 450
transition metal prooxidants, 351

P

packaging. *See also* storage
containers, glass vs. plastic, 61, 61t
oxygen exclusion, 313
packed bed reactor, 233
packing parameter, Israelachvili's, 281
palm oil, 222, 460
palm olein, 226t, 426
palm stearin
in cocoa-butter equivalents, 217
fat emulsions, oxidative stability, 228t
short-path distillation, 231
with soybean oil in various reactors, 233
for *trans*-free fats, 219, 225
Pangasius species of fish, dietary antioxidants, 366
partial hydrogenation, 130
partition coefficients, octanol/water, 245, 247
partitioning. *See also* cut-off theory; polar paradox
α-tocopherol, 145
of antioxidants, 141, 144, 235, 255, 262, 283
chlorigenic acid alkyl chain length, 277f, 279
cut-off effect and self-aggregation, 282–284

partitioning *(continued)*
emulsion phases, 134–135
filtration and, 282–283
hydrophobicity and, 274
interfacial oxidative affinity, 157–160, 158f, 159f, 245
multiple emulsions, 133
phenolic, and scavenging, 303–304
polarity-based, 140, 250, 262, 473–474
proteins into continuous phase, 141
quantification of emulsifier partitioning, 181, 185–187, 185f
rosemary phenolic behavior, 421
as statistical distribution, 284
surface area and oxidation nonlinearity, 136
by surfactant micelles, 143
surfactants into continuous phase, 143
patents
chewing gum with tea polyphenols, 434
ergothioneine, 391
peanut oil, 107, 421
peptides, as food system antioxidants, 235
peroxidase
as endogenous antioxidant in fish, 352
fish muscle lipid oxidation, 351–352
flavor development in cook chicken, 107
peroxyl radicals (LOO•)
additions, 17–18, 39t, 40
autoxidation initiation, 325
β-elimination of oxygen, 9–11, 10f
carnosic acid as scavenger, 420–421
catechin radical scavenging, 429
characteristics of, 8–9
cyclization, 14, 15–17, 39–40, 39t
disproportionation, 18–19, 39t, 40
ergothioneine and, 389f
H abstraction, 2, 5, 11–15, 38–39, 39t
from hydroperoxides and transition metals, 135, 351
phenolic prooxidant reaction mechanism, 306f
propagation, 4f, 6f, 155, 346–347
rates of reaction vs. alkoxyl, 19, 20, 38, 39t
peroxynitrite
autoxidation initiation by, 325
BODIPY-based probes, 335, 336f
ergothioneine as antioxidant, 387–388
PeroxySafe™ assay, 55t, 72, 77t
PG
FDA limits and tolerances, 443–444t
as synthetic antioxidant, 441
pH
conjugated diene formation, 190, 190f
dispersion aqueous phase, 262
emulsion droplet charge, 132, 134, 137, 159
emulsion droplet ζ-potential, 134
food system antioxidant oxidation, 327
liposomal formation, 280
phenolic metal-catalyzed oxidation, 305, 308, 313–314
phenolic radical scavenging, 302
phenolics as prooxidants, 304–305
polar head- and tailgroup sizes, 183
protein and
conformation of protein, 191
emulsion droplet charge, 137
pH and pI, 141, 169, 182, 187–188, 189f, 190, 190f
surface loads, 179
as reaction condition, 72, 181
self-aggregation, 281
soil pH for rosemary plants, 418
transition metal properties, 308
Tween 20 and, 180, 191
phase inversion temperature (PIT), 132
phenolics, 297–314
absorbance wavelength, 328
as amphiphilic antioxidants, 235
as antioxidant, minor components, 231
antioxidant activity comparisons, 247, 270, 298, 301
as antioxidant in emulsions, 183
antioxidant mechanisms, 299–300, 300f
as antioxidants of food lipids, 297–298, 382
as bioactives, 298–299
in canola oil
free-radical scavenging, 457
hydrophilic, 462–463, 462f
lipophilic, 463–467, 464t
sinapic acid. *See* sinapic acid
chelation by, 299–300, 300f, 304, 374
color changes by, 309
combinations of, 300
as free-radical scavengers, 300, 300f, 301–304, 301f

metal-catalyzed oxidation, 305–308, 306f, 309–314, 310f, 312f
as natural plant extracts
antioxidant mechanisms, 299–304, 300f, 301f
combinations of phenolics, 300
mushrooms, distribution in, 359, 360t, 361f, 361t, 362
mushrooms, mechanisms underlying, 373–374, 373f
as nonsynthetic antioxidants, 297–298, 299t
prooxidant mechanisms, 304–309, 305t, 306f
prooxidant minimization, 309–314, 310f, 312f
structure, 297, 298f
olive-leaf phenolics, 329
oxidation products and quality loss, 309
phenolic diterpenes, 418, 420–422, 420f. *See also* carnosic acid; carnosol
phenolic glycosides, 302
phenolipids and cut-off theory, 256, 268, 269
polyphenols. *See* polyphenols
prooxidant activity, 304–305, 305t, 306f
prooxidant minimization, 309–314, 310f, 312f
proteins and, 311–312, 382
Rancimat method of stabilization, 251–252, 252f
reactive oxygen species generation, 308–309
in rosemary, 418, 420–421
structure of, 240f, 249f, 297, 298f, 420f, 440f, 462f
surface-active phenolic derivatives, 303–304
phenolipids
critical micelle concentration, 280–281
cut-off theory and, 256, 268–271, 275
internalization hypothesis, 278–279
self-aggregation, 280–284
phosphatidic acid, in canola oil, 472–473
phosphatidycholine with oleic acid (DOPC), 145–147, 146f
phospholipids (PLs)
α-tocopherol and, 144–145
as antioxidant, 141, 144–145, 197, 201, 352
canola seed preheating, 463
chelation by, 141
in crude canola oil, 472–474
emulsion droplets and, 141
fish postmortem, 348
hydrophilic stabilization, 276
oxidation susceptibility, 67
phenolics as antioxidants, 299t
phospholipid emulsifiers and polar paradox, 266
as polar lipids, 347–348, 472–474
as prooxidant, 141
protein/phospholipid interfacial layers, 201–203, 201f, 202f
reverse micelles, 145–146, 146f
phosphorescence probes, 331, 331–333f
photo-induced electron transfer (PeT), 336–337
photosensitized oxidation, vs. autoxidation, 22
phyto-factories, 442
Pierce kit, 55t, 77t
pine nut oil, 226t, 227
plant lipases, 212, 213t, 216
plant-derived extracts. *See* natural plant extracts
plastic fats, oxidative stability, 225, 226t, 227
plastic packaging, permeability of, 61, 61t
plum, as food system antioxidant, 444–445, 445t
polar paradox, 243–258. *See also* cut-off theory
antioxidant location in emulsion, 139–140, 245
in bulk oil vs. oil-and-water emulsions, 159–160, 264
concentration and, 244–245, 244t, 246, 257
critical chain length. *See* critical chain length
cut-off effect, 254–255, 263, 267–272, 268f, 270f, 271f. *See also* cut-off theory
definition, 243–244, 263–264
EDTA, 249–250
experimental contradictions to, 265–267
food matrix and polarity, 248
FRAP, 253–254
initiation and evaluation of oxidation, 248–249
lipophilic vs. hydrophilic antioxidants, 246–248, 247f, 248t
nonlinearity of results, 264–265, 265f
Porter, William, 243–246, 258, 264

polar paradox *(continued)*
- Rancimat test, 251–252, 254
- structure–activity studies, 251–255, 252–254f
- surfactant effectiveness of antioxidants, 256
- TEAC, 253–254

polarity. *See also* polar paradox
- of antioxidants
  - bulk oil vs. emulsions, 247
  - food matrix complication, 248
  - interface and antioxidant partitioning, 157
  - lipid dispersions, 262
  - no quantitative relationship, 247, 248t
  - nonlinear relationship, 160, 250, 264–265
  - structure–activity relationships, 253
- catechins and liposomes vs. emulsions, 429
- crude canola oil polar lipids, 472–474
- HPLC–DPPH for polar and nonpolar, 403
- molecule distribution driven by, 156, 250
- partitioning as polarity-based, 140, 250, 262, 473–474
- scission, 32–34
- of solvents
  - H abstraction, 20–21, 20t, 38, 39t
  - peroxyl radical reactions, 14, 18–19, 22, 38, 39t
  - scission, 33, 38, 39t
- and structure, rosemary extracts, 249f

polymers
- assays commonly used, 57t
- as lipid oxidation products, 4f, 6f, 81f
- toxicity of, 43, 82

polyphenols
- antioxidant activity, 266
- chelation by, 304
- cocoa polyphenols and cardiovascular disease, 298
- definition, 301
- green tea antioxidant mechanism, 428–430
- in green tea extract chemistry, 426–428, 427f, 428f
- matrix pH and stability, 308
- radical scavenging, 303, 311

polysaccharides, in emulsions, 142–143, 178

polysorbate 20. *See* Tween 20

polysorbate 80
- regulation of, 453–454, 453t
- Tween 80, 186, 187, 276

polyunsaturated fatty acids (PUFA). *See also* higher polyunsaturated fatty acids (HPUFAs)
- deodorization of, 230
- oxidation susceptibility, 222
- oxidation toxicity, 177

pork
- ergothioneine content of, 386t
- ergothioneine HPLC chromatograms, 395f
- lard. *See* lard
- natural plant extract uses, 444–445, 445t, 446t, 448–449
- rosemary extract as antioxidant, 425
- sausage and phenolics, 299f
- synthetic antioxidant USDA limits, 444t

Porter, William, 243–246, 258, 264

positional distribution, of FAs, 217

potato chips, rosemary as antioxidant, 426

potato flakes, 425, 443t

poultry meat. *See* chicken; turkey

preheated seed canola oil (PSCO). *See* canola seeds, preheated

pressurized fluid extraction (PFE). *See* accelerated solvent extraction (ASE)

pressurized solvent extraction (PSE). *See* accelerated solvent extraction (ASE)

preventative antioxidant, 249

products of lipid oxidation. *See also specific products*
- assays, standardized, 54, 55–57t, 60
- autoxidation vs. photosensitized oxidation, 22
- beyond classic chain reactions, 2–3
- closed vs. open sample analysis, 94
- environment of reaction on, 38–40
- epoxide rare mention, 23
- first stable products, 34
- lipid alcohols, 3, 20
- multiple product analysis, 78–84, 110–111
- time curves of products, 41–42, 42f
- typical products analyzed, 80, 81f
- which to monitor, 53–54, 59

prooxidants
- antioxidants as, 267, 327
- catechins as, 304, 305t, 429
- EC as, 305t
- ECG as, 304, 305t

EDTA as, 313
EGC as, 305t
EGCG as, 304, 305t
in emulsions, 134–135, 136, 142–143
in fish meat, 349–352, 350f, 374
gallic acid as, 304, 305t
grape pomace extract as, 305t
green tea extracts as, 304, 305t, 429
heme protein as, 350–351, 374
lipoxygenases as, 349
metal chelators as, 250
minimization of, 309–314, 310f, 312f
myricetin as, 304
phenolics as, 304–305, 305t, 306f
phospholipids as, 141
polysaccharides in emulsions, 142–143
reverse micelles as, 257
rosemary extracts as, 305t
SDS as, 136
singlet oxygen as, 144, 325, 327, 348–349, 349f, 421
propagation of lipid oxidation
alternate reactions focus, 2
quinones, 306
by radicals, 4f, 6f, 155, 346–347, 348
stage identification, 1
propyl gallate, 235, 267
proteins
aggregation and denaturation of, 200–201
aggregation by dityrosine, 195
aggregation by quinones, 309
aldehydes and crosslinking, 43
analysis effects, 57, 58, 80
as antioxidants in emulsions, 141–142, 169–170, 182
as antioxidants in lipid dispersions, 311
cooxidation, 96–98, 97f, 99–101t
denaturation of, 200–201, 356–357
double role in oxidation, 191–195, 193f, 194f, 196f
emulsion droplet effects, 137, 139, 141
epoxide rapid reaction with, 23, 43, 81–82
fluorescence, 191–192, 193f, 195, 326
as food-grade emulsifiers, 178
heme proteins in aquatic foods, 350–351
ice storage of aquatic products, 357
on interface of emulsion, 179, 180
interface packing vs. Tween 20, 179
modification of, 191–192, 193f, 195
peroxide scavenging, 311
phenolics and, 311–312, 382
protein/phospholipid interfacial layers, 201–203, 201f, 202f
vs. surfactant for stabilized emulsions, 183–184, 187–191, 189f, 190f, 197
unadsorbed emulsifiers
concentrations of, 183–184, 184f
emulsions designed for study, 185–187, 185f
minimal vs. excess, 195, 196f
physical hinderance of interactions, 459–460
protein double role, 191–192
water-holding capacity, 356
PSCO (preheated seed canola oil). *See* canola seeds, preheated
purge and trap (P/T) thermal desorption GC
as dynamic headspace analysis, 86–87
sensitivity of, 83
as standard assay, 57t, 86

Q
quercetin
antioxidant activity in lipid systems, 299t
as antioxidant for plastic fats, 226t, 227
chelation by, 304
phenolic combinations, 300
phenolic glycosides effects, 302
polar paradox, 248, 264
quinones
furthering propagation, 306
nucleophilic amino acid reactions, 309
pH and, 308
superoxide dismutase (SOD) and, 307

R
radical recombinations, vs. scission, 3
radical scavengers. *See* free-radical scavengers
rancidity
aldehydes as source, 43
aquatic foods, oxidative deterioration, 346, 353, 354f
in dispersions, 262
driving product analysis, 80
lipid-protein adducts, 139
natural plant extracts retarding, 430t, 449

rancidity *(continued)*
scission reactions, 3, 25
soybean oil and tea polyphenols, 434
transition metals in emulsions, 135
unsaturated FAs, 458
Rancimat method
olive oil, 251–252, 252f
for oxygen consumption rate, 328
tyrosol lipophilic ester derivatives, 253, 253f
rapeseed oil. *See also* canola oil
vs. canola oil, 460
chemometrics to distinguish, 107, 110
designed emulsions for study, 186
erucic acid in, 460
interesterified SLs and HMFS, oxidative stability, 223t, 232
mayonnaise and EDTA, 250
Nieto "virgin" rapeseed oil, 467
oilseed rape, 460
phenolics as antioxidants, 299t
plastic fats, oxidative stability, 226t
production levels, 460
short-path distillation, 231
reactions of lipid oxidation. *See* mechanisms of lipid oxidation
reactive oxygen species (ROS)
autoxidation initiation, 325
BODIPY-based fluorescent probes, 335
generation of, 308–309, 350, 350f
phenolic metal-catalyzed oxidation, 305, 308
phenolic prooxidant activity, 305
short lifetimes of, 327
recombinations
of peroxyl radicals, 19
recombinations vs. scission, 3
reduction potential
of α-tocopherol, 145, 244
antioxidant activity and, 139
tocopherol methyl substitution, 244
regulation
of antioxidants in foods, 362, 382, 442, 443–444t
as natural plant extract challenge, 453–454, 453t
rosemary extract regulatory status, 422, 423t
resorcinolic lipids, amphiphilicity of, 280
resveratrol, structure of, 298f
reverse micelles
as association colloids, 144–147, 146f
bulk oil formation, 256, 257f
lipophilized antioxidants, 276, 281, 284
reversed phase HPLC
antioxidant evaluation, 404–405
aqueous reversed-phase HPLC columns, 91, 93
challenges of, 92
phenolic diterpene quantification, 422
riboflavin, as prooxidant, 144
rice-bran oil
oxidative stability, 223t, 226t
short-path distillation, 231
for *trans*-free fats, 225
ROS. *See* reactive oxygen species (ROS)
rosemary, 417–426
as antioxidant in food systems, 235, 299t, 382, 439
antioxidant mechanism, 420–421
bulk vs. emulsion oxidation, 452, 452t
extract quality/price determination, 421
extraction process, 418
flavor as additive challenge, 451
food application, 422–426, 423–425t
vs. green tea extract, 432–433, 432f, 433f
HPLC–DPPH-UV-NMR, 404
incorporation methods, 419, 419f, 420f
maximizing performance of, 450, 451t
meat product storage, 424–425, 424t
natural plant extract uses, 444–450, 445–447t
phenolic diterpene quantification, 421–422
polar paradox antioxidant, 247, 247f
prooxidant activity, 305t
regulatory status of, 422, 423t, 453
rosemary plant characteristics, 417–418, 422
synergism with synthetic antioxidants, 442
vs. TBHQ and ascorbic acid, 446–447, 447t
rosmarinic acid
bulk vs. emulsion oxidation, 452, 452t
ester alkyl chain variation
cut-off effect, 255, 268, 268f, 269–270, 270f, 271
phenolic conjugation, 303
polar paradox challenge, 140
polarity and antioxidant activity, 160
reduced mobility hypothesis, 286
HPLC–DPPH assay, 403
as hydrophilic, 452

internalization hypothesis and esters, 276, 277–278
in lemon balm, 440
metal-chelation and activity, 248–249
in oregano, 448
polarity and activity, 160, 247, 248t
in rosemary extract, 418, 421, 440
self-aggregation, 281, 282, 283
in spearmint, 440
structure of, 249f, 440f
surfactant concentration, 283

S

safety of foods, and odor, 43, 81–82
safflower oil
gallate esters and polar paradox, 244–245, 244t
sugar alcohols as antioxidants, 143
transesterified SLs and HMFS, oxidative stability, 224t
salad dressings
EDTA as antioxidant, 249
emulsion droplet size, 136
fat emulsions, oxidative stability, 228t, 229
neat oil as delivery system, 234
as o/w emulsion, 458
phenolics as antioxidants, 299t
tocopherols as antioxidants, 235
4-VS recovery for, 467
salmon meat, 366, 368, 447–448
salmon oil
metal chelators as prooxidants, 250
phenolic prooxidant activity, 305t
surfactant head-group size, 138
whey-protein fractions as antioxidants, 142
saturation. *See* concentration
sausage
ergothioneine content, 386t
natural plant extract uses, 444–446, 445t, 446t, 449
pork sausage and phenolics, 299t
rosemary EU regulatory limits, 423t
synthetic antioxidant USDA limits, 444t
scission
alkoxyl radicals, 24–34, 39t, 40
α or β, 24–25, 32
facilitation, 39t, 40
linoleic acid, 10f, 28–29f
linolenic acid, 30–31f
oleic acid, 26–27f
vs. radical recombination, 3
SDS. *See* sodium dodecyl sulfate (SDS)
SDS-polyacrylamide gel electrophoresis (SDS-PAGE), 191, 192, 194f
seal blubber oil (SBO), 433, 434
sensory impact
alkoxyl scissions, 25
antioxidant flavor, 362
browning of food, 43, 467
canola seeds, preheated, 458
chemometrics for flavor patterns, 102, 107, 110
color changes by phenolic oxidation, 309. *See also* color
difference-from-control test, 451–452
discoloration of meat, 353, 450
epoxides, 43
flavor. *See* flavor; off-flavors; rancidity
green tea vs. rosemary, 433
lipid content correlation, 353
natural antioxidants, 447t, 448–449, 449t
odor. *See* off-odors
as oxidative status marker, 43, 222, 223t, 345
phenolics and proteins, 382
plastic fats and fat emulsions, 225, 229
potato flakes and rosemary extract, 425
singular importance as fallacy, 43, 81–82, 95
triangle test, 451–452
turkey burgers and green tea extract, 432t
4-VS in frying oils, 466–467
sesame oil, and interesterified SLs, 223–224t
sheep meat
ergothioneine content of, 386t
fat, rosemary EU regulatory limits, 423t
fresh mutton and green tea extract, 434
shelf life, of foods
accelerated shelf life studies, 42–43, 90
analysis challenges, 78
analysis frequency, 41
aquatic products, 346, 366
green tea extract extending, 430t, 434
oxidative deterioration and, 357
plastic fats, 225
PUFA challenges, 1, 129, 328
rosemary extending, 425
short-path distillation (SPD), 230–231, 232–233, 467

shrimp meats
ergothioneine toxicology studies, 399–400
mushrooms as antioxidants, 390, 399
synthetic antioxidant FDA limits, 443t
sinapic acid
in *Brassica* oilseeds, 457, 462
structure of, 462f
4-VS production, 457–458, 464–465, 464t, 474
singlet oxygen, as prooxidant, 144, 325, 327, 348–349, 349f, 421
SL. *See* structured lipids (SLs)
sodium caseinate. *See also* caseins
4-VS and oil-in-water emulsions, 467
chelation by, 459
as emulsion stabilizer, 459
hydrogen peroxide scavenging by, 311
oxidative stability vs. other proteins, 137
packing structure of interface, 474
sodium dodecyl sulfate (SDS)
emulsion droplet charge, 136–137, 157, 158f, 159
gallate alkyl ester diffusion, 285–286
micelles and alkyl chain length, 278
micelles and α-tocopherol, 285
vs. other emulsifiers, 137, 159
as prooxidant, 136
SDS-polyacrylamide gel electrophoresis (SDS-PAGE), 191, 192, 194f
self-diffusivity and concentration, 289
solid phase microextraction (SPME)
quantitation issues, 83
SPME-GC with chemometrics, 107, 110
4-VS and o/w emulsion oxidative deterioration, 468
solubility, as natural plant extract challenge, 452–453, 452t
solvents
alkoxyl radical reactions
cyclization, 21–23, 23t
H abstraction, 20, 20t, 33t
pathway facilitation, 38–40, 39t
scission, 33–34, 33t
H abstraction and, 20, 20t, 33, 33t
headspace, 64
lipase amphiphilic helix peptide sequence, 212
peroxyl radical reactions, 15, 19, 38–40, 39t
polarity of
H abstraction, 20–21, 20t, 38, 39t
peroxyl radical reactions, 14, 18–19, 22, 38, 39t
scission, 33, 38, 39t
solubility of gases, 61, 62f, 63t
solvent peroxides, 64
stability of, 61, 64
Soxhlet extractions, 65, 68, 474
soy protein isolate
as antioxidant in emulsions, 141, 170, 184
oxidative stability vs. other proteins, 137
soybean oil
casein-lipid oxidation adducts, 139
chemometrics to distinguish, 106f, 108f
interesterified SLs and HMFS, oxidative stability, 223–224t
microchannel emulsification, 166f
with palm stearin in various reactors, 233
plastic fats, oxidative stability, 226t
production levels, 460
reverse micelles, 146f, 257
rosemary as bulk oil antioxidant, 421, 426
rosmarinates vs. rosmarinic acid, 268
short-path distillation, 231
Soxhlet extractions, 65
soybean liposomes, 269, 278–279, 287–288, 287f
TAG emulsion droplet size, 167, 168f
TAG oxidative stability, 166f
tea polyphenols and rancidity, 434
for *trans*-free fats, 225
Span 20, 197–200
sparging, 60–61, 66, 70
SPD. *See* short-path distillation (SPD)
spearmint, rosmarinic acid in, 440
spectrophotometry
correlation with HPLC–DPPH, 403
cut-off theory, 278, 283
of ergothioneine, 391–392, 394–395, 401–402, 405
of unadsorbed proteins, 186
spices/herbs
as antioxidants in food systems, 235, 382
deodorization of, 382, 419f
squid miso, 403, 404f
static headspace analysis, 56t, 83, 85–86
stationary phase HPLC, 92
stearic acid

in cocoa-butter equivalents, 219
human milk TAG, 222
oxidative stability vs. capric/oleic, 221
sterols, as antioxidant minor component, 231
stoichiometry of assay, 59, 71–73, 73t
storage
analysis results skewed, 78, 84, 94
aquatic products
antioxidant dietary supplementation, 366, 368, 369f, 370f
lipid content and oxidation, 353, 355f, 390
loss of nutrients, 357, 358–359t
metMb formation, 356, 356f, 362, 391
storage temperature, 352–353
green tea extract and meat products, 430–432, 431f, 432–433, 432f, 432t, 433f, 434
natural plant extracts and meat products, 444–450, 445–447t, 449t, 451t
plastic packaging permeability, 61, 61t
rosemary and meat and poultry, 424–425, 424t, 432–433, 432f, 433f
tissue postharvest oxidation, 347, 348, 390
structured lipids (SLs)
functional foods containing, 229–230
interesterified lipids, 222–229, 223–224t, 226t, 228t
MLM-type SLs, 217, 220–222
styrene, 337
sugar alcohols, 143
sunflower oil
chemometrics for quality, 106f
chlorigenic acid partitioning, 277f, 279
fat emulsions, oxidative stability, 228t
fatty alcohols and reverse micelles, 145
gallic acid as prooxidant, 304
interesterified SLs, oxidative stability, 224t
mayonnaise and milk drink SLs, 230
pH and conjugated dienes, 190f
phenolics as antioxidants, 299t
phenolics as prooxidants, 304, 305t
plastic fats, oxidative stability, 226t
polar paradox cut-off effect, 255
supercritical fluid extraction (SFE), 66–67, 421, 422
superoxide
vs. AAPH, 302
catechin radical scavenging, 429
membranes and electron transport, 348
superoxide dismutase (SOD), 307, 352, 354f
surface activity
bulk oil vs. emulsions, antioxidant polarity, 160
driving molecule distribution, 156, 250
enhanced by esterification, 255
enhanced by whey protein, 200
nonpolar antioxidants at interface, 159
reverse micelles and polar antioxidants, 256–257, 257f
Tween 20 and phospholipids, 141
surface area
bulk oil oxidation as surface effect, 243, 246, 459
of emulsions, 135, 136, 178, 181, 182f, 459
lipid dispersion oxidation susceptibility, 262
low surface-to-volume (LSV) ratio, 243
membrane oxidation susceptibility, 348
surface load, 179
surface tension
CMC of hydroxytyrosol alkyl esters, 281
emulsifiers at interface, 178
polar paradox interfacial effects, 245
surface tension at CMC (γCMC), 281
surfactants
concentration and emulsifier packing, 168
concentration and self-aggregation hypothesis, 283
designed emulsions for study, 186, 187
diffusion and micelles, 286
as food-grade emulsifiers, 178
head- and tail-group sizes, 138, 183, 197
in interface of emulsions, 179–180
mixtures of, 197–200, 198f, 199f
polarity and antioxidant diffusion, 276
vs. protein for stabilized emulsions, 183–184, 187–191, 189f, 190f, 197
surface load and, 179
surfactant effectiveness of antioxidants, 256, 256f
surfactant micelles, 143–144
synergistic reactions
canola oil phenolics, 463, 465
catechins, ascorbic acid, α-tocopherol, 429
phenolic combinations, 300
phenolics and proteins, 311–312
polar paradox antioxidants, 246–247

synergistic reactions *(continued)*
rosemary with synthetic antioxidants, 442
wine and phenolic oxidation, 308
synthetic antioxidants
BHA. *See* butylated hydroxyanisole (BHA)
BHT. *See* butylated hydroxytoluene (BHT)
decreased use of, 442
regulation of, 362, 382, 442, 443–444t
TBHQ. *See* TBHQ (*tert*-butylhydroquinone)
toxicity of, 382
synthetic flavorings. *See* flavoring agents

T
TAG. *See* triacylglycerols (TAGs)
tallow
DART–MS, lard differentiation, 89
interesterified and water content, 232
rosemary EU regulatory limits, 423t
synthetic antioxidant USDA limits, 444t
TBARS (thiobarbituric acid-reactive substances)
fat emulsions, 228t
fish meat ice storage, 357
green tea extract and turkey burgers, 431f
mushroom extract and tuna, 365f
natural plant extracts as antioxidants, 445, 446–450, 446t, 451t
oxyMb autoxidation inhibition, 363f
peroxide-quenching mechanisms, 312f
plastic fats, 225
polar paradox, 250
rosemary extract and meat and poultry storage, 424t, 425
structured lipids, 224t
TBHQ (*tert*-butylhydroquinone)
as antioxidant for structured lipids, 229
FDA limits and tolerances, 443–444t
vs. green tea extract, 434
in plastic fats and fat emulsions, 225
vs. rosemary, 446–447, 447t
as synthetic, 441
tea. *See also* green tea extract
black, green, and oolong, 426, 427–428
catechins and oxidation potential, 302
catechins and prooxidant behavior, 304
catechins and protein-rich foods, 311
catechins in green tea extract chemistry, 426–428, 427f, 428f
fermentation of, 426, 427, 428
green tea extract and lipid systems, 299t
polyphenol health benefits, 298
tea extract prooxidant activity, 305t
TEAC (Trolox equivalent antioxidant capacity), 253, 254, 331, 332–333f, 334
temperature
accelerated shelf life studies, 42–43
alkoxyl radical reactions, 21, 22, 34
anthocyanin color changes, 309
aquatic products storage, 352–353, 355f, 356, 356f, 357, 358–359t, 362
bond dissociation energy (BDE) and, 15, 35, 39
canola oil from preheated seeds, 458, 467
canola oil headspace volatiles, 468
canola seed preheating. *See* canola seeds, preheated
casein conformation, 459
catechin content of tea and, 427
chemical interesterification, 211
coagels to gels/micelles, 282
denatured, partially aggregated proteins, 200–201
deodorization, 230
enzymatic lipid modification, 233
glycerolysis, 220
heat-sensitive antioxidants, 382
homolytic decomposition, 35–36
lipase activity and heating rapeseed, 463
peroxyl radical reactions, 15, 18
phenolic metal-catalyzed oxidation, 313–314
Rancimat method of stabilization, 251
reaction pathway effects, 38–40
rosemary extract and, 420
self-aggregation, 281
short-path distillation, 230
of SL purification process, 229
unsaturated FAs, 220
4-VS thermal instability, 467
teratogenicity, synthetic antioxidants, 382
termination of lipid oxidation
alkyl radicals and, 8
alternate reactions, 2, 18
peroxide formation, 18, 19, 347
protein-lipid oxidative interactions, 196f
stage identification, 1, 4f
thermal degradation in oils, assay for, 87
thermodynamics, 275, 276–277
thin-layer chromatography (TLC), 391, 396

thyme, as food system antioxidant, 235, 382
time of flight spectrometers, 87
tissue. *See* meat products
tocopherols. *See also* α-tocopherol; γ-tocopherol
  amphilicity and association colloids, 280
  antioxidant dietary supplementation, 366
  as antioxidants, 221–222, 231, 235, 299t, 352, 354f, 366, 382, 445t, 449, 461
  ascorbate and tocopherols, 300
  bulk oil ORAC assay, 331
  in canola oil, 457, 461, 469f
  metal-catalyzed oxidation, 306
  methyl substitution and lipophilicity, 244
  as oxidation marker
    interesterified SLs and HMFS, 223–224t
    SLs vs. unmodified, 222, 225
    stearic vs. caprylic vs. oleic acids, 221
  oxidation/reduction potentials, 244
  vs. phenolics, 298
  polar vs. nonpolar, 160
  polarity and surface activity, 250, 251f
  short-path distillation, 231, 233
  structure of, 298f, 462f
tocotrienols, 231
toxicity. *See also* health advantages
  of antioxidants
    regulation of, 362, 382
    of synthetic antioxidants, 382
    toxicological studies and ADI levels, 399–400
  of erucic acid, 460
  of oxidation
    bioactive high doses, 299
    free radical health damage, 381
    ingestion of oxidation products, 82, 129
    multiple pathways challenge, 43
    oxidative stress and chronic disease, 323
    polyunsaturated fat oxidation, 177
    quinones and DNA, 309
    sensory cues for, 43, 81–82
  oxidative stability and heart disease, 130
  of traditional rapeseed oil, 460
tragacanth gum, 142
transesterification, enzymatic, 218f, 219–220
trans-free fats
  coconut oil for, 225
  margarines, 225
  palm stearin for, 219, 225
  partial hydrogenation and, 130, 262
  rice-bran oil for, 225
  soybean oil for, 225
transition metals
  biological system oxidation, 349–351, 350f
  charge of oil droplet and, 159, 159f
  chelation. *See* chelation
  in classic chain reaction, 3, 4f
  cyclization of alkoxyl radicals, 22
  EDTA effects, 249
  electron transfer as antioxidant, 5
  in emulsions, 135, 136, 159, 159f, 170, 182, 182f, 249
  hydroperoxide decomposition, 35
  hydroperoxide reduction stoichiometry, 71–72
  metal redox cycling disruption, 310, 310f, 312–313
  pH and properties, 308
  phenolic oxidation, 305–308, 306f, 309–314, 310f, 312f
  polar lipids and, 473–474
  Trolox and α-tocopherol depletion, 146–147
triacylglycerols (TAGs)
  assays for, 92, 107, 329
  chemical interesterification, 211
  emulsion droplet size, 166f, 167, 168f
  enzymatic lipase reactions, 217–220, 218–219f
  esterase hydrolysis, 212
  vs. free fatty acids in aqueous dispersion, 161
  human milk TAG structure, 222
  hydrophile-lipophile balance (HLB), 169
  lipase hydolysis, 212
  microchannel emulsification, 166f
  microwave extractions, 66
  in o/w emulsions, 160
  oxidation susceptibility, 67
  Soxhlet extraction, 65
  tallow vs. lard differentiation, 89
triglycerides (TGs)
  α-tocopherol vs. Trolox, 248
  fish postmortem, 348
  lipid peroxide quantification, 337
  as modified lipids, 217, 219
  as neutral lipids, 347
  as oil-in-water emulsion particles, 161

Trolox
vs. α-tocopherol, 146, 146f, 161, 248, 257, 274
BODIPY fluorophore (B-TOH) and, 336–337, 337f
conjugated autoxidizable triene (CAT) assay, 330f
vs. ergothioneine, 373, 388, 389f
lipophilic vs. hydrophilic antioxidants, 246
micelle charge and, 157–158, 158f
ORAC assay, 326
TEAC (Trolox equivalent antioxidant capacity), 253, 254, 331, 332–333f, 334
trout meat. *See also* fish meat
antioxidant dietary supplementation, 368
enzymatic oxidation, 351–352
ergothioneine content of, 386t
natural vs. synthetic antioxidants, 446–447, 446t, 447t
tuna meat. *See also* fish meat
discoloration from oxidation, 353, 354f, 356
ice storage and PUFAs, 357, 358–359t
mushrooms and oxidation, 362, 365f, 366, 367f, 373f, 390–391
myoglobin content, 362
tuna oil
lipid peroxide quantification, 338
preheated seed canola oil as antioxidant, 474–475, 475f
4-VS and o/w emulsions, 468t, 470, 471f
tung oil
chlorogenate esters and cut-off effect, 255, 267–268, 268f
conjugated autoxidizable triene (CAT) assay, 329, 330f
rosmarinate esters and cut-off effect, 268f, 276, 286
turkey
green tea extract as antioxidant, 430–431, 431f
phenolics as antioxidants, 299t
rosemary extract as antioxidant, 424, 424t
Tween 20
designed emulsion for study, 186
droplet cross-linked proteins, 139
hydrophilic stabilization, 276
interface packing vs. proteins, 179
phosphatidylcholine and ζ-potential, 141
vs. proteins for compactness, 197
proteins vs. surfactants, 187–191, 189f, 190f
vs. SDS and DTAB, 159
steric and electrostatic repulsions, 180, 199–200
surfactant mixtures, 197–200, 198f, 199f
Trolox vs. α-tocopherol, 248
Tween 80
designed emulsion for study, 186
hydrophilic stabilization, 276
polysorbate 80, regulation of, 453–454, 453t
proteins vs. surfactants, 187

U
ubiquinols
as endogenous antioxidant in fish, 352
membrane mobility, 285, 287–288, 287f
membrane systems and cut-off effect, 269
ultrasound extractions, 66
ultraviolet (UV) radiation, oxidative damage, 388–389
U.S. Department of Agriculture (USDA)
oxygen radical absorbance capacity (ORAC) assay, 326
synthetic antioxidant limits, 444t
U.S. Food and Drug Administration (FDA)
food-grade EDTA, 249
polysorbate 80 regulations, 453t
rosemary extract as GRAS, 422
synthetic antioxidant limits and tolerances, 443t
trans-fat labeling, 130
uric acid, vs. ergothioneine, 373, 388, 389f

V
4-vinylsyringol (4-VS; canolol)
as antioxidant, 458, 463, 466, 467, 470, 474
case study, oil-in-water emulsions, 467–476
methods, 467–470, 468t, 469f
oxidative deterioration, 470, 472f
preheating temperature, 470–471
tuna emulsion fortification, 458, 470
4-VS concentration in emulsion, 470, 471–472, 471f, 474
water degumming, 472–474, 473f
as lipophilic, 458, 466, 467

production of, 457–458, 464–465, 467, 474
structure of, 462f
virgin canola oil oxidative stability, 466
viruses, green tea as antiviral, 434
viscosity
aldehydes increasing, 43
H abstraction rates by peroxyl radicals, 15, 39
lipid dispersion lipid droplets, 262
oxypolymers increasing, 17–18
polysaccharides in emulsions, 142–143
reaction pathway effects, 39–40
of rosemary oil, 418
texture modifiers in emulsions, 178
vitamin C. *See* ascorbic acid
vitamin E
antioxidant-substrate molar ratio, 329
BODIPY fluorescence, 336
phenolic compounds vs., 374
Trolox as water-soluble analogue, 326
volatile products
assays commonly used, 56–57t
canola oil, heated, 468
cryotrapping for highly volatile products, 94
matrix binding of, 83, 86
rancidity in postharvest aquatic foods, 353
4-VS and o/w emulsions, 468, 470, 472, 472f, 473f

W
warmed-over flavor (WOF), 424–425, 432t
water
ice crystals accelerating oxidation, 352
modified lipid product stability, 232
permeability of plastics, 61, 61t
as reaction condition, 40
solubility of gases, 61, 62f
water-degumming
oil from preheated rapeseed, 463
4-VS and oil-in-water emulsions, 472–474, 473f
water-in-oil emulsions
vs. oil-in-water emulsion, oxidation, 227
oxidative stability, 227–229, 228t
water-in-oil-in-water emulsions, 132–133, 133f
whey protein. *See also* β-lactoglobulin (BLG)
as antioxidants in emulsions, 141, 142, 169, 184, 227
oxidative stability vs. other proteins, 137
peroxide scavenging, 311
residual denatured and aggregates, 200
for unadsorbed emulsifiers, 184
whey-protein fractions (WPF), 142
whey protein isolate (WPI). *See* whey protein
wine, 308
Wöhler, Friedrich, 243

X
xanthan gum, 142, 183
xylenol orange assay
detection range for hydroperoxide, 77t, 83
molecular structure, 74
nonlinear response to saturation, 76, 77f
oxygen interference, 70
product concentration effects, 78, 83
sensitivity of, 83
spectral overlap, 69
as standard assay, 55t
stoichiometry, 71–72, 83

Y
yeast
polysorbate 80 FDA/EU regulatory limits, 453t
synthetic antioxidant FDA limits, 443t
yellowtail meat. *See also* fish meat
antioxidant dietary supplementation, 366, 368, 370f, 372, 396, 396f, 399
autoxidation inhibition, 363f
ergothioneine HPLC chromatograms, 395f
ergothioneine toxicology studies, 399–400
ice storage and hydroperoxides, 355f
mushrooms as antioxidants, 390
yogurt
neat oil as delivery system, 234
as oil-in-water emulsion, 458

Z
ζ-potential, 134

FSC
www.fsc.org
MIX
Papier | Fördert
gute Waldnutzung
FSC® C083411

Kolibri 360 GmbH
Ferdinand-Jühlke-Straße 7
99095 Erfurt, Deutschland
produktsicherheit@kolibri360.de